29 $\frac{80}{3}$

Life Strategies,
Human Evolution,
Environmental Design

Valerius Geist

Life Strategies, Human Evolution, Environmental Design

Toward a Biological Theory of Health

Springer–Verlag

New York   Heidelberg   Berlin

Valerius Geist
Faculty of Environmental Design
University of Calgary
Calgary, Alberta
Canada

QP
82
.G37

Library of Congress Cataloging in Publication Data

Geist, Valerius.
  Life strategies, human evolution, environmental design.

  Bibliography: p.
  Includes indexes.
  1. Adaptation (Physiology) 2. Life (Biology)
3. Human evolution. 4. Environmental health.
5. Human Behavior. I. Title.
QP82.G37   573.2   78-10807

Printed in the United States of America.

9 8 7 6 5 4 3 2 1

ISBN 0-387-90363-1 Springer-Verlag   New York

ISBN 3-540-90363-1 Springer-Verlag   Berlin   Heidelberg

To Ian McTaggart-Cowan and J. B. Cragg

Acknowledgments

This book took shape in response to many discussions with colleagues, students, and especially my wife, Renate. I am in debt to them all and would like to thank those who took time to discuss ideas of mutual interest. The following colleagues read and commented on various chapters:

Drs. R. S. Sainsbury, R. Lein, P. K. Anderson, H. Buchwald, A. Fullerton, D. D. Detomasi, W. T. Perks, and T. Howarth.

I owe a large measure of gratitude to the dedicated work on the manuscript by Muriel Enock who sheparded it through years of toil and seemingly unending alterations.

Preface

Consider that you were asked how to ensure human survival. Where would you begin?

Conservation of resources jumps to mind. We need to conserve resources in order that economic activities may continue. Alas, this is a false start. Resources are always defined by a given economic system, and only it determines what is and what is not a resource. Therefore, conserving resources implies only the perpetuation of the appropriate economic system. Conservation of resources as we know them has nothing to do with the survival of mankind, but it has very much to do with the survival of the industrial system and society we live in today. We have to start, therefore, at a more basic level.

This level, some may argue, is addressed by ensuring for human beings "clean genes." Again, this is a mistaken beginning. It is thoroughly mistaken—for reasons of science. It is a false start because malfunctioning organs and morphological structures are not only due to deleterious hereditary factors but particularly due to unfavorable environments during early growth and development. Moreover, eugenics is not acceptable to any but a small fraction of society. Eugenics may not be irrelevant to our future, but is premature and should be of little concern until we understand how human genes and environment interact.

Where do we begin? How do we ensure our survival? We in industrialized nations need answers to this question, urgently, since our life style does not bear the mark of permanence as our social critics have made eminently clear. The outlook for our existence in the future is not encouraging. How do we move toward a more durable, survivable, life style? What does such a life style look like? The dogmas of nineteenth century politics and economics are under attack, as implied by such notions as the postindustrial society, the political ecological or "green" movements and their successes at the polls, the "relevant" technologies and "economics as if people mattered." Unfortunately, there is no assurance that these movements carry any more of the seeds of a durable, humane, survivable life style than the systems they are intended to replace. We must begin at very basic levels indeed!

It is self-evident that for a society to thrive its individuals must be healthy. Each generation succeeding the last must be composed of individuals of equal or superior quality to the preceding one. If the term "quality" is objectionable, let me replace it with the next best, "competence." Note that the emphasis is not on numerical or demographic aspects but on the maintenance of competence, on the individual's ability to deal with the social and physical milieu he is born into. However, competence is very closely related to an individual human being's environment, particularly during its long juvenile development. Thus, we must search for an environment that maximizes competence in individuals. Of course, this ought also to be an environment that supports health, for this is an essential ingredient of survival. However, by a stroke of good luck, it turns out that to maximize health *is* to maximize competence. Thus, the very first concern in ensuring survival is to know how to structure environments that maximize the health and competence of individuals. This is a requirement mandatory to all socioeconomic systems laying claim to viability.

However, it is no simple matter to visualize a life style maximizing health, for there is no theory of health. When I began to concern myself with human adaptations, this discovery came as a rude surprise. I was teaching graduate students entering urbanism, architecture, and environmental science and who required an understanding of how to maximize health environmentally. This book deals with my quest for such a theory and with the details needed for its understanding.

A theory of health is a simple predictive statement that can be put into just one sentence. *Health is maximized when diagnostic features of a species are maximized phenotypically.* It is also obvious that this simple sentence is neither lucid nor self-evident, nor very useful even though it expresses a principle and can be verified. To show how it is developed, I had to return to fundamentals, which are dealt with in the first part of the book and dwell heavily on very new notions arising from the field studies of large mammals, plus I had to tackle theoretical problems, at times with heretical conclusions.

I wrote this book not without trepidation, not in fear of being wrong, but afraid of the misunderstanding it can generate. Let me state from the outset that I aim to convince nobody. I am after a theory of health that can be disproven; that is all I can or should do as a scientist. A scientist's task is not to do what novelists and poets or artists do but to develop concepts, that can be disproven beyond reasonable doubt. I am, thus, not an advocate but do hope that the conclusions I come to can be subjected to critical tests that I cannot perform.

This book represents an attempt at an interdisciplinary synthesis. It ought to be obvious that in dealing with the understanding of an organism's life history one is always dealing with an interdisciplinary understanding. Synthesis, in contrast to analysis, takes place at the expense of depth of knowledge; synthesis, by its very nature, is interdisciplinary and based on the assumption that he who synthesizes adequately grasps the essence of the disciplines he trespasses on. Credibility and authority, however, rest within disciplines. Therefore, can the authority of science be invoked for the results of an *interdisciplinary* synthesis? There is danger enough that a synthesis reaches the stage where scholarship gives way to something less. How does one avoid the pitfalls of synthesis?

I am not aware of a formal procedure to reduce error in interdisciplinary work, nor do I know a shortcut to the slow process of mastering different disciplines one by one. Nor do I believe any more that a gathering of minds from different disciplines can be more than pooled ignorance unless each member of the interdisciplinary team knows the essence of all other participants' disciplines, and all speak the same intellectual language. Interdisciplinarity rests with the individual, be he alone or in a group. However, interdisciplinarity, despite appearing to be modern is quite old. It is the essence of such disciplines as ornithology, mammalogy, and ichthyology, and above all, it should be the essence of anthropology. Scientists in these fields do go about synthesizing information from diverse fields and are no less scholarly for doing so. One can, in fact, recognize some disciplines as mainly analytical, such as physiology, chemistry, and physics, and others more prone to synthesis, such as those mentioned previously. Still, the question I raised is pertinent: How does one verify one's information in an interdisciplinary study?

Consider this: Many papers and books outside one's discipline cannot be read at the same level of critical appraisal as writings within one's own discipline. Therefore one cannot give the expected assurance that each paper cited has been fully understood, and that one supports its conclusions from an examination of the methods and the relevant theoretical background. This is a very serious limitation. One can, however, partly surmount this handicap by asking the following.

(1) Is the fact or concept one wishes to use commonly referred to and accepted in the given discipline? If so, it is probably valid.

(2) Is the paper published in a refereed publication? If so, the chances are the substance has merit.

(3) Is the fact or concept logical in the context of one's own theoretical knowledge? Can the fact or concept be deduced from a different line of reasoning than that used by the author? If so, it is likely to be valid.

(4) Is the author of the paper a reputable scholar tested by peer review? If so, his work is likely to be valid.

(5) In case of serious doubt, discuss the paper with a *good friend*. If it is discussed with anyone else, one may give the impression of an amateur questioning a field and no useful discussion ensues.

(6) If the new information is used, will the theoretical construct then derived predict correctly? If so, chances are that the information is valid, especially if it predicts rather diverse phenomena correctly.

In working on a synthesis one is usually concerned with arranging diverse facts in a logical network into a system, and often into a particular kind of system, an adaptive syndrome. To do this, one must often imagine missing facts or logical links, and then work as if these assumptions were valid, construct the conceptual framework, and test it by its predictions. One must be prepared to work with diverse assumptions. This, however, is not peculiar to interdisciplinary research alone. The important point is to continually test by looking for the most unlikely prediction.

Regardless of how poorly or how well a synthesis has been achieved, a review of chapters by colleagues showed me that their content was often difficult for the reader to evaluate. In large part this is due to the peculiar mix of disciplines included. This is one more reason why I ask the reader for some patience. Please, neither reject nor too readily accept these writings. I can assure you that what I am writing about is dreadfully important in the sense that this is knowledge that ought to be heeded by professionals who are the builders of our society and thus of your environment and mine. However, this knowledge requires critical appraisal and development.

I might be permitted to give a brief description of how this book came about. It started in the winter landscapes of the northern Canadian Rockies during my investigation of the ecology and behaviour of large mammals. I became suspicious during those studies that some of my insights were relevant to an understanding of humans, and my early excursions into anthropological literature only strengthened my suspicions. Whether much would have come of it, I do not know, had I not accepted a position with the Faculty of Environmental Design at the University of Calgary. Here I was confronted by diverse academics and professionals, urbanists, architects, planners, economists, lawyers, scientists, all of whom perceived themselves as interveners into human environment and, at least implicitly, as benefactors of humanity. One of my tasks was to teach graduate students of highly diverse backgrounds those aspects of human biology relevant to environmental design. It was here that I ran into problems.

I focused on health. However, by focusing on human health I was confronted on one hand by the inadequacy of presently available literature and on the other by the insights that I as a zoologist had that were relevant to the subject. We had to answer the question of what environments we ought to design toward without robbing individuals of health. Some, like Boyden (1972, 1973), have answered that we ought to look at primitive human cultures of hunter–gathers and learn from them the determinants of optimum health. That, however, is a mistake in principle. As a zoologist, I am aware that natural populations of mammals may be far from healthy; in fact, animals are healthy only under exceptional circumstances. Also, the data on parasites, pathogens, and development of individuals in some of the primitive cultures presented no appealing picture. A model of man the hunter–gatherer was, for these reasons and others, quite useless as a baseline concept for environmental management, design, or preventive medicine. Which model was useful?

The literature is not a great help in answering this question. At one extreme, we have the spirited works of social critics who point out what is wrong with our society. There is Gidion's (1948) revolt against mechanization, Rachel Carson's (1962) historic attack on pesticides, Hall's (1974) critique of the food industry, René Dubos' (1968, 1972) concerns with professional tunnel vision, as well as those of reformers; Ivan Illich's (1975) war on our education and medical establishments, Galbraith's (1958, 1967) critique of the economics of the industrial state, Schumacher's (1973) crusade against conventional economics, or Edward Goldsmith's perpetual volleys at the establishment in the *Ecologist*. We have with us the symptoms of justified discontent in the form of the environmental movement, the consumer movement, advocacy planning, an upswing in public hearings on major decisions once left to government and bureaucracies, the United Nations conferences on environment and housing, and also the biting sarcasm of Parkinson (1957), Peter (1972), or Boren (1972) criticizing big bureaucracies. Yet a mere critique of society leaves unanswered the question—what is a better alternative? To do the antithesis of what is criticized is no guarantee of betterment. There have been brave and most constructive attempts to provide alternatives such as the *Blueprint for Survival*.[1] There are environmentally oriented political parties afoot to work on and get a vote on alternatives to the present socioeconomic systems. There is no shortage of suggestions for improvement from within, as well as without, the establishment. What is lacking is a tangible vision of the desirable future, and criteria that tell us if and when we have reached that state. One aspect of a desirable future that transcends political views because it is so fundamental is health. What are the criteria that tell us that we are on the right road toward that goal?

[1]*Ecologist* **2(1)**, January 1972.

It may appear that with so fundamental a question many would have given answers to it. A review of the pertinent literature shows differently. There is no shortage of books on man and the environment, nor on preventive medicine. However, the disciplines of environmental health and preventive medicine are a collection—a very detailed, very well-organized, collection—of the crises generated by modern societies and how to deal with them. We have an account of the dangers we face in our artificial self-made environments. We learn what not to do, but not how to anticipate what not to do. There is no theoretical guidance. Preventive medicine and environmental health are essentially reactive and therefore highly relevant for the past and useful for the future in so far as the past repeats itself. It does not tell us how we ought to live to maximize health, except by telling us what not to do.

This criticism does not make the works of environmental health and preventive health irrelevant, but only shows that they are incomplete. Professionals in modern society would be well advised to concern themselves with environmental health. Nor are the conceptual gaps in preventive medicine or environmental health simply filled by examining conventional aspects of human biology. It assumes that a student familiar with human biology is not likely to make serious errors in his professional life. Such an approach to teaching about man was done by J. Z. Young (1971) and recorded in a truly wonderful book, *An Introduction to the Study of Man*. This is the fruit of a life-long study and teaching of dental and medical students. Yet it leaves a gap between biology and environmental design; one cannot readily distill from Young's approach normative criteria, and one cannot expect students to do what Young himself does not.

An exposure to urbanists and architects makes one aware of the long history in these professions of a deep concern with environmental matters pertaining to human health. Fresh air; water; sunshine; the control of sewage; the availability of green spaces and of nature close to habitation; and localities for family life, socializing, and recreation have been very much on their minds, in their writings as well as in their designs. It is quite astonishing how much these professionals concerned themselves with the writings of Konrad Lorenz, Robert Ardrey, Desmond Morris, and Lionel Tiger. They sought the collaboration of sociologists, medical experts, psychologists, and ecologists, and stimulated the present man/environment field with their concerns. As critical as I am of its writings, I recognize the deep humanistic concern that gave rise to it. In general, though, these design professions often had a very healthy common-sense approach to providing built environments to house people. Yet again, the question raises its head, "What are the critiera for health?" Common sense will not answer it, nor the man/environment field, as I have indicated in the final chapter. To gain an appreciation of the concern for a supportive functional environment, I might briefly point at the tome edited by Buchwald and Engelhardt (1973) on landscape management, a four-volume work of some sixty authors, an account of the German experience.

To answer the question of what environment will maximize health, one must return to basics. One must know what an organism is designed to do by natural selection. One must therefore know the adaptive strategies of a species, and also how they came to be. One must know in detail the environment in which a species arose. Without such knowledge, no adaptive strategy can be discerned. This, however, is a rather young field of science not only for humans but for organisms in general. This is one reason why this book is addressed in large part to evolutionary biologists, ethologists, and mammalogists. However, to understand the design of an organism is only one necessary prerequisite to answering the question of what environment will maximize health.

Another prerequisite is to know under what conditions health is maximized in natural populations of organisms. Studies by my students and I accidentally opened a window on this matter. We had studied the phenotypic response of sheep to different environments and compiled a syndrome of characteristics of individuals living in different environments. Individuals from good habitats differed conspicuously from those in less favorable habitats in size, maturation rates, reproductive output, social behavior, and also in incidence of parasitism and congenital deformities. We could therefore predict in principle the health of individuals from their phenotypic development, including their behavior. I have discussed this in Chapter 6. Now it was also clear why classical medicine could not possibly develop a theory of health; it had no way to investigate the correlates of health in natural populations of large mammals, but investigated their dysfunctions at best in laboratories. The investigations among zoologists of phenotypic variabilities as related to the environment were in their infancy and no theory existed to explain the significance of the phenotypic responses. Such a theory had to be developed and it provided additional surprises.

To understand human adaptive syndromes is to understand human evolution. On the face of it, that would seem to require little more than a reading of textbooks of anthropology. However, the accounts of anthropologists simply would not square with the findings students of large mammals had developed, nor did I find explanations for many human characteristics, such as our social system—so different from that of other primates—or our vocal and visual mimicry, humor and laughter, superb self-control or self-discipline, altruism, or our very intellectual competence, etc. Yet it was very important to know how these features—diagnostic features—arose, since a theory of phenotypic response to environment stated that health was maximized when the diagnostic features of a species were maximized. Here I reached a no-man's-land.

I could not rely on the anthropological account of human evolution. It was, from my perspective as an ethologist, but especially as a student of large mammals, as an ecologist, and as a student of evolutionary biology, not adequate. I was puzzled by statements such as "the volar pads on our feet cannot be explained." Had no anthropologist ever taken off his shoes and stalked a deer? Apparently not. Had they done this in earnest they could hardly have missed the significance of volar pads or, for that matter, would they embrace the view that feet evolved for walking. What to an ungulate behaviorist was so obvious—that a naked human being with his padded feet and superb control over his leg muscles is capable of getting closer to shy wild ungulates undetected than any carnivore—was not so obvious to others. Or take the hypothesis that early robust hominids were seed eaters. This view conflicts with their body structure, for to feed on this small, highly digestible, fraction of the plants' biomass requires a small body built for roaming, as we know from ungulate studies. Or take the hypothesis that the discovery of relatively advanced hominds millions of years prior to the extinction of the australopithecines disqualifies the latter from being the ancestor of hominds. As a student of large mammals, I am only too aware that very primitive ancestral forms not only can survive side by side with advanced ones, but may even *outlive* them! In no way do the recent Lake Rudolf finds disturb the ancestral position of the australopithecines. One has to know something about evolution and extinction of large mammals in general before proceeding to an interpretation of human fossils. I had little choice but to bring such understanding to bear on human evolution, and the resulting picture departs from the presently accepted one. Alas, there is no guarantee that I have not made large blunders myself. Regardless of this, it is important that an understanding of large mammal evolution be brought to bear on human evolution. Whether I succeeded is another point.

In the studies of mammalian evolution, I used a conceptual approach which can be labeled today as a "systems approach," albeit one without computers. One simply asks, what ecological niche or profession and what adaptive strategies are compatible with the empirical data about the animal and its environment? If one formulates correctly a given adaptive strategy, then it will predict correctly the behavior, morphology, etc., of a species. One formulates adaptive strategies until every adaptation is accounted for that is logically related to the systems of actions and organs by means of which a species extracts its share of resources from the habitat, escapes predators, finds mates, etc. One can also use the very same approach on extinct forms, albeit with greater uncertainties. I have illustrated this approach in Chapter 7, giving an account of the transformation of reptilian features into mammalian ones, using one—and only one—major adaptive strategy. I used the same approach in reconstructing the adaptive strategies of successive hominid species, adaptive strategies that are compatible with the known facts about these forms. Each explanation in this process must be compatible with earlier explanations, but the more complex the system of interlocking, logically related, explanations the more unlikely their predictions. Therefore, if the predictions are verified, the less likely are the explanations to be invalid. Moreover, the adaptive strategies of an earlier form must be smoothly transmuted into those of a later form by being placed in the environment of the new form. A perfectly acceptable adaptive strategy for a primitive form that fails to transmute logically into that of an advanced form in the latter's environment is invalid. Thus, postulated adaptive strategies not only must be internally logically consistent and harmonize with all known facts about a form, and be compatible with basic rules of adaptation, but must also pass the test of transmutation. Thus, the adaptive syndrome of *Australopithecus* had to change into that of an advanced *Homo* if it encountered the environmental condition of the next most advanced hominid. The adaptive syndrome of a *Homo erectus* had to transmute into that of *H. sapiens,* etc. The approach I used is thus a hypothetical deductive model building of increasing complexity and improbability of prediction.

It surprised me many times. I began with the hypothesis that *Australopithecus* was human, and discovered that he could not be. I assumed that *Homo erectus* was not human, but found this assumption untenable. Human characteristics that were unexplained found a logical explanation. Humor, laughter, dancing, role playing, music, altruism, self-discipline, the nuclear and extended families, the waxing and waning of sexual dimorphism, all fell out as logical consequences, as important parts of adaptive syndromes during various stages of human evolution. To a mammalogist, it was not surprising that the course of human evolution paralleled that of other mammals, with an ancestry in the tropics and an adaptive radiation, first into increasingly xeric habitats and thence, by preadaptation, into temperate, cold, periglacial, and arctic ones. Other lineages of mammals that produced cold-adapted bizarre giants during the ice ages had done much the same.

The biggest—and inescapable—surprise arose when it became evident that, at least since the beginning of the last glaciation, and probably earlier, human beings must have artificially intervened in the natural process of their population growth to ensure a maximum of resources for development of individuals during ontogeny. If this is valid, then we are in large part a self-made species, and the superb physical development of late Paleolithic humans, is an expression of their wilful, directed, and guided manipulations. I was incredulous, because this did not fit with the conventional haughty picture we have painted of us in relation to "primitive" civilizations, let alone ancestral forms or the "aberrant" Neanderthal man.

Although I dwelt here on humans and health, as this is the central concern, I must point out that to get to a theory of health I had to develop a number of new perspectives on the following matters: the basic rules organisms live by, as deduced from the concept of reproductive fitness; a new conception of animal communication; a comprehensive theory of aggression, based on individual selection; a theory on the biology of art and culture; a theory of phenotype development through communication between genes and environment; a theory of mammalian origins explaining those features we share with other mammals; and a theory about how glaciers shaped the ecology in their vicinity to form the rich periglacial ecosystems, our final evolutionary home. Some of what I have elaborated on here has appeared in print. In particular, I point to Geist (1975a), a popular book in which I broached some of the subjects developed here, and Geist (1975b), a paper that outlines the theory of health I develop here.

Some of my conclusions bothered me, such as the need to abandon the neo-Darwinian paradigm of natural selection in favor of the Waddingtonian. So did the recognition that the fate of health was not at all in the hands of the medical profession, and the recognition that the social structure, phenotype, and health of individuals, in both animal and human societies, are a close consequence of their economic system. Only in the last chapter of the book could I deal with the implications of the preceding fourteen chapters for environmental management and design. Even here, I cannot go very far, for, if health is the consequence of an economic strategy, then it can be maximized only by shaping economic systems and social values to become compatible with this goal. Here I am completely out of my depth. I cannot make more than the vaguest suggestions about how to achieve a durable, humane, healthy life style, but can only indicate what ciriteria it has to meet. I am left with a brooding suspicion that, to achieve such a life style, fundamental changes in presently accepted economic strategies, and therefore in social values, are required. However, I do know that by focusing on a policy of maximizing individual competence we ought to come close to durable healthy human life styles.

If I saw wrongly, dear colleague, what is the alternative?

Alberta, Canada
1979
                                                                    Valerius Geist

Contents

Chapter 14 / From Periglacial to Artificial Environments   354

Chapter 15 / Health, Professionals, and Creature Comforts 400

References   436

Index   481

Chapter 1

Organism Theory: The Dictates of Genes, the Meanings of Environment

Introduction

If we are to answer the question of who we are, if we are to explain why we are built the way we are, and why we function and form societies, institutions, and cultures the way we do, then we can begin by examining some fundamental rules that all creatures are subject to. No serious examination of ourselves ought to bypass these considerations because it is these rules that tie us to life on earth and make us a part, a spectacular part, of it. The rules I refer to are logical derivatives of the theory of evolution as we presently understand it; they are rules that encompass the causal system of our existence like the limits of a universe beyond which none can penetrate. These rules are the ultimate causes shaping an organism's historical development and we hit upon them if we trace back the chains and branches of causal links from whatever point in the biology of a creature we care to start with; they are the rules by which organisms extract their share of fixed solar energy and nutrients from plants, and by which organisms maintain and replace themselves.

To understand these rules, we must begin with a fundamental generalization of evolution. Individuals act on behalf of their genes; this is a clear derivative from the concept of *reproductive fitness* which defines the individual's biological success (see Mayr 1966). To maximize its reproductive fitness an individual maximizes the proportion of individuals endowed with its own genes in the succeeding generation. Reproductive fitness thus measures the effective reproductive output of an individual *after* natural selection has acted. Clearly, reproductive fitness and not Spencer's "survival of the fittest" is the driving force of evolution, and Charles Darwin, in the sixth edition of his book on the *Origin of Species* (1972, p. 52), was aware of just this, even in the absence of our modern theory of heredity. But in maximizing reproductive fitness an individual must clearly act on behalf of his genes, not on behalf of itself as an individual, and at best on behalf of other individuals only if they share a significant fraction of the same genes. This notion is termed "inclusive fitness'; an individual contributes to the success of other individuals in proportion to their relatedness (see Hamilton 1964).

If the driving force of evolution is indeed reproductive fitness, then genes are subject to conditions that can best be summarized as grand strategy, strategy and tactics of survival. Now, genes are not, of course, conscious individuals adjusting their aims and means to achieve certain goals. Yet such an analogy is useful as it leads to quite unexpected insights, and an excursion into strategy, along the lines of military strategy discussed by Liddell Hart (1967), or into the mechanisms of the complex industrial market society defined by Kenneth Galbraith (1967), can be most enlightening. Genes, directing very complex individuals as executers of their goals, appear to run—really not very surprisingly—according to rules similar to those of other teleonomic complex systems, be they armed forces or corporate enterprises. Had Liddell Hart been with us still, he might have relished the thought that organisms in their existence and evolution are masters of the *indirect* approach to strategy.

The policy of the gene must be to ensure its existence and its grand strategy, on one hand to secure its gains against competitors and the vagaries of the physical environment, and on the other hand to expand, either geographically or into new resource exploitation opportunities, whenever the opportunity presents itself. However, genes act through individuals. What must an individual do to maximize its reporductive fitness? It ought to do the following.

(1) Minimize expenditures on maintenance so as to save a maximum of resources for reproduction.
(2) Maintain physiological homeostasis. That is, keep healthy and unharmed.
(3) Create and maintain access to scarce resources essential for reproduction.
(4) Reduce directly the reproductive fitness of others.
(5) Support individuals with similar gene compositions in relation to their relatedness.
(6) Grasp every opportunity to expand into unutilized resources.
(7) Mate with individuals equal or superior to itself in maximizing reproductive fitness.

These rules apply to all individuals regardless of what species, provided they are reproducing sexually. They apply simultaneously, so that a compromise must be struck by an individual to the extent that it follows the dictates of one rule rather than another. Therefore, these rules cannot be deterministic (philosophers, please note!). Furthermore, individuals can and do opt out of reproduction and then act contrary to at least some of these rules until the opportunity arises for them to enter into reproduction. These rules imply the existence of very basic mechanisms of monitoring and evaluating alternatives and adjudicating between then. We shall dwell on these in the second chapter and again in the last. Here we shall examine in detail only some of the rules derived from reproductive fitness. We shall also examine some rules generated by the mechanisms of gene transmission and expression, as well as the logic of gene dispersal. Ultimately, this is vital to the development of a theory of health.

Primary Rules of Reproductive Fitness

Let us begin examining the rules deduced from reproductive fitness by analyzing the dictate that individuals ought to minimize expenditures on maintenance in order to maximize reproductive fitness, thereby sparing maximum resources for reproduction. *This is the Law of Least Effort* (Zipf 1949, Arnheim 1971). An examination of nutrition and bioenergetics soon reveals a reason for this law.

To living organisms, energy is precious and fleeting. Relatively little of it is fixed by green plants from the sunlight; energy is usually hard to come by and often hard to liberate from the food; it is easily lost, since the energy cost of living is surprisingly high and energy is converted to work or tissue with a low efficiency. For every calorie of work performed, the animal loses, in addition, from three to five calories of heat. Thus, no effort must be spared by the animal, no opportunity missed, to conserve ingested energy. It must be a niggardly miser, a clever investor, a good housekeeper, an opportunist, for only in being so, will it spare—with some luck—sufficient resources for reproduction to ensure the continuity, let alone the dominance, of its genotype. For a calorie saved from maintenance is a calorie available for investment in reproduction.

The evidence for the foregoing comes from the textbooks of bioenergetics which summarize the research done primarily on man and domestic animals for the benefit of medicine, agriculture, and also nutrition and physical education. The great classics are the books by Samuel Brody (1945) and Max Kleiber (1961), which remain to this day veritable treasures. In recent years, bioenergetics has begun to recover from the neglect it suffered in the postwar era, and has begun to extend itself into zoology, ecology, environmental science, and wildlife management. Moen's (1973) book can be taken as a sign of this interest, as can be Schmidt-Neilson's (1972a) and Watt's (1973). Much basic information is also found in text books of physiology (e.g., Hoar 1966) and medicine (e.g., Guyton 1971). Let us turn to examining the costs of life, for they quickly explain the rationale for the law of least effort.

In choosing the following examples, I hope I will be pardoned for sticking so closely to large mammals, in particular to ruminants, but also dogs and, only to a lesser extent, human beings. Animals illustrate the points to be made quite adequately; there is good data available for them; moreover, one of the overriding principles discovered by bioenergetics is that animals use energy in their physical system, not in accordance with their uniqueness as a species, but in accordance with their mass, surface area, temperature, etc., that is, in accordance with their physical parameters.

The *energy cost of activities for humans* are as follows. The basal metabolism, the lowest metabolic rate measurable under a set of specified conditions which need not concern us here, is about 70 $kcal/wt^{0.75}$; resting already costs 90 $kcal/wt^{0.75}$; standing costs about 100 $kcal/wt^{0.75}$; walking doubles that to 200 $kcal/wt^{0.75}$; moderate work increases it to 240 $kcal/wt^{0.75}$; swimming costs twice as much as moderate work,[1] or 500 $kcal/wt^{0.75}$; hard labor such as that performed by loggers in the olden days with axes and hand-pulled cross-cut saws costs about 560 $kcal/wt^{0.75}$; running costs a bit more than that, or 570 $kcal/wt^{0.75}$; hard running, of which we quickly tire, costs about 1050 $kcal/wt^{0.75}$; climbing increases that to about 1100 $kcal/wt^{0.75}$; during exertion, when we go into oxygen debt, a man can put out for a few seconds up to 9000 $kcal/wt^{0.75}$. Since 1 hp is 10.7 $kcal/min$, a 70 kg person at peak exertion produces about 0.58 $hp/kg\ wt^{0.75}$ of heat, that is, some 14.0 hp.

Note the above figures referred to kilocalories of *heat* produced per 24h. per metabolic kilogram (weight in kilograms raised to the power of 0.75). The above

---

[1]However, per unit of distance covered, swimming is about five times as costly as the same distance covered by walking (Schmidt-Nielsen 1972b).

figures suggest that activities can usefully be measured either as multiples of basal metabolic rate or of resting metabolic rate. Resting heat production requires some 95 kcal/wt$^{0.75}$ of metabolizable energy (for convenience sake, rounded off to 100 kcal/wt$^{0.75}$). This is a useful shorthand; thus the cost of standing tends to be $1.1-1.3$ times basal, work tends to cost $3-8$ times basal, running and sprinting have upper ranges of $15-20$ times basal; exertion costs some $40-100$ times basal or more. Most activities, however, are at the low-cost end of the spectrum; walking— depending on gradient—costs some $1.5-2.5$, basal, foraging by sheep, cattle, and deer costs $1.3-2.0$ basal, and ruminating costs around 1.3 times basal.

The foregoing showed the cost of various activities, and it can easily be seen that costs accelerate drastically with an increase in activity. Let us now turn to see the fate of food on its way to becoming useful work. Let us assume a cow, sheep, or deer ingests 100 kcal of good quality hay; this would amount to about 22g. in weight. Its digestibility is taken somewhat arbitrarily at 56.6%.

(1) From 100 cal of the *gross energy* ingested, 43.4 cal are lost in the feces.

(2) Of the remaining 56.6 kcal, which are termed *apparent digested energy,* some 10.6 kcal are lost in the urine and as methane gas belched from the animal. This leaves 46.0 kcal of *metabolizable energy* in the body.

(3) The energy cost of simply metabolizing this energy, called the *caloric effect,* and in this case amounting to some 10% of the remaining energy, must now be deduced. This leaves 41.4 kcal of *net energy* in the animal.

(4) If the 41.4 cal of net energy are used for work, then at the most, 10.3 kcal of work will be accomplished. Top efficiency of converting net energy to work is about 25% (see Brody 1945, Kleiber 1961, Schmidt-Nielsen 1972b).

(5) However, the above is an optimistic figure; after working, there is a recovery cost to be paid which varies between 25 and 10% of basal metabolism (Kleiber 1961). This, of course, reduces the overall efficiency of converting energy to work. We find from Kleiber that the efficiency will be reduced from 25 to some 21%. Thus, 41.4 cal of net energy will produce, at best, 8.7 kcal of work. Since at low rates of work the instantaneous efficiency of energy conversion is less than 25% the above is most optimistic.

(6) Let us assume that the digested net energy could not be used at once for work, but was stored for future use, then 41.4 kcal of net energy will convert into about 20.7 kcal of fat, while the remaining 20.7 kcal will be lost as heat. This is the cost of lipogenesis, or fat production; it costs $50-70\%$ of net energy, depending on forage type (Blaxter 1960, Kleiber 1961). The 20.7 kcal of fat will amount to about 2.2g. of body fat.

(7) Should the animal now mobilize the 20.7 kcal of stored energy for work, it will accomplish at best 5.2 cal of work and lose 15.5 kcal as heat, not counting the cost of recovery. If recovery cost is included, then at best 4.3 kcal of work will be accomplished.

Thus, 100 kcal of good hay will yield only some 8.7 cal of useful work and, if the energy is first put through a fat storage depot, the animal will get only some 4.3 kcal of work, due to the high cost of lipogenesis.

The foregoing has made it evident that, not only does the energy cost accelerate greatly with the increase in activity, but that energy is processed in the body with a very low efficiency. Moreover, as we shall see next, the animal's ability to ingest and store food is very limited so that very little excess energy is available.

How much energy does an animal take in daily? A roe deer *(Capreolus capreolus)* on good forage in summer ingests only some 197 ± 53 kcal/wt$^{0.75}$ of digestible energy; in winter it is only $70-102$ kcal/wt$^{0.75}$ (Drozdz and Osiecki 1973). Since 197 kcal of apparent digestible energy are equivalent to $197 \times 0.81 = 159$ kcal of metabolizable energy, without calculating the caloric effect this will permit the animal some 45 kcal/wt$^{0.75}$ in work without going into negative energy balance. In winter, however, the roe deer would be working with an energy deficit which he would have to balance by drawing on his fat depots.

The above figures for roe deer are within the ranges of energy uptake and expenditure reported for various domestic and wild ruminants. Maintenance costs for sheep on good and poor pastures range from $1.1-2.0$ times basal according to Lambourn (1961), Graham (1964), and Young and Corbett (1968); for 2-year-old cows, B.A. Young (1971) reports maintenance values of some 114 kcal/wt$^{0.75}$, and for cows raising calves from $140-189$ kcal/wt$^{0.75}$. For white-tailed deer *(Odocoileus virginianus)*, Moen (1973) reports expenditures of energy varying between 1.24 and 1.70 times basal. Only under experimental conditions on selected forages of high quality do ruminants ingest up to $350-400$ kcal/wt$^{0.75}$ of apparent digested energy (Blaxter et al 1961, Kleiber 1961, Graham 1969). For nonruminants, the intake values appear to lie somewhat higher judging from Kleiber's (1961) figures, and for carnivores we expect them to lie higher still, since their adaptations require greater energy expenditures per unit of food gained, which is the reason they convert energy and nutrients to protoplasm from the herbivore layer with a lower efficiency than do herbivores converting green plants (see Watt 1973).

These figures indicate that ruminants, for instance, take in only slightly more than they expend—at best—and that they may easily slip into the red and have to fall back on stored body fat. This is, in fact, how deer must live in cold climates; they do not take in sufficient food in winter to cover the cost of living, and depend on their fat stores from the excesses of summer to survive the winter; this can be deduced in part from the great weight and fat losses they incur during winter (Hall 1973, Drozdz and Osiecki 1973).

Earlier on, I drew attention to the high cost of accelerated activity. However, an animal can easily go bankrupt expending small amounts of energy frequently, such as when it gets excited about something in its environment, be it a predator, conspecific, or some strange object or situation. Energy cost of excitation is hard to average out, since it can easily shoot up to 200 kcal/wt$^{0.75}$ and quickly decline again, but chronic excitation costs somewhere between 20 and 50% above maintenance, and can easily last for days (see Graham in Blaxter 1962, Webster and Blaxter 1966). Somewhat similar results have been reported for the increased energy expenditure of mice in groups by Gorecki (1968), and by Moen (1973) for deer.

Lactation and gestation are both energetically expensive, and are also not easy to average, since they vary with time, gestation cost increasing up to parturition, and lactation cost declining after an initial upsurge following parturition as the young grow older. Kleiber (1961) cites data for dairy cows, indicating a milk production of about 25 kcal/wt$^{0.75}$. Since the energy cost of milk production is some 1.6 times that of the energy in the milk, the cost of milk production for these cattle is about 40 kcal/wt$^{0.75}$ above the cost of maintenance. This checks nicely with the figures given by B. A. Young (1971) that cows raising calves had a cost of $140-189$ kcal/wt$^{0.75}$.

What is the consequence of disturbance to animals, say to a herd of wintering caribou repeatedly disturbed by aircraft? First, the energy expended beyond the daily intake must come from the precious energy stores the animals accumulated in fall and summer. The bulls have mostly lost these during the rutting season, anyway. In these animals, as in other northern ungulates, the mortality of males probably increases if they lose their fat depots (see Heptner et al 1961, Bannikov et al 1961, Geist 1971a). Therefore, fat loss means increased adult male mortality.

In the females, the fat is required to nourish the growth of the fetus, particularly in late winter. Disturbance, regardless of source, leads to loss in body weight in reindeer (Preobrazhenskii 1961, Mokriden 1961), domestic goats (Liddell 1961), and domestic fowl (Allee and Guhl 1942). It leads to increased metabolism (Graham in Blaxter 1962, Webster and Blaxter 1966) and reduced food intake (Allee and Guhl 1942); it precipitates in domestic sheep psychological stress and pregnancy toxemia (Reid and Miles 1962); it raises the level of adrenocortical hormones and affects not only unborn young (Thompson 1957), but also the young of unborn young, that is, its effect is transmitted across generations (Denenberg and Rosenberg 1967, Moore 1968); it may cause disintegration of muscle fiber due to extensive running, *overstrain* disease (Young and Bronkhorst 1971) and reduction of horn growth in goats (Liddell 1961). If reindeer females lose their fat depots, they may either begin to resorb the fetus, which begins happening when their body weight decreases by more than 17% (Preobrazhenskii 1961), or they may give birth to a small, poor-quality fawn that is destined to die at birth or shortly thereafter. As reviewed in Geist (1971a), small calves tend to be not only poorly developed, but also, as we know from studies on sheep, where this point has received particular attention, may have poorly ossified joints, and have so little energy stored in the form of fetal fat that they fail to survive to the first suckle. Many of these calves are not expected to rise. If the energy expenditure of the female is incurred during lactation, it reduces milk production and consequently the growth and development of her young. Exertion during late pregnancy can lead to abortion in reindeer and mares (Nikolaevskii 1961, Nishikawa and Hafez 1968); it also makes the animals susceptible to diseases (Preobrazhenskii 1961), and in late winter to emphysema (Nikolaevskii 1961). Saiga antelope may also die of pulmonary edema due to chasing by cars (Bannikov et al 1961). Severe running of ungulates can also lead to death through overstraining disease as described by Young and Bronkhorst (1971). How severely reindeer are affected by hard work is illustrated by the fact that a herder requires several mounts a day in order to do his work, just as cowboys did when ponies were fed mainly on grasses. Moreover, reindeer need to be rested for several days after working, and females go barren from work unless fed concentrates (Nikolaevskii 1961, Popov 1961). Clearly, it is best for the animal not only to maximize energy intake, but also to avoid excessive expenditures. It must ration and parcel out its energy reserves very carefully, and it is this phenomenon that accounts for the law of least effort.

The foregoing illustrated that energy is indeed difficult to obtain, costly to store, and very easy to lose, and that organisms must indeed conserve it as much as possible or not have a sufficient amount for reproduction. We find the law of least effort wherever we may look in the morphology, physiology, and behavior of organisms. One expression of this law, and a very striking and important one for the explanation of the structure of organisms, is that organisms closely follow physical laws in their structure and function. D'Arcy Thompsom (1917) explored this phenomenon in his justly renowned book *On Growth and Form,* but he apparently did not recognize that, in principle, taking advantage of the laws of mechanics was the cheapest way to exist, because it costs energy to act against the forces of gravity, surface tension, etc. In the growth and functioning of their forms, animals have captured and incorporated to their advantage the normal properties of matter; they have not fought but rather "parasitized" mechanical rules in order to reduce the cost of maintenance. We shall meet this tendency to parasitize existing opportunities again and again, such as in combat, parasitizing basic tendencies to withdraw from stimuli distorting the body surface or causing trauma, female or juvenile mimicry by subordinates parasitizing the inhibitions of dominant males to strike females or children, courting males mimicking juveniles and parasitizing maternal responses of females, and dominance displays parasitizing the mechanism of cognitive pattern matching to generate high arousal (fear). In short, the organism takes advantage of existing structures or environmental dictates and fits itself to them, rather than developing redundant systems *de novo* or expending energy uselessly against environmental dictates. Some examples may be relevant.

If the form of many cells resembles that of a drop of liquid whose form is determined by surface tension, then the similarity is partly a result of the cell's incorporating the physical property of surface tension into its form. Surface tension is a convenient, free, shape-giving force in the size range of cells. No energy need be expended to assume that shape, but energy must be expended to deviate from the spherical shape. The law of least effort dictates that the cell, unless its function precludes it, utilizes the freely available force of surface tension in its structure (see Thompson 1961, p. 49).

The size of heads varies greatly but predictably with body size; the larger the animal, generally, the smaller its head, relatively speaking. This observation is again explained by the law of least effort. Because the metabolism of an animal is proportional to its surface area, which varies as the square of its linear dimension, but its mass increases as the cube of its linear dimension, it follows that metabolism per unit of body mass must decline as the body of an animal increases in size. Therefore, relative to its body size, it requires less food. Therefore, relative to its body size, it chews less food. Therefore, relative to its body size, it requires a smaller set of jaws and teeth than a closely related small-bodied form. Therefore, large-bodied animals have relatively smaller heads than small-bodied animals. Or, conversely, a mouse compared to a giraffe requires a much larger head relative to body size to hold relatively larger jaws and teeth to permit it to process more food to fuel its high rate of metabolism. The head is thus adjusted in size to the metabolic needs of the organism, unless the head serves additional functions in combat, for instance, when it requires additional mass.

Small mammals such as mice, squirrels, rabbits, duikers, small antelope, and deer tend to have relatively large hindquarters and small front quarters, while in large mammals such as large antelope, deer, giraffes, or elephants the front quarters may be as heavy as the hindquarters or even heavier. Again we find here organisms obeying the rules of mechanics, and in so doing they are obeying the law of least effort. The large hindquarters in small-bodied mammals permit sequential jumping as a form of running. According to Thompson (1961, p. 27) we note that, whereas the work done in leaping is proportional to the mass of the body and the height to which the body is raised, but the power available to do the job is proportional to the muscle mass of the body, with increasing body mass animals of similar build will tend to jump much the same absolute height or distance. Relative to body size, therefore, a small animal jumps much further than a large one. As body size increases, a point is soon reached when jumping barely propels the animal up or forward at all, while at greater size still the mass of the animal becomes too large to permit its muscles to lift it off the ground in a jump, as is the case with elephants. Clearly, large animals must adopt a form of locomotion different from jumping, since not only is the cost of jumping high, but it is inefficient in rapidly gaining distance as well. Hence the change from hopping to a cross-walk or run. One also suspects that the energy liberated in hopping would, in a large-bodied animal, be so high that heat dissipation might be a problem, since it does have a smaller surface area relative to its muscle mass.

These examples may suffice to support the point I am trying to make, that in their morphology organisms obey the laws of mechanics closely, as it is the best way to minimize energy expenditure for maintenance and maximize that available for reproduction. We find the same in examining their behavior.

If we note the activities of an animal during the daily cycle in order of their frequency of occurrence, we see at once that this corresponds to the order of increasing energy cost; the more frequent an activity, by and large, the less expensive in energy it is, and vice versa. This is comprehensible at once from the law of least effort, because it is clearly in the interest of the animal to reduce expensive activities to the very minimum and replace them with activities that achieve the same but cost less.

If we look at the trail system of mountain sheep or mountain goats, we are struck by the fact that they tend to run horizontally along the contours of hills; in fact, they are rather pleasant to hike on. This is also understandable at once from the law of least effort. To a domestic sheep, for instance, the cost of movement on the level is about 0.59 ± 0.63 cal/horizontal meter; the cost of moving vertically, however, is 6.36 ± 0.32 cal/vertical meter (Clapperton 1964, Graham 1964, see Moen 1973, p. 349). Much the same values apply to humans, cows, horses, and dogs where they have been measured (see Moen 1973, p. 349). That is, climbing costs some 12 times as much as walking on one level. Clearly, there is little point in climbing unless it cannot be avoided.

The above also explains another phenomenon every fieldworker has encountered: game trails converge on saddles and diverge in flats. The cost of climbing thus converges trails; where there is no such cost, trails are free to wander. Also, it can be noticed again and again that in crossing some strange country for a second time one finds one's own footsteps. In part this is also due to following the lay of the land so as to do a *minimum* of climbing.

The law of least effort is exemplified multifold by various big game animals living in the winter landscape, and their techniques can be readily adapted by humans who must survive in winter. Heat is lost through long-wave radiation, convection (particularly when it is windy), conduction by touching the substrate and evaporation of water from the body. To avoid radiative heat loss to the clear cold sky, particularly the nocturnal winter sky, a radiation shield above the animal is essential. Mountain sheep go into caves in the coldest months of winter, where they have not only a radiation shield above them but also a reflector and a spot with reduced convective heat losses. Long-wave radiation leaves the animal's body, but bounces back from the cave walls and returns to the body again; furthermore, a group of sheep in a cave absorb each other's long-wave radiation. Deer move under heavy conifer cover at night, which acts as radiation shielding (Moen 1968) and also reradiates the heat it absorbed during the day. This may also explain why sheep frequent the cliffs at night in late winter; rocks absorb heat and give it off at night to a much greater extent than the moist or snow-covered mountain slopes.

To reduce convection heat loss, the animal should get out of the wind, or at least search out areas with minimum wind speed. This is found in the forest, of course, or behind obstacles, but also just below the crest of hills. Here one sees cattle huddled into masses during blizzards; here one finds deer during strong winds.

Conductive heat losses occur primarily if the animal is in contact with a medium that has a high conductance of heat, e.g., water or stones. To reduce conductance, the animal requires a layer on which to rest that has a very low conductance, and must avoid resting in water—unless it is very hot and heat loss is aimed for rather than heat conservation. Usually the animal's hair provides this layer of low heat conductance. In addition, animals prefer dry places with duff, litter, or a layer of their own dry fecal matter; moose slump into deep powdery snow, which reduces conductance (Des Meules 1964).

Another way to retard heat loss is through a decrease in the surface-to-mass ratio of the animal. The more an animal resembles a sphere in shape, the less heat it will lose; conversely, the more its surface area deviates from that of a sphere, the more heat it will lose. If we give an animal a standard food ration, but vary its surface area, we can plot the temperature at which it goes into negative energy balance. For instance, a decrease in relative surface area from 1.67 (the relative surface area of a cow[2]) to 1.4 results in a saving of heat energy of 18.4% (Geist 1972). This makes it evident why dogs, deer, and pronghorn antelope curl up when resting in the cold, and why snakes clump in masses during hibernation.

Mammals can and do make use of direct solar radiation. In early winter, moose are found in northern Canada notably above timberline on slopes exposed to the weak rays of the winter sun. Do they really gain anything from it? Research on cattle demonstrated that for every megacalorie of heat that falls on a black cow, the cow saves one-third of a megacalorie of fat from oxidation (Webster 1971). Granted the large size of moose, they must be able to save an appreciable amount of fat by exposing themselves to the sun.

Selection of microclimates such as little solar bowls in which the topography of the land concetrates the radiant energy of the sun, pushing activity into the warmest part of the day in winter (Geist 1971a), pawing only in soft and not in hard soggy snow or through hard snow crusts, and staying in the inversion layer in the mountain where temperatures are higher than in the valley or in the peaks are all examples explained by the law of least effort; we shall meet many more in the course of this book. The law of least effort states the kind of solution the animals should strive toward. The law of least effort has a logical inverse, the law of maximum gain, and occasionally it is useful to think of adaptive strategies from this point of view.

Since animals do exert themselves greatly at times and appear to squander resources, such as male deer during rutting, it must be emphasized that the law of least effort applies only in so far as it maximizes reproductive fitness. Applied without this important qualitification, it leads to ambiguities, for all our behavior, clearly, is not governed by the law of least effort alone. According to students of Freud (see Arnheim 1971), Freud recognized the law of least effort as important in human behavior. Arnheim's (1971) uneasiness in accepting this appears to stem from his unawareness that the law of least effort in organisms is subject to the

---

[2]The surface area of a standing cow is about 1.67 times that of a sphere of equal mass.

dictates of reproductive fitness. The law of least effort is not paramount, but operates in so far as it maximizes reproductive fitness by ensuring that no more energy and nutrients are spent on maintenance than will ensure survival and that the remainder is channeled into reproduction.

The second rule organisms must obey is that *individuals must maintain homeostasis*. They must, therefore, maintain the body in good functional repair and parcel out the ingested energy and nutrients appropriately.

When we investigate how resources used for maintenance are expended, we enter a broad, diverse, and complex field of investigation. These resources first pay the cost of physiological functions encompassed by basal metabolism, the cost of growing and renewing tissues and organs, and the cost of repairing the body, all of which are the domain of physiology, animal science, and medicine. The remaining resources, beyond the immediate needs of the body, are invested in various activities, such as the gathering of resources acted out in diverse *strategies of resource exploitation*, protecting the individual by diverse *strategies of predator avoidance*, reducing the contact with and effect of *parasites and pathogens*, and moving the individual on the basis of daily, seasonal, or situational demands into different *microclimates*. Not only do species differ in these strategies, but so do individuals of different sex and age, or individuals under different ecological conditions. Moreover, the strategies interfere with one another so that one dominant strategy, say of predator avoidance, affects the strategy of resource exploitation, or vice versa. This means that one strategy can, and does, severely limit the options of evolution. Once a given strategy of resource exploitation is adopted, the species begins to conform to the dictates of that strategy in a rather predictable fashion. For instance, if a species of ruminant exploits the highly digestible fraction of the plant biomass, then natural selection keeps for it a very small body size, according to the reasons expounded by Bell (1971). Besides, such a species can attain only a very low biomass owing to the low biomass of the highly digestible parts of plants (fruits, flowers, buds, shoots) (Jarman 1973). Therefore, such a species may have a very low density of individuals per unit area of habitat. Small body size makes hiding or flight mandatory for such a ruminant, whereas large-bodied forms can confront predators, and are then altered so as to confront predators successfully (see Geist 1974c). In forests, hiding is a good strategy that calls for minimum visual and vocal stimuli by the individual, and this selects for solitary existence, as well as inconspicuousness while active (Geist 1974a, d). This, in turn, puts a premium on olfactory communication, which is not only functional in the dark, but also separates the communicator from the medium, as in territorial marking. Dusk and darkness permit hiding, hence the greater activity of forms that hide from predators at that time, and the concomitant development of good hearing and rather poor eyesight. The above feeding strategy also promotes pair territories, and species opting for this also opt for reduced sexual dimorphism (Geist 1974a) and for small damaging combat organs and poor defences (Geist 1978a).

It can be seen from the foregoing example that a food strategy, if dominant, locks a species into adapting into one of a few options only. The same appears valid for antipredator strategies. The greater the diversity and density of predators a given species faces, the greater the security measures it requires. Very large-bodies species compared to very small ones are likely to require the fewest security measures. Thus we would expect elephants, for instance, to shape their home ranges closely to food availability, while small rodents may shape theirs first to maximize security and only then to food intake. If so, large forms are free to evolve diversity in adaptive strategies, but small forms are not. This may explain the surprising frequency of small mammals fitting into the "rat" or "mouse" syndrome, while the giants tend to fill the ranks of the unique life forms (Chapter 8). Small bodied animals, in maximizing security, may converge into a very few but eminently viable life forms; large bodied animals are free to shape themselves to exploit diverse food sources, almost unhampered by predation. Large body size appears to be the prerequisite for evolutionary innovation.

Antipredator strategies may take the form of imposing obstacles of the habitat between the prey and the predator, as we see in the mountain bovids that run into cliffs upon the appearance of predators. Antipredator strategies may also be independent—within reason—of terrain features, such as the clumping of individuals into large groups on flat, open landscapes. In either case, the antipredator strategy powerfully shapes the body proportions, social behavior, and weapons of the species (Geist, 1971a, 1978a).

The energy available for reproduction cannot be spent haphazardly to maximize reproductive fitness. Individuals must spend their resources on reproduction in such a fashion as to maximize the proportion of viable successful individuals per unit of resources invested in the following generation. We are dealing here primarily with *parental investment,* a subject formally and elegantly treated by Trivers (1972, 1974) and Trivers and Willard (1973) (see also Wilson 1975). Trivers' model is a powerful predictor of the behaviors of males and females according to the amount of resources each sex invests in offspring; it permits one to evaluate the costs and benefits of various strategies of reproduction and predict the actions of the male and female, as well as that of the offspring, and even the sex ratios of offspring most likely to be produced by individuals.

The effect that one strategy has on other adaptations is also labeled as the "multiplier effect" by Wilson (1975, p. 11), for a small change in one dominant adaptation may have rather great repercussions on other adaptations. We therefore deal with a hierarchy of dominance among the adaptive strategies of a species. Although it would go too far in this sketch to list the diverse rules generated by adaptive strategies in their interactions, the reader may want to peruse some papers on this subject, which has only recently begun to blossom in the literature on animal behavior. Most workers have been concerned with the relation between the strategies of resource exploitation and the concomitant development of social compromises between adaptive strategies dealing with the maintenance of homeostasis. I point to the studies on birds by Brown (1964, 1974), Crook (1965, 1970a), Verner and Wilson (1966), Simmons (1968), Orians (1969), Brereton (1971), Carothers (1974), Wiley (1974); also, on mammals, by Eisenberg (1966), Crook (1968b), Itô (1970), Geist (1971a, 1974a,b,c, 1978a), Eisenberg et al (1969), Jarman (1973), Estes (1974), Kaufmann (1974a,b), Barash (1974). Also interesting are two symposia dealing with adaptative strategies in relation to environment (Banks and Wilson 1974, Geist and Walther 1974), and a book by Wilson (1975). The above list is, unfortunately, more a reflection of my ignorance than my knowledge; however, in recent years, this type of study has blossomed and, if the above is not an exhaustive list of relevant studies, it is at least an introduction to the literature.

Not all resources essential to reproduction are abundant; some are scarce, and there is not enough for successful reproduction if all individuals in a population get an equal share of such resources. The individuals that manage to reproduce and raise viable offspring are clearly those who obtained more than an equal share of the scarce resources—despite the activities of other members of the population. This leads us to the third rule individuals must obey in order to maximize their reproductive fitness.

Individuals must maintain access to scarce, limiting resources and obtain a share large enough to ensure reproduction. The individual must therefore obtain more than an equal share of scarce resources in order to reproduce and obtain all he needs for maximum reproductive fitness. This rule generates direct and indirect *competition* for resources; such a scarce resource might be a mate, as may be seen by the contest for mates by individuals of the same sex.

Indirect competition is a contest for the same resource by ever improving the means of obtaining, processing, and using that resource. However, if the resource is defendable (Brown 1964), then individuals enter into confrontation and maintain access to the scarce resource by means of aggression. This in itself is a very comprehensive subject as it deals not only with the means of direct or overt aggression, but also with cheaper alternatives to overt aggression and strategies of resource acquisition and defence through aggression. This subject encompasses weapons and defences, strategies of combat, selection for good and poor defences, retaliation as a control on the evolution of means of overt aggression, threats, deception, dominance displays, territorial defence, kin selection, display organs, and strategies of aggression based on the costs and benefits involved. Chapters 4 and 5 deal with this subject.

The fourth rule is that individuals ought not only to maximize their reproductive fitness by competition for resources or by their frugal, efficient, and strategic utilization of energy, but *they should also reduce the reproductive fitness of other individuals directly.* After all, individuals must not necessarily maximize the absolute number, but rather the *proportion* of offspring in the following generation. This subject is dealt with in Chapter 4; there are diverse means by which this rule can be followed from outright murder of the offspring of others to a directed reduction in the growth and development of playmates.

The fifth rule is that *individuals ought to support individuals with identical genes in their efforts to maximize reproductive fitness.* Wilson (1975) considers this rule the central problem in sociobiology and he identifies a trend toward altruism in all lineages of life. This rule is presently under much scrutiny in sociobiology. For a detailed account, I recommend E. O. Wilson's (1975) discussions on this subject. We shall meet it occasionally in this book, particularly when discussing kin selection and its products.

The sixth rule, that *individuals ought to grasp every opportunity to expand into unutilized resources,* will be the subject of the following subchapter.

The seventh rule is that *the individual must select mates to raise offspring that are its equal or superior in reproductive fitness, unless the cost of reproducing with an individual of lesser competence is negligible.* Thus, where male and female share equally in the support of their offspring, the male must select the best possible mother for his children and be most reluctant to bond and mate with any female, while the female must select the best possible provider or defender of resources she requires to raise and safeguard her children. In polygynous species, in which males have a short reproductive period and inseminate many females each reproductive season, the female must select the most dominant male to mate with. The male is permitted to be less choosy since his direct contribution per offspring can be as little as the cost of inseminating the female. The greater his contribution, however directly, such as defence of resources required by the females to raise their mutual offspring, the more choosy he must be in accepting a female.

Although a polygynous male may have a low direct cost in reproduction, his indirect costs may be very great indeed. In order to choose the most dominant male, the female must be able to compare the activities and appearances of males. Thus, males must advertize themselves to females, and prevent others from advertizing themselves. The greater the amount of advertizing, the greater the amount of resources the male is able to accumulate above and beyond the needs of maintenance. Therefore, the greater the amount of advertizing, the more efficient the male's indirect means of competition. However, if the male has the best ability to produce resources, obviously he is more likely to sire children competent in procuring resources. Therefore, a female maximizes her reproductive fitness by mating with such a male. Females obviously ought to act in such a fashion as to maximize competition between males for polygynous species; in monogamous species, females ought to reduce competition for their mate, since such competition jeopardizes their intrauterine or suckling offspring. As in the case of the rule stating that individuals must maintain access to limiting resources, so this rule depends on mechanisms of communication. Here too we are dealing with dominance displays, as these become not only inexpensive means of attaining the aims of overt aggression but also symbols of dominance attractive to the female. The ramifications of this rule are discussed in further chapters of this book.

Grand Strategies of Maintenance and Spread

In order to maintain itself, the grand strategy of the gene must be, on one hand, to be able to maintain itself wherever it gains a foothold and, on the other (rule 6), to be able to exploit instantly any opportunity for spread—be this geographically or into new resources. Let us explore simply the logic of this grand strategy, an attempt I have not seen performed to date and, of necessity, a very speculative one.

It is self-evident that the problems faced by an individual during colonization of an area previously unexploited by its genome are quite different from those faced by an individual in an established population. We can deduce from this *a priori* that a different phenotype is required for colonization than for existence in a settled population. In the latter, emphasis is on competition with conspecifics for resources; in the former, the emphasis is on dealing with the many uncertainties of colonization. A *dispersal phenotype* is thus required during colonization, as is discussed in greater detail in Chapter 6. This, in turn, leads to the conclusion that genes can control phenotype expression dependant on the environment the individual encounters and that genes communicate with the environment. By what means they communicate we do not know. Thus, we conceive the genes as hovering within *maintenance phenotypes* in an established population ready to exploit any breakthrough. We shall see later why this, a hold strategy, is better for maximizing dispersion than tackling competitors, directly trying to increase one's share of available energy and nutrients within a given ecosystem.

When individuals have succeeded in breaking through to, say, an area uninhabited by the species, their offspring become dispersal phenotypes and, as discussed in Chapter 6, they are characterized by greater reproductive output, greater exploratory activity and mobility, apparently a better immune mechanism, and larger body size. These are all features that deal with the uncertainties inherent in exploring an area hitherto unexploited by the species and fill the area rapidly. In addition, dispersing phenotypes may encounter conditions in which it is not adaptive to follow tastes and thus perform activities that werre formerly adaptive. Conversely, they may well have to do some unpleasant, even painful, activities in order to survive or prosper. Therefore, they must be able to easily suppress "desires" as well as "aversions"; we can also say they must be able to practice exceptional "self-discipline."

Because phenotypic development is maximized by heterozygosity (Mayr 1966), clearly dispersal phenotypes, entering somewhat unpredictable environments, ought to select mates with genomes unlike their own. The converse may be true for maintenance phenotypes.

By virtue of enhanced exploration, play and roaming, the dispersal phenotype quickly adjusts behaviorally to the problems it encounters. Behavior is the cheapest means of adjustment and—following the law of least effort—is thus the first line of defence against environmental demands, followed by physiological and then morphological adjustment of the body. C. H. Waddington (1957) recognized this. A constant environment demand would lead to behavioral exertion by all individuals to fulfill this demand. If selection is strongly in favor of this behavioral activity, and individuals exert themselves to achieve it, we soon have selection of those individuals that are genetically best endowed to perform that activity. The following evolution is then based on chromosomal rearrangements and mutations appearing that favor the trait and enhance both the actions and the phenotypic growth of the organs involved in the activity (Waddington 1957, 1960, 1975). Therefore, the colonizing population begins to diverge genetically from the parent population. Each deviation from the parent population becomes a *diagnostic feature* of the colonizing population.

The best invasion strategy is clearly to disperse widely and build up numerically when the opportunity permits it, while continuing the advance. Initially the dispersal phenotype of the colonizers carries the advantage over competitors in the area colonized, because it is more plastic than the maintenance phenotype of the competitor. Moreover, by disturbing the delicate balance of coexistence of its competitors it can gain resources at their expense, thereby broadening its original ecological niche. This ought to lead to a saturation phase off the colonizing population in which plastic adaptable dispersal phenotypes in large numbers fill and probe all potential opportunities of a landscape.

This situation can, however, be only temporary. If, for any reason, the colonizing population suffers a numerical setback and the resources its competitors formerly exploited lie unused, the latter may surge back. As an abundance of resources leads to dispersal phenotype development, the now-rallied competitors also change to plastic dispersal phenotypes and face the colonizers as much more formidable opponents. We can label this expected development the "countersurge." The countersurge has the best chance of wiping out the colonizers at the point of breakthrough because here the colonizers are least changed from the parent population or least adapted to deal with the new competition. Moreover, the countersurge is likely to be formed by several species and thus be formidable, so that it has a good chance of eradicating the invader at least close to the point of breakthrough. Therefore, if the invader survives the countersurge, it is likely to be some geographic distance from the point of original breakthrough.

We can see also that the invader ought to maintain a dispersal phenotype over many generations, even when the area has been colonized, because of the inherent plasticity of this phenotype. This permits individuals to continue adjustment to competition and other environmental contingencies. Therefore, we expect a quick response from maintenance to dispersal phenotypes by a population at breakthrough, but a slow return to maintenance phenotypes after colonization.

The maintenance phenotype characterizes the holding phase of a population. Clearly, it must be designed to preclude any opportunity for interspecific competitors to expand at their expense. This can best be achieved by the removal of all resources within its ecological niche (ecological profession). This amounts, in essence, to a scorched earth policy. It is best achieved by maximizing the number of individuals per unit of exploitable resource so as to maximize coverage of the landscape. This can be done by reducing individuals to the smallest possible body sizes. Also, it pays to diversify, ensuring that all possible resources are indeed utilized. In addition, the maintenance phenotype should feed on lower-quality food than the dispersal phenotype, as this also ensures removal of a maximum of resources. Also, it may pay the maintenance phenotype to select against genetically distant individuals in mate choice. This would reduce heterosis. Heterosis is bioenergetically expensive, and such a cost may be charged against the cost of reproduction of the offspring, lowering its reproductive fitness. This prediction requires investigation, but it fits with the observation that heterozygocity within populations may have low heterosis (Mayr 1966, p. 231). As direct competition using aggression is bioenergetically very costly, the cheapest strategy to maximize a species' dispersion geographically or ecologically is to adopt an opportunistic approach of hold and disperse when the opportunity arises. That is, the direct assault on competitors is avoided and an indirect approach to the objective is given preference.

An axiom arising from the hold strategy is that a balance is struck between competitors that results in coexistence of populations of different species. It can be predicted that, in a tightly packed community, each species keeps the other in check by depriving it of critical resources, or of access to them, with minimum expenditure. It also follows that in such communities each population refines its competitive ability by specializing in exploiting resources where the opposition is weakest, which leads to a net decline in competition. Because the success of individuals depends on a narrow margin of resources acquired and saved from maintenance for reproduction, it also follows that a great penalty will be paid by the individual should it mate with a conspecific of a genotype shaped elsewhere and introduced by dispersal. In densely packed communities, this should lead to the rapid isolation of genetic pools adapted to local situations. Therefore, densely packed communities will be communities of specialists and loosely packed ones will be of generalists. This, in turn, predicts that generalists will not disperse into tightly packed communities of specialists, as there they would face several specialists in their respective niches that collectively will deprive the generalists of the resources needed for existence. The converse expansion is quite likely, however. Therefore, zoogeographic dispersal is likely to be a one way street from the tropics to the arctic, from the humid to the dry, from the productive landscapes to less productive ones.

The foregoing is a sketch, no more and no less, and I have not even attempted to reference its predictions to show whether it is valid. This is done, in part, later in this volume. I emphasize that all the deductions lead ultimately to reproductive fitness. The sketch of gene strategy in military terms ought to serve not only to present an unconventional view of biology but also to illustrate the morality of the gene, a subject Wilson (1975) touched on but did not elaborate. Let us face it: The morality of the gene is that of the marketplace, in which the ends always justify the means. Moreover, the ultimate aim of each competing gene pool is not simply to enlarge its slice of the available energy and nutrients but to engulf it all. We must also be aware that genes are ruthless dictators, and at times short-sighted ones, and the boneyards of paleontology appear to bear this out. We have to know and understand genes if we are to make good use of their potential in maximizing human health—a major aim of this book. We must know them just as any general must know his opponent, being aware of what is going on "on the other side of the hill," as Wellington said prior to the last great battle which finished his great adversary, Napoleon. By fair means or foul, genes will maintain themselves or disperse, and if altruism is an aid toward their goal, so be it. Altruism cannot serve, therefore, as a vindication of the morality of the gene; it does not have any.

Rules of Conduct Dictated by the Mechanisms of Gene Transmission and Expression

In the preceding section, we dealt briefly with the dictates from reproductive fitness. We also noted that genes protect themselves against environmental variability by communicating with the environment and shaping the phenotype of the individual in accordance with the information received. Phenotype plasticity is thus the manifestation of one insurance mechanism of the genes. There are others, as discussed at length by Ernest Mayr (1966), such as polymorphism, heterozygosity, recessiveness, pliotrophy, and at times, heterosis. The latter, however, is problematic.

Heterosis enhances the phenotypic development of individuals so that in similar environments the heterozygote may be better developed and larger, and may reproduce more than the homozygote. In highly variable environments, in which individuals may be called upon to do many things well, heterosis is, therefore, highly advantageous, as it preadapts the individual to handle and overcome environmental constraints. Therefore, in environments with great variability we expect selection to favor individuals that roam and mate with distantly related genotypes, maximizing heterosis.

For populations living under conditions of resource scarcity, this argument no longer holds. Enhanced growth and vigor of the resulting offspring of distantly related genotypes is a costly process. Growth is very costly in energy and nutrients. As we have seen, energy is hard to procure, is inefficiently processed and stored, and can be quickly lost. Individuals must follow the law of least effort so as to maximize resources available for reproduction. Therefore, under conditions of resource scarcity, such as in very stable environments with a maximum number of individuals at carrying capacity, heterosis may be disadvantageous as it raises the cost of maintenance and reduces the resources available for reproduction, without necessarily giving an individual a significant competitive advantage. This must be true particularly for demes of kin-selected individuals in which selection favors a maximum number of adult defenders per unit of defendable resource. Here, heterosis would reduce the scarce energy and nutrients available for reproduction. It would fly in the face of selection for altruism as well.

We noted under rule five that a given gene ought to support its copy in a second individual rather than an allele. Thus, individuals closely related genetically may enhance the reproductive fitness of their common genes by mutual aid, provided this results in greater net reproduction than could be achieved by acting alone (Hamilton 1964, Wilson 1975). Therefore, under those conditions, i.e., kin-selection, individuals with closely related genomes ought to mate. How, then, does an individual recognize the degree of genetic relatedness in other individuals? Without this, it cannot possibly choose a partner on the basis of genetic relatedness, in part to avoid heterosis.

It appears obvious that the closer two genotypes are related, particularly if they are raised in common environments, the more similar their tastes and aversions. Individuals with similar likes and dislikes are likely to act in concert; they will be more frequently in agreement than two individuals of widely divergent genotypes. Therefore, harmony of responses through identity of responses to common stimuli can function as a recognition of genetic closeness. Therefore, under kin selection, individuals with identical response ought to stick together, while individuals behaving differently ought to be expelled. If this argument is valid, identical twins should get along better than fraternal twins or siblings, as the former are of closer genetic identity than the latter; or, relatives ought to support each other in social interactions in proportion to their relatedness. This was shown by Massey (1977) for pigtail macaques. Also, in kin-selected populations, there ought to be group actions in which individuals perform actions in harmony.

The harmonious performance of individuals in schools of fish or flights of birds should not be confused with the above-mentioned performances. Schools and dense swarms of birds appear to be adaptations against predators (see Wilson 1975, p. 38); perfect close-order drill, and thus acting as a unit, is advantageous to individuals, since in so doing they avoid acting discordantly and being at once selected and pursued by a predator. The more likely a predator is to focus on discrepant individuals, the greater will be the selection for practice flights by birds and daily exercise of close-quarters flying. Predators do select for discrepant individuals (Mueller 1971). This explanation of coordinated group activity is quite likely to fit the many examples collected by Wynne-Edwards (1962). Thus, coordinated behaviors in groups could serve two different functions: one, to select and expel deviant or distantly related members in kin-selected populations, and, two, as a training exercise by which individuals moving with a group reduce their chances of being selected and caught by a predator.

The discussion on the foregoing topic began with a proposition that heterosis may not be adaptive in populations under resource scarcity, and that it would be least adaptive under conditions of kin selection. One predicts, therefore, that in closely inbred species heterozygosity and heterosis need not be related, as is indeed the case, at least in some instances, as discussed by Ernest Mayr (1966, p. 231). We should also note, as was indicated in the preceding subchapter, that during dispersal it is clearly adaptive to maximize heterosis so as to maximize an individual's ability to resist environmental variations. Conversely, we expect, in populations at carrying capacity and short of energy and nutrients, some selection against mating of divergent genotypes. The more a population approaches extremes in developing individuals of maintenance phenotypes, the less individuals ought to roam, select distantly related mates, or tolerate strangers.

Clearly, the actions of individuals are, therefore, dictated not only by the rules we deduced from the concept of reproductive fitness, but also by rules generated by the mechanisms of gene expression. Roaming and tolerance of strangers is dictated under conditions of resource abundance, in variable habitats or under conditions of dispersal, and the converse under conditions of stable habitats and resource shortages. Yet in the former condition the individual must not mate so indiscriminately that it mates with genotypes so distant that the offspring suffer loss of competence. When this condition exists, reproductive isolation is selected for, as Mayr (1966) discussed in detail.

We must also note that in stable environments there must be great selection for phenotype redundancy, or canalization, as C. H. Waddington (1957) called it. That is, entirely different genes produce much the same phenotypic expression in individuals. Mechanisms of canalization, whatever they may be, maximize reproductive fitness under such conditions, because they overcome mutant genes or irregular strong signals from the environment that might otherwise lead to the development of an uncompetitive phenotype. Phenotype redundancy ought to be maximal in long-lived organisms.

The Environment

The term "environment" is global in meaning and therefore quite useless. It can denote anything a speaker wishes it to, unless qualifiers are added and its meaning is severly restricted. Unless this is done, communication cannot proceed unhampered by excessive ambiguity. One can conceive of many ways of defining and categorizing "environment"; the following is the manner in which I intend to use the word in this book.

I will define "environment" as the sum total of factors impinging on an organism. Thus, clearly, a first division is into the *external environment* and the *internal environment:* the latter is Claude Bernard's *"milieu interieure,"* the environment that bathes the cells of our bodies. Conversely, the external environment is that which impinges on the exterior of our bodies; that exterior logically includes the surfaces of our gut, lungs, and nasal oropharyngeal region. In discussing the internal environment, we deal with the substances that cross the body surfaces into the body; the chemicals that circulate within us; the ionic, water, and acid−base balance; the many regulatory mechanisms; the fuels and breakdown products of metabolism; the repair mechanisms and defense mechanisms; the transport mechanisms; excretion, etc. The internal environment is the domain of physiology and its many branches. It will not concern us greatly in this book.

The *external environment* is still too vast to be talked about meaningfully. It can be broken down into three components: the *surficial environment,* which is the environment of the body surfaces; the *physical environment,* which I equate with the term *habitat,* encompassing the climate, geology, and land form in which a species lives, as well as its food, cover, commensals, and predators; and the *social environment,* encompassing the relation of each individual of a species with its conspecifics.

The *surficial environment* need not detain us long. It deals with the adaptations of the body surfaces and that, of course, means with the interplay between the animal's exterior and the factors in proximity to it. Thus, parasites that cling to the body's exterior and the manner in which the host deals with them are encompassed by this division. So are all behaviors and activities concerned with maintaining the body surface in a functional state, be it through maintenance or repair. The maintenance behaviors, or comfort movements studied by ethologists fall into this division. It is self-evident, of course, that the surficial environment cannot be discussed in isolation from the external or internal environment, yet it is a useful division since it does help organize thought about the biology of an organism and it emphasizes, in so doing, that adjustments and adaptive strategies are fusions of elements from the realms of behavior, physiology, and morphology. The surficial environment, again, will not concern us greatly; it is the domain of parasitologists, ethologists, pathologists, and also physiologists.

The *physical environment* or *habitat* is a useful concept. As indicated, it refers to the relationship of the organism to everything except the relationships with conspecifics and those excluded by the internal and surficial environment. The habitat is the subject of study by autecologists, that is, ecologists dealing with the ecology of single species. Whatever is studied in the physical environment or habitat of a species, it is always with reference to the adaptive strategies of a species. This conception makes the term "physical environment" a global one for the human species and includes, therefore, the structured and altered environments of man. Thus, urbanists, planners, transportation engineers, and architects, to cite examples, are practicing autecologists of sorts. Clearly, when talking about human beings and their physical environment, one must use terms that denote subdivisions of that physical environment. For the purposes of this book, however, I need not draw up an appropriate taxonomy of human physical environments, be it spatial or conceptual, because in my discussions I shall remain largely within the natural environment that houses human beings. This brings us to points of view from which the environment may be considered.

The term *natural environment* is problematic. In everyday conversations it is used in different ways, so much so that the context in which it is used denotes its meaning. For instance, it may be synonymous with "nature" or "out of doors," as in the sentence: "We enjoyed the natural environment on our trip to the mountains." It may also denote a value judgement, such as illustrated in this sentence: "The city is the natural environment of man." Here "natural environment" denotes an obligation; one "ought" to support cities because humans fit best in them. The term "natural environment" may be contrasted to "artificial environment." Here, "natural" denotes in a vague way both "nature," as indicated earlier, as well as a "good fit between organism and environment." Used in that context, without further specification, the term "natural environment" is also value laden, and the term "artificial environment" becomes pejorative. This is unfortunate since the term "artifical environment" could stand as an equivalent for "man-made, created, altered, or influenced environment." For the purposes of this book, I have tried to avoid the terms "natural" and "artificial" environments. Where the term "natural environment" does appear in relation to man, it is used in the following artificially and arbitrarily limited, meaning: It denotes the physical environment man lives in and exploits at the level of and below that of a Paleolithic technology. Thus, by definition, the early agriculturists or, for that matter, the Erigbaagtsa

Indians of the Amazon basin practicing slash-and-burn cultivation, are not living in a natural environment. By contrast, the Caribou Eskimos of Canada are.

In the literature, the term "natural environment" has become linked in an unfortunate manner to "optimum" and "evolutionary" environment. Thus, Boyden (1972) treats these terms as synonymous, in arguing that hunter—gatherer cultures from subtropical zones are from environments that man is best adapted to, and that we should use as the basis of comparison for all other environments, and, I presume, that we should imitate in structuring future human environments. Although I agree very much with Boyden's (1972, 1973) aims, I disagree both with his choice of the "standard environment" and, even more, with the equating of "natural," "optimum," and "evolutionary" environments.

"Natural environment" I defined as that in which man exploits his physical environment with cultural means no more advanced than that of Paleolithic cultures. It is the environment of hunters and gatherers wherever they are found, and it may or may not be an optimum environment—or, for that matter, an evolutionary one!

The term *optimum environment* is more difficult to define than appears at first sight. For instance, if we define an optimum environment to be that in which a species' genetic potential can be unfolded, then we are probably speaking nonsense. If there are epigenetic mechanisms that alter the phenotypic expression of a genotype according to the environment experienced, then there cannot be an optimum environment, but only optimum adaptations of individuals to a given environment. The dispersal phenotype is likely to be ill-suited to life under conditions of resource shortages, while the phenotype adaptive under conditions of resource shortages is hopelessly inadequate when resources are abundant (see Chapter 6). The genetic potential unfolded ontogenetically under conditions of resource shortage is different from that unfolded under conditions of resource abundance. If we speak of an optimum environment, we betray a value judgment as to which type of phenotypic development we prefer. For instance, if we speak of the optimum environment as that in which the individual suffers a minimum loss of physical, intellectual, and social development, then we clearly opt for, and consider as desirable, individuals highly competent in diverse competition and, with reference to ourselves, individuals who are on the whole unruly and difficult to control, but at the same time highly capable of looking after themselves. If someone should value docile individuals of low competence, then his definition of an optimum environment is that which maximizes the number of sexually reproducing individuals. I am biased. I opt for the former definition; that is, an optimum environment is that which maximizes physical, intellectual, and social competence and minimizes physical, mental, and social breakdown. Technically, such an environment can be detected by the degree to which tissues of low growth priority are developed. If this sounds like gobbledy-gook to the uninitiated, as it must, I would refer him to Chapter 6.

Once we have decided what characteristics of individuals we desire, the relevant attributes of the environment can be discovered empirically, that is, by observation, experiment, and trial and error. For the optimum human environment, Boyden (1972, 1973) correctly seeks a shortcut to that haphazard process, as I do. We are searching for a theoretical model that will assist us to ask relevant questions and gain guidance and economy of effort in our search. Yet, in so doing, we are at once confronted by the value-laden question: What do we mean by "high physical, intellectual, or social development"? What criteria are to be used? What are acceptable measures of social breakdown? Is divorce? Is failure to finish high school? Is it the number of institutionalized children and old people? We have here the choice of discussing such "criteria," and consequently facing inevitable long, complex, and usually futile discussions, or avoiding getting bogged down and searching for criteria most of us can agree upon. I propose to restrict myself to a definition of "optimum environment" that rests on measurable physical fitness, on whose desirability most people can agree. *The optimum environment for man is that which during ontogeny maximizes body size and still produces a disease-free organ system at the termination of body growth in early adulthood.* Not an appealing definition, is it? Yet it is a useful one, since it avoids futile discussion about social and intellectual criteria, while including intellectual and social development. The definition states that we cannot just maximize body size, for a very good reason: As will be shown later, conditions that foster exceptional body growth can also lead to diseased organs, leading to early physical breakdown and the death of individuals. Also, to reach large body size with healthy organs, one must pay meticulous attention during ontogeny to psychological, social, and cultural factors, for these do affect ontogenetic development. More of this later.

From the foregoing it is evident that the optimum environment, in theory, need not be a natural environment. It is possible to develop organisms artificially that fit the criteria I have stated. In searching for a theoretical basis for the optium environment, we come upon the *evolutionary environment*.

The evolutionary environment of a species can be defined as the sum total of environments experienced by populations forming that species. This is a useless definition, as it says virtually nothing. Evolution denotes change. Therefore, an evolutionary environment must be, by definition, an environment that produces noticeable genetic change. This does not imply that, in an environment in which no noticeable change takes place in a species, there is no natural selection. On the contrary. Natural selection may be severe, and return the gene pool of the population to the very same values with each generation (Mayr 1966). It is possible to have natural selection that produces no evolutionary change; natural selection and evolution are not synonymous. Contrary to a widely accepted belief, evolution appears to be a rare occurrence, and most populations live in environments that produce no noticeable change in the gene pool. Evolution can fall asleep, and usually does; natural selection never sleeps (see Chapter 6).

Evolution deals with genetic change that occurs under quite specific conditions. There is accidental evolution which one finds in small populations, be they decreasing because of deteriorating environmental conditions, or increasing as a consequence of colonizing an area. In suchs small populations, genetic information is usually lost through genetic drift, inbreeding and reduced viability of offspring, and the production of phenodeviants, and colonizing populations may differ from parent populations by virtue of the founder effect; the latter term refers to the accidental loss of genetic information because of the accident by which individuals became colonizers and formed a new population. This type of evolution has been of considerable concern to students of evolution (see Mayr 1966, Wilson 1975), yet it is of no great concern to us here; directed evolution is.

Directed evolution takes place whenever the genetic variance of a trait under selection exceeds the environmental variance, and it is most rapid when the environmental variance of a selected trait is nil. This occurs under two conditions: first, when resources are superabundant and the genetic potential of a given trait can be expressed maximally; secondly, when resources are so scarce that there is competition among organ systems for the available energy and nutrients for growth. We already touched on the former conditions when discussing the strategies and tactics of dispersal and the subsequent selection for new traits in the colonizing population, which were termed "diagnostic features." Under conditions of resource scarcity, a trait under selection must be enhanced in its development through multiple gene actions and pleiotrophy; it is then "canalized," as C. H. Waddington (1957) termed it. Selection acting on individuals suffering shortages of resources develops growth priorities among organ systems, thus explaining the existence of growth priorities long studied by animal scientists. Note that in the same species, entirely different adaptations will be selected for under conditions of resource scarcity than under conditions of resource abundance. Thus, the genome of any species obviously must produce different phenotypes under each condition.

From the foregoing, we obtain the following definition of the evolutionary environment in which new adaptations are evolved. *The evolutionary environment is that which maximizes the diagnostic features of a species.* The diagnostic features are permitted maximum expression, be it through the selection of conditions favorable to their development or selective intake of nutrition or specific activities of individuals that enhance an organ; the diagnostic feature is maximized environmentally, thus maximizing the genetic expression of the trait. Only then is the genetic variance between individuals exposed, and genes favoring said diagnostic trait are selected for. For instance, cold increases the vertebral count of fish during ontogeny, but not surprisingly, individuals from cold waters have a genetic tendency toward a higher number of vertebrae (see Mayr 1966, p. 145). This is exactly what we expect.

I used the term "diagnostic features." It may be noted that taxonomists are concerned with diagnostic features and distinguish species on this basis. Clearly, diagnostic features are the very ones selected for, creating the distinct species. If we know how to maximize environmentally the diagnostic features of any species, then we are keeping that species in its evolutionary environment. We have then learned how to maintain it. Our definition of the evolutionary environment then becomes a key to the understanding of how to create environments fitted for a given species to live in. This is particularly relevant for species that develop their diagnostic features maximally under conditions of resource abundance. We shall see later that the diagnostic features of *Homo sapiens,* those features that we rightly cherish, are developed maximally under conditions of resource abundance. Moreover, an environment that maximizes phenotypically our diagnostic features, such as language development; musical ability; visual mimicry (dancing, sports); intellect; manual dexterity; large body size relative to our parent species; large frontal, occipital, and parietal lobes of the brain; blind altruism; powerful control and discipline over voluntary and even involuntary muscular movements; laughter; crying; extended sexual receptivity; and culture also appears to maximize physical health. Therefore, the evolutionary environment under conditions of abundance of resources is the same as the one maximizing health. Here is the link between evolution and health. This is one reason why an understanding of the human evolutionary environment is so fundamental to a theory of preventive medicine. There are other reasons, such as the concept of "creature comforts," which I discuss in the last chapter, or the concept of relict adaptations, which we shall encounter repeatedly in this book. An understanding of the evolutionary environment thus gives us a basis for comparison and for meaningful research, provided we agree that our diagnostic features are worth maintaining. To discover and describe the evolutionary environment of our species is, therefore, one task of this book.

The evolutionary environment of a population, however, need not be the evolutionary environment of a species. The environment of a given population may retard certain attributes otherwise diagnostic of the species and enhance others diagnostic of the population. In defining the evolutionary environment of a species, we may, in fact, be defining that of the extinct ancestral group, not the conditions of any living population. The diagnostic features of a population may be superimposed over those of the species, and there may be a conflict between an adaptive trait of the species and one of the population. This has, of course, the implication that a model of the evolutionary environment of our species may only be used subject to an understanding and consideration of the environment of a given living population. This caution is reinforced by the fact that humans can develop—and lose— specialized physical (body) adaptations—a magnificent attribute, since it permits our bodies to adjust to seasonal or local environmental demands, and yet spend only the resources needed to maintain these adjustments as long as they are needed. This insight we gain from exercise physiology.

After this discussion of natural, optimum, and evolutionary environments, we turn to other basic definitions of environmental compartments. One of these is the *social environment*. It contains the factors that govern relationships among individuals. It is useful to conceive of the social environment from the viewpoint of specific individuals, e.g., the social environment of a neonate, or a male adolescent in a specific social setting. Clearly, with the term *social environment* we denote a collection of exceedingly complex interpersonal relationships. Another manner of compartmentalizing the social environment is to think of the differences between cultures. Thus each individual exists in a given cultural milieu or environment. The social environment is not comparable to von Uexküll's notion of *"Umwelt"* which is broadly defined as "environment." It is possible, of course, to form divisions of the social environment almost *ad infinitum*. These, to be useful, must be related to some use in which these definitions are an aid to communication. For the present, I shall use only the divisions based on sex–age class, social class, and culture.

Chapter 2

Cognition—Predictability

Introduction

There is a fundamental rule governing animal behavior: In order to increase reproductive fitness, the individual must live in an environment predictable to itself, and
one it can cope with. This rule is a consequences of the laws of least effort and the
maintenance of homeostasis. To live in a predictable social and physical milieu is to
be prepared and fittingly equipped to cope with almost any situation that may arise,
at least under normal conditions. Each situation arising, each problem confronting
the individuals, is countered with an appropriate proven solution that preserves the
animal from harm, maintains homeostasis, and costs the animal a minimum of
scarce energy and nutrients. Conversely, if the organism is constantly confronted by
situations that are new (unpredictable environment), it must quickly develop appropriate responses *de novo*. This, of course, increases the chances of formulating
inappropriate expenditure, or becoming a victim of learned helplessness (Seligman
1975). In order to maximize its reproductive fitness, the animal must strive to live
where, firstly, its adaptations and, secondly, its experience fit closely with its
environment, permitting it to live at the cheapest cost in energy and nutrients, as
well as extending its period of reproduction maximally. This applies to all life,
including ourselves, and we cannot escape its consequences; it is, for instance,
fundamental to an understanding of why neither humans nor animals can accept a
constantly changing environment (see Denenberg et al 1962). This is a point championed by Alvin Toffler in his well-known book *Future Shock* (1970), how stress is
generated in the psychic realm, or why individuals of different cultures respond
differently to the same environment.

One can illustrate this principle by visualizing what could happen to an animal if
it were to violate the above fundamental rule. If it were to wander into strange
habitat, it would not know, for instance, of the best escape routes, terrain, and
cover, and if confronted by a predator it would waste time deciding where best to
run, as well as taking the chance of running into terrain quite unsuitable for escape.

Its chances to maintain homeostasis, or even to keep intact, would be slim under such circumstances. There is some experimental data to back up this contention (Metzgar 1967, Seligman 1975).

In the social milieu, our animal would be at a disadvantage not knowing the rank and combat potential of its conspecifics. Conflict with a physical superior would quickly ensue and the wounds inflicted by combat, plus the high energy cost of exertion and body repair, are not compatible either with the law of least effort or the rule to maintain homeostasis.

It is clearly advantageous for an animal to have mechanisms by which it makes the social and physical environment predictable; gains, stores, retrieves, and alters knowledge; and structures responses to fit each problem arising. It is self-evident that only by continuously monitoring, by comparing manifestations as they are against how they have been, does the organism know when a new situation has arisen to which attention must be paid; stimuli not found in its habitat and social environment are irrelevant to the animal's biology, and it should not surprise us if the creature is not prepared to handle them and ignores them; stimuli that partially match with the experience of the animal should elicit strong responses from it, since they are partially, although not entirely, known and not proven harmless; stimuli that match the knowledge of the animal should elicit no response unless they are of a kind appropriate to be acted on for various immediate needs.

When the animal is confronted by known harmless stimuli, clearly it need not respond; when it is confronted by unique stimuli that do not occur on its range normally, it may or may not respond; it may ignore the stimulus if it falls well outside its range of possible experiences or semblences thereof. If, however, it has semblance to known stimuli, in particular those signaling danger, the animal should become alterted and investigate the stimulus. To run at once would deprive the animal of ingested energy and nutrients, as well as of the opportunities it was in the process of seeking; to ignore the stimulus could also lead to high costs of energy and nutrients in a belated escape, or to injury and death. Preparation for all eventualities is a better strategy, that is, some investment of time, energy, and nutrients must be made to clarify the strange stimulus, as well as keeping the body fired up and ready for the exertion of running or fighting that may or may not follow. This syndrome we call excitation combined with curiosity. It is important to note that it does cost energy and nutrients, and if practiced too often is a drain on the animal's resources and, ultimately, those of its home range. Furthermore, chronic excitation is equivalent to stress. Thus, excitation ought to be an inverted U-shaped function on stimuli ranking from known to totally unknown and unmatchable ones. When the strange stimulus has been identified, that is, matched with previous experience, there should be a signal telling the animal that it is so, thus releasing it from excitation and exploration either to return to its previous activity or to flee. This signal is probably the "Eureka" effect we all experience. We can regard the matching of expected and perceived stimuli as either a neutral or a pleasurable emotional event, depending on whether the event is very common and the animal is habituated, or whether the event is rare and the animal was searching for it; the unpleasant, disturbing, and ultimately damaging condition is the one in which expectation cannot be matched against reality. Thus, we are discussing here a fundamental mechanism within the animal that permits it to dwell where its experience is most relevant and it is best prepared to act or, conversely, that permits it to sense that it is in a dangerous situation and withdraw to a known and less dangerous one.

We have so far pointed to a system of internal "pattern matching" and stored perceptions and experienced perceptions, a system signaling that matching occurred, a system alarming the animal if the patterns fail to match and triggering the consequent behavioral responses of searching for stimuli that will fit the internal perceptions. Crude as this little model is, it is fundamental to an understanding of behavior and many other events in the life history and biology of mammals—including ourselves. Moreover, it appears, as we shall see, in various disguises in various disciplines—ethology, psychology, sociology, linguistics—and goes by various names, i.e., discrepancy principle, theory of emotionality, *auslösendes Schema*.

We are becoming aware of the physiological nature of the pattern-matching process. It entails electrical activity in the brain, which matches with a given stimulus—or the expectation thereof; the flow of stimuli is broken into units, of which about three to seven can be held in short-term memory, depending on the length or size of the units; long-term memory coding takes place apparently through RNA structuring; specific learning tasks and exercises leave their mark as increased brain or central nervous mechanisms parts; thus, learning depends not only on a proper inflow of stimuli and experiences, but also on adequate nutrition to supply the building blocks for the memory stores and activity centers.

The origin of the pattern-matching system is quite likely a mechanism operating at the spinal level, the reaference principle. We can suspect this to be so on the basis of the law of least effort; modifying and enlarging an existing system is easier and cheaper than developing one *de novo*. This is also in line with the general observation that new organs arise from old ones and may change dramatically in function, e.g., the reptilian articular—quadrate joint of the jaw becoming the malleus—incus joint of the mammalian middle ear.

The physiological correlate of how we acquire, process, store, and act on knowledge is important to understand, not only because in our lives it plays so significant a role, but also because the physiological base can be so severely damaged during ontogeny. Since humans have a very long ontogeny, the slow structuring of the physiological base for our behavior is very much subject to inappropriate environmental conditions, which inhibit and stunt its development. Moreover, a knowledge of it permits insights into the process of communication and its limitations, and sheds some light on environments we can and cannot cope with. It gives insight into goal-directed behavior, as well as the old nature—nurture controversy, and lays the groundwork for the following chapters dealing with aggression and status displays.

The following will not be an exhaustive review of the speculative subject of the physiology or, better still, "natural history" of cognition. Rather, it will be a sketchy review with emphasis on aspects that pertain directly to a better understanding of the rules governing the life of higher creatures such as ourselves. In a recent book, Konrad Lorenz (1973) attempted to sketch a natural history of cognition. The purpose of this chapter is not to do this, and so it will be sketchy and peculiarly lopsided to the expert.

Predictable Environment

In order to extend its chances of survival maximally, an organism must not be confronted with unique problems. If this were the case, it would have to develop solutions *de novo,* and discard these after use. This would be a very risky life indeed, for the chances of quickly finding and applying the correct solutions to unique problems are, of course, smaller than those of implementing solutions to problems repeatedly encountered. Thus, the more limited the range of problems confronted by an organism, the more likely it will be to possess and implement proven solutions, cheap solutions, and thus maximize its reproductive fitness. Conversely, if the range of problems encountered by an organism is very broad during its adult life, then it requires an extensive period of preparation and practice in confronting and solving such problems; it needs a programing period prior to confronting the environment as an adult, or reproducing, individual.

From this *a priori* exercise, it is evident that it is in the individual's best interest to reduce the number and kind of problems it will encounter during its life. We expect to find that the animals are structured to search out a minimum of unique experiences (and hence problems). They should go to environments in which their adaptations and experience match closely with the environment. We can expect that animals will strive to find environments they are able to handle if displaced from familiar environments; we expect to find mechanisms that permit animals to gain familiarity, effectively and cheaply, with strange environments; we expect to find that animals from highly variable, changing environments confront new, unique situations at their own choosing, since in so doing they remain prepared for the eventualities of their changing environment.

The foregoing has been a strictly hypothetical deduction from the rules that an individual must maintain homeostasis and obey the law of least effort in order to maximize reproductive fitness. Yet it predicts precisely the one generalization we may make about animal behavior: *Animals strive to live in an environment predictable to themselves.* Conversely, we can claim that animals are structured by natural selection to live in predictable environments.

It is difficult to identify who first formed the conclusion that animals strive to live in predictable environments; various authors are aware of this. It is certain only that this conclusion did not generate major debates in ethology, nor was it fully developed, despite its relevance to human existence, as is illustrated by Toffler's popular book (1970). This is not to say that psychologists, ethologists, sociologists, animal scientists, and medical scientists—and even designers, e.g., Perin (1970)—have not been aware of some aspects of this conclusion. Studies in psychological stress deal with the damaging effects of unpredictable social environments (e.g., Chitty and Southern 1954, Barnett 1955, Merton 1957, Brady 1958); the change of life scales of Rahe and co-workers (Rahe et al 1967, Masuda and Holmes 1967, Wyler et al. 1968, Cassel 1971), as related to morbidity, deal with the same; animal scientists know only too well that disturbances cause animals to reduce weight gain and egg or milk production (Allee and Guhl 1942, Liddell 1954); psychologists and ethologists dealt voluminously with frustration-generated behaviors, e.g., aggression (see Ewer 1968, Hinde 1970); textbooks in psychology deal with it, if not explicitly (see Hebb 1966). There is for the searching, therefore, a large literature to draw from to buttress the contention that animals perform optimally in physical and social environments predictable to themselves. One can go one step further and claim that most of the social and habitat behavior of organisms serves the function of creating and maintaining a predictable environment, be it through play, exploration, dominance displays, aggression, courtship, maternal behavior, etc. (Geist 1966a, 1971a). This conclusion, based on rather different observations, would hardly be new to students of bureaucracies (see Boren 1972).

The notion of the predictable environment requires some refinement. Animals strive to live in responsive environments, environments they can manipulate to their own satisfaction. If they cannot do it they learn not to try, to become helpless, and to suffer the damaging consequences to their bodies, so well discussed by Seligman (1975). Secondly, prior to being fully sensitive to even slight environmental changes in adulthood, animals must go through juvenile stages in which they are not very sensitive to new unexperienced stimuli (see Hebb 1966). Only by being so do they learn to sort the deleterious from the pleasant and useful. Clearly, the old animal must be more "emotional" than the young, or learning could not take place.

As indicated, we expect that animals are constructed to live in a predictable responsive environment; at the least, this means a steady inflow of stimuli to their sense organs, as can be inferred from studies on sensory deprivation (Berlyne 1966). What happens if they are forced to live in an unpredictable, unresponsive environment? We are in a position to generalize: Unpredictable environments cause a spectrum, depending on the rates of change, from mild excitation to aggression, stress, increased frequency and intensity of illness, and ultimately death. The unpredictable environment can manifest itself in weight loss, severely disturbed metabolic functions, reduction or cessation of egg-laying in hens and milk production in cows, increase in sickness, increased overt aggression, neglect of offspring, cessation of reproduction, and neurosis, as well as early and even quick death (see Chapter 6).

It is necessary to point out several expectations arising from the relationship between the environment and the animal's means of dealing with it. First, the more simple, repetitive, and unvarying the environment (or the adaptive strategy), the more stereotyped, genetically rather than environmentally, will be the behavior of a species, and the less it will be found to play and explore. The reverse is, of course, also expected. Furthermore, since diverse activity stimulates the growth and development of diverse brain centers (Cummins et al 1973, Greenough 1975), we expect that mammalian species dealing with a wide range of diverse environmental problems will possess larger, more complex, brains than species from relatively simple environments. This hypothesis is in need of verification, but there is some data to support it and none against, as can be seen in the data pertaining to brain size in mammals of tropical and cold-temperate origin (p. 188). Second, the more varied the environment, the more mechanisms an animal must have to preadapt itself to diverse problems. This requires not only the larger memory storage implied in a more complex brain, but also a drive to explore the environment and budget a significant amount of time for this activity (see Hinde 1970), a sophisticated system of problem classification, a plastic motor response, refined ability to discriminate between stimuli, a long period of time and varied milieu in which to deal with problems in their various nuances as encountered in the environment, and consequently a long life expectancy for the adult. Hence we expect a shift toward K-adaptations in species with complex adaptation syndromes. One point is of particular importance: A long ontogenetic process is required for animals evolved to handle complex social and physical environments. Thus, it is not surprising that the tendency to explore is greater in carnivores than in herbivores, and greater in large-brained mammals than in small-brained mammals (Glickman and Sroges 1966). The ultimate development, of course, is the complex exploratory activity of gaining knowledge as found in humans, which Berlyne (1966) describes as epistemic behavior; it may be that fantasizing "autistic thinking," that is, conceptual play, is also a direct consequence of selection for play, since fantasizing has been found to be a most important element in discovering solutions to problems, and is consciously used in group problem-solving strategies (Gordon 1961, Crosby 1968, Singer 1974).

It can rightly be asked what one can consider a simple and what a complex social and physicial milieu. In general, I would argue that the milieu of a social animal is more complex than that of a solitary one (I exclude here asexual individuals such as are found in insect societies), that species occupying one plant community live in a simpler physical environment than those occupying the ecotone between two or more communities, and that species from regions with stable climates live in a simpler physical milieu than those from fluctuating climates. Paradoxically, the biotically diverse tropical rain forest would be, from the viewpoint of the animal's *Umwelt,* a simpler environment than the subarctic boreal forest. Granted its great temporal stability, and severe competition between species (see Amadon 1973), the tropical rain forest permits the species to adapt to a narrow range of foods, pre-dators, and climatic conditions and practice these all year round; this is not possible for a permanent resident of the boreal forest who has to deal with a seasonal succession of foods, predators, and environmental conditions. Likewise, a solitary species may specialize in territorial defence and live year-round on the territory; these requirements clearly are simpler than those imposed on an individual who lives in the domiannce hierarchy of a group and depends on learning complex migratory routes to establish its home ranges. As self-evident as these deductions may seem, they are, at the present state of knowledge, only deductions and hypoth-eses.

Despite much of the speculative nature of the foregoing, it is important to examine the physiological basis of the behavioral phenomenon that animals strive to live in a predictable environment. It has important ramifications for ourselves.

Not to be forgotten is the fact that species may respond differently to different environments in their home-range behavior. White-tailed deer, which form small stable home ranges in a highly structured habitat such as forests, become drifters with no fixed home range in the flat, open, nearly coverless, plains of South Dakota (Sparrow and Springer 1970). The poverty of features makes the habitat predictable, even in areas an individual has not visited.

Relation Between Pattern Matching and the Reafference Principle

If we are to go to the root of the physiological system that forms the basis for the phenomenon that animals strive to live in a predictable milieu, then we must look at the monitoring systems in an animal's body. To maintain homeostasis a body requires information and integration mechanisms by which it can orchestrate the functions of organs, as well as obtain resources and fulfil other requirements of life; it must monitor its internal and external milieu. Much of its internal monitoring depends on a multitude of more or less sophisticated negative feedback loops between control center and target organ. This includes feedback loops based on simple reflexes, hormonal circuits, and also loops which contain behavioral systems interspersed between sensor and effector organs, such as kinesis, phototaxis, to-potaxis, innate responses containing complex behaviors, behavioral responses based on quick irreversible learning in specific environments, and behavior responses formed by conditioning, learning, and memory.

In addition to the negative feedback loop is another type of loop, which controls comples functions. This type is found in ethology under the heading of "reafferenz principle" (von Holst and Mittelstaedt 1950, von Holst 1954, see Hinde 1970, p. 98). It differs from the simple negative feedback loop by the addition of a second command loop, the efference copy. The returning afference (reafference—carrying information as to how the command (efference) was executed by the motor organ— is read against the efference copy. Thus the system can determine whether there is a discrepancy between the afference and the efference copy. If the difference is zero, obviously the afference is exactly that commanded by the efference, and the order has been perfectly executed. If the afference is greater than the efference copy, obviously there was a greater response by the motor organ than had been commanded. The efference copy thus sets the "expected" extent of motor organs' performance.

It can be seen from the foregoing that the efference copy is the means by which an animal monitors its action. If the reafference mechanisms are found at the level of the spinal cord, then they are clearly the precursors of the complex cognitive pattern-matching mechanisms that have been repeatedly postulated for the brain, as we shall discuss later. This hypothesis is strictly in line both with our knowledge of how biological mechanisms on organs evolve and with the law of least effort. Biological mechanisms are developed or evolve from existing simple structures into complex ones; granted the existence of spinal reafference mechanisms, the law of least effort dictates that they be used as the basis monitoring mechanisms for complex cognitive processes.

In order to function adequately at the level of the exterior milieu, a storage of efference copies is required for monitoring. This we call memory. In addition, it requires a mechanisms to sort and select efference copies for comparison with reafference (now the sensory input about the *milieu exterieure),* and a signal system that indicates whether efference copy and reafference match. Some sort of a reward mechanism is required for matching, with a punishment mechanism for a failure to match. The familiar "Eureka" effect may be the extreme emotional form of the former, fear and anxiety the extreme form of the latter.

Clearly, the more complex the behavior of an organism the more storage space, and the better a sorting mechanism, it requires for its efference copies in the CNS; the more predictable and variable the environment, the more it must rely on efference copies formed from learning rather than by genetic programing with a minimum of environmental input. Thus, for external monitoring, the organism requires an increasingly sophisticated CNS system with an increasingly complex environment. We shall see that this deduction indeed does hold. Granted a sufficiently large number of efference copies, the animal is in a position to constantly compare incoming stimuli (afference) against internally stored cognitive patterns (efference copy), and thus note at once *deviations* from the expected. Secondly, it can search for an *appropriate afference* using its efference copy as a guide. This would manifest itself as goal-directed behavior.

Pattern Matching

From the study of higher mammals and humans, we can regard goal-directed behavior as a function of *cognitive patternmatching*. Although this concept has been widely recognized by diverse behavioral disciplines, it is by no means easy to identify it in their diverse languages and problem orientations. Judging from the reviews of Kimble and Perlmuter (1970) and de Long (1972), patternmatching has been the essence of theories advanced by philosophers, psychologists, and linguists beginning as early as William James. The concept is strongly implied in the founding works of ethology by Tinbergen (1951), von Holst (1954), and Lorenz (at least as early as his earlier neurophysiological formulations), although Lorenz (1973) writes as if he is indebted to Campbell (1966) for the concept, not only the term "pattern matching." Without making claims for completeness of coverage, as this is well beyond the scope of this work, I shall make reference to a sampling of authors who used the concept of pattern matching, although it was in some cases only implied, or named "schema," "sense of competence," "discrepancy principle," or "auslösendes Schema cognitive dissonance" (Flesh 1951 and Walter 1957 in Shapiro 1968, Festinger 1957, Sokolov 1963 in Gray 1971, White 1963, Pribram 1963, 1967, Flavel 1963, Schleidt 1962, Perin 1970, Spitz 1963, Barnett 1967, Magoun 1969, Hinde 1970, Tobach 1970, Hinde and Stevenson 1970, Salzen 1970, Gibson 1970, Beck 1971, Cunningham 1972).

The evidence for cognitive patterns that are formed, stored, and recalled in the central nervous system comes from several sources. John (1972), in a review of theories of learning and memory formation, reports that in humans specific memories conform to specific electric patterns recorded from the brain as reported by numerous studies. Noton and Stark (1971) showed that for visual configurational stimuli the scanning movements of the eye during learning of configuration were matched by those scanning the stimulus during recall. They postulated that the brain stored in its memory the initial attention-catching points, and the angle and the direction at which the eye must move from these points. This is some of the evidence which points more directly, rather than indirectly, to cognitive patterns based on brain activity, or at least is closest to the activities of the brain, as exemplified by the work of Noton and Stark (1971). Rich indirect evidence for the existence of cognitive patterns comes from the language and image learning of children (see Elkin 1975 and Cunningham 1972).

Evidence that, in dealing with cognitive patterns, we are dealing with a structural coding of biochemical material also comes from other sources. Cummins et al (1973) showed that rats trained to do specific tasks have heavier brains than do controls, and also show marked differences in the structure and biochemistry of the brain; Passingham (1975) showed that the ratio of the volumes of the neurocortex to the medulla increased with the learning of set performance and the frequency of response to novel objects. Krech et al (1960) showed that exposure to complex environments increased the brain cholinesterase activity in rats. Zimmerberg et al (1974) showed a small hemispheric asymmetry in the brain chemistry of rats that was associated with trained left or right space preference. During learning, cholinergic nerve endings in the brain become more sensitive and synaptic conductance is altered (Deutsch 1971). Greenough (1975), in a review of experimental modifications of developing brains, indicated that environmental effects, such as learning tasks, increase the length and branching of dendrites of cortical neurons, increase the number and size of synapses, increase the size of the cell bodies and the number of glial cells, and enlarge the RNA/DNA ratio. Rats trained to fight developed alterations in the norepinephrine uptake mechanisms in the brain and increased catecholamine synthesis in the brain (Lamprecht et al 1972, Henley et al

1973). Stimulation of the brain with catecholamines in rats led to increased killings and aggressive behavior (Smith et al 1970, Igić et al 1970). The subject of neurohumors and their relation to aggressive behavior in mice is treated by Welch and Welch (1970). Noteworthy in this regard is the finding of neuroanatomists that, in different species, brain areas controlling specific functions are enlarged if these functions are commonly performed (see Hofer 1972). Gray (1971) reports the work of Sokolov and his colleagues, which demonstrated the existence in the visual cortex, reticular formation, caudate nucleus, and hippocampus of novelty or comparator neurons, that is, neurons responding to mismatches of actual and expected stimuli. Last, but certainly not least, are the now-classic findings of Penfield (1959) that stimuli from the current of an electrode reawaken long-forgotten memories that last as long as the experimenter maintains the stimulus. This work showed that memories are deposited as if on tapes and—as indicated—are subject to recall by a low dosage of electric current. Memory—we deal both with long-term and short-term memory (see Kesner and Conner 1972)—is evoked in the brain as specific spatial and temporal patterns of electric activity in the brain, but is not confined to any single locus (John 1972). This, of course, explains Lashley's (see Zangwill 1961) classic findings that learned tasks in rats were not abolished by brain lesions, but were progressively impaired with an increase in brain lesions.

A number of studies give us further information about the nature of cognitive patterns. The very existence of such patterns is supported by the findings that animals classify their sensory experiences and recognize them as discrete units (see Gibson 1970). These units are agglutinated and held as "chunks" in the short-term memory (Simon 1974). Studies on humans and higher vertebrates indicate that an upper limit of about nine chunks can be held in short-term memory; the upper limit in humans is usually seven (Koehler 1952). It was shown by Simon (1974) that chunk size is not unrelated to chunk number. If a syllable were counted as a chunk, Simon could remember seven syllables; if words and sentences were "chunked," he could remember about 26 syllables, represented by three to five chunks. The larger the chunks, the fewer one can store in short-term memory. Simon showed also that chunking of information depended on coding, and that coding was indeed a labor-saving mechanism. Thus, for nonsense syllables the time required for learning was 27.9 sec/unit learned; for digits, 25.5 sec/unit; for prose, 7.2 sec/unit; for poetry, 3.0 sec/unit. The ratio of chunk content between words and nonsense syllables was 2.5 to 1; 2.5 words could be retained per nonsense word, owing to the redundancy inherent in the words, as well as the "meaning" that existed in the specific linking of syllables into meaningful words.

It is evident from *a priori* deductions that an animal requires a mechanism whereby recognition or nonrecognition is signaled, that is, matching and nonmatching of cognitive patterns, one derived from memory and the other from the ongoing efference of the animal. Such a signal apparently occurs from an activation of "reward centers" in the brain which are brought into an excited state during goal-oriented behavior, and extinguished by the appropriate efference (see Olds 1967). Recognition is thus triggered as a reward for each reduction in uncertainty; recognition is reinforcing. We may also speak here of reduction in arousal. In humans, such "reward centers" appear to consist of adrenergic neurons that lie deep in the brain stem, but whose axons ascend to the hypothalamus, limbic system, and cortex. These axons lie opposite the reward neurons and extinguish them by means of norepinephrine. A destruction of these adrenergic neurons leads to a deficit in goal-directed and reward-directed behavior and losses of consumatory acts. Such a destruction appears to take place in schizophrenics, since they produce an autotoxin, 6-hydroxydopamine, which interferes with the transfers of noradrenalin between the adrenergic neurons and the reward neurons; in the long run it leads to a degeneration of these neurons, sympathetic nervous endings, and catecholamine-containing endings in the brain. Linked with the reward neurons are also systems of activating and deactivating neurons, that is, there are stimulatory and inhibiting centers, which may be kept in balance by simple neurohumural negative feedback systems (see Stein and Wiese 1971). These centers can be deduced from selective destruction of brain parts by surgical or chemical means. Thus, hypersexuality may be achieved by the destruction of the inhibitory center; conversely, sexual extinction is the consequence of destruction of the stimulatory center. In contrast to the "reward centers," there are also "punishment centers," whose activation brings ongoing behavior to a stop. If, in goal-directed behavior or during the animal's activites, perceived patterns cannot be matched against internal cognitive ones, that is, a strange perception is noted, arousal of the organism is the first and mildest; consequence of pattern mismatch. From the work of Kagan (1970) on the attention span of children in relation to known and unknown visual configurations, it appears that arousal may be an inverted U-shaped function on unfamiliarity. That is, arousal or attention span is least with well-known familiar stimuli, as well as with totally unknown ones; arousal is greatest with stimuli of which some elements only are familiar. It is also pertinent to note that the response to stimuli is a function of the level of arousal. Berlyne (1966) found that aroused rats tend to search for more familiar stimuli; rats with a low arousal level search for more novel stimuli.

Patterns are recognized by serial inspection of components; hence it takes longest to discriminate known patterns from irrelevant ones, since incongruency is quickly detected and congruency is confirmed only after total inspection (Noton and Stark 1971). Clearly, where congruency is only partial, a child apparently tries repeatedly to fit the perceived pattern with the one in memory. The concept of the inverted U-shaped function of arousal at strangeness does not fit entirely with the generalization that strange objects, sounds, smells, etc., arouse higher vertebrates, as has been found repeatedly (Berlyne 1966). Clarification in this area is still required. Effects more potent than arousal due to mismatch of perceived and stored cognitive patterns will be discussed below.

Theory of Emotionality

The foregoing can be structured into what has been referred to as the theory of emotionality. It is no new concept, as is evident from papers such as Hebb (1946, 1966), Thorpe (1963), Tobach (1970), and Gray (1971), or Hinde's (1970) comprehensive textbook in ethology. The following is an elaborated model, since it is based on ethological data as well as psychological ones, and cast into the context of evolutionary theory. As we shall see, it predicts correctly a rather diverse set of phenomena in mammalian and human behavior. The theory runs as follows:

(1) The adult individual screens the stimuli that impinge on it for unexpected strange stimuli by constantly comparing the stream of inflowing patterns with those from its memory.

(2) As long as the perceived stream of stimuli matches the stream of patterns generated internally, that is, actuality matches expectation, the individual continues with whatever task he is doing.

(3) When the perceived patterns fail to match the internally generated ones, arousal is triggered and the individual interrupts its task. The degree of arousal tends to be directly related to the novelty or degree of discrepancy between patterns (Berlyne 1966), and to the stage of arousal the animal is in.

(4) The aroused animal experiences the beginning of general adaptation syndrome, as classically described by Selye (1950); after all, the organism is preparing itself physiologically for violent activity so that it is prepared to flee or to fight. These activities liberate a lot of energy in short order, and the body must be physiologically prepared to deliver such activity.

(5) The individual, if mildly aroused, begins to act in a fashion we consider as expressing "curiosity" (Berlyne 1966). It begins to explore the strange stimulus but remains cautious, and is prepared to flee, as often indicated by intention movements to withdraw. It must be emphasized that exploration may be exceedingly tenacious (Hinde 1970). Should the strange object or pattern disappear, the individual may flee owing to the unpredictable situation this generates; if the unfamiliar pattern persists, and in the course of exploration remains harmless, the animal gains familiarity with it and ignores it ultimately. It is not surprising that aroused animals tend to search out familiar stimuli, whereas quiet, unexcited animals tend to seek more novel stimuli (Berlyne 1966); mild arousal is "rewarding" but extreme arousal is not. Thus the animal avoids nonarousal as much as extreme arousal; we can conclude that there is an optimum level of arousal.

(6) If the strange situation persists and frustrates the individual's effort at resolution, it may experience anxiety, use inappropriate behavior, become neurotic, suffer from learned helplessness, and ultimately suffer from depression (Akiskal and McKinney 1973, Seligman 1975). It is this situation that sets in when the frequency of unknown situations relative to known ones becomes large, and frustrates the animal's attempt to cope. It is here that we find the roots of "future shock" (Toffler 1970). Effects of stress will be discussed under Population Quality in Chapter 6.

(7) If the strange situation persists despite attempts by the individual to resolve it, a phenomenon occurs that deserves to be singled out: It is *helplessness*, as discussed by Seligman (1975). When the environment is no longer responsive to the acts of the individuals, it motivation to respond drops, it generalizes and does not even try in other situations, which of course reinforces the vicious cycle, and the consequence of this is apathy, depression, loss of vigor, impairment of the immune system, increased susceptibility to diseases, even to cancer, and very likely a premature death. Thus an environment may be predictable, but that does little good if the individual is incapable of dealing with it, and once this is learned it may give up trying and vegetate to death. The psychic deaths of men and animals can be attributed to this state, such as death following a severely traumatic event. This Seligman (1975) elaborates on in a most worthwhile book.

Physiological preparatory activity, that is, symptoms similar to the general adaptive syndrome, may begin very early in anticipation of unpleasant tasks. Dogs may begin to develop high blood pressure and brachycardia about one hour before commencement of a regular unpleasant task (Anderson and Brady 1971).

Brachycardia is also developed by persons preparing for overt or covert activity (Schwartz and Higgins 1971), and by infants exposed to novel sight and sound stimuli (see Kagan 1970). Referring back to the law of least effort and the cost of excitation, it is quite evident that an individual living in an environment that calls for unpleasant tasks, and exposes it to unexpected stimuli, will pay a considerable amount in energy and nutrients for living in such an environment, as well as in impaired reproduction and health. Rats raised in an environment in which they could control some aspects of it were far less emotional in open field tests than rats that had not been able to control aspects of their environment (Galle et al 1972); the ability to make the habitat predictable, therefore, does have a beneficial effect in the sense that the animal is less aroused by novel situations and less likely to suffer helplessness (Seligman 1975).

The foregoing theory also explains why persons of different ethnic backgrounds may be less than content with the cultural environment of a different group (Clark 1968), or why older people find modern concepts trying. It explains why individuals selectively screen out uncomfortable information and "hear what they want to hear," why they remember the information congruent with their own belief and forget the other, why they attempt to see issues in "black and white" terms and ignore the "gray areas" of doubt and uncertainty (Strauss and Sayles 1972). It explains why in verbal communication words that have a "bad" connotation take longer to be recognized than words having a "good" connotation (Newbigging 1961). We strive for pattern matching, particularly those with a positive reinforcment. In "friendly" discussions we "harmonize" with others on known experiences, subjects, knowledge, etc. Thus, we search out the company of our peers, since here harmonizing is posssible. Thus the attraction of likes, provided they need not compete. Thus the cohesiveness of groups of persons with a homogeneous background (Strauss and Sayles, 1972, p. 85), or of small size in which, of course, a familiar conceptual milieu can be established. It also explains under what conditions apathy, morbidity, and increased mortality can be expected in persons, namely, when circumstances have conspired to deprive them of the minimum satisfaction of being successful in helping themselves. In the face of incongruency they give up, having learned the futility of responding, and now also lack a belief in self-worth. An unpredictable and unresponsive environment spells doom.

Not to be forgotten in this discussion is another effect of incongruency: The animal may *attack* the "unusual." This I have observed repeatedly when healthy mountain sheep attacked sick companions (Geist 1971a); it is also reported for chimpanzees by von Lawick-Goodall (in Hinde 1970). Scott (1966), in a review of agonistic behavior in mice shows that "unlike" mice were attacked more than "like" mice; thus unusualness or strangeness increased the frequency of aggression. In human females, Hess (1965) recorded involuntary aversive responses to crippled or cross-eyed children, but positive ones to healthy babies, or mothers with babies. He also found the same negative response to modern abstract paintings and cited the work of Shachnowich, who found this response appeared when subjects were presented with unfamiliar geometric patterns.

Intelligence, Creativity, Memory

At this point it can be pointed out that three attributes of intellectual competence can be understood within the framework of the pattern-matching theory. *Intelligence* can be equated with the rapidity and economy of pattern matching; the quicker a person can detect and match cognitive patterns, the more intelligent he or she is. This attribute of intellectual competence can be enhanced greatly by training, and is discussed later more extensively (see p. 370). *Creativity* can be equated with the rate of cognitive pattern dissolution and reconstruction. The more creative a person, the less he or she sees the world in stereotypes, since such a person continually beholds the same situation in the light of different visualizations. Cognitive patterns, old knowledge, are constantly restructured and compared against incoming stimuli; problems are constantly found and solved. Here some very basic mechanism is at work at the molecular level dissolving and restoring patterns (see p. 31). *Memory* becomes pattern storage and recall, in essence, the filing system of perceived information.

Structuring and Ontogeny of Cognitive Patterns

It is evident from the foregoing that the cognitive patterns of the individual are, in essence, its knowledge. It is these that generate expectations about the outside world. Some such knowledge may be structured into the organism, even the human organism, without prior experience with appropriate situations in which such cognitive patterns could be structured. That is, such knowledge is "innate." This is a point greatly stressed by ethologists (e.g., Eibl-Eibesfeldt 1967, Ruwet 1972) and supported for lower animals by elegant, convincing experiments (e.g.., Eibl-Eibesfeldt 1955). In humans, the response to the "visual cliff," or the mechanisms responsible for "size constancy" of an object seen at different distances, appears to be based on innate knowledge (Gibson 1970); children begin to develop cognitive patterns soon after birth (Kagan 1970), but they are biased from the start by selecting sharply contoured stimuli, complex rather than simple stimuli, and intermittently presented rather than steadily presented stimuli (Kagan 1970, Barlow et al 1972, see also Berlyne 1966). The attention paid by infants to a complex visual field is apparently a function of the square root of the absolute amount of sharp borders (Karmel 1969); this also fits with Noton and Stark's (1971) findings that attention span lengthens with the number of corner, angles, and sharp curves in the infant visual field. Further work showed that it was not the complexity of the visual stimulus but the amount of border that fixed attention; in addition there was also a preference for moderately long borders by infants over very long and very short ones. The visual system of the child is thus already structured at birth to pay attention to certain stimuli over other ones. The auditory system may also be preadjusted, according to Mattingly (1972), who argues that speech may be innately recognized. If we put these preferences together, we note as Kagan (1970) points out that the infant is born with the minimum requirements to organize a visual schema "face" and abstract it from the visual surrounding. Mothers move about, have complex faces, speak, and have sharply contoured exteriors. The infant's biases confine it to looking at "relevant" stimuli, namely its mother.

The cognitive pattern "face" is thus formed from a minimum, inherited "knowledge" and is built up and completed during the infant's ontogeny. There are sensitive periods in which specific thing are learned. Thus, attention to "face" drops after the infant's fourth month. However, after 12 months it again rises sharply and continues to do so until 36 months of age. Now the child apparently differentiates between "face" and "individual faces." Prior to the fourth month of age, the child spends an equal amount of time scanning symmetrical and asymmetrical forms, but thereafter prefers symmetrical forms. It also prefers increasingly complex over simple patterns (see Berlyne 1966) and acquires conditioned responses faster (Kagan 1970). During early childhood, "field effects" such as dominance of continued line or formation of "good form" as described by gestalt psychologists dominate the child's perceptual abilities. As it grows older, the field effects are progressively dismantled and perceptual regulatory patterns arise that permit perception patterns to be joined, inverted, reversed, coordinated, fixed as object or as field, held for comparison, and integrated with temporal or spatial features. The development of such regulatory processes can be enhanced through training; the field effects appear to be the individual's largely innate component of perception (Elkin 1975). On the whole, investigators dealing with child development conceive of the child as unfolding its inherited capacities—the environment permitting!

It is important to note that linguists studying the ontogeny of human language echo the same conclusion (Lenneberg 1969, Bronowski and Bellugi 1970, Mattingly 1972). Language development is tied closely to motor development and brain maturation; the child's basic language structure is essentially formed at 4.5 years of age. It is difficult to learn language thereafter, particularly after the teenage formative period. This is well illustrated by studies on brain-damaged individuals.

Language is formed neuronally in the left cerebral hemisphere in the frontal and temporal lobes in most persons, primarily in the Broca's and Wernicke's areas (Lenneberg 1969, Geschwind 1970). Injuries to the left hemisphere lead to aphasia (impairment of ability to use words) in adults; in children, before the age of two years, injury to the left lobe causes no impairment of ability to learn language; between the ages of three and four years, injury causes transient aphasia and such children repeat the stages of language acquisition, albeit they do it quicker. If language retardation is present and is not abolished before the early teens, nothing can be done afterward, since language structure is formed beyond major modification. If children become deaf accidentally at three to four years of age, they lose the language rapidly; however, a residue of acquired cognitive structure apparently remains, since such children do better in schools for the deaf than do children deaf from birth. In humans, therefore, it appears that cognitive structures are shaped in critical periods over a long ontogenetic period from a small innate base; moreover, the ability to continue forming certain classes of cognitive structures continues throughout life, as is evidenced by learning ability. Which cognitive structures are of early formation and which late, which has a large and which a small innate basic structure, which are irreversibly fixed and which are readily modified, requires further research. The important point here is that we are particularly subject to a long formative period, during which an improper environment can raise havoc with us and our capabilities as adults. I shall discuss this under Population Quality (Chapter 6).

Prediction with the Theory of Emotionality and Pattern Matching: From Sheep in Pastures to Conventional Wisdom

The theories of emotionality and pattern matching are very handy in explaining apparently diverse behavioral phenomena in animals and men. The extensive play activities of young mammals and their curiosity are not mysterious, as Hinde (1970, p. 359) indicates they are, but appear to be an obviously necessary process in programming the individual to cope with adult life, as is the tenacity with which adult animals stick to familiar home ranges, or pursue their next subordinates and by various means attempt to keep them subordinated. The development of emotionality during ontogeny, so well described by psychologists (Hebb 1966), is a means of permitting learning and maintaining adults in an adaptable milieu. Humans readily develop rigid spatial use patterns, and are upset if a routine is upset, and they avoid the unusual, such as keeping distance from crippled, oddly behaving, persons (Sommer 1969, Hebb 1966). Behavior can be understood as a means by which the organism shapes a predictable environment or shapes itself so that this environment contains a minimum of surprises for it (Geist 1971a). The theories of emotionality and pattern matching explain in detail behaviors that without such theories remain obscure.

Note, for instance, that sheep released into large paddocks in Australia defy the expectations of agriculturists. The number of sheep released into such paddocks is based on the amount of forage contained in the pasture. There is an underlying assumption that sheep will graze the area uniformly, and in so doing achieve a satisfactory weight gain and keep the pasture lightly and evenly grazed. Yet sheep do not act that way. Instead of moving across the pasture, they cluster at the gate. Here they graze out the area closest to the gate and reluctantly move beyond. When they do begin to move out onto the pasture because of food shortage, they stick closely to the fence lines and graze out the perimeters of the pasture. The center of the pasture is last to be grazed (McBride et al 1967; Arnold, personal communications). Puzzling? Not at all!

In a strange environment the animal will stick to the site it was liberated at or rush into the closest escape terrain or cover. Any move from this now-known area carries it to an unknown and emotionally upsetting site. Hence, as we know from many studies (Hinde 1970, p. 353), the animal will oscillate from the *known* to the *unknown* and will initially spend more time at the known site. In the above example, sheep find at the gate a cluster of companions that they have been with previously. Thus the gate is the most familiar territory to them. The presence of known companions is quieting rather than upsetting to the animals, just as Liddell (1958, 1961) demonstrated for lambs, Harlow and Zimmerman (1959) for monkeys, and Morrison and Hill (1967) for rats. If the sheep moves, it will move along some border, the fence, since this border is connected to the known site and is thus a part but not a substitute for the known site. This explains movement along fence lines and trails; thus occupation of the pasture's periphery and lastly its center.

I used the above principle in the taming of wild mountain sheep, both in approaching them and, even more so, in conditioning them to accept my hand on their faces and the subsequent manipulation of their ears in which small metal tags were embedded (see Geist 1971a). Thus, once I could touch them on the nose I moved my fingers down the face and back to the spot where I touched them first. My fingers thus moved from the area on the head where touching with my fingers was familiar to the animal to the areas where it was not familiar. In principle this is an oscillation from the known to the unknown. However, the above explains much more than the spatial behavior of animals exploring new environments, such as mice exploring cages or lambs exploring new rooms; it also explains the actions of humans, but in the area of conceptual rather than physical space. It explains well how human beings deal with new knowledge that confronts their conventional wisdom.

A student of animal behavior may find himself in surprisingly familiar intellectual terrain when reading John Kenneth Galbraith's essay on conventional wisdom in the second chapter of his book, *The Affluent Society* (Galbraith 1958). It is true that he is dealing with more exalted beings than mere animals, although neither the evidence Galbraith marshalls nor his arguments are particularly convincing on this point. Conventional wisdoms are familiar, accepted, intellectual milestones or, better, pillars that support the norms of behavior of our society by their very acceptability and association with wealth, power, and hence prestige. They are our intellectual home ranges, the known, familiar, accepted conceptual terrain. So as not to lose the spice of life—or an "optimum level of arousal"—they are reiterated, in somewhat novel versions lest boredom be our lot. It takes effort to learn them, to discuss their intricacies; it calls for association with the dominants of society; it conveys a measure of privilege or status; it satisfied the soul, since there are many adherents, and so reduces loneliness; it has all of the functions of a good home range, for it leaves the bearer of conventional wisdom eminently prepared for any discourse he may want to call discussion. Then, the two bearers of conventional wisdom harmonize on their common knowledge, as well as competing by making little form changes in their conventional wisdom, and they may even have a heated debate about some trivial aspect. This latter part need not concern us here; it is the legitimate domain of Chapter 5 on dominance and ego display. We are concerned, however, with the tenacity with which conventional wisdom is clung to, the psychological rewards we experience in recognizing conventional wisdoms, the hesitant departures we make into new intellectual realms and the quick return to conventional wisdoms, as well as the tendency to regard those of different theoretical convictions as different species, of low worth—if of any (see also Strauss and Sayles 1972). This only indicates that in our intellectual discussions we do not deviate in principle from the actions of other higher mammals, but follow animal tendencies with comforting regularity. A zoologist would expect to find no different. Although the substance of our discussions may be cultural, the form in which we change from one idea to another appears to be homologous to animal exploration. From the basis of neurophysiology there should be no difference, and this is supported by the law of least effort. There should be no difference in the principle of how the central nervous system handles *spatial* knowledge and *cultural* knowledge. This is suggested by the theory of pattern matching and emotionality.

Even without the benefits of the neurophysiological model, those dealing with humans in a professional capacity recognized the effects of the law of least effort in humans. Allport (1954) recognized correctly that our tendency to avoid trouble and emotional arousal, the need to learn and master inconveniences, expressed itself in persons choosing for company others having similar backgrounds. Lofland (1973) indicated that residential segregation is a function of human craving for predictability. He recognized that symbols in dress, behavior, and manner of speaking are means to sustain predictability in the social milieu; symbols permit us to form categories of organized knowledge. Scuttles (1972) suggests that people choose defended neighborhoods as a means of coping with heterogeneity. One could also say, as a means of coping with an overload of social problems.

Thus the "irrational" behavior of rational man is not at all surprising when strong conflict of cognitive patterns arises. We can even detect the familiar inverted U-shaped function of response frequency plotted against "strangeness": Utter familiarity is dull and not worthy of discussion; the partially familiar, and thus puzzling; occupies us most as we strive to match knowledge; the unmatchable leads to no discussion of any length since no common basis is found. Of course there are some who have recognized this principle without the benefit of science. Anderson (1952), a pioneer in the study of agricultural plants, mentioned that scientists who are one step ahead of everybody else are considered great, those two steps ahead are considered eccentric and those three steps ahead are ignored!

Chapter 3

Communication

Origin, Components, and Definitions

Whereas predictability in the physical environment, or habitat, is generated through the explorations of the individual, in the social environment predictability is created and maintained through social behavior. The habitat characteristics are learned, but the social milieu is not only learned, but also created by the individual. To the individual a reduction of ambiguity or the creation of predictability increases its reproductive fitness by permitting it to follow the course least likely to be costly in precious nutrients and energy and least likely to lead to the death of the individual. It can also be adaptive to an individual—as we shall see—to reduce predictability to another and benefit by the mistakes of its aroused, indecisive, opponents. Predictability is generated by the individual's precise knowledge of each companion's combat potential, as well as the signals emitted by each companion, so as to correctly anticipate and avoid or, alternatively, to provoke specific actions from them. Individuals must know when active competition is futile and alternative means of maximizing reproduction are more profitable. Predictability of the social milieu permits the developing animal to maximize its growth and development by developing strategies that enable it to get maximum nutrition and maternal care, even though it is in the parent's interest to pay out as little as possible in support and still have a viable offspring (Trivers 1974). The predictability of the social milieu is created by social behavior, through the mechanisms of cognitive pattern matching discussed earlier. Thus, communication is the evolutionary solution to the law of least effort, the rule that individuals must maintain homeostasis, create access to scarce resources, and choose the best possible—or the greatest number—of mating partners.

Communication is a global subject debated by different disciplines, each having different, often untransferable, concepts and sharing at best a common interest in the subject, so Cherry (1957) points out. There is no shortage of discussion about communication, and yet, despite such extensive treatments as those of Sebeok (1968), Altmann (1967), Hinde (1974), or Wilson (1975), one may be permitted another attempt at ordering this subject. Some important concepts need to be included, while the aura of complexity has obscured the elementary simplicity and elegance of animal communication. In addition, some rather questionable and poorly founded ethological theory can be ignored with profit in the discussion of communication.

Communication can be defined as a process adaptive to the genome of the *sender* of signals; it transmits information between sender and receiver by means of signs and signals whose semantic or pragmatic components are encoded as messages. Only exceptionally, namely, under conditions of kin selection in which natural selection maximizes the number of adults per unit of defensible resource, is it the receiver of signals—rather than the sender—whose reproductive fitness may be enhanced. Under this condition, however, the genomes of the sender and receiver are identical or exceedingly similar. Only under conditions of kin selection do I follow Wilson's (1975) lead of confining communication to signals that enhance the reproductive fitness of sender *and* receiver. Some examples illustrate the limitation of the above definition: Plants, in attracting insect pollinators, do indeed communicate with the insect, but not with the herbivore who is also attracted by the flower but comes and eats the plant. This is despite the fact that the plant's shape, color and smell, were recognized as a message by the herbivore. Thus, communication excludes information derived from prey, forage, inanimate objects, commensals, or predators. In the latter case, for instance, the organism does obtain, of course, valid and vital information and may react vigorously, but the information derived is incidental to the predator's activities, and the predator in no way enhances its reproductive fitness by sending it. Had the predator, however, sent information that attracted prey, then communication would indeed have taken place. It would have been "communication to" the prey, rather than "with" the prey, as an angler fish would perform, but it would be communication just the same, as it enhances the reproductive fitness of the sender.

Information can be defined as that which, in a message, causes organisms to order their priorities in behavior; it affects the probability of behavior occurring. Information has thus a wider scope than communication; all communication carries information, but not all information is derived from communication by the organism. Information is a function of experience stored or inherited by an organism in its central nervous system; it is not a function of messages communicated or of inanimate objects, commensals, or companions.

The above refers to the semantic meaning of information, which is not to be confused with the mathematical concept of information as expressed in "information theory" by the Shannon–Wiener formula. Here information is a mathematical expression defined in binary bits measuring their statistical rarity (Cherry 1957, Wilson 1975 pp. 195–200). The mathematical definition of information speaks about the content in purely quantitative terms; it says nothing about what the sender coded or the receiver understood. Information theory—if applied to organisms— tells us at best how, not what, they communicate. Information theory, disregarding momentarily the great methodological difficulties in applying it to animals, tells us something of the sophistication achieved in the process of communication. Wilson (1975) discussed some of the difficulties of applying this quantitative approach to animal communication, and indeed, it has been a rather unexciting, not very fruitful pursuit by zoologists.

A most important point has so far been missed in the discussion of information theory, namely, that it runs afoul of biological realities. Thus, according to information theory, a rare or unique event carries little or no information. Information is carried only by multiple patterned events. Yet, it is the rare event that animals are programed to respond to most readily; they respond with alarm (Berlyne 1966, p. 33). Rare events in nature are, on principle, dangerous. Predators may be recognized by their rarity (Schleidt 1961). Although rare events may only arouse animals, they are nevertheless, for that very reason, most important in animal communication and are being used for that very purpose—to arouse. One needs only to consider the rarest of behavior patterns used in communication to illustrate the point, such as the very exceptional roar of mountain goat males in intense dominance displays prior to combat, or the very rare and terribly effective roar of moose when confronted in deep hard snow, or more commonly the assumption of the alert posture by mountain sheep, deer, or elk—a relatively rare behavior to which others respond with alarm (Geist 1971a). If one experimentally makes an individual act in a rare and unusual manner, such as by sticking a paper bag in the mouth of an inquisitive mountain sheep upon which it will buck, bounce, and twist on the spot, one can observe at once the alarm that it causes.

Signals that, according to information theory, would convey little information owing to their enormous redundancy and yet are continually performed, and that appear to aim at generating arousal and anxiety, are the undirected dominance displays. I shall argue this point later. Here it need only be emphasized that communication mechanisms need not aim at transferring only semantic information, but also its opposite—making the immediate social milieu unpredictable—thereby transferring a benefit to the sender.

Semantic information, in contrast to information mathematically defined, is thus a function of cognitive pattern matching, as discussed in detail in the preceding chapter. Information has been transferred when the receiver is able to match internal patterns—be they learned or innate—with the patterns sent by the sender, and as a consequence he can perform specific appropriate acts. However, it is worth emphasizing again that communication may aim not only at cognitive pattern matching, but also at the prevention of cognitive pattern matching, thereby generating arousal in the receiver, and this in turn increases his cost of existence and causes flight or abandonment of the locality.

Signs and signals are defined as adaptations by which the sender sends messages that act as stimuli on the receiver. The signal is the syntactic form of the message, that is, the form itself, such as a visual *configurational gestalt* of the animal; it may have the same message, for instance, but differ in syntactic form from a *chemical released by the sender simultaneously.*

The message is the package of information sent by the sender and received by the receiver. It has a *semantic* form, that is, the meaning encoded by the sender, and a *pragmatic* form, that is, the meaning decoded by the receiver (Cherry 1957). It can come in different syntactic forms across different *channels* of communication, the visual, auditory, olfactory, tactile, and even the gustatory and proprioceptive, as will be discussed later; it can also come in different *syntactic* forms across one channel. For instance, as will be discussed later, the rush threat and weapon threat are two different syntactic forms of the same message transmitted along the visual channel.

Classifying Meaning

The semantics or meaning of a message is aided in transmission by the work horses of communication, the *message modifiers*. Before these can be discussed, one must turn to a classification of meaning. There have been a number of attempts, notably those of Sebeok (1962), Marler (1961), and Smith (1969), and these have been critically discussed by Wilson (1975), but maybe not critically enough. Sebeok's (1962) six basic functions of communication, emotive, phatic, cognitive, connative, metacommunication and poetic are held together by no common criteria. Yet, this is a prime requirement in classification. For instance, the phatic function (maintenance of contact between sender and receiver) may be either a consequence of the arousal of emotions or of a cognitive process; the poetic, metacommunicative and connative functions may be consequences of cognitive processes. Thus, a simplification of Sebeok's (1962) classification is possible, as I show below.

The classifications of Marler (1961) and Smith (1969) are those of information, not of the broader-category meaning. Signals may *withhold* information and thereby have meaning in the sense that animals respond with arousal. Also, the meaning of a message to an individual may be unrelated to its encoded information; classical conditioning experiments make this plain. By and large, the meaning of a signal is the behavioral response of the receiver. A classification of meaning thus depends on some theory of how a message is processed internally and its effect is manifested in the behavior of an organism. The theory I am proposing to use is that of cognitive pattern matching, discussed in the preceding chapter; three classes of meaning can be identified.

(1) Messages that bypass the cognitive evaluative apparatus entirely and result at once in a response. These are the reflexes and Sebeok's emotive responses and the social behaviors released by social releasers (see Hinde 1974). Messages whose syntactic form emphasizes stimulus contrast fit this category, such as many alarm calls and threats. Messages that trigger maternal responses, such as infantile distress calls, belong here. Messages transmitted by pheromones that unlock goal-oriented behavior, such as the estrus pheromones of female Rhesus monkeys which trigger courtship activities by males (Michael and Keverne 1968), belong here as well.

(2) Messages that exercise the cognitive apparatus, leading to pattern matching; this means messages imparting objective information, which harmonizes with the animal's experience and permits it to choose a course of action on the basis of that experience. Here iconic messages belong, based on intention movements, as well as abstract messages, which gain meaning through association.

(3) The third class of messages also exercises the cognitive valuative apparatus, but they withhold information, or they reduce information so as to generate an emotive state in the receiver, or they permit the receiver to use the signals as a standard to judge other signals against. Two examples may illustrate this.

In the dominance displays of large ungulates, we observe activities that are not actually directed at, but are addressed to, the receiver. The dominant may slowly approach at a tangent, look away from the subordinate, halt in his vicinity, and maybe horn the ground or a shrub. No specific intent is signaled. No appropriate counterresponse by the receiver can be formulated. No prediction can be made as to whether an attack is forthcoming or not, nor what exact form it might take, nor when it might take place. Here the dominant is withholding information that is crucial to the defence of the receiver. Yet the sender is continuously implying, albeit not denoting, imminent danger. To the receiver, the social milieu is now unpredictable, and according to theory it should respond with arousal and/or flight. Here the dominance display appears to aim at generating fear through creation of discrepancy in cognitive images.

The third class of messages may also generate the opposite of arousal, namely quiescence and relaxation, so that individuals can go about their normal activities. Such messages are apparently essential where large numbers of individuals gather, such as in herds of ungulates or swarms of birds, who by their very presence generate much noise and obscure vision, and thus disarm the individuals, in part, against predators. The larger the group, the less likely that an individual can perceive what is going on on the other side of the group; nor can an individual caribou, for instance, readily detect the difference between a conspecific stepping on a branch and a wolf, when it is surrounded by tall dwarf birch or in a willow swamp. Here, there is a need for a continuous signal that denotes conspecific identity, location, and "normality" and that can be ignored as long as it lasts. But let the flow of signals stop, and at once everyone knows that danger has been perceived. In groups, it is to the individual's advantage to signal "normality," lest it accidentally alarm other individuals, then become alarmed in turn and so waste precious energy and nutrients. Here, not the communicated signal, but its absence—silence—is the "novel" and alarming event. The many contact sounds of social birds and ungulates, or the clicking sounds of the carpals and phalanges of caribou and Pére David's deer (*Elaphurus davidanus*), two highly gregarious cervids that often live, or have been hypothesized to live, in thickets, appear to fit this category.

It may be noted here, that the necessity for this type of communication is recognized and applied by herders of domestic ungulates. They have developed herd calls that are associated with rest and calm grazing, which can calm nervous herds in times of dnager or when moving the animals through strange places (Baskin 1970, 1974).

The foregoing makes it evident that redundancy (to be discussed later) can be generated not only by repetition of a message, by coding the message in different syntactic form, by transmitting the message over several, rather than one, sensory channel, but also by coding messages to satisfy all three meaning classes. Even at a glance, it is obvious that the rush threat of an attacking mule deer buck does just that. It incorporates *looming* (see p. 59), vocalization based on stimulus contrast, and hard slapping of the ground with the hooves—which is also a signal of stimulus contrast. The hard slapping is an intention movement and a prelude to attack, as is the fast approach. This should harmonize with the conspecific receiver's past experience; the rush generates uncertainty, for the rush threat may terminate in an attack—or it may not.

The foregoing classification of messages according to meaning classes suffers from a serious, if not fatal, flaw. The criterion for putting a signal into one class or another depends on the neurophysiological processing of that signal, at present a matter of educated, but rather imprecise, guesswork. No signal can be readily classified, except by reference to investigations, which show—or at least make it likely—that the signal in question is either causing a reflexive response, or harmonizing with the experience of the receiver. Even if this can be plausibly accomplished, and it often can, one is still left asking the obvious question, "How many neural processes are there that handle signals beside those used to form the classification?" Still, I believe the foregoing did illustrate the heuristic values of the classification, as it does point to previously unrecognized processes in communication, such as signals structured to withhold information, to create redundancy by structuring the same messsge to fit the enumerated meaning classes, or to become meaningful by their *absence,* as in "Stimmfühlunglaute."

Core Messages

The classification of information in messages, attempted by Marler (1961) and Smith (1969), is an open-ended process and for this reason is not likely to be terribly successful, as Wilson (1975, p. 216) argues, I believe correctly. It must always be arbitrary, unless some criterion of classification can be established, and this is not evident in the above classifications. What unequivocal criterion can be gleaned from the diversity of signals transmitted between sender and receiver, and from the multitude of responses both generate concurrently? I believe we can identify only one, the attraction of sender to receiver as expressed by their repulsion or cohesion. This means that there are only two dyadic superclasses of information or core messages transmitted, namely "stay/come" and "go away." The core messages, however, are modified by message modifiers and thereby gain specificity and effectiveness. In addition, there is auxiliary information, that is, information that the receiver obtains that is incidental to the act of communicating by the sender, and signals devoid of semantic or pragmatic meaning which arouse, such as alarm and contact calls.

Note that this proposed classification comes close to the approach/withdrawal conception of animal behavior by Schnierla (1965); the core messages may correspond to two basic motivation sets. These core messages are now modified by various means toward various ends.

Signal and Message Modifiers

Message modifiers operate on the structure of signals so as to facilitate their transmission. Signal and message modifiers are as follows.

1. Duration.   The length of time a signal is emitted is varied; the semantics remain unchanged.

2. Repetition.   The signal is repeated so that the message is transmitted repeatedly, increasing redundancy and increasing comprehension of the message.

3. Modulation.   The signal is varied in intensity during the course of signaling; this signal modifier may act against habituation to the signal by the receiver.

4. Direction. The signal may be directed at no one in particular, or it may be modified so as to apply to one individual only; the signal may be addressed. Altmann (1967) suggested several ways in which a signal may be directed: by reducing output through the channel of signal transmission so that the message is recieved only by the nearest conspecific; to use a private code known only to a select few; to add an auxiliary component to the signal denoting whom the signal is directed to. The latter may take the form of the sender's facing the receiver when directing the signal at him, or presenting to him a specific part of the body.

The signal may also contain structural elements that permit quick localization of the sender, and there are signals that are structured to do the opposite. Here we are dealing not with the enhancement of the semantics of a signal, but with the enhancement of what is normally auxilliary information. Sounds that have a sharp onset and termination can be easily localized in space. The process by which this is done is known as angular sound localization (Diebold 1968, p. 554, Marler and Hamilton 1961). Predator warning calls, however, may have a gradual beginning and termination, making their localization quite difficult.

5. Frequency alterations. Since sounds of high frequency are more easily absorbed, and thus muffled by objects, than are sounds of low frequency, it follows that species living in forest, where they are surrounded by foliage, ought to use vocal signals of lower frequency than species living in open habitats. Using this argument, Altmann (1967) showed that forest-dwelling primates used not only loud vocalization for long-distance communication, but—as in the case of the howler monkey (*Alouatta palliata*)—low frequency sounds as well. The same principle is illustrated in Morton's (1975) work on tropical birds.

6. Grading. This is not so much a signal modifier as a message modifier for, in varying the intensity of the display, grading provides an infinite shading of meaning and permits the receiver to judge the probability of overt actions. Not all signals can be, or are graded; some are discrete (Wilson 1975, p. 178).

7. Obscuring. This is a message modifier. Here the animal overlaps the intention movements for one act with other acts, in order to camouflage or hide its intentions; the animal plainly lies about what it will do next. We find an example of this in bighorn sheep, during dominance fights. Prior to clashing, the attacking ram moves away from the defending ram. This is a clear signal that a clash will follow, and the actions of the defending ram, who gets into position to meet the clash, make it plain that the turning and moving away is understood this way. Quite often, however, the attacking ram will stop, paw vegetation and feed, move a few steps, feed, then suddenly whirl, rise, and charge at the defending ram. The pawing and feeding can be considered to be functional elements of behavior that reduce the predictability of the time of the clash for the defending ram. The attacking ram *lies* about his intent. The advantage gained by the attacking ram, should the deception succeed, is to throw his oponent off guard for a split second, thus depriving him of time to ready his defences while the attacker crashes into the unprepared opponent. This would throw the defending ram downhill and would very likely injure him as well. The question of whether rams "lie" consciously is irrelevant to the discussion, since they could be innately programed to mix feeding behavior with actions just prior to the clash. It may be noted that defending rams continued to watch very carefully, despite pawing and feeding attempts by the attacking ram, and showed clear indications of expecting a clash.

The foregoing raises the question of whether some of the celebrated "displacement activities" scattered in ethological papers might not be functional in obscuring intentions (for detailed discussion, see Otte 1974).

8. Restricting.   Here, signal modifiers reduce the ambiguity of a message so that limited, specific, information is transmitted. A good example is the *attention-guiding adaptations* of ungulates during display, which I have described in detail in Chapter 5. Here, the receiver's attention is restricted to certain parts of the sender and the signal, presumably, gets across with greater clarity. In restricting, we deal with means to reduce the signal to its essence; the striking simplicity of social releasers (Hinde 1974) can be said to be caused by restricting.

9. Iconicizing.   This refers to a message modifier that denotes specifically of intended acts by means of iconic signals. Iconic signals are those in which the sender abstracts from the complete overt behavior a small but diagnostic part; in performing that part, the whole overt behavior is clearly recognizable. Good examples of iconic signals are the weapon threats (Geist 1965, 1971a, Altmann 1967), the tongue flicking of courting elk, denoting licking of the female, and the waggle dance of the honey bee (*Apis mellifica*) (Frisch 1959).

10. Coupling.   In coupling, arbitrary acts, or at least behaviors without an iconic communication component, are linked by conditioning to specific meanings. Thus an arbitrary signal gains "semanticity," to use Altmann's (1967) language. The term "arbitrary" denotes here that no relation exists between the form and the meaning of a signal, not that any signal can arbitrarily become, say, the alarm signal of a species. Good examples of arbitrary signals gaining meaning through conditioning are probably status and dominance signals, which individuals learn to associate with the combat potential of individuals. Here, not only behaviors, but also display organs gain symbolic significance, as do horns in bighorns (Geist 1966a) and antlers in axis deer (*Axis axis*) (Schaller 1967) or red deer (Lincoln 1972). The same can be said of the status signals of Rhesus monkeys and other cercopithecines, which Altmann (1967) regards as examples of metacommunication. This is communication about communication. In the case of cercophithecines, Altmann (1967), suggests that status signals communicate about aggressive behavior. One can consider, as an alternative, that status signals convey not so much information about aggressive behavior, but are memory props about all the experiences to be expected by subordinates from dominants. This includes more than aggression. In the latter case, status signals are not metacommunication, since they do not specifically communicate about a specific act or even class of acts, namely aggression. To consider otherwise is to stretch the term metacommunitication so broadly as to include in it all signals whose meaning is learned. Nor do I agree with Altmann (1967) that the address prior to a message is communication about communication, except in a very trivial sense; the method of attention provoking is thus a signal modifier permitting clearer passage of the message. Play invitation, considered by Bateson (1968) and Wilson (1975) to be metacommunication, is iconic communication, and can be metacommunication only in the trivial sense that all play behavior is communication and thus an unending row of signals. I know of no convincing example of metacommunication in mammals using the narrow definition of this term proposed above, and thus disagree with Bateson (1968), Altmann (1967), and Wilson (1975) that metacommunication exists in mammals. I concur with Sebeok (1962) that it is a concept that is restricted to the human, thanks to his use of language.

11. Context.   Placing the same signal in different contexts, in which the signal changes its meaning, is message modification. Smith (1965, 1969) stresses this point and gives examples from birds; Wilson (1975, pp. 192−193) discusses the change of meaning of a signal in different contexts using, in addition, examples from vertebrates and insects. I do not, however, agree with Wilson that mammals use contextual information extensively, but use it they do. For instance, staring by a mountain sheep within a band of feeding sheep signals danger; the starer has almost certainly discovered something unusual, most likely a predator. However, starting by a sheep to a distant point just prior to migration signals intent and readiness to move to that distant point and other sheep join in. Here, staring becomes a signal to follow the lead animal. In mule deer, what is submissive behavior by a subordinate may be appeasement behavior by the dominant.

12. Syntax.   Ordering of signals so that their order, not the signal *per se,* transmits a message, is using syntax for communication. It is unreported in animal communication, and may be restricted to human language. Experiments with chimps on their language ability also have shown it in this species (Kellogg 1968, Gardner and Gardner 1969, Premack 1976).

13. Multichannel Transmission   This is a form of increasing redundancy and reducing ambiguity about a message; another process is to send the same message in different syntactic forms across the same channel. This is discussed later.

The foregoing list of signal and message modifiers of the semantic form of the message, that is the form emitted by the sender, may not be exhaustive. We must also consider the modifiers of the pragmatic form of the message. The list here is shorter. Message modifiers appear to be the experience (age), sex, dominance rank, health, and motivational (hormonal, physiological) state of the receiver relative to the sender. The receiver also has means of enhancing or reducing the transmission of the message. For instance, in dominance fights of mountain sheep, the subordinate ram-to-be close its eyes during the horn displays of the dominant (Geist 1971a). This is a modification of the pragmatics of a message by reducing its redundancy. Modifiers, which would increase the correspondence between the semantics and pragmatics of a message, would be enhanced attention on the part of the receiver, movement to a vantage point to reduce obstructions between sender and receiver, or approach by the receiver to reduce distance and maximize reception on all channels.

The experience of the receiver can clearly affect the pragmatics of a message, particularly if the signals are arbitrary rather than iconic. The arbitrary signals are linked through conditioning to certain experiences; if these linkages are not very similar for sender and receiver, misunderstanding must inevitably result. At present, this is a hypothetical consideration.

Auxiliary Information

The sender can transmit in a message not only its semantic meaning, but also information that is incidental to the act of communication. We may term such information *auxiliary information*. This information may inform the receiver about the sender's direction, distance from the receiver—maybe his exact location, identity, and sex-age class, as well as the activities he may be engaged in. Such information may at times be as important as the semantic component of the message, in the case of distress calls, for instance. As noted earlier, angular sound localization enhances or diminishes the information about the direction of the sender relative to the receiver.

Granted three meaning classes, a minimum of 13 signal and message modifiers of the semantics of the message, a minimum of 8 or more modifiers of the pragmatics of the message, plus all the auxiliary information, then two core messages can have great diversity of information with an infinite shading of meaning. For this reason, a classification of the information animals can communicate must always fall short of being unambiguous or even complete.

A number of authors (e.g., Smith 1965, Diebold 1968) state that messages give information about the internal state of the sender to the receiver. One may be sceptical of this. Would not messages aim at releasing memories and conditioned responses in the receiver? Even if the sender is "angry," his message should evoke at best the proper response in the receiver based on the receiver's experience with "angry" conspecifics. To invoke an intellectual insight into the sender's "mind" by the receiver would violate Occam's razor.

The systems of communication of interest here are variously called paralanguage (Diebold 1968), kinesiology, defined as "the study of the visual aspects of nonverbal, interpersonal communication" by Birdwhistell (1960), and proxemics (Hall 1959, 1963), which covers much the same area as kinesiology but emphasizes the use of space. These terms refer to human behavior; however, they would include the nonverbal, nonlanguage communication processes common to higher mammals.

Some Principles of Communication: The Law of Least Effort

The first rule of communication is that information should be imparted in the most economic manner, that information of maximum clarity should be delivered at the lowest cost. Messages are to be coded with a minimum of elements compatible with the avoidance of ambiguity; the more energy required to form a signal, the less likely it is to be used, and the more likely is its replacement by an element costing less energy to form and transmit. This may be the ultimate cause for the abstraction of signals. The patterning of simple signals is favored over the display of a complex signal, not only because the latter may be more costly in energy, but it may also affect the pragmatics of the message by its rarity. This is a point rightly emphasized by Smith (1969) and Moynihan (1970) in trying to explain why each species commands only a limited set of signals, some 15 to 45. The more signals there are, the more rare is their individual use and the greater is their novelty. The greater the novelty of signals, however, the more likely the elicitation of arousal and a startling response in the receiver. Hence, communication must deal with signals that are familiar enough not to trigger startling or arousal.

The principle that the simpler the signal, the less costly it is of energy, the more frequently it will be used, is termed Zipf's law (Cherry 1957). If we plot the log of rank−frequency against the log of frequency of individual patterns (signals), we obtain usually a straight declining line, indicating that frequency declines with rank.

Log rank is closely related to the complexity of the signal or the energy cost of the message. I have illustrated this in Figure 3-1 using some data from mountain sheep. It may be noted (in Figure 3-2) that behavior patterns very costly of energy, such as clashing, butting, etc., tend to be used less than their corresponding threat patterns, which are in turn less frequently used than the simpler patterns such as the horn displays.

Zipf's law was determined primarily from human communication, but it does work where it has been applied to animal communication, such as by Poulter (1968) for the vocalization of killer whales (*Orcina orca L*) or in my own work to the behavior patterns of mountain sheep (Geist 1968a, 1971a) and my work on mule deer, as yet unpublished.

Imitation, Automimicry

In a second expression of the law of least effort, communication takes advantage of, or "parasitizes," existing response systems in conspecifics. This explains why intraspecific mimicry should not only exist, but even be common, and why threat and combat should exploit existing mechanisms of withdrawal behavior (Geist 1971a).

In intraspecific (and interspecific) mimicry there is a model which is imitated by

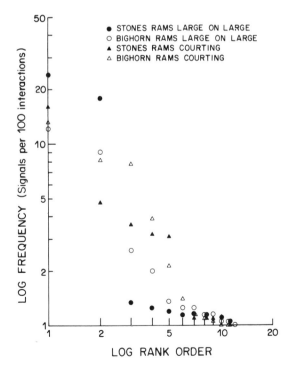

Figure 3-1.   The logarithm of rank order of social behavior patterns of various mountain sheep plotted against the logarithm of the frequency with which each behavior pattern occurred within 100 interactions. (The abscissa is marked off in decimal units.)

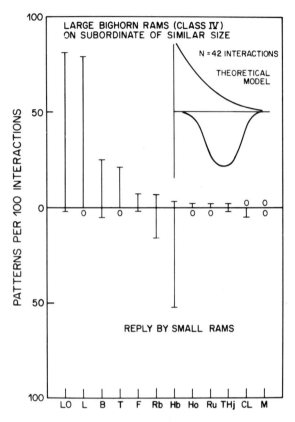

Figure 3-2.   Rank order of social behavior patterns in bighorn sheep rams against the frequency of these patterns, ranked according to the the behavior of the dominant. The upper graph shows the actions of the dominant; the lower, the actions of the subordinate. Note that the behaviors most frequently performed by the dominant are infrequently performed by the subordinate. Behaviors indicative of overt combat and therefore strenuous, are not performed commonly by neither dominant nor subordinate. This graph and the theoretical model depict settled social conditions between rams. (Lo = low stretch, L = Laufeinschlag or front-kick, B = butt, T = twist, F = lipcurl or flehmen, Rb = rubs body, Hb = horns body, Ho = horn threat, Ru = rush, Thj = threat jump, Cl = clash, M = mount. After Geist, 1971a supp. 132–146).

the mimic for its sole benefit. Wickler, who has been one of the most industrious and far-ranging investigators of mimicry in recent times, also calls mimicry the "evolution of lies" (Wickler 1970). Indeed, mimicry aims at deception. For a detailed description of mimicry, I refer to Wickler's book (1968) and to Otte (1974).

In intraspecific mimicry, we find messages structured on the basis of food, sex, and child rearing. The use of food mimicry is well illustrated by a small fish, the swordtail characin (*Corynopoma riisei*). The male has a long, thin outgrowth on its gill covers, which terminates in a small dark bulb. In courtship, the male swims ahead of the female and twitches the outgrowth in front of her. The bulb now mimics a water flea. The female is attracted alongside the male and later may commence copulation. The same type of structure, but originating from a scale above the pectoral fin, is found in another fish (*Plerobrycon*) and is used in much the same fashion as in the swordtail characin (Wickler 1970). Here, use is made of the neural releasers structured to respond to food items; we note a conservation of systems.

Female mimicry is found in a few highly social vertebrates such as baboons (Wickler 1963, 1967), mountain sheep (Geist 1966b, 1968a, 1971a), and cichlid fishes (e.g., *Tropheus,* Wickler 1970). The subordinate assumes the appearance and behavior of a female when in the presence of, or confronted by a dominant. The rationale is that the dominant male who attacks female stimuli will attack real females as well as mimics. In so doing, he will select against his reproductive success. Clearly, males should not attack conspecifics characterized by female stimuli.

To us, it seems odd that in higher mammals a male should fail to distinguish male from female as we can so easily do. In most mammals, however, the evidence suggests that individuals perceive others as signal patterns (*releasers,* in ethological jargon; see Ewer 1968). Hence, a male assuming the female's signal patterns can be mistaken for a female, or at least the female signals are so overpowering as to inhibit attack. In this context, we must note that human children in their first years of life are very much at the mercy of "field effects," that is, perceptual mechanisms such as dominance of continuous line or formation of good form discovered by gestalt psychologists. Only as they get older do they develop perceptual regulatory mechanisms that overcome field effects in late childhood (Elkin 1975). It is tempting to suggest that the field effects are the same dominant perceptual mechanisms recognized by ethologists as releasers; field effects appear to be the hereditary base from which human perception develops.

Female mimicry can also act in a very different context. Struhsaker (1967a), in his studies of vervet monkeys (*Cercopithecus aethiops*), found that dominant males, *not* the subordinates, displayed the female's estrus markings (red, white, and blue display). When dominants displayed to subordinates in such a fashion, the subordinates would sit down and, in so doing, apparently hide their colorful rear ends from the dominant. The display increased in frequency during the mating season. Quite evidently, this does not fit with Wickler's (1963, 1967) views. It is, nevertheless, an excellent example of female mimicry, albeit its interpretation is quite different from that valid for baboons or mountain sheep.

I suggest the following. It is evident that in a vervet troop, dominated by a large male, all males see the red, white, and blue display in association with the assertive behavior of the dominant. Thus they are being continually conditioned to associate the display, and therefore the estrus colors of the female, with fear. Only the dominant male escapes such conditioning. Therefore, females coming into estrus carry on their bottoms "danger" signals to the subordinate males. This would, for the dominant male, reduce competition with subordinate males for estrus females and would reduce the chances of fighting, injury, and loss of reproductive fitness. The subordinates would be subject to "psychological castration," so called by Guhl (1941), who showed that punishment by dominant cocks greatly reduced the frequency of mounting by subordinate cocks, even if these were tested in the absence of other males.

Male mimicry is the imitation of male behavior and appearance by the female. It is found in a good many vertebrates. I described the selection process responsible for its appearance in some detail elsewhere (Geist 1974a). It is found where the female defends resources, with the male against other conspecifics, or against the male. In pair forming, territorial forms in which both sexes defend the territory, male mimicry increases the redundancy of the message to the intruder that the territory is occupied, it also increases ambiguity for the intruder deciding who to attack, since the intruder must attack the territory owner of its own sex in order to gain territory and mate. If it attacks the territory owner of the opposite sex, it drastically reduces its chances of getting that territory since, in essence, it must do twice the fighting. In highly gregarious species, male mimicry has another function. Here it permits the female to compete more effectively against the male and reduce the cost of the harassment imposed by males. In ungulates and parrots, we find male mimicry thus confined to pair-forming territorial forms and highly gregarious, migratory species (Geist 1974a).

In infant mimicry, a subordinate adult plays "baby" as a means of staying close to a dominant or avoiding retribution from a dominant. We find it in species in which male and female raise their young together. The rationale: The adult who beats children will not be likely to raise offspring of its own. Again, this can work well only in species that are far less capable of abstracting than we are, and that respond to conspecifics as if they were bundles of releasers. Not surprisingly, we find infant mimicry widespread in birds, and also in social mammalian carnivores such as wolves and hunting dogs. A trace of it can be recognized in ourselves. An interesting observation was made by Kummer (1967) on Hamadryas baboons. Here a subordinate subadult male may, if confronted by adults, pick up an infant, put it on its back, and act like a female.

Infant mimicry may also be used by the male in courtship to stop the female, or at least to slow her down. Here the male "parasitizes" the maternal emotions of the female. It has been identified by Ewer (1968) in the courtship of some rodents and carnivores. In ungulates, it is found vividly in the mule deer, as I indicated earlier (Geist 1971a).

Another form of parasitizing an infantile reflex is shown by the meercat (*Suricata suricata*), a small South Africal viverid, as described by Ewer (1968). In the young, one finds a relaxation reflex; it is triggered when the female picks up the young by the scruff of the neck. The young then hangs limply from the sharp teeth of the mother, and by not struggling avoids the mother's having to bite harder to hold on and thus possibly injuring the young. This relaxation reflex persists into adult life. One can grasp a hissing, snapping meercat by the scruff of the neck and at once it hangs limp and peaceful from one's fingers. This the male takes advantage of. Courtship—if one can call it that—is rough. When the male succeeds in getting hold of the female by the scruff of the neck, she tones down, relaxes, and copulation begins.

In essence, communication systems take advantage of existing response mechanisms; it is inconceivable how or why two separate mechanisms could or should evolve in a species to do the same job. Moreover, in the communication systems, a change in the sender system requires a corresponding change in the receiver system. Thus, changes in the CNS mechanisms dealing with communication can be expected to be slow. This explains Edinger's principle, that brain morphology is most conservative in its evolution, as discussed by Hofer (1972); it also explains Krumbiegel's law (Krumbiegel 1954), that behavioral signals may be more conservative than corresponding morphological structures.

Some Rules Governing Semantic and Pragmatic Structure of Signals: Novelty

As discussed earlier, animals are structured by natural selection to live in a predictable social and physical milieu. An individual who perceives the surroundings as familiar, and who can anticipate the actions of its companions, is calm and can go on with the business of maximizing energy gain and conservation. The moment it perceives something "unusual" or "novel" it becomes aroused. The background against which the animal judges the "unusualness" of events are the common every-day activities of its companions. For instance, in a group of wild sheep the normal feeding, resting, ruminating, walking, and standing postures are the usual background. The very moment one animal stops and does *not* move a hair, it not only attracts attention but also alarms the whole group. A walking sheep coming to a dead stop and standing rigidly is an *unusual* event, and a very alarming one (Geist 1971a). It means something exciting and probably dangerous has been spotted. In communication, we thus find that signals come to the attention by rising above the common sounds, motions, and postures simply by being relatively rare, and novelty, as we noted earlier, created arousal and even flight, or fight in the extreme (Schleidt 1961, Moynihan 1970).

We can also consider the construction of bowers by courting bower birds an example of the use of "novelty" or "unusualness" in the process of communication. Here the stimuli are bound, not to an individual, but to a locality. It also follows that in principle the communication process would be enhanced if some of its manifestations were tied to unusual or uncommon localities. This would greatly facilitate, for instance, the meeting of sex partners or rivals. Granted the great ability of birds and mammals to locate and relocate specific localities, the formation of leks in arbitrary localities can take place. To my knowledge, the formation of leks in Uganda kob, for instance (Buechner 1963, Leuthold 1966), is not tied to any *a priori* recognizable locality.

Rare events by themselves convey no message, but simply arouse the animal and force it to pay attention. It is evident that "novelty" must be an integral part of the structure of all signals lest the receiver fail to note the signal through habituation. To put it more strongly, "novelty" in signals maximizes attention in the receiver and circumvents habituation to the signal. Signals are distinguished by deviation from the normal events. They are structured to exaggerate motion (either unusually slow or unusually fast), orientation of the sender to the receiver (by making the orientation an unusual one), visual configuration of the sender's body and appendages (exaggerating postures) and vocal patterns, as well as by emission of strange or rare odors (Geist 1965, 1966b, 1971a). The foregoing thus explains some of the puzzling and most conspicuous attributes of animal behavior.

Antithesis or the Law of the Opposite

This rule of communication was first discovered by Darwin (1872) and came to the attention of ethologists relatively late (Marler and Hamilton 1966, p. 178, also Tembrock 1959, Eisenberg 1962, Geist 1965, 1971a, Wilson 1975). The rule of antithesis states that the signals with opposite meanings are syntactically structured in opposite directions.

According to Tembrock (1959), many mammalian sounds can be classified into socially positive and socially negative sounds. These have the following characteristics. Socially negative sounds are loud, harsh, nonharmonic, nonrhythmic, and have a sudden beginning, thus contrasting sharply with the background noise. Socially positive sounds are structured in exactly the opposite direction. They are of low intensity, they sound soft, they are harmonic (are pure tones), usually rhythmic, and have a gradual beginning. They thus rise less strongly above the background noise.

Socially negative and positive sounds are used in antithetical or opposite social situations: Thus the male mountain goat uses the opposite behavior pattern in courtship from that in his dominance displays (Geist 1965); the elk uses behavior in courting the female antithetical to that when herding females back to his harem (Geist 1966c). In general, socially positive sounds are used in the mother−young context, in mating, and in submission, while socially negative sounds are used in threat, in displays of dominance or status, or against predators. A perceptive observer will note that humans have not escaped this rule in their communication.

The principle of antithesis also applies to touch. For instance, contrast combat against grooming. The socially negative tactile stimuli would have the following characteristics: generation of a large amount of mechanical energy, sudden application of energy, energy application usually nonrhythmic. Socially positive tactile stimuli would differ in the opposite direction, of course.

The principle of antithesis expresses itself also in the postures, movements, and relative frequency of behavior patterns used by animals. Darwin (1872) noted that the attack and appeasement postures of his dog were the very opposite. In the threat or socially negative stance, the dog stands high and his movements are stiff, his hair is erect, as are the ears and the tail. Thus the posture of the dog emphasizes size. When appeasing, or in the socially positive posture, the dog's posture is depressed, the dog is wriggly—that is, it shows many small, quick, relaxed motions—the ears are down and so are the tail and the body hair. Thus, in appeasement, the dog uses the very opposite motions and body configuration from those in threat.

It is of interest to note that the principle of antithesis is well illustrated in the electric communication of fish, as found in the mormyriform and gymnotid teleosts. Hopkins (1974) describes that an arresting of electric discharges served in *Gymnotus, Gymnarchus,* and *Gnathonemus* as a means of reducing attacks by larger bodied conspecifics, that is, as a submissive behavior. Electric silence renders the subordinate relatively inconspicuous, just as vocal silence and depressed crouched postures, and engaging in common activities such as feeding, does for ungulates. Hopkins (1974) recognized the antithetical nature of electric silence as an appeasement behavior.

It is important to note that we cannot predict *a priori* which posture is socially positive and which is socially negative. There is no close relationship, and often no relationship at all, between the syntax and semantics of a signal, that is, between its form and its meaning. A good many authors emphasize this point for both human and animal communication (Diebold 1968, Hockett and Altmann 1968, Smith 1965, 1969, DeLong 1972). In the house cat, for instance, the agonistic defence posture and the submissive posture are also antithetical, but in almost exactly the opposite directions from that of the dog (Darwin 1872, Leyhausen 1956). We can, however, anticipate that, granted the law of least effort, once a given threat behavior has been adopted some socially positive behavior will be its antithesis.

Since we have touched on the subject of *a priori* predictability, it may be of value to digress a little and show that solutions to a common communication problem are not *a priori* predictable, differ widely, and are still consistent with the rules of communication. I refer to the approach of the male to the female in courtship, as illustrated by some American deer and bovids. In the mountain goat, the courtship posture and behavior of the male is the *antithesis to the dominance display* of the male; here the male uses, in essence, submissive behavior to court a very aggressive and dangerous female. In the mountain sheep, the male uses *dominance displays,* advertizes himself to a skittish female, and sticks close to her, despite her efforts to shake him off; the male is thoroughly dominant and can lose out only if he is too rough, chases the female too much, and attracts the attention of larger rivals. The elk uses the *antithesis of his herding posture* in courting the female; the herding posture is a threat posture. In the mule deer, the main type of courtship (there are two kinds) is based on *juvenile mimicry* in which the buck presumably parasitizes the maternal emotions of the female. The white-tailed deer uses, in his approach to the female, initial elements of the suckling run of the fawn, but changes this into a charge in which he chases the female and keeps close on her heels; when the female is in estrus, the male may use a posture akin to submission in approaching her. The mule deer also uses the rush courtship, albeit rarely, but when he does use it it is far more extravagant than that of the white-tailed deer. In moose, we find males use a *dominance display* less to advertize than to intimidate a very aggressive, but not very dangerous, female; when the female becomes more or less tractable, the bull uses antithesis to the dominance display only in his voice. This soft call is also used by submissive cows and bulls. I hope this sketch illustrates how very diverse the solutions to a problem may be, despite the fact that none violates the rules of communication.

Antithesis is also expressed quantitatively in the behavior of animals. Let us once more use the mountain sheep as an example. If we order by rank frequency the behavior patterns performed by dominant to subordinate males, we get the graph show in Figure 3-2. If below we graph the frequencies of the same behavior patterns performed by the subordinates towarrd dominants, we obtain the lower graph of Figure 3-2. It can be seen at once that those patterns performed most frequently by the dominants are performed less frequently by the subordinates and vice versa; in addition, there are behavior patterns performed infrequently by both. To a considerable extent—although not completely—the behavioral repertoire of dominant and subordinate are antithetical.

Let a diversion be permitted at this point: In Fig. 3-3, the behaviors of two males are plotted as they actually occurred from Geist (1971a). Each acted toward the other in the same manner, in this instance as dominants. Clearly, this is an untenable situation theoretically, as it was practically, for these rams were in a dominance fight and both acted as if they were the larger. The situation shown in Fig. 3-2 is found only when dominance orders are established.

In Figure 3-4, I show a theoretical possibility; each animal acts as subordinate to the other, one appeasing the other. Appeasement here is the use of submissive patterns by dominants toward subordinates as a means of holding them, a point well discussed by Barrette (1977a). In mountain sheep, appeasement in that sense is

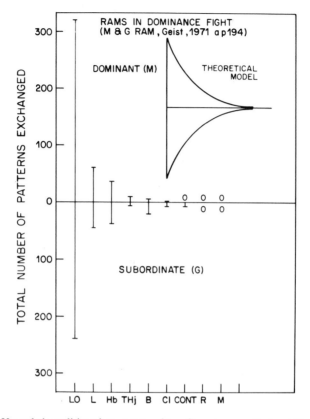

Figure 3-3.   Unsettled conditions between two large Stone's rams. Both act like dominants. The eventual loser, G−ram, butts and clashes more frequently than does M−ram.

not found between males; in some canids I expect it, as I do in human beings. The graph in essence shows *courtesy* which, in principle, is acting like an inferior to a superior. This simple rank-ordering method, therefore, shows quantitatively in graph form such states as enmity and friendship.

Stimulus Contrast

It is widely acknowledged in the study of animal behavior that stimuli that appear suddenly, and contrast sharply with their background, attract attention (see Hinde 1970, Marler and Hamilton 1966, Thorpe 1963). They provoke attention not only by virtue of deviating from the normal as explained earlier, and thus having the property of novelty, but also because they markedly stimulate, if not "offend," the sense organs. Mammalian visual systems, as well as those of other vertebrates, contain edge receptors that essentially enhance borders and fire strongly in response to moving edges. For discussion of these properties, I refer to Hinde (1970, pp. 101−118). Thus stimulus contrast, in relation to these receptors, would be a multip-

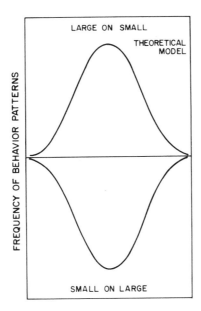

Figure 3-4. A theoretical possibility: The behavior of two "friends", each acting like a subordinate toward each other. The rank order versus frequency graphs permit one to express quantitatively the actions of antagonists, friends and normal social relationships.

lication of edges and movements. Visual receptors also ought to respond intensely to an object rapidly increasing in size, because of triggering more and more edge receptors in a short period of time. Psychologists call such a stimulus *"looming,"* and humans as well as a host of animals respond by withdrawal on first experiencing it (Gibson 1970, Schiff 1965, Schiff et al 1962, Hayes and Sariff 1967). The "fear" response to looming appears to be innate; however, it can be explained—as I did above—as a special case of stimulus contrast. For the auditory and olfactory or tactile systems, stimulus contrast is created by stimuli that quickly excite many receptors and increase receptor response, such as sudden loud noises, sudden intense odors, or blows. In principle, stimulus contrast bodes no good and it is adaptive to respond with arousal, that is, to be prepared for the eventuality of flight. This property of the receptor systems and the CNS mechanisms, triggering arousal, is exploited by the communication systems in order to maximize the attention-arousing properties of signals. I shall discuss this property under attention-guiding adaptations (Chapter 5).

Redundancy

Since the aim of communication is to transfer information, we expect to find attributes of communication systems that reduce the risk that a message may be misunderstood. Misunderstanding is less likely if the message is repeated in any one channel, and coded so as to satisfy more than one meaning class (see p. 49). We call this property of communication systems "redundancy." It is encountered in the sign languages of mammals repeatedly; note in Figure 3-1 how often and varied display patterns are. Here we have, in the visual channel alone, the same message sent in different forrms. We also find that dominance displays contain visual, auditory, olfactory, and even tactile components; in short, they are exceedingly redundant. For a discussion of redundancy in human language, I refer to Cherry (1957).

Stimulus contrast, antithesis, and redundancy are all mechanisms that reduce ambiguity of messages to a minimum; how a reduction in message ambiguity emhances the reproductive fitness of an individual I discussed in Geist, 1971a, pp. 235−236, and in Chapter 5.

Intention Movements—Iconic Signals

The semantic structure of messages is formed in a large number of behavior patterns or signals by intention movements. The sender signals a total action by performing part of it, which allows the receiver to anticipate and prepare for the complete behavioral sequence to follow. Since the signal is related directly to the message, signals embodying intention movements are iconic signals. Darwin (1872) recognized this category of signals, and they are well recognized in ethology (Marler and Hamilton 1966, Altmann 1967, Hinde 1970, Wilson 1975).

Threat behavior, in contrast to dominance displays, is differentiated from the latter by being formed from movements of intention to use specific weapons; threats are thus iconic, and displays arbitrary, signals (Geist 1965, 1971a). In the category of weapon threats, exposed fangs, lowered horns, and raised and cocked appendages may signal specific kinds of attacks. Many examples of this are enumerated and discussed by Marler (1968).

Another type of threat, the *rush threat,* may not contain a discernible weapon threat, but it does signal the intention to close quarters to do violent action and it may terminate in a weapon threat. This is found in moose or mule deer where the animal may strike the ground audibly with the front legs following a rush threat. A rush threat and a real attack are either difficult or impossible to distinguish—which creates uncertainty, which "frightens," and thus adds to the effectiveness of the threat. In addition, as indicated earlier, it is a special case of stimulus contrast, since a rapidly enlarging object should rapidly activate many edge receptors in the CNS; withdrawal from such a stimulus appears to be innate and has been studied under the term "looming" by psychologists (Gibson 1970).

Even in submissive or appeasement behavior, we find iconic signals. In ungulates, grazing may act as a submissive or nonaggressive gesture. In the majority of Old World deer that I studied this grazing appears to be ritualized into rapid chewing motions *in vacuo;* in the small primitive muntjac (*Muntiacus*), the subordinate may pick up nonfood items and chew them loudly in front of the dominant (Barrette 1977a).

The foregoing dealt exclusively with a communication system we use quite unconsciously ourselves; however, its conscious application can enhance our verbal communication, and forms the basis of good oratory. From paralanguage to human language, however, is a jump I must refrain from taking, even though it is evident that the weight of evidence is in favor of the concept that human language is based in part on a genetic program whose unfolding is environmentally determined; thus grammar has a biochemical structure in the CNS, and we can safely predict that it is formed from some precursor common to subhuman animals. This we anticipate from our studies of structure and behavior in relation to evolution. This view is indeed supported by the recent studies of communication by way of learned signs in chimpanzees, as discussed by Bronowski and Bellugi (1970), Jolly (1972) or Premack (1976). The language acquisition of children is characterized by their own experimentation with language and cognitive restructuring. They do it in such a fashion that leaves little doubt that they have inherited the rules of language formation. Unfortunately these rules are outside my competence to discuss and I must close regretfully at this point without as yet conceiving how grammatical rules could have evolved.

Levels of Communication

Tavolga (1968) has proposed a useful scheme for classifying communication by its specificity and level of abstraction. At the *vegetative* level, little beyond the presence of the organism is communicated; at the *tonic levels,* byproducts of ongoing activities, which serve primarily to maintain physiological homeostasis, communicate the location, activity, and possible sex of the sender, the kind of information I termed auxilliary information; at the *phasic level* tonic processes are organized in discrete bundles of information and are emitted at the appropriate situation in a general, rather than specific, fashion. One can, however, detect elements of the next level already, the *signal level*. Here information is emitted by specific structures in the service of communication, and the messages are of great specificity; in mammalian communication, we deal from this level upward. The *symbolic level* encompasses signals whose syntactic structure has no relation to their semantics or pragmatics, nor have the prroperty of signals that act as releases. Much of mammalian communication proceeds at this level. The *language level* is the level of extreme abstraction and complexification of communication as found in man.

Chapter 4

Aggression

Competition

To maximize its reproductive fitness, an individual must obtain all the scarce resources essential for reproduction. Such resources must of necessity be limiting resources. Since even a minimum share of such resources may be more than an equal share, and since scarcity of essential resources is a function of their removal by conspecifics, the successful individual obviously is one who obtains more than an equal share of scarce resources despite the activities of other members of his population. Such an individual is competing for scarce resources with other members of his or other species. Competition can thus be defined as a process in which an individual uses all means at his disposal to obtain more than an equal or proportional share of scarce, limiting, resources compared to the shares obtained by other individuals, be they conspecifics or not.

We must pause here for a moment. An individual enhances his reproductive fitness, not only by successfully competing for scarce resources, but also by reducing the reproductive fitness of other individuals. The concept of reproductive fitness dictates that the individual leave behind in the following generation the greatest proportion of individuals blessed with its genes, not the greatest possible number. Therefore there is a payoff to the successful individual both if it maximizes its own reproductive output and if it minimizes that of others. This can be done by a great diversity of means: destruction of the competitor's nest, eggs, fry in some fish (Collias 1944), or young in langurs (Sugiyama 1965, Hrdy, 1974) or in lions (Schaller 1972, Bertram, 1975), interference with mating or the rearing of young (Collias 1944); keeping competitors in a state of costly excitation and stress (Barnett 1958, 1967, Bronson and Eleftheriou 1964, 1965), which in chickens may result in psychological castration (Guhl 1941) and in mice reduce the sexual competence of males by reducing the size of seminal vesicles, or which results in the lowering of resistance to infection (Vandenberg 1960, Stein et al. 1976) or causes loss of body weight and faulty development of body insulation (Allee and Guhl 1942) or reduction of social activities and thus their chance of breeding (Archer 1970); delaying the mating or nesting of competitors so that their young are raised at an unfavorable time, as may be the case in prairie chickens (Robel 1970); disturbance of the offspring in the intrauterine environment so that they are destined to be less viable and socially competent (Thompson 1957, Lieberman 1963, Weltman et al. 1966); inflicting of costly body repairs on subordinates through frequent wounding (Scott 1966), or inflicting so much stress that the subordinates ultimately die of the unpredictable infrequent attacks (Welch and Welch 1970) or become an easy prey for predators (Robel 1970). Clearly, one can "work both sides of the fence" to increase one's reproductive fitness, as is obviously being done. Yet, in raising the cost of reproduction to others, the individual must do so only if it can be done at little cost to himself.

Competition can take many forms. It need not be dramatic, as is active competition by aggressive behavior; it may be quite inconspicuous, as would appear to be the rule for passive competition. In passive competition, individuals compete primarily through efficiency in sensing and consuming the scarce resources for reproduction. We expect passive competition when scarce resources cannot be defended (Brown, 1964). The distinction between active and passive competition has been made by ecologists for some time and has been called *interference* (active competition) and *exploitation* (passive competition) by Miller (1967).

Although competition is normally conceived as an activity of adults, some second thoughts show that it need not be. One must not wait until adulthood to win access to scarce resources essential to reproduction. Competition may begin and be deadly among embryoes as O'Gara (1969) has shown for the pronghorned antelope (*Antilocapra*). Obviously, it pays to follow strategies that maximize one's gains in adulthood by controlling the future. In a species with a high probability that individuals born together will also live together, it pays not simply to defeat prospective competitiors early in ontogeny but to impair growth, development, and competitive abilities as adults as well. This is most urgent if adult competitiors are armed with dangerous weapons and overt aggression is extremely risky. A future competitor's competence can be undercut by interfering with its acquisition or assimilation of resources needed for growth and development. If this hypothesis is valid, we expect some weapons in competing juveniles, and we expect a rigid dominance hierarchy among juveniles that is maintained in adult life.

Such weapons are indeed found in suckling pigs (*Sus*), where they exist as short, sharp, temporary canine teeth. These are used to settle which piglet occupies which teat along the maternal body. The further forward an individual can establish itself, the better the milk supply; therefore the better its body growth will be and the smaller the risk of being trampled on by the female's hind legs (Hafez and Signoret 1969). Juvenile aggression and specialized weapons pay off in larger body size, and therefore probably in increased combat ability as an adult and greater access to scarce resources. The establishment of rigid dominance hierarchies in juveniles that apparently last throughout adulthood have been suggested for canids (Bekoff 1974) and the females of pronghorn antelope (*Antilocapra*) (Bromley 1967, Kitchen 1974). From the foregoing it is obvious that "play" in juveniles may be less innocent than it appears and that it may have to be reexamined from this new perspective.

Individuals may compete not only alone, but also with the aid of those who share a common interest. Individuals, in an abstract sense, do not work for themselves but for their genotypes, and may therefore work with closely related individuals in spreading their genes; *cooperation* and even *altruism* arise as a consequence (Hamilton 1964, Wilson 1975). Parents working on behalf of common offspring, sibling groups of male turkeys working on behalf of the most dominant brother (Watts and Stokes 1971) or relatives, as in a wolf pack helping to raise the pups of the dominant female, are all examples of the principle that closely related individuals may jointly compete to spread their common genes. These concepts are treated in some detail by Wilson (1975).

The concept of cooperation in animal societies is largely the domain of American biologists (Collias 1944, Allee 1951); it has often been ignored even in recent textbooks on animal behavior. This may be due in part to such difficulties as comprehending how predator-prey or herbivore-herb relationships can be termed "cooperative" (Collias 1944), a difficulty I also have failed to master.

Whether a resource is subject to active or passive competition is quite evidently dependent on the distribution of the resource in space and time and on the density of competitors. These parameters determine whether the resource is *defendable* (Brown 1964). If it is defendable active competition will ensue, provided the gain in resources exceeds the cost of active competition to the individual. If food, for instance, is widely scattered and found in small pieces so that each piece is relatively low in energy and nutrients, then it does not pay to enter into active competition for each piece. Any aggressor contesting for food will expend more energy on the preparation for aggression than he will ingest; the better strategy is to maximize time spent feeding and minimize all expenditures in doing so, as well as increasing efficiency in finding and gathering food. Marine plankton is one example of such a dispersed food source, at least to larger teleosts, marine birds, and mammals; another is a grassy plain to grazing herbivores or the usual browse sources to deer, elk, and moose.

An example in which a spatially highly concentrated food source is best exploited by passive, rather than active, competition is described by Kruuk (1972) for hyenas. Instead of fighting over a downed prey as lions are wont do to (Schaller 1972), hyenas feed as rapidly and thoroughly as possible, ingesting a maximum amount of meat in a minimum amount of time. Each hyena deprives its neighbor of food, not by fighting but by eating as quickly as possible from a limited amount of food accessible to both. If a hyena were to attack its neighbor, it would first incur the direct cost of fighting and the indirect ones, such as the cost of wound healing and fighting pathogenic organisms that infect wounds. Secondly, it would deprive itself of the opportunity to feed while fighting, and thus would lose out in obtaining the maximum amount of food, since other hyenas would in the meantime be rapidly depleting the available meat supply. Every second lost in fighting can be costly, since hyenas dismember and ingest a carcass—depending on its size—in minutes at the most.

In the foregoing example, the resource may be localized and defendable, but the density of equal competitors makes its defence too costly for any individual. Thus, a highly concentrated resource that attracts a large number of competitors promotes selection that ultimately reduces sex and age dimorphisms, creating a society of lookalikes (Geist 1974a). If food sources are concentrated into limited areas and such areas are dispesed widely, the social groups move as units between these areas; if the food source cannot be predicted in space and time, such social systems appear to develop group dispersal as a mechanism for finding new food sources. We find this in highly social parrots (Brereton 1971) or crossbills (Newton 1970). Thus, improved means of passive competition are not the only consequence of a species exploiting dispersed patches of highly concentrated food; there are secondary and tertiary effects that then mould the biology of the species, only some of which I have mentioned.

From the foregoing one can make a number of self-evident predictions about agonistic adaptations in relation to ecology. Active competition will be absent or insignificant in species in which no resource essential to reproduction is defendable. If neither food, nor mates, nor shelter are defendable, then natural selection will form a species void of intraspecific aggressive behavior. We expect this situation in pelagic plankton-feeding schooling fish, for instance, with broadcast rather than internal fertilization. We also expect no courtship behavior in such fish; this is apparently what is found (Norman in Collias 1944, Noble 1938). As a consequence, we expect no morphological correlates of agonistic or courtship behavior, since such would be superflous in such anonymous societies. Rather, reminiscent of adaptations of parasites, we expect a dearth of adaptations except those essential for individual survival, gamete ejection, and reproductive synchrony. In such species all, or virtually all, mechanisms of competition can be expected to be passive ones; many marine teleosts and plankton-feeding intertidal invertebrates, such as mussels, clams, and barnacles, are examples.

If, however, the egg cluster of a female becomes defendable because it is clustered, localized, and immobile as when resting on a substrate, then males will enter into active competition in order to fertilize the eggs; the same is true for species in which females are fertilized internally. Thus, a mandatory consequence of internal fertilization is agonistic behavior and courtship, and it is a probable result if eggs are fertilized externally, provided the eggs are clustered, localized, and immobile. Granted only one scarce defendable resource essential to reproduction in the ecological niche of a species, then mechanisms of active competition will be found. However, the frequency or degree of importance of agonistic adaptations cannot be judged from the frequency of defendable resources a species depends on, for too many factors—as we shall see—affect the evolution of agonistic behavior and its morphological and physiological counterparts.

Granted individual rather than mass fertilization of gametes, it is at once in the interest of the female's reproductive fitness to select the most successful or ''best'' male to fertilize her eggs, while it is in the interest of the male not to waste his energy and gametes on any female who is an adaptive failure. This requires that adaptive success be advertised so as to permit individuals to select and compete for the best possible partner. Thus, signals facilitating active competition for mates are required. Here means of active competition assume a secondary function, namely that of rank symbols, which have the dual purpose of repelling individuals of the same sex while attracting individuals of the other. In addition, we expect various means to enhance the signal quality of active competition itself; we shall discuss this later.

The foregoing deductions, though self-evident and partially supported, require further verification. It requires a close quantitative investigation of the causal links between the ecological niche and the behavior of a species, identifying the many goal-oriented or teleonomic strategies, measuring the energy and nutrient cost of activities and their benefits, the latter in terms of reproductive fitness.

Agonistic Adaptations

By agonistic behavior we understand, following Scott and Fredricson (1951), the behavior associated with active competition, that is, aggression in its widest sense, including combat, threat, dominance displays, and the strategies of their application, as well as submission and appeasement behaviors. It encompasses all activities aimed at displacing conspecifics as well as those resisting displacement. Thus, one finds on one hand attack and dominance displays (direct and indirect forms of displacement) and defence and submission (direct and indirect forms of resisting displacement), as well as appeasement behavior which attracts, or at least permits, the dominant to approach a subordinate.

In this essay on aggression, I shall very frequently come back to examples from ungulates, those generally large herbivorous mammals of diverse lineages. There is no accident in that. In this group, through a quirk of nature that I shall shortly explain, there is frequently a segregation of weapons from feeding adaptations, while in most mammals the teeth double splendidly as weapons. Therefore many ungulates, but primarily the ruminants, have evolved organs that function strictly as weapons. We can study the effects of adaptive strategies on weapons without having any feeding functions imposed on them. To add to this good fortune is the great size range of ungulates. This enables us to study the effects of similitude on combat and weapons. The great range of ecological adaptations and geographic dispersion in ungulates permits insights into the relationship of weapons to feeding and antipredation strategies. Furthermore, the weapons of ungulates, by and large, are well preserved in the fossil record, and that allows us precious insights. All in all, they are the best animal group available for a study of aggression.

The "quirk of nature" I referred to? It is simple. Many ungulates and all living ruminants lack upper incisors. This makes the mouth an impotent weapon. Some do have canines, it is true, but most cannot afford even those, since canines interfere with effective mastication by limiting the movement of the jaw. With the mouth becoming ineffective as a weapon, hornlike organs evolved as weapons. And why, one may ask, do so many ungulates—ruminants especially—lack upper incisors? The answer appears to be as follows: Since ruminants cannot eat more than they can digest, they must eat the most digestible forage possible in order to maximize the intake of food (Blaxter et al. 1961). The digestibility of plants is very closely related to their fiber content. Therefore ruminants must try to ingest plant food with a minimum fiber content. This can be done, among other things, by feeding on the meristematic, that is, growing, parts of sedges and grasses. All that is required is to pull the blades free rather than snapping or biting them off. This requires that the blades be firmly grasped, but not too firmly, lest they be weakened by the bite and snap off. Therefore, in ruminants, these blades are pressed between the lower incisors and the palate and then pulled out, so that the meristem is withdrawn with many blades. This is the succulent, tender, protein-rich part of the plant. By selecting tender parts of plants, ruminants require less food than do monogastrics, such as horses, and can generate a larger biomass for the same amount of food. Note that the rise of ruminants parallels the spread of grasslands during Tertiary times, which indirectly supports the hypothesis that the meristems of grasses and sedges are related to the evolution of the ruminants' feeding apparatus and digestive system.

Definition of Aggression

Much bitter, pointless, and insulting discussion has taken place because of conflicting definitions of aggression and careless reading of the literature. All authors initially define aggression by *effect*. That is perfectly acceptable. However, some authors choose for their effects phenomena that impose a *narrow* definition on aggression, while others choose effects that impose a *wide* definition. Hinde (1974) is an example of the former. He defines aggression as that behavior which is "directed toward causing physical injury to another individual." Note that Hinde has at once equated aggression with *violence*. He has excluded from aggression even threats, that is, behaviors signalling violence. He has also excluded pushing and shoving—in fact, any bodily displacement of an individual by another. Clearly, Hinde's definition of aggression is untenable, since it excludes much of the behavior we readily observe even in the fights of animals.

If we choose for our effects to define aggression processes that physically displace individuals during interactions, we have chosen a *wide* definition. It encompasses under aggression, violence, unritualized fights, dominance fights, threats of all categories, dominance displays (bravado displays, present threat, Geist 1965, 1971a; display threat, Lent 1965; *Imponierverhalten* in German, see Walther 1974), as well as ego displays and self-assertion in humans, as I shall argue in Chapter 5. Note that workers using aggression in the *wide* sense (as many ethologists, psychologists, and psychiatrists do) talk about different matters than do individuals using a *narrow* definition of aggression. However, worse is to follow.

Hinde (1974) claims that limitations are necessary in order that a definition may be useful. Agreed. Now, Hinde limits aggression to damaging conflict between individuals of the same species. Hence, according to this limitation, a cow moose attacking a wolf would not be acting aggressively. Never mind that she is using the same weapon system, the same short changes with loud hoof slapping of the ground, and the same threat postures that a subordinate moose uses against a dominant (Geist 1963). Clearly, this limitation is unacceptable.

The converse limitation, to exclude from aggression the attack of a wolf on a moose, or more generally speaking the attack of a predator on its prey, is less absurd but still untenable. This limitation was strongly championed by Lorenz (1963) and—unfortunately—widely accepted (Hinde 1970, 1974). Lorenz argues that a predator attacks and kills its prey for entirely different reasons from those it has for attacking a conspecific. In the first instance, food is involved in motivating the individual; in the second instance, it may be a factor of territorial, dominance, or mate defence. The latter imply different motives from the former.

It should be noted that if the predator species is also cannibalistic, this segregation by motivation falls. What Lorenz (1963) was not aware of, and neither—apparently—is Hinde (1974), is that large carnivores are cannibalistic. Lorenz argued for a separation of cannibalism and carnivory. He arued that well-armed canivores do not kill each other. History has long since contradicted Lorenz. Large carnivores are indeed cannibals and may hunt and kill weaker and disabled conspecifics for food, or at least eat conspecifics killed in fighting. We know this from lions (Schaller 1972, Schenkel 1966b), mountain lions (Hornocker 1967), grizzly bears (Pearson 1972, Miller 1972, Payne 1972, Russell 1972, Larsen 1972), polar bears (Jonkel, personal communication), black bears (Jonkel and Cowan 1971), wolves (Rausch 1967, Jordan et al. 1967, Mech 1970), hyenas (van Lawick and van Lawick-Goodall 1971, Kruuk 1972), and exceptionally in cheetahs (Eaton 1974). Cannibalism is not rampant in carnivores and may be missing, as it aparently is in hunting dogs (Schaller 1972, van Lawick and van Lawick-Goodall 1971); yet it is present in many forms and may even be well developed. Hence, segregating certain acts from aggression on the basis of motivation—disregarding momentarily the fogginess of that notion—appears ill advised, particularly since in interspecies attacks both opponents may use species-specific weapons and defences of a morphological, physiological, and behavioral nature. Hence a definition of aggression should include fights in which species-specific weapons and defences are used. The logical conclusion of this concept is that some types of overt aggression in carnivores are a prelude to feeding behavior. Wilson (1975, p. 243) makes much the same point. Separating feeding and aggressive behavior may become a moot point, as is also pointed out by Southwick (1970a).

Hinde (1974) is to some extent a captive of his concept of motivation. For example, he treats threats and displays as products of two conflicting motivations, attack and withdrawal. As I indicate later (Chapter 5), displays are treated as impurities, not as behaviors in their own right, with their own evolutionary history and discrete functions. If one wishes to interpret behavior as the product of conflicting drives, a narrow definition of aggression is desirable. This is not quite good enough, because it usurps the term "aggression" for the perfectly good term "violence"; it is worse to misread or ignore pertinent literature that argues differently, as is evident in the claim that most students of animal behavior define aggression narrowly (Hinde 1974, p. 251). I read differently and point to Collias (1944), Tinbergen (1951), Lorenz (1963), Davis (1964), Hall (1964), Marler and Hamilton (1966), Scott (e.g., 1966), Eibl-Eibesfeldt (e.g., 1967), Walther (e.g., 1958, 1974), Schaller (1967), Estes (1969), Goodall (1971); in fact, it is difficult to find ethologists who restrict the term "aggression" to violence. It appears that Crook (1968b) uses it in the narrow sense. As a student of mammalian aggression, I have used a wide definition (e.g., Geist 1965) and am perfectly happy with the definition of aggression as adopted by psychiatrists (e.g., Gilula and Daniels 1969).

Aggressive behavior in man and animals is recognized as such by the same criteria. Please note I said "criteria," that is, in this instance, external manifestations accessible to the senses. I am neither concerned, nor even interested, whether aggressive behavior is homologous or analogous between higher mammals and man; I cannot determine this with any degree of certainty between different species of higher mammals, or even between individuals in the same population. After all, it may be that the neural structures controlling aggression in two individuals are quite different owing to genetic, as well as learned, differences, and in the strictest sense they are, therefore, not homologous. Evolutionary theory dictates that, under similar conditions solutions with similar effects will appear for similar problems: Evolutionary theory says nothing about the internal structuring of the physiological mechanisms controlling aggression.

Of course, Hinde is perfectly welcome to use any definition of aggression. However, when discussing the works of other authors he should at least point out in all fairness to them that his definition of aggression and theirs differ significantly. For instance, in reference to Lorenz (1963), Hinde (1974) asks, "Is aggression a valuable human characteristic?" Granted Hinde's definition of aggression, this is a rhetorical question, surely! No person in his right mind would consider violence a valuable human characteristic, especially in regard to the last six decades. Granted the wide definition used by Lorenz, aggression becomes a very valuable characteristic of humans. Indeed, without it the very best in cultural achievements would have been impossible, as indeed would have been love. Quite contrary to Hinde's suggestion (1974, p. 275), neither Lorenz nor Storr reach the height of absurdity by suggesting this. The foregoing illustrates that differences in defining a term can lead to different points of view and to unenlightening debates.

Another definition is advanced by Wilson (1975, p. 242). He contends that the common English usage conveys a definition that is hard to improve on: Aggression is the abridgement of the right of others by whatever means. I can concur with this definition when it is confined to human aggression, assuming of course the existence of universal rights; I have difficulty accepting it for animals, since I can see no universal rights prescribed by nature guaranteeing access to resources, reproduction, and survival, be it for individuals or species. This is not to dispute Wilson's valid point that aggression encompasses a great variety of acts with rather different functions.

I propose to define aggression as social behavior that displaces individuals and/or bars their access to scarce resources, as well as those actions that protect the individual from bodily harm. This definition of aggression makes it part of *agonistic behavior,* namely, that excluding submission, appeasement, and withdrawal. *Overt aggression* refers to combat in which weapons and defences are brought into play and opponents interact physically. It includes fighting, combat, dominance fights, attacks, defence, unritualized and ritualized fighting, etc. *Fighting* is defined less broadly than overt aggression; it excludes attacks, since here only one partner may use weapons while the other flees. Fighting implies that both partners exchanged attacks and defended themselves. *Combat* refers to intensive fighting between conspecifics or nonconspecifics. *Dominance fight* refers to extensive fighting between individuals of the same sex and age; it is composed of elements of attack and defence, plus threats and displays. *Attack* and *defence* are groups of behavior patterns that inflict damage or trauma, respectively, and prevent the infliction of these in fighting; included under attack and defence are *tactics* which maximize their respective efficiencies. *Ritualized combat* or fighting is the same as dominance fights, but emphasizes the reduction or absence of behavior patterns that inflict damage to the opponents. *Unritualized fights* refer to fights between conspecifics in which damaging behavior patterns are brought into play, and from which dominance displays—but not threats—are absent. Such fights are usually of short duration, since either they terminate or they are quickly escalated into full-blown dominance fights. *Defensive fighting* can occur between conspecifics or nonconspecifics and refers to the actions of the smaller subordinate or the prey. It is characterized by extensive threats, no dominance displays, and a minimum of overt aggression. Its converse would be *aggressive fighting* in which a dominant, larger, or predatory animal—without preliminary threats or displays and with a minimum of defensive behavior—applies its weapons quickly and efficiently. *Threats* are iconic signals recognizable by individuals bringing specific weapons systems into a position for instant use; threats signal intentions to use specific weapons. *Displays* are abstract signals recognizable by criteria described in Chapter 5; these include apparent enlargement of body or weapon size by various means, including orientation to conspecific, use of attention-guiding organs and optical illusions, attention-attracting motions, and an indirect approach apparently designed to generate arousal by robbing the opponent of predictability. This is discussed in detail in the following chapter.

Other words often used in discussion of aggression are 'dominant' and 'subordinate.' A *dominant* is defined as an individual who is capable of displacing others in contests over resources. A *subordinate* is defined, of course, conversely, as an individual who withdraws from contests leaving contested resources to the dominant.

Submissive behavior is not based in its naming on human analogy in an attempt to anthropomorphize animal behavior, as some critics have contended (e.g., Leach 1966), but in an attempt to communicate. As I shall argue below, the similarity, be it analogous or homologous to human behavior, is exceedingly crucial to an understanding of its function; its motivational origins are not. Lorenz (1974) makes a similar argument. Submissive behavior is defined functionally; it reduces aggression by conspecifics and is neither a threat nor a display as defined below, nor a withdrawal.

The Phylogeny Fallacy

The ultimate purpose of writing about aggression is, of course, to state some meaningful conclusions about human aggression which are based on knowing aggression as a phenomenon in its own right. This brings us to a question often heard but nowhere put more succinctly than in Boulding's (1968) critique of Konrad Lorenz's (1966) book *On Aggression*. What can the study of animal aggression possibly teach us about human aggression?

In trying to answer this question, Boulding (1968) made a number of errors which, however, even biologists make. He asserted that the evolution of aggression cannot be studied, since the ancestors of present forms are dead and fossils tell us nothing. Fortunately, the study of evolution does not depend on fossils. Moreover, fossils do tell us something about the morphological counterparts of aggression, the organs of attack and defence. So we can learn something about the evolution of aggression even from fossils (G. Clark 1957, p. 101, Geist 1971b, 1974a, 1978a, Galton 1971, Barghusen 1975, Hopson 1975, Stanley 1974). Boulding also pointed out, correctly this time, that the study of animal aggression as a simpler form of human aggression may still tell us little or nothing, for the behavior of complex systems cannot be predicted from that of simpler ones, nor even from isolated parts of the complex system. In a later book, Lorenz (1973) emphasized that an amalgamation of simpler systems leads to properties that cannot be predicted from an examination of each simpler system; he referred to this development of new properties as "fulguration."

However, Boulding committed a far deeper error. He assumed that present-day forms have systems of aggression because their ancestors had them, in the sense that the adaptations of today result from phylogenetic determinism directly derived from the adaptations of yesterday. If this were so, indeed, we should study primitive forms to understand advanced ones. Yet this is an error. I call it the *phylogeny fallacy*. Anthropology, in embracing primate studies, has already been guilty of it. Schaller (1972, p. 378) makes the same point in a more diplomatic fashion.

To understand the phylogeny fallacy is to understand the statement that a horse does not look like a horse because its ancestors looked like horses; it looks like a horse because the physical and—in particular—the social environment of the modern horse (which shaped the modern horse) are not too different from those that shaped the ancestors of the horses. The rhino's ancestors, for instance, looked like primitive horses, not like rhinos. Moreover, rhinolike animals have appeared repeatedly in the history of mammals, as I shall discuss in detail in Chapter 8. Thus, evolutionary theory dictates that similarity is due to convergent selection in similar ecological professions (niches), not necessarily due to genetic similarity. The origin of a system is thus irrelevant to an understanding of the function of a system.

Let me illustrate these important points. In order to understand the function of the mammalian ear, a knowledge of the Reichert–Gaupp theory, which states that the malleus and incus are derived from the quadrate and articular of the reptilian jaw, is irrelevant. In a similar vein, the study of aggression in primates will tell us very little about human aggression; in fact, aggression as practices in carnivores and in ungulates teaches us considerably more, as I hope to show. To an understanding of a process or structure, a knowledge, not of its origin, but of the environmental circumstance that shaped and maintain it, is crucial. Phylogenetic knowledge may be most enlightening when we look at the embryology of the middle ear (see Portmann 1959), but it tells us little more of an organ's function than the trivial.

If this is so, then similarity or dissimilarity, not the origin of structures, is very significant to an understanding of their role in the adaptive strategies of organisms. As indicated, it is irrelevant to discuss whether these structures are analogous or homologous, and the caution of conventional wisdom that similarity does not imply identity of origin, process of development or, in the case of behavior, motive, is irrelevant. A similar behavior may have the very same purpose in both animal and man, but may, of course, be of very different origins. Therefore, similarity is indeed worthy of attention, never mind analogy or homology, which create more difficulty when applied to behavior than the effort is worth (Atz 1970). This is dictated by evolutionary theory.

The study of animal aggression thus serves to create a general knowledge framework of aggression and permit us to contrast the specific subject of human aggression within the general framework of aggression. We learn by contrasting the specific against the general and in so doing are able to make predictive statements about humans, and in testing these predictions we can determine which lessons from animal aggression apply for our species and which do not.

The physiologic structure underlying agonistic behavior and its ontogeny are signifcant if, by interfering with the physiological base, alteration in the aggressive behavior of organisms is desired. Then a similar and phylogenetically closely related system is preferable to a similar but phylogenetically remote system as a model, provided the subject system cannot be studied directly. This applies to humans, of course, and for ethical reasons animal models must substitute where necessary. This necessity, however, must not be expanded into a general faith that the study of organisms phylogenetically related to man will tell us all that is to be known of man. Such a belief is based on the phylogeny fallacy.

Misadventures in Aggression

The phylogeny fallacy is not the only intellectual misadventure in aggression; there are more. Thus, a relatively simple subject has been transformed into one of complexity, thanks largely to faulty observations and at times thoroughly muddled thinking. Lorenz (1963) wrote a most provocative treatise on aggression that stimulated fierce debate, but the outcome was more smoke than fire.

At that time the study of large mammals in their natural environments had only just begun. Lorenz, and ethologists as a whole, were familiar primarily with small-bodied organisms of a great diversity, and failed to realize that they had not yet grasped some fundamental processes. For instance, the findings that combat could be divided into attack and defence and that there were morphological adaptations against weapons systems, as well as behavioral strategies of offence and defence, were only published after Lorenz's book appeared (Geist 1966a, Schenkel 1967). These papers, however, failed to penetrate to the attention of ethologists, as an examination of recent literature and textbooks will show.

The reason such fundamental distinctions went unnoticed is easy to explain: Given the speed with which small-bodied animals act, the behavioral morphology of fighting is easily overlooked. Furthermore, in short-lived creatures—which small-bodied birds and mammals usually are—we expect little emphasis on *defensive* adaptations and a predominance of overt damaging aggression (Geist 1974a). Again, this does not make it easy to study the morphology of aggressive behavior. Large bodied forms can be better observed and, owing to their mass, use diverse means of combat and multiple weapon systems, all of which clarifies the nature of aggression (Geist 1971a).

Second, Lorenz made several misobservations that led him to erroneous conclusions. He claimed that dogs expose their necks in fighting so that the dominant can kill a subordinate so displaying. Scott (1967) denied that such behavior exists in dogs, while Schenkel (1967) reported that the dominant, not the subordinate, may display his neck, but that the subordinate simply will not snap since he would be thrown off, overpowered, and probably bitten in a moment. Lorenz's misobservations and erroneous interpretation of submissive behavior led to the ethological dogma that predators do not kill conspecifics and that carnivory and cannibalism are different entities, while in principle animals fight so as *not* to maim each other (Lorenz 1963, Tinbergen 1953, 1968, Matthews 1964, Eibl-Eibesfeldt 1963, 1967, Southwick 1970a). Injurious combat or the death of combatants was considered an unnatural occurrence, one easily explained as a consequence of exceptional circumstances, unusual environments, or accidents at the worst. Species equipped with lethal weapons had also evolved reliable means of blocking these weapons in intraspecific combat; only species without such weapons had no inhibitions against using them. Studies that indicated the contrary received little if any attention, since one can read the same general theory of aggression in relatively recent textbooks of ethology (e.g., Eibl-Eibesfeldt 1967, Ruwet 1972, even in Wilson 1975) or in research papers (e.g., Maynard Smith and Price 1973).

Studies of large mammals during the decade following Lorenz's book (1963) have shown clearly that death through interspecific combat is by no means rare, and that injury during such combat is reduced less by inhibitions against using weapons than by skillful uses of defence strategies, behaviors, and morphological adaptations. For carnivores, I gave earlier references to intraspecific killing and fighting; for ungulates, I point to several of my reviews (Geist 1966a, 1971a, 1978a); references to intraspecific killings and injury can also be found in Collias (1944), Scott (1966), Southwick (1970a), Jolly (1972), Bygott (1972), and Wilson (1975). Thus, in the last decade, animal aggression was robbed of much of the innocence that ethologists had given it, in part to contrast it against human aggression. It is still true that killings among animals are not rampant, but we are able to understand the reasons for it today, as I shall show, and these are not that animals in combat disdain to injure each other.

To confound the already unfortunate theoretical conception of animal aggression, the argument was advanced that dominants must not injure the subordinate in the interest of the *species*(!). This was proclaimed by Eibl-Eibesfeldt (1963) as if there were a disadvantage to the reproductive fitness of a dominant if he killed an unrelated subordinate. It reinforced the contention that combatants should not injure or kill each other, albeit for an illogical reason, as I have shown elsewhere (Geist 1971a, pp. 231−36).

Eibl-Eibesfeldt (1963) and others argued that ritualization of combat robbed it of its injurious effects. That is, *show* replaced *force*. An analysis of the "ritual" of combat in fencing, as well as in the dominance fights of mountain sheep, showed quite the opposite: The "ritual" consisted of postures, stances, and motions that *increased* the severity of blows, that *maximized* the chances of inflicting injury on the opponent, and that minimized the chance of the opponent succeeding in his attack (Geist 1971a). The "ritual" effect was thus not *less* damaging combat on the part of the overt aggressor but *more* damaging combat. However, the ritual also developed partly as a means of defence (see also p. 75).

A further reason for reconstructing the theory of aggression as held until recently by ethologists, is the confounding of observation and interpretation in discussing aggressive behavior. Thus, threats and displays are still conceived as results of antagonistic *motivations* (e.g., Manning 1967, Hinde 1970, 1974) instead of signal structures in their own right. The failure to separate threats from dominance displays appears to be one reason the latter have not been recognized in animals, or their counterparts in man. If such a separation is made, both can be seen as signal structures and not accidental by-products of conflicting tendencies. Critics of the ethological position on animal aggression also failed to notice that there were fundamental differences between threats and dominance displays. It was the students of large mammals who differentiated threats from displays and emphasized these differences (e.g., Antonious 1939, Walther 1958, 1966, 1974, Schloeth 1961, Geist 1965, 1966a, 1971a); they were formally defined for the first time by myself (Geist 1965).

Other difficulties that hindered, rather than advanced, a clarification of the subject of aggression were a critique of the ethologists' position, but using aggression defined strictly as violence (e.g., Crook 1968a) (ethologists—Lorenz in particular—discuss aggression as I define it above, i.e., widely) and the formulation of aggression so defined as a response to aversive stimuli or unusual conditions (Crook 1968b, Scott 1966, Southwick 1970a). This was then followed by a denial that aggression was spontaneous and had appetitive behavior (Scott 1966, Crook 1968b). Again this was an unfortunate formulation. It does fit animals in which combat is injurious owing to good weapons and poor defences; it does not fit at all where defence mechanisms are excellent, as in many horned ungulates. Here individuals search out fights and develop these into playful combat which can serve as an excellent indicator of their health, vigor, and well-being (Petocz 1973). In short, to illustrate aggression as Scott (1966) and Crook (1968b) did is totally inadequate for such ungulates as cattle (Schloeth 1961), mountain sheep and their relatives (Geist 1966a, 1968, 1971a), horses (Klingel 1967, 1974), gnus (Estes 1969), and gazelles (Walther 1958, 1966, 1965); it would be perfectly adequate, however, for the mountain goat (Geist 1965, Petocz 1973), and with some difficulty for the chamois (Kramer 1969), black rhino (Schenkel and Schenkel-Hulliger 1969), and white rhino (Owen-Smith 1974). Thus, with increased examination of different species of animals a different conception of aggressive behavior is warranted.

Weapons and Defenses

The following is an abstract of a theory of aggression as a function of individual selection which I published earlier (Geist 1966a, 1971a,b, 1974b, 1978a). Some new material is added here.

Assuming a defendable resource essential for reproduction, overt aggression is the ultimate, albeit most expensive and risky, means of maintaining access to it. In its simplest form, overt aggression is pushing, shoving, or wrestling, that is, the physical displacement of one individual by another. We find such behavior in frogs competing over females or territories (Noble 1938, Emlen 1968).

At the next stage of development, overt aggression aims at triggering withdrawal responses by the opponent; it parasitizes the withdrawal responses of organisms to aversive stimuli by mimicking them. It triggers withdrawal by causing trauma or causing "pain"; the latter is caused by distorting the body, particularly the body surface. Trauma is caused by robbing the opponent of control over his own body, such as the desert tortoise (*Gopherus agassizi*) turning its opponent on its back or a large stag pushing a smaller opponent effortlessly across the landscape. Triggering an opponent's withdrawal is, of course, cheaper in energy than lifting him off bodily. Furthermore, the larger an opponent the more difficult it becomes to move him bodily. Therefore, in large-bodied species the emphasis in fighting shifts increasingly to mechanisms causing pain or trauma and, therefore, reflexlike withdrawal. Fighting has not escaped the law of least effort.

The organs evolved to maximize the inflicting of pain or trauma are weapons. Since we know from medical evidence that wounds to the body surface are more painful than cuts to the body core (Guyton 1971), it follows that pain is maximized by maximizing surface damage. Therefore, weapons that maximize pain through damage to the body surface are quite inappropriate for deep penetration. Weapons that maximize surface damage may act like clubs maximizing bruising or like knives which cause multiple punctures or lacerations to the skin. The horns and manner of fighting of the giraffe are an example of the former (Backhaus 1961, Spinage 1968), and the claws of cats used to rake the opponent (Leyhausen 1956) are an example of the latter. Note that a weapon penetrating the body surface causes about the same amount of pain as one penetrating into a vital organ. However, such a weapon places the aggressor in danger, since withdrawal of a long stabbing weapon is far more difficult than the withdrawal of a short one. Thus, long horns stuck deep in an opponent are likely to cause a skull or neck fracture in their owner owing to the victim's struggle to free itself. Therefore, one does not expect weapons to evolve to kill opponents quickly, but rather they evolve to cause a large amount of damage to the body surface.

In order to inflict trauma, the aggressor must be able to grasp and hold on to his opponent so as to exert maximum force. Teeth, claws, grasping organs, or horns and antlers evolved as locking organs are essential to this manner of fighting. Thus wrestling arises secondarily as opponents, holding onto each other's bodies, attempt to make each other lose control over their bodies. We therefore expect two principal types of weapon in animals, namely, those that maximize damage to the body surface and can be quickly withdrawn and those that permit opponents to lock onto each other. The points, bumps, spirals and surface texture of antlers and horns of many ungulates are means of gripping the opponent's weapons (Walther 1966, Geist 1966a).

It is, of course, not in the individual's interest to be displaced from a scarce essential resource. It can resist displacement by force, and use means that make the attempts of the aggressor to displace it by causing pain and trauma entirely futile. These means we call "defenses." There are defense organs, strategies, and tactics to neutralize attacks. These include evasion of attack, reducing the target area exposed to attack, catching the opponent's blow against a heavy dermal shield on the body or against an armoured head, deflecting the opponent's weapons, denying the opponent the use of his weapons, grasping the opponent's weapons and holding on to them, thereby effectively disarming him. There are now a good many examples reported in the literature of the defenses mentioned. We note that dermal shields are grown where weapons are most likely to make contact with the body surface: on the face and front of the mountain sheep (Geist 1971a), on the rear end of the mountain goat (Geist 1967a, 1971a), on the neck of the male impala (Jarman 1972), on the shoulders of the wild boar male (Frädrich 1967, 1974). We also note that adaptive strategies of catching, blocking, or holding onto an opponent's weapon exclude elaborate dermal shields, and that horns and antlers act as guards, locking organs and weapons (see Krämer 1969, Harkness 1971, Cumming 1975, Kitchen and Bromley 1974, Sowls 1974, Barrette 1977c, Lincoln 1972).

We find, therefore, a correlation between the structure of weapons and defenses and their manner of use. Since in combat each individual alternately maximizes the efficiency of attack and of defense, depending on circumstances, it is evident that for any given system of armament only a limited number of strategies will maximize the efficiency of attack and defense. The interplay of these limited numbers of strategies of attack and defense gives rise to the "ritualized" species-specific fight.

An important factor limiting the evolution of weapons and defences is body size. Since the mass increases as the cube of the linear dimension, and increase in linear dimension permits an increase in the speed of movement, acceleration and deceleration soon reach the point where animals will fracture bones, tear ligaments, or rupture organs if they should jump at each other, collide, and roll on the ground like small-bodied animals do. We find, therefore, that in large bodied mammals appendages are used as weapons; the very mass of these appendages, accelerated rapidly, is sufficient to cause large surface bruises or penetration of the body by hornlike organs. The head is commonly used to deliver blows, but occasionally the tail is used and studded with weapons as exemplified in the extinct glyptodonts or the ankylosaurs and stegosaurs. Small-bodied mammals have opted mainly for fighting weapons that penetrate the body surface, in part because their low body mass precludes the generation of sufficient force by their appendages to inflict bruises on their opponent's body.

Defences are also influenced in their evolution by the body mass typical of the species. If the mass of the body is low, the individual can successfully evade blows by jumping aside; in large bodied species their mass and the greater distance to go to evade blows precludes this. We therefore find dermal body shields primarily on large-bodied mammals. I discussed this and some exceptions recently (Geist 1978a).

One can also recognize a relationship between the reproductive strategy of a species and its weapons and defences. In species in which individuals rarely reproduce more than once, and do so during a short reproductive season—such as various salmon—we find in the males very elaborate toothed jaws that can inflict significant damage, but we find no obvious means of defence. Clearly, injured individuals do not live long enough to heal wounds and reproduce thereafter; there is little advantage in elaborate defences, only in effective weapons. We find a similar situation in the short-lived saiga antelope (*Saiga tatarica*) in which combat is often lethal during the rutting season (Bannikov et al 1961). Conversely, in long-lived species in which individuals live through numerous reproductive seasons, we find well developed defences, nowhere better illustrated than in the genus *Ovis* (Geist 1971a).

Retaliation as a Control on the Evolution of Aggression

Some of the differences in the agonistic behavior between species are a direct consequence of their weapons and defences. The heavy armor on the skulls of caprid males permits frequent long-lasting, or even severe, fighting without noticeable injury to the opponents; it permits "sporting combat" that flourishes when forage conditions are favorable (Petocz 1973). Yet in an animal with weapons capable of causing severe wounds, with poor defences, such as the mountain goat, the incidence of fighting is negligible and even threats decline with improved forage conditions (Petocz 1973); in this species we find no sparring matches, that is, "sporting combat." The rare instance of fighting in species with sharp weapons can be explained as a consequence of retaliation (Geist 1966a, 1971a, 1974a); it is also an explanation of how individual selection acts to limit the development of arms and defences.

It is known that pain causes various birds and mammals to attack companions or inanimate objects (Ulrich and Azrin 1962, Ulrich et al 1965, Azrin et al 1965, Galef 1970). The frequency of attacks, however, is an inverted U-shaped function of the intensity of the stimulus causing attacks (Azrin 1964). Thus a blow, unless it is so severe as to be totally debilitating, will trigger an attack on the aggressor by his victim. Given damaging weapons, a counterattack by the victim is likely to inflict injury on the aggressor, particularly in an all-out attack. Therefore, both the aggressor and the victim are likely to suffer damage and the aggressor is not likely to be successful in reproduction, particularly in long-lived species in which individuals reproduce in more than one reproductive season. Only the instant death of the victim, or its inability to retaliate, or some means of escaping retaliation, will permit damaging aggression to flourish. If the biological weapons are selected to maximize pain through surface damage, retaliation is certain and there is selection against frequent combat in species with poor defences, since the victim punishes the aggressor.

A similar idea, that retaliation was an adaptive strategy to maximize gain in a zero-sum game, was developed independently by the late G. R. Price and formalized mathematically by Maynard Smith (1973). This earlier paper suffered from a number of defects, as I pointed out (Geist 1974b).

The foregoing makes it evident that the only way for an aggressor to circumvent prompt retaliation is to stun or kill the opponent outright on the first attack. This, however, is virtually impossible even to the well-armed ungulates and large carnivores. Even lethally wounded opponents do not die instantly but succumb to wounds later; they are, therefore, often in a position to retaliate. Only humans, by virtue of cultural weapons and defences, can escape prompt retaliation either by killing the opponent outright or, if not by killing outright, by ducking behind some protective shield. This hypothesis suggests that humans have not evolved biological inhibitions against using cultural weapons (as is only too obvious). I tend to agree with authors such as Goode (1974) who suggest that humans have no "killer instinct"; humans have social restraints against cultural weapons. They probably do have biological inhibitions against using their teeth in combat, which inflict most painful wounds and may be poisonous as well.

The retaliation hypothesis solves another riddle: Why do some poisonous snakes wrestle in combat rather than bite (Shaw 1948)? If A bites and B retaliates, clearly both are dead.

Although the skillful use of defences, a high threshold for overt aggression and the principle that weapons maximize damage to the body surface and not its core may greatly reduce combat mortality in animals, such mortality exists, nevertheless (Geist 1966a, 1971a). If an opponent loses his defences, he can be injured. If the aggressor uses deception by inserting into combat activities inappropriate to combat—once called displacement activities—it may throw his opponent off guard and land a telling blow. The behavior of mountain sheep in their dominance fights gives ample evidence of this. A ram not quite ready to parry the clash of the attacker may be thrown a long way downhill; in such an event the fight is over and the deceived ram submits. If combat is costly, alternatives to combat are expected to evolve that have the same effect but not its price, such as dominance displays (to be discussed in the next chapter) or strategies such as juvenile aggression, which I mentioned earlier, or sparring matches that pit individuals of settled dominance rank in "sporting" engagements against each other, as can be seen in the mule deer (*Odocoileus hemionus*).

Wilson (1975, p. 247) proposes another control on the expression of overt aggression, namely, kin selection. The benefit to a genotype dominating a relative is less than dominating unrelated genotypes. The benefit of dominating a relative is least where kin selection maximizes the number of individuals with a common descent which defend a resource territory as a group. When this condition does not apply, aggression against relatives may be quite adaptive, particularly in species exploiting seral rather than climax plant communities. Aggression may serve to disperse individuals so they may find available habitat that is unpredictable in time and space (e.g., moose, Geist 1967b, 1971a).

Armament and Ecological Strategy

Weapons and defences are not selected for independently of strategies and tactics of aggression or of threats and displays. Rather they are part of an agonistic syndrome, and it in turn appears to be closely related to the ecological strategy of a species. Our point of departure will be species from highly productive tropical environments, since these appear to be ancestral to the highly evolved forms from temperate and cold zones. When comparing the weapons of species, it is soon apparent that primitive forms tend to have sharp piercing weapons, while many highly evolved forms have locking-type weapons which are often elaborate and used in displays. A number of hypotheses were developed relating weapons to ecology and these require further testing, although at present they fit the available facts; a detailed discussion is found in Geist (1977, 1978a,b,c).

(1) *Defence of resources needed for maintenance, reproduction (excluding mates), and growth selects in both sexes for weapons maximizing surface damage and for defence by evasion. If body mass or combat technique preclude evasion, dermal armor is selected for.*

If resources that are required daily are defended, then clearly natural selection will maximize the effectiveness of the defender per unit of defendable resource. Since resources must be maximized for reproduction, combat that is expensive and risky must be minimized in frequency and duration. Weapons that maximize surface damage in violent instantaneous attacks and inflict pain well beyond the threshold of maximum retaliation (Azrin 1964) are most advantageous, since they ought to quickly cause an intruder to flee. We also expect a diversity of deterents to evolve, such as group displays, olfactory marking, surprise attacks, mass attacks, and a reduction in sexual dimorphism to confuse intruders (Geist 1974a). All these factors would reduce predictability to the intruder, catch him unprepared, reduce his retaliation, and condition him negatively to the area of attack. Why small-bodied species will not evolve armor but large-bodied species will if they fight with damaging weapons, I explained earlier. The hypothesis advanced here is also supported by the observation that in the defence of estrous females, ungulate males, regardless of weapons system they have, attack in such a manner as to inflict a maximum of surface damage on the opponent. For the short duration of estrous, when withdrawal of the male can be equated with the loss of reproduction, all means must be marshalled to retain access to this essential resource. Attacks that inflict pain maximize the defender's time for vigilance, courtship, and breeding; the more attacks deter, the more assured is the defending male's reproductive success.

(2) *When resource defence is not possible owing to low density or significant annual fluctuation of the resource in space and time, then females tend to compete passively and lose or reduce their weapons, while males evolve strong defences, elaborate displays, or low-intensity interactions.*

Here we are dealing with somewhat opportunistic species with extended mating seasons that do not reproduce continuously and therefore must extend longevity. This selects against intense combat in the males and for good defences, such as catching and holding the opponent's weapons or dermal shields, depending on body size. Defence is all-important in a strategy of extended opportunistic breeding. The females, quite unable to compete over resources actively, have little reason to be aggressive and consequently lose whatever armament they possessed. Field studies on muntjac (Barrette 1977b), warthog (Cumming 1975), bushbuck (Leuthold 1974, Walther 1964a), reed buck (Jungius 1971), and wild board (Oloff 1951, Heptner et al. 1961, Eisenberg and Lockhart 1972, Frädrich 1967, 1974, Gundlach 1968) suggest these species possess the above adaptive syndrome.

(3) *If gregariousness evolves as a strategy against predation, weapons are selected for that minimize surface damage and retaliation. Weapons that permit contests based on strength are selected for.*

In open landscapes that do not permit hiding, where individuals maximize security by grouping (see Wilson 1975), it is clearly a disadvantage to possess weapons that inflict severe pain and trigger retaliation in kind. Severe pain would cause the attacked conspecific to leave the group, thereby reducing security for the aggressor; if group members avoid the aggressor, he would probably often find himself at the edge of the group and therefore in the most dangerous zone, for it is here that predators are most likely to strike. Retaliation would probably cause wounds and make the aggressor susceptible to predation (Estes 1974). Weapons and behaviors that disrupt gregariousness are clearly incompatible with an antipredator strategy based on gregariousness. The above hypothesis explains why we find wrestling-type weapons in gregarious and highly social ungulates, why there is little evidence for body attacks in these forms (Walther 1966), and why we find very little dermal armor. At first glance, the camel described by Gauthier-Pilters (1974) appears to be an exception. The males have sharp incisiform canines and they attack readily and even kill rivals. We must note, however, that the above hypothesis assumes heavy predation pressure, so that solitary individuals are selected against. For the large-bodied, fleet, well-armed camel living in a highly unproductive landscape, this assumption does not apply, at least not to the same extent. Owing to its size, its wide scanning horizon and concomitant requirement of less time for scanning, and because it rarely meets predators, the camel ought to live reasonably successfully alone. Therefore it can maintain sharp surface-damaging weapons and determined attacks, which would be selected against in smaller-bodied species living on more productive open plains.

The foregoing three hypotheses explain the basic weapon and defence structures found in ungulates; there are, in addition, a series of secondary hypotheses which I discussed earlier (Geist 1974a, 1978a,c). These explain diverse attributes of the primary weapon systems. Thus, a high degree of gregariousness of both sexes selects against sexual dimorphism in species from open plains because females enter into competition with males and mimic them; conversely, in pair-forming forest forms that defend resources jointly, we also expect the female to mimic the male, and this leads to a reduction in sexual dimorphism. It is in the ecotone between forest and plain that we find maximum sexual dimorphism in ungulates. Should males be servely exhausted by the rut, as is common in northern ungulates, they may escape predation after the rut by joining females, provided they mimic females (Bromley 1977). In practice, this means casting horns and antlers shortly after the rut. Males may also hide by withdrawing to small seasonal ranges which they defend for exclusive use. In this case, we expect weapons causing surface damage. Males may also withdraw to a different habitat and exploit it gregariously, in which case they require wrestling-type weapons. These and other concepts explain the biology of diverse ungulates (Geist 1977, 1978a,c).

The weapons of carnivores are a case apart. These animals use the weapons on their prey as on conspecifics. Since these weapons are very damaging, carnivores must either develop complex defences or reduce fighting to a minimum, as dictated by the concept of retaliation. Heavy dermal armor, however, is clearly incompatible with the speed and agility so essential in bringing down prey and avoiding its defensive attacks. This leaves virtually only one avenue open to carnivores to reduce damage to themselves, namely, a high threshold against overt aggression. This appears to be found in mountain lions (Hornocker 1967) and lions (Schaller 1972), as well as in social carnivores such as wolves, hunting dogs, and hyenas, at least within the social group (Scott 1967, Mech 1970, van Lawick and van Lawick-Goodall 1971, Kruuk 1972). Curiously enough, we do find a dermal armor on the neck of grizzly and polar bears. In the black bear, it is less developed (Jonkel, personal communication). Here the extent of dermal armor appears to correlate with the aggressiveness and combat ability of the species, being least in the small timid black bear and greatest in the large aggressive polar bear. On the whole, neither in carnivores nor in most mammals do we find distinct combat organs as we find in ungulates, although the fossil record does suggest exceptions, such as the horned armadillo *Peltephilus* or horned gopher *Epigaulus* (Scott 1937). It appears that where capable incisors or canines were present, or where limbs terminated in large claws, they were secondarily functional as effective weapons and were only rarely enlarged for combat purposes.

Finally, we must briefly consider the evolution of weapons over time in several lineages of mammals. Where separate organs of aggression and defence do exist, they show a progressive enlargement and complexity, provided the weapons function to parry, block, and hold onto an opponent during wrestling. As we go from the middle Tertiary to the Pleistocene, or from the tropics to the arctic, or from the ecological generalist to the specialist, we find this trend. I investigated this primarily on caprids (Geist 1971a) and old world deer (Geist 1971b), but I also showed that sufficient evidence exists to identify it in other animal lineages (Geist 1977). I explained by means of the "dispersal theory" how the external appearance, including weapons, changes as a consequence of dispersal, and I showed how this process differed in the glaciated zones from the tropical areas. In general, the more specialized a species ecologically the greater the likelihood of its weapons complexifying, its combat specializing, its display organs becoming more ornate, and its body size enlarging. Since ecological specialists are more likely to be subject to extinction than are generalists, it is not surprising why the boneyards of paleontology feature so many bizarre giants, while some tropical small-bodied mammals appear to have survived for a very long time indeed. Nor is it surprising that frequent major ecological changes are a great stimulus to the evolution of bizarre giants such as flourished or still flourish in the Pleistocene and the present.

Circumstances and Aggression: Some Rules

The subject of aggression is complex, since so very many environmental factors affect it. Overt aggression is subject to aversive stimuli, pain, frustration, novelty, isolation, crowding, the appearance of strangers, unnatural environments, nutrition, environmental quality, and prenatal and postnatal factors affecting ontogeny (Collias, 1944, Calhoun 1963, Scott 1958, 1966, 1967b, Weltman et al. 1966, Southwick 1970a, Ginsburg 1968, Crook 1968b, Welch and Welch 1970, Archer 1970, Geist 1971a, Petocz 1973, Shackleton 1973, Halas et al. 1975). In attempting to order these factors, it has also become apparent that species with excellent morphological defences will respond quite differently from species with injurious combat to similar conditions. Thus Petocz (1973) found that mountain sheep and goats in the same area under the same climatic conditions in winter showed opposite responses in their overt aggression to the same environmental factors.

In general, the following appears valid. When the population experiences an environmental optimum, by virtue of relatively low density and abundant food and cover, species in which overt aggression is highly injurious will show a minimum of aggressive behavior, while species in which overt aggression is rarely injurious will show a maximum of aggressive behavior. Thus, when populations of voles, primates, or mountain goats are expanding or at low density in excellent habitat we will expect little aggressive behavior, while for the males of most antelope, gazelle, sheep, goats, and deer we expect much aggressive behavior. When nutrients and energy can be readily replaced, it obviously is to the advantage of individuals to inflict defeats on actual or potential competitors that they will remember, and the memory of which will impair their competitive ability at some time in the future. When the physical and social environment imposes stress, we expect a rise in overt aggression in species with injurious combat and a drop in those with noninjurious combat; in the latter the cost of inflicting significant pain is very high and probably too costly in relation to any benefits obtained. However, it was found by Petocz (1973) that, although overt aggression did decline in mountain sheep males with increased severity of wintering conditions, the number of sharp butts and jabs associated with competition over feeding sites rose.

The foregoing emphasizes that, in species with noninjurious combat due to excellent defences, one can recognize three kinds of fighting. One kind functions to settle the dominance ranks; the second is of playful nature, as in sheep, ibex, or deer, and may serve to maintain rank hierarchies; the third settles priority to food, salt, or water whenever such competition arises, and is in the nature of a short sharp fight.

In species with injurious overt aggression, the frequency of aggression is apparently an inverted U-shaped function of the deviation from the optimum ontogenetic environment, as well as the familiar environment experienced by the individual. Thus, overt aggression would be lowest in individuals raised or living at the environmental optimum, and again when the environment deviated drastically from the optimum, so that either a behavioral sink (Calhoun 1962) is created or individuals simply ignore the stimulus that normally would spark attack. Under conditions of abundant resources, aggressive behavior is simply a cost charged against the reproductive cost of the individual; in conditions of resource scarcity, the individual husbands resources it cannot readily replace and uses strategies other than aggression to maximize reproductive fitness. Behavioral sinks develop because of food deprivation (Southwick 1970b) or severe crowding (Calhoun 1962, Crowcroft 1966, Craig and Bhagwat 1974). However, not all reduction of aggression with crowding can be ascribed to a behavioral sink (see Welch and Welch 1970). Increases in aggression occur among children as group size increases above the optimum (see Chapter 14), so that they act much as other vertebrates studied do (Hutt and Vaizey 1966).

Another example of aggression triggered by stimuli signaling deviation from optimum is overt aggression triggered against individuals that act "abnormally," be it due to injury, disease, or congenital deformities. Here the conspecific remains inconspicuous by virtue of conforming in its behavior to the norm experienced by all. Once it deviates, it may panic conspecifics or cause them to attack (see Chapter 2). In its refined form, the discrepancy between a normal familiar individual and a normal but unfamiliar one causes attacks on the unfamiliar one. The common observation that individuals or groups trespassing into each other's territory enter into conflict is a special case of the rule that deviation from the environmental optimum increases aggression but decreases it if the deviation goes very far. This is because territorial defence breaks down under increased challenge as population density increases (Southwick 1970b). In essence, that which increases stress, up to a point also increases aggression, so that the latter becomes a nonspecific response to a diversity of factors. Note, for instance, the very high frequency of injuries and painful diseases found in natural populations of wild primates, as discussed by Schultz (1961); these surely must keep many individuals cantankerous and ready to assault at the slightest provocation. It may be noticed that the concepts just discussed fit closely with the "discrepancy principle" treated under pattern matching (Chapter 2). Thus, empirical evidence from the study of aggressive behavior supports the concept of the discrepancy principle developed from very different data and theoretical bases.

Although the frequency of overt aggression appears to be an inverted U-shaped function on environmental quality, a good case can be made that it is a function of the nutritional state of the animal. Riney (1954) noted that starving red deer were irritable. In a poor environment, the individual's nutritional state may be quite poor owing to excessive energy and nutrient expenditures and inability to feed adequately despite an abundant food supply; conversely, an individual may be fattening on seasonally superabundant food and yet be constantly on the look-out for food and "hungry," owing to rapid storage of energy and nutrients liberated during digestion. The individual's nutritional state, I hypothesize, is linked to the neuronal mechanisms controlling overt unritualized aggression in such a fashion that it is lowest when the individual is in a good nutritional state and again when it is in a very poor nutritional state. It ought to be more gregarious, therefore, when its demands for forage are low, since it is least likely to be aggressive. This simple hypothesis explains a diversity of phenomena.

Caribou ought to be least gregarious in summer when their nutritional demands are very high, and this is found (Bergerud 1974). Moose ought to be most frequently solitary when food shortages exist, as during winter; cow moose, because of their high nutritional demand when pregnant and lactating, ought to be less gregarious than barren cows and more aggressive; bull moose ought to be least gregarious in summer when fattening and growing antlers and they should be more gregarious as antlers cease to grow and their food intakes declines, and again after the rut when their rut antagonism is decreased, and they should progressively become more solitary as winter progresses and forage becomes scarce but should form social groups in spring and early summer when nutrition is excellent and their antlers have barely begun to grow. All this is indeed found (Geist 1963, Dodds 1958). Barren ewes ought to be more gregarious than those in late pregnancy; they are (Geist 1971b). Agression of the unritualized kind ought to increase during inclement winter weather in sheep, and it does (Geist 1971a, Petocz 1973). Aggressive individuals ought to raise fewer young and of lower quality than nonaggressive individuals, and this is found to be valid for voles (Krebs et al. 1973) and red grouse (Watson and Moss 1972), but has not yet been verified for ungulates.

Thus it appears that somehow nutrition can be translated into levels of overt aggression, which in turn can act to disperse individuals, thereby controlling group size and density. Not only aggression, but aversion, could be subject to a similar mechanism; poorly "fueled" or fed individuals are least likely to be gregarious, explaining why starving or ill individuals prefer seclusion to company. One can thus explain the correlation between food availability and density of individuals on the basis of behavioral mechanisms, albeit due to different ones from those postulated by Wynne-Edwards (1962). He visualized agonistic behavior as a means to reduce excess energy taken in by individuals in the form of behavioral conventions; I see agonistic behavior and aversion as a consequence of the nutritional state of the individual.

The influence of environment on the occurrence of aggression is exceedingly great. Even in species such as mice, in which the motor patterns of fighting develop innately without experience of fighting, as in classical deprivation experiments—a fact emphasized by Welch and Welch (1970) in their extensive studies of aggression in mice—attacks on conspecifics can still be regarded as a *habit*. To form the habit of attacking conspecifics a reward is necessary—the reward of winning. Thus, mice that attack do so because they have learned to attack, and mice responding to "frustration" with attack do so because they are in the habit of doing so (Scott 1958), since "frustration" can lead to behaviors other than attack in mice with different backgrounds.

This, of course, raises the question as to why chasing an opponent away should be reinforcing, and the conclusion appears to be that any possibility of performing an action without being blocked and frustrated is reinforcing. This is suggested by the concepts of pattern matching and emotionality discussed earlier (see Chapter 2). Thus, after the fight the winning mouse can move wherever it "decides" without interference. Thus, using the theory of pattern matching and emotionality, we can state that the expectancy of the mouse and the experienced execution of the behavior matched—acting as a reward (Hinde 1970, p. 594). The reward of pattern matching is reduction in arousal if arousal is high, as one expects it to be shortly after a fight (Berlyne 1966). Secondly, the mouse, in moving at the defeated opponent, sees it withdraw. This it experiences as the opposite of "looming" (see Chapter 2). Looming causes severe arousal; the antithesis to looming should cause a reduction in arousal. The victorious mouse should notice this and form "expectations" about its move at the opponent. It would test the expectations at once. In reaping rewards, once through pattern matching and the concomitant reduction in arousal and once through arousal reduction by the opponent's moving away (the antithesis to looming), the naive inexperienced mouse links its aggression with arousal reduction and thus with reward. This little thought exercise simply illustrates that we need to assume only very general and basic innate neural processes in explaining why attack can be reinforcing to a naive mouse.

The foregoing explanation matches well with the finding that in human subjects the opportunity to perform overt aggression against those individuals causing frustration is accompanied by a reduction in blood pressure (which I interpret as a reduction in arousal due to a parasympathetic internal stimulus), provided the experimental subject believed the acts of overt aggression caused pain and discomfort to the person causing the frustration (Hokanson and Shetler 1961). The performance of overt aggression should thus act to reinforce overt aggression if it is perceived as successful. It is evident that where the subject experienced the expected response from his tormentor, where "pattern matching" occurred, a reduction in arousal was triggered. Further work by Hokansen and Burgess (1962) showed that blood pressure, which rose in response to verbal insult, dropped if the subject responded in kind; verbal and overt aggression thus had similar physiological effects. Yet responses by the subject were not necessary to lower blood pressure; if its blood pressure were raised by viewing violence on film, all that it had to do was to watch someone receive just punishment. Williams and Eichelman (1971) found that rats experiencing electric shocks in pairs, and able to attack and fight upon being shocked, experienced a drop in blood pressure. The blood pressure returned to normal levels within four hours. Rats shocked singly, however, retained their high blood pressure. Fighting may, therefore, indeed reduce "tension" and relieve "frustration."

Contrary to the case of such species as domestic mice and rats (Scott 1966), we must postulate for some ungulates *appetitive* behavior for aggression and an internal reward system for aggression, so that the execution of combat movements serves as a consumatory act. One cannot term the behavior with which a mule deer buck solicits sparring matches anything except appetitive behavior (Geist, in press 1978d). The same can be said of rams racing to an erosion site prior to long-lasting huddles and clashing sprees (Geist 1971a). In the absence of sparring partners that are cooperating in sparring matches, mule deer bucks horn shrubs so extensively and vigorously that the exercise may last up to 20 minutes. In relict sheep populations with a low density of rams, large rams may "attack" cliff walls and heavy trees (Welles and Welles 1961), a phenomenon only exceptionally observed where many rams are present. This is very circumstantial evidence indeed, but it leaves us no option but to assume that fighting is a "pleasurable" experience sought out for its own sake. The reward may be not only arousal reduction, but also a stimulation of sexual rewards, since in ungulates erection accompanies aggressive behavior in many species as well as other aspects of sexual behavior, such as mounting and ejaculation (Buecher and Schloeth 1965, Geist 1971a). It is also evident why practice in fighting is so very important to these ungulates: Granted the short rutting seasons and intense competition of males prior to the rut, then the experience gained in judging opponents in fighting, plus developing the appropriate musculature through exercise, must all be enormously adaptive to small and large alike; to the former in extended life expectancy through the conservation of energy, and to the latter in increased rate of mating at minimum energy cost (see Geist 1971a, 1974a).

The foregoing indicates that overt aggression can take very different forms in different species, and that familiarity with groups of species in which overt aggression is damaging and rare, such as rodents or primates, may lead to a different understanding of overt aggression than would a familiarity with ungulate species in which excellent morphological and behavioral defences protest the fighters. There are human parallels: Heavily protected armored knights were also free to exercise their weapons on each other in fighting rituals and jousting. From my reading of the European hero sagas, it appears that combatants were more likely to sink to the ground from exhaustion than from loss of blood.

The frequency of overt aggression is, of course, also a function of hormones, in particular, sex hormone secretion, as this is controlled through environmental stimuli. This subject underlines that neural structures that control aggression are subject to hormonal influences, and that aggression is quite easily triggered at appropriate seasons, and less so at others (Hinde 1970, Marler and Hamilton 1966, Lincoln 1972, Wilson 1975).

There is a curious link between overt aggression and sexual behavior, as I indicated previously (Geist 1971a). We find, for instance, that in some cervids and bovids females in heat not only behave aggressively toward other females, or even toward small males, but may also act as if displaying to or attacking the breeding male. The latter appears as a part of the female's courtship behavior and may function in arousing the male. In old world deer and primates, dominance displays of males may incorporate a showy erect phallus, a link between dominance and sexuality. In mountain sheep, I showed that the levels of overt aggressive and sexual behavior were linked during the maturation of the males (Geist 1971a); when overt aggression was high so was overt sexual behavior, and vice versa. Since sexual behavior can normally only be performed after successful competition against conspecifics of equal or near-equal status—a condition that may even apply on rare occasions to females in polygynous ungulates—it appears logical that an individual prepared to copulate must also be prepared to do combat. Thus neural mechanisms controlling sexual and agonistic behavior may have to be equally susceptible to any given level of arousal.

The development of overt aggression or violence in humans has been well reviewed by Hinde (1974, pp. 288−292). I refer the reader to this review. The subject of human aggression, however, cannot be understood from a study of violence; one must also consider aggressive strategies, threats, and the rich indirect means of aggression, such as dominance displays and self-assertion. These will be discussed in the next chapter.

Chapter 5

Dominance Displays: The Biology of Art, Pride—and Materialism

Dominance Displays: Their *Raison d'Etre*

Although overt aggression is the ultimate means of forcing access to scarce resources essential to reproduction, it is a costly and dangerous means. It not only uses precious energy and nutrients at a very high rate, but also risks the loss of dominance, wounding and secondary infections, as well as death of the aggressor. In some situations overt aggression would be futile. Thus the minimum number of agonistic interactions increases exponentially with group size, in order to have a dominance hierarchy based on individual recognition ($x = (n^2 - n)/2$, where $x =$ number of fights, $n =$ size of group). Clearly, one soon reaches a group size in which individuals do little more than fight. The laws of least effort and maintenance of homeostasis dictate that a cheaper means than fighting be found which has the same effects, but not the same consequences, as overt aggression. The same argument applies to agonistic behavior in large-bodied, long-lived species, as compared to small-bodied, short-lived ones, or K-selected against r-selected species (Geist 1974a). If the reproductive effort of individuals must be spaced over very long periods of time, then overt fighting will be reduced to a minimum per unit of time and replaced by a cheaper substitute. The cheaper substitute for overt aggression can be identified as the diverse showy, and very common, dominance displays and the strategies with which these displays are used; in our own species these displays appear to be the root of art and pride, maybe of culture itself. That displays save energy was suggested quite early in the study of animal behavior (Collias 1944), but I am not aware that this has been verified to date.

However, the reduction in resource expenditures to achieve the very aims of overt aggression are not the only reason for the evolution of dominance displays. By displaying and preventing the displays of others, a male advertises that he is the most dominant. It is adaptive for a female to select such an individual for insemination, since she confers on her male offspring the genes of this successful male and increases their chances of becoming dominant males in their turn (Orians 1969).

Hence, dominance displays would be selected for by females, as well as by males, owing to their intimidating effect on the less competent. We may ask, further, what benefits does a dominant displaying father pass on to his daughters, compared to one who is a poor competitor in the sexual realm? There are a number of benefits. First, the most dominant male has not only competed successfully with other males for resources but also spared a relatively greater share from maintenance for reproduction. This may be due not only to a superior strategy of passive competition or active competition over resources but also to an efficient physiological machinery that required relatively less resources for maintenance. Thus, enhanced expenditure on, say, heavy elaborate antlers and time spent displaying betrays indirectly competitive superiority. Females, by "falling for" such males, ensure for their daughters the very same superior mechanisms for competition and maintenance of homeostasis (Geist 1977).

A male that is very vigorous and spends his fat resources freely during the rutting season increases his chances of death during winter, be there predator present or not (e.g., saiga antelope, Bannikov et al 1961; chamois, Knaus and Schröder 1975). A male that acts vigorously in front of females, that has large antlers, and that has survived a number of years is obviously competent to escape predators even if exhausted after the rut, and has an excellent capability for foraging during the season of reduced forage availability, namely winter. In "falling for" the vigorous large-antlered stag, a female ensures for her offspring superior abilities in both escaping predators and foraging during winter. Thus the large-antlered, but calm, stag ought not to be as attractive to females as the large-antlered and very active one. The males become the testing ground of antipredator and foraging adaptations. Dominance displays can thus enter into courtship, permitting females a choice and an opportunity to select the male with the highest potential reproductive fitness.

In classical ethology, dominance displays have not been identified as displays in their own right, but rather as hybrids of conflicting motivations to attack and escape (Hinde 1970). They were thus not seen as displays of aggression with an evolutionary history of their own, with their own rules, but as impurities in the ethogram of animals. Students of large mammals, however, and particularly those concerned with ungulates, saw it differently. They argued early that dominance displays were displays in their own right, and that the predictions of the "motivation theory of displays" did not hold (Walther 1960, 1974, Geist 1965, 1971a). Thus a change in the classification of agonistic behavior permits one to adopt a new view of its organization.

Characteristics of Dominance Displays

The following deals with the characteristics of dominance displays in mammals abstracted from many studies. These characteristics apply also to the dominance displays of reptiles, where descriptions permit one to identify dominance displays, and probably to all vertebrates. If we want to identify dominance displays, we look for the following characteristics.

(1) They are performed mainly by the larger (or dominant) toward the smaller (or subordinate). The separation is not absolute, as can be seen from my statistical data for mountain sheep (Geist 1968a, 1971a); in some species,

however, like the mountain goat, subordinates may display exceedingly rarely. Dominants may punish displaying subordinates and thus terminate their displays.

(2) Displays can be differentiated from threats in that the latter show a species-specific weapon system in the initial stages or preparation for use. Weapons may play a very great role in dominance displays, but they are not shown in the intention of use (excepting some primitive species that combine threat and display elements, as I shall discuss below). Weapons are displayed in such a fashion as to be clearly seen by the opponent, and are displayed at their maximum size (Geist 1965, 1966a).

(3) In dominance displays in which body size is maximized, apparent size is enhanced by the erection of hairs, by rising to a maximal height, by a broadside orientation of the displayer to the opponent, and by the raising of the head and tail so as to increase apparent height or length (as in the yak, for instance), but also by the use of a *diversity of attention-guiding adaptations* which function to guide the onlooker's eye over a maximum body surface. This I shall discuss in detail below. By far the most common dominance display in mammals entails a broadside orientation of the displayer to the opponent, which often leads to a parallel orientation of the opponents if both display. The "faking" of larger than actual body size probably has roots in the following principle: The mass of an animal increases as the cube of its linear dimension, so that a small increase in linear size greatly increases the mass and weight of an animal. Thus, an increase in the linear dimension of only 10% leads to a nearly 30% increase in mass; an increase of 20% in the linear dimension leads to a doubling of body mass. This makes it evident that individuals ought to be sensitive to size differences in conspecifics, since small increases in external size are associated with disproportionately greater strength or power. Power happens to be proportional to mass (Thompson 1961).

(4) Dominance displays are characterized by being relatively rare among the normal everyday behaviors, although they may be common social behaviors. They are conspicuous by their unusual body postures, sounds, odors, and, in particular, unusual movement patterns. Movements may be unusually slow, as in most dominance displays (e.g., mountain goat, Geist 1965; lion, Schaller 1972), unusually fast, as in red deer or wapiti (Geist 1966c), or of peculiar rhythms, as in the "parade march" of the fallow deer when herding his harem. Thus *rarity* or *novelty* is used as a means of attracting attention (see Chapter 3, p. 55).

(5) In dominance displays, the displayer usually approaches his opponent at a *tangent*, not directly as in a threat.

(6) In dominance displays the dominant displayer tends to avert his eyes from the subordinate; he acts as if looking away or ignoring the subordinate. This is clearly seen in the long-necked ungulates, and less clearly in small, short-necked mammals such as the Norway rat, but the element is present just the same. On closer observation one will notice the dominant glancing at the subordinate or looking at him, provided the latter is not looking at the dominant. The displaying animal acts as if he were calm and unexcited despite the threats of the subordinate. The displayer retains "composure";

he acts as if "fearless." The same quality is not only valued in humans of high rank, but the show of fear may be punished, even with the death of an individual in primitive tribes (see Henry 1970, p. 134). We can consider this attribute of dominance displays among animals as a deception, denying the opponent any evidence as to the state of arousal of the displayer, and thus denying him any evidence of his effectiveness in threatening. The opponent's aggressive behavior thus results in no feedback and no reinforcement of the aggression.

We can understand the significance of this attribute of dominance displays from the *theory of helplessness* (Seligman 1975) that I have alluded to repeatedly. The dominant, by *not* responding to the subordinate's threat, teaches him that threat is in vain, and teaches him to be helpless. The threat of violence will not rouse the dominant, yet the dominant may attack at unpredictable intervals. A subordinate so treated can only regain reinforcement from his own threat by attacking his subordinates repeatedly, which leads to the well-known phenomenon of redirected attacks. The more a dominant can teach a subordinate the futility of challenge, the more helpless the subordinate, the less his motivation to rebel, and the more likely the chances of morbidity and mortality in the subordinate. For this reason, the dominant gains by displaying often, reinforcing in the subordinate again and again the experience of futility of threats.

(7) Dominance displays include acts of vigor and weapon play executed by the larger in the vicinity of the smaller, but not obviously directed at the smaller. This includes, in bison and cattle, wallowing, pawing, and horning of soil so that clouds of dust arise; in deer, it includes horning of shrubs and trees, which may be literally shredded.

(8) Dominance displays usually contain no element of iconic communication; they contain no intention movements and one cannot predict when and whether dominance displays will switch to overt aggression, grazing, or other behavior. Note in this context the aversion of eyes, the tangential rather than direct approach, the often slow rather than quick approach, the averting of weapons. And yet the dominant displayer closes distance, in contradiction to the aversion of weapons, or looking away, or the apparent direction of his movements. With some experience, one can tell, of course, approximately what is likely to happen, but this ability to predict is derived from associative learning and not from signals emitting from the displayer. There is thus a considerable amount of *uncertainty* about the course of action of a displaying dominant, owing to the contradictions of the components of the display. I believe it is valid to state that dominance displays aim at precisely that, namely, the generation of *arousal* in the subordinate owing to its inability to predict the outcome of a dominance approach by a dominant. Contradiction in communication can thus be used to generate "fear" in others and cause them to leave. Here ambiguity is seen as a means to disadvantage an opponent, parasitizing the cognitive pattern-matching apparatus to generate arousal.

(9) Dominance displays are generally performed by males.

(10) Dominance displays may be the most common behavior patterns performed by dominants.

(11) Dominance displays function both in aggression and courtship (p. 102, 86–87).

(12) Dominance displays precede dominance fights.

(13) Dominance displays tend to maximize sensory stimulation by filling all channels of communication (visual, auditory, olfactory, tactile), intensifying stimulation in one or more channels, and using attention-catching and -holding mechanisms, thus probably preventing habituation by the onlooker; the displayers select vantage points that maximize signal transmission and evolve mechanisms that maximize signal transmission, as well as develop tactics—such as the temporal ordering of different dominance displays—that mutually reinforce each other's effect. A large stag not only displays visually but is noisy and marks frequently; olfactory marking, just like other displays, increases with status (Ralls 1971).

(14) In some species dominance displays are associated with phallic displays in the male. Examples are found among Old World and New World monkeys (Jolly 1972); phallic displays are not absent in human beings. We can regard these displays as signals of supreme confidence, and as such a symbol of a very competent dangerous opponent. A male ready to copulate is likely to defend a female far more vigorously than a male that is not. Thus, an erection in the dominance display signals that state of readiness for exertion and sacrifice; it is symbolic of a totally fearless, determined, ruthless opponent. If males interact aggressively about estrous females, then by associative learning young males ought to associate erected phalli and the concomitant scent with the attacks, fear, and pain they suffered on such occasions. Clearly, phallic displays can evolve fully only in polygynous species of mammals, in which copulations occur in groups and are readily visible to all.

From the foregoing it should be evident that dominance displays are showy and common, and they generate arousal through uncertainty, as well as by stimuli that facilitate associative learning and are difficult to "understand" except by the initiated. It is this latter characteristic, a consequence of the absence of iconic components and the tendency of the displayer to "look away," that has caused death and injury to zoo wardens who failed to grasp that the peculiar antics of the ruminant males were directed at them and could easily snap into quick attacks. Whereas threats are easily understood across the species boundary, dominance displays are not.

At this point we must look at a very important attribute of displays and social communication, namely, the principles of *attention guidance*.

Attention-Guiding Organs: The Biology of Art

Some time ago, I pointed out explicitly (Geist 1966a), and Walther (1966) implicitly, that the coat patterns of some ungulates incorporated the very principles of attention fixation and guidance used by artists. This idea was unfortunately not developed further by someone better trained in art and aesthetics than myself, and for this reason I shall develop it here in some detail. In addition, we shall note that the "designs" in ungulates or large carnivores harmonize well with the principles of gestalt psychology, as summarized by Katz (1948, see also Elkin 1975), with some ethological generalizations (e.g., law of heterogenous summation), as well as with physiological investigations of perception. However, one cannot recognize that principles of attention guidance are involved unless one knows the characteristics of dominance displays as I have listed them, and assumes that displays are purposeful, that is, that displays aim at maximizing apparent body size, weapon size, etc. Some of the principles of attention fixation and guidance used by artists in painting are described by Munro (1970) in his introduction to aesthetic morphology entitled *Form and Style in the Arts,* as well as in Arnheim's (1966) book which discusses aspects of gestalt psychology in relation to art. The principles I discuss below are stated more often implicitly than explicitly by the above authors. Nevertheless, we have here a clear link between ethology and the scientific branch of aesthetics, aesthetic morphology.

A SINGLE SPOT WITH SHARP BOUNDARIES CONTRASTING WITH THE BACKGROUND FIXES ATTENTION ONTO ITSELF. Thus the eye wanders to the spot and returns to it during scanning. Such single spots evolved on the wings of butterflies, and function as predator deflector adaptations. That is, birds peck at these conspicuous spots rather than the camouflaged body of the butterfly (see Tinbergen 1958). This principle is illustrated in Figure 5-1.

THE VANISHING POINT ATTRACTS ATTENTION. Intersecting lines or parallel lines leading to infinity produce a "vanishing point." The attention of the onlooker is focused at the point where the lines intersect. One can, of course, increase the attention of a subject by combining a contrasting spot with a vanishing point. Figure 5-2 illustrates an example from nature. We are again looking at what is obviously a sophisticated predator deflection mechanism in which vanishing points aim at the contrasting dummy head.

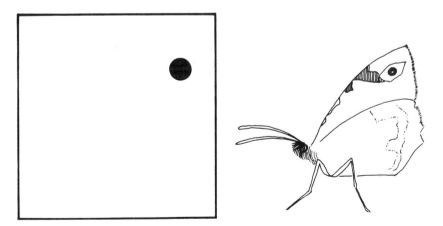

Figure 5-1.   The simplest manner of capturing attention within a picture plane, according to artistic theory, is to place a sharp-edged, contrasting spot within it. We see this principle of attention guidance illustrated by the Grayling butterfly (*Eumenis semele*)—the dark wing spot acts as a deflection mark for predators (Tinbergen 1958, p. 166).

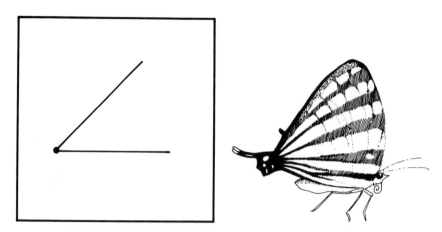

Figure 5-2.   Attention focuses on the intersection of two lines, known as a vanishing point. Its effectiveness is enhanced by placing a small dot on the intercept. This principle is illustrated by the tropical butterfly (*Thecla togarna*)—the vanishing point focuses on a fake head and antennae (Wickler 1968 p. 77).

SPOTS IN CLOSE PROXIMITY TO ONE ANOTHER LET THE EYE JUMP FROM SPOT TO SPOT, CON-NECTING THE SPOTS INTO A FORM. BY ARRANGING FOCI OF INTEREST, ATTENTION CAN BE DRAWN TO A "GESTALT" FORMED BY SPOTS.   It is this mechanism that is most fruitful in explaining markings on many social ungulates, as I shall show shortly (Figure 5-3).

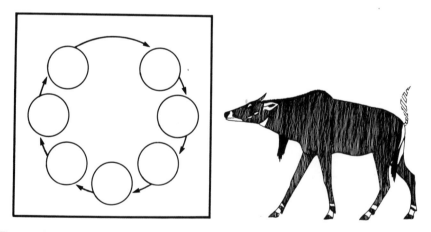

Figure 5-3.   Attention can be guided within the picture plane by objects forming a *gestalt*; the eye jumps the interval between objects (or foci of attention) and remains within the picture plane. This principle is illustrated by the dominance display of the Indian nilgai antelope bull (*Boselaphus tragocamelus*); the foci of attention are the contrasting spots on the nose, cheeks, throat and legs, and the "turkey beard" on the throat. A periodic flick of the tail tears the attention from the head and makes it rotate over the whole broadside (and mass!) of the bull (after Walther 1966 p. 130).

Figure 5-4.   Attention can be retained within a picture plane by framing it. This principle is illustrated by the dominance display—a broadside display as in the nilgai—of the oryx antelope (*Oryx gazella*).

STRONG BORDERS, BOUNDEDNESS, OR DETACHMENT OF THE FORM FROM THE BACKGROUND, HOLD THE ATTENTION OF THE ONLOOKER AND GUIDE IT ALONG THE BORDER. Thus, ''framing'' holds attention on the picture plane; the eye wanders to the border of the picture plane and returns again to the center of the plane. This principle is illustrated in Figure 5-4. In addition, the width of the border framing the body can generate an illusion about the size of the body; I am referring here to the Delboeuf illusion (Day 1972). A strong border surrounding an object creates the illusion that the body is larger than it actually is. This illusion is based on the perception of distance cues; a strong, thin border around a circle is perceived as nearer, and hence larger, than a wide border around the same body.

TO DISTRIBUTE ''ATTENTION'' OVER THE WHOLE PICTURE PLANE, THE PICTURE PLANE MUST BE ''BALANCED.'' Thus, if it is indeed the function of the broadside dominance display to maximize the apparent size of the animal, it is imperative that the onlooker's eye wander over the *whole* body of the displayer. In order to achieve this, there must be attention-catching mechanisms all round the body, or attention must oscillate between the extremes of body length. If the latter is the case, attention-catching mechanisms of *equal* effectiveness must be at each end of the body. Therefore the body plane must be symmetrical in its distribution of attention-guiding mechanisms. Figure 5-5 illustrates this concept—in part, the case of the broadside display of the markhor (*Capra falconeri*). Note that the horns act as *vanishing points* and draw attention from the body; the body plane is unbalanced by the large dark horns on the head; ''balance'' is restored by the *wiggling of the tail*. It should be noted also that *C. falconeri* is the only goat with a well-developed broadside display—and a tail wiggle; the domestic goat still has vestiges of a broadside display—and a simultaneous tail wiggle. Furthermore, the raised tail that is quivered, jerked, bobbed, etc., is part of all the broadside displays in ungulates known to me, except for the mountain goat.

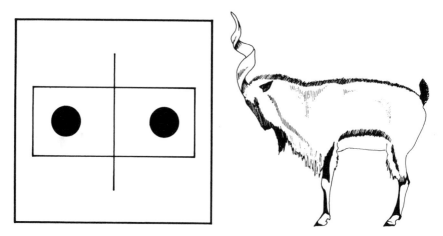

Figure 5-5.   Unless the picture plane is balanced, the onlooker's attention neglects a part of the picture plane. If the picture plane presented to an opponent by a displaying ungulate were unbalanced, the opponent would fail to note part of the mass of the displayer. Here we see this principle in the markhor (*Capra falconeri*). The heavy head and horns deflect attention from the body. Attention is drawn from front to rear by the flicker of a quickly wagging tail. Thus "balance" is restored.

Bilateralism of displays, as well as color segregation, are also shown in the bowers of bower birds, in particularly striking fashion in that of the orange-crested gardener (*Amblyornis subalaris*). Here the structure of the bower, and the fruits, flowers, snail shells, and iridescent beetle carapaces, strike us as very "artistic" because of the exacting bilateral arrangement and segregation of objects. One also notes that the whole bower display is "framed," so that the viewer's attention is kept to the colorful and contrasting interior of the bower. Moreover, during the frequent restoration sessions (the bower is often damaged by pouring rain), the male repeatedly retreats a little from the bower, spends some time viewing it, and then returns to move a flower or shell elsewhere. The bowers of other bower birds also show clear evidence of bilateralism. I refer to Frisch (1974) for illustrations and further descriptions.

ATTENTION IS CAPTURED BY QUICK LOCALIZED MOVEMENTS AND BY CONTRASTING LINES RAPIDLY CROSSING ONE ANOTHER.   This mechanism is supported by physiological investigations into the functioning of brain cells in perception (Grüsser et al 1964, Hubel 1963, Mize and Murphy 1973). It explains at once the quivering or flicking tails in the displays of various ruminants and the dark stripes along the rump patches. Note that in the display of the impala (*Aepyceras melamphus*) (Schenkel 1966a) and in that of the Grant's gazelle (*Gazella granti*) (Walther 1965), the tail is flicked on the side on which the opponent stands; the tail cuts across the dark band along the rump patch (Figure 5-6).

EXTENSION OF BODY APPENDAGES INCREASES APPARENT SIZE OF THE BODY.   In itself this is not a principle of attention fixation, but it is functional only with the aid of attention-fixating mechanisms. Figure 5-7 illustrates this principle. Note that the

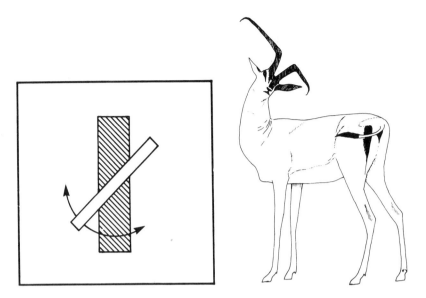

Figure 5-6.   A flicker may be created by wagging the tail (white) across a rump-patch border (black) as is illustrated by the dominance display of the Grant's gazelle (*Gazella granti*). Here attention bounces between the complexly structured head and the flickering tail (after Walther 1965, 1968 p. 67).

display makes use of a common optical illusion, the Müller-Leyer illusion (Pressey 1970), and thus parasitizes the rules of perception as much as the earlier attention-fixating rules.

In its classic form, the Müller-Leyer illusion increases apparent size some three- to fourfold; it would be folly, of course, to extrapolate this to, say, a displaying mule deer and expect the same three- to fourfold increase. No one knows at present how much or whether a subordinate mule deer experiences a size increase in the displaying dominant. The illusion is based on a sensory illusion of distance cues (Day 1972). The smaller the size of the terminal elements on the line (in Figure 5-7 the terminal elements are arrows) the longer or larger the apparent size of the line, since the small terminal elements give the illusion of greater distance. This suggests that the extension to the body, i.e., the tail, ought to be relatively *small* in a displaying ungulate in order to increase the apparent size of the body. A *large* tail ought to reduce the efficiency of the Müller-Leyer illusion because a large tail would create the illusion of nearness, and consequently the body would appear small. The foregoing does explain why the terminal tail brush in *Bison* or in mule deer is relatively small. As noted above, large tails, as in horses, gnu, or yak, may function to extend the apparent length and size of the body.

Let us now interpret the dominance broadside displays of the nilgai bull (*Boselaphus*), a peculiar antelope from India, as described by Walther (1966). Note in Figure 5-3 that the body, neck, and head form a vanishing point during the display. The *unusually* slow movements of the displayer are "unusual" compared to its normal walking movements and thus capture attention. The onlooker's eye should move to the head, as predicted by the vanishing point mechanism. Here

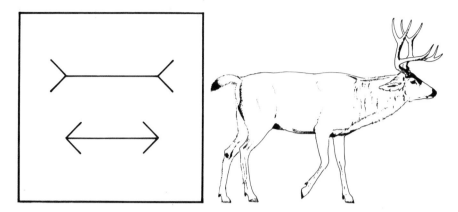

Figure 5-7.   Apparent body size is not only increased by focusing attention on the whole picture plane shown to the opponent in a dominance display, but also by the use of optical illusions. The displaying mule deer (*Odocoileus hemionus*) appears to take advantage of the Müller-Leyer illusion by extending head and tail linearly away from the body.

attention is held by the white spots contrasting with the bluish dark gray of the head. If attention were to remain fixed here it would defeat the purpose of the broadside display, since it would *distract* the onlooker from viewing the body mass of the displayer. Somehow, attention must be shifted from the head. This is achieved by the bull by quickly flashing the tail upward. At once the onlooker's attention jumps to the animal's rear in response to the quick motion. The tail, however, has disappeared from sight, since it is now held pressed down between the hind legs. The rounded lines of the rear permit no fixation of attention, so that attention jumps to the next attention-catching adaptation, the feet. Note the two clearly set-off white rings above the hooves. These remain visible under natural conditions, since the nilgai inhabits desert and scrub desert, areas with relatively little ground cover (Schaller 1967). During the display the legs move, and by the movement of contrasting, sharp borders, they should activate edge receptors in the onlooker's brain. Granted the large amount of black/white border and its distribution over about 1.5 meters, it ought to catch attention. Again, attention would be fixed on the legs, if it were not competing with the dark body mass above, and the jerking "turkey beard" on the nilgai bull's neck. Once it jumps to the turkey beard, the vanishing point principle quickly slides the onlooker's attention to the head, where it remains fixed until the tail flick jerks it to the back of the animal, and the cycle of attention guidance around the animal begins anew. Thus the picture plane of the broadside display of the nilgai bull is "balanced" between the vanishing point and contrasting dots on the head and the tail flick at the rear; the apparent size of the animal is increased by the extension of the head and the movement of attention to the legs, rather than to the belly line, which one would normally expect. The attention-fixating mechanisms are thus arranged to distribute attention *around* the displayer's body mass, to continuously rotate the onlooker's attention and not permit the onlooker's attention to escape from the displayer's body. These mechanisms are arranged around the body in a manner compatible with the interval principle. Note, the nilgai's display is explained by assuming that the principles of attention-fixation as elaborated by artists apply not only to humans but also to large mammals. We can put it a little more precisely; art is apparently based on mechanisms of perception that human beings share with other large mammals. In this form, the assumption is less startling, in fact it is reasonable.

Here we are probably dealing with mechanisms of perception such as are found in children, and which are termed "field effects" by students of perceptual development in children (Elkin 1975). These produce such phenomena as dominance of continuous line (note "framing"), formation of "good form" (note the interval principle), etc. Children have a perception early in their ontogeny dominated by field effects; it is likely that field effects dominate the perception of most mammals intensely. Note also that the explanation assumes that the function of the broadside display is to maximize the apparent size of the animal, and in so doing intimidate the opponent. Without these two assumptions, the broadside displays and the arrangement of the markings we find on ungulates remain unexplained.

Using the foregoing principles of attention fixation, we can explain most of the markings and concomitant behaviors of ungulates. The broadside display of the mountain goat (Figure 5-8) depends for symmetry on the hump of long stiff hair growing from the top of the rump; the raised, wagging, jerking, quivering, nodding tails in the broadside displays of mule deer, white-tailed deer, bison, cattle, kudus, markhor, and many gazelles are now seen as mechanisms to jerk attention away from the front and distribute it more evenly over the displayer's body; the many-tined antlers of caribou, red deer, elk, moose, or barasingha are now seen as organs that during displays create many edges, and thus maximally stimulate the edge receptors, not only fixing attention on the antlers, but also generating so massive a stimulation of edge receptors that this in itself becomes a message; in the caribou (Figure 5-9) we suspect that the balance of vanishing point and massive antlers makes for a frequent return of attention and "relooking" by the opponent. If this interpretation is correct, we witness a far more simple and elegant solution than that of the nilgai bull.

Figure 5-8.   Picture plane balance in the displaying mountain goat (*Oreamnos americanus*) depends on the rough symmetry of front and rear halves of the body. The display of this ungulate incorporates a weapon-threat in that the head is held low and away from the opponent in readiness to strike (Geist 1965).

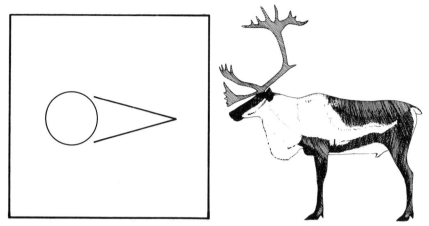

Figure 5-9.   Picture plane balance in the caribou (*Rangifer tarandus*) appears to be maintained by the contradiction of juxtaposed massive attention focus (head and antlers) and a vanishing point formed by the rearward extension of the neck and shoulder cape.

In principle, the above is testable experimentally. It predicts that an impairment of attention-fixing and -guiding adaptations leads to an increase in energy expenditure in maintaining social position and increases the chances of wounding, loss of status, and reduced reproduction.

Noting the high frequency of dominance displays and the ornateness and size of display organs in the form of hornlike organs, coat patterns, and color schemes, we can safely assume that they are a cheaper solution for capturing and maintaining high social rank than is overt aggression. Some of these organs appear ''excessive'' to us (Portmann 1959); I explain this excessiveness as a function of ecological conditions and the species' evolutionary history (p. 87).

Quantitative Rules in Dominance Displays

Dominance displays may be far and away the most common social behavior pattern of dominant males. A detailed analysis of such is found in Geist (1971a) for mountain sheep. Figures 3-1 to 3-3 illustrate the relative frequencies of behavior patterns in mountain sheep. From thirteen behavior patterns quantified, three are dominance displays; the fourth dominance display, the ''present,'' was not quantified by difficulty of identifying it at a distance. Therefore, the quantitative picture is biased in recording fewer dominance displays than actually took place. Granted even this bias, 89.2% ($N = 549$ recorded patterns) of the behavior patterns of dominant Stone's rams were dominance displays, and 73.9% ($N = 976$) of dominant bighorns. Shackleton (1973), using a slightly different and more detailed classification, obtained for bighorns much the same results; dominance displays made up 78% of the behavior patterns of dominant class IV rams (from Shackleton 1973, Figure 25), 70% and 76% of dominant class III rams from two populations (from Shackleton 1973, Figures 25, 29), and 72% of dominant class II rams (from Shackleton 1973, Figure 26). It is evident that dominance displays are far and away the most common behavior of dominants, but they are not absent from the repertoire of subordinates as Figure 3-2 shows. Shank (1972) recorded much the same for feral domestic goats, Schaller and Mirza (1974) for Punjab urials (*Ovis orientalis*).

As discussed earlier in Chapter 3, dominance displays and submissive behavior are inversely related in the interactions of dominant and subordinate. What one does frequently, the other does rarely. These behaviors are thus antithetical.

To date, little quantitative information on dominance displays is available. However, granted that dominance displays are the principal means of gaining the same objectives in strategies as overt aggression, but at a reduced cost, the following is clearly expected: Dominance displays ought to be elaborate, diverse, and frequent in societies in which males form subgroups that exploit areas or habitats different from females, in polygynous as against monogamous species, in species with open as against closed societies, in species with short reproductive seasons, in species with poor defences against the conspecific's weapons, and in species with a high as against a low threshold of arousal.

The relative frequency of dominance displays is also a function of the nutritional state of the animal. This conclusion is based on circumstantial evidence; mountain sheep males from high-quality populations—that is, those on a better forage base—tend to show more dominance displays (Geist 1971a, Shackleton 1973); males that in winter experience mild weather and good forage availability show more dominance displays than during conditions of deep snow and great cold (Petocz 1973).

Ontogeny of Dominance Displays

In polygynous ungulates, where dominance displays have mainly been studied to date, these behaviors are subject to maturation and appear relatively late in the male's ontogeny (Geist 1966a, 1968b, 1971a, Shank 1972, Shackleton 1973, Walther 1974). They appear first in young males at about six months of age, accelerate in relative frequency during the ages of one to four years, and continue to increase in relative frequency as long as the male continues growing. This may take seven to nine years. Accelerated growth accelerates the relative frequency of dominance displays (Shackleton 1973).

That dominance displays are subject to hormonal influence is indicated by their virtual absence in the behavior of females and prepuberal males. As was shown in mountain sheep, females and prepuberal males are exceedingly similar in behavior and body form; these similarities decline and vanish upon the sexual maturation of the male (Geist 1968b, 1971a).

Evolutionary Directions in Dominance Displays

We may ask, what is a primitive and what is an advanced dominance display? When we compare different lineages of mammals, it appears that the body display is more primitive than the weapon display. This is shown clearly in caprids (Geist 1971a), it is noticeable in the lineage of Old World deer (Geist 1971b), and it is evident in the comparison of primitive ungulates as compared with the higher evolved forms. The difficulty one encounters is that weapon displays are comparatively rare, and body displays are all too common.

The great frequency of body displays with a broadside orientation may be due to poor visual acuity in most mammals, excepting primates (Grzimek 1952, Backhaus 1959, Walls in G. Altmann 1966). Poor visual acuity may necessitate a massive superficial enlargement of the body, and grant little selective advantage, if any, to the refined and relatively small weapon displays. It may well be that an increase in visual acuity permits a reduction in size and conspicuousness of display organs, since small organs are now just as noticeable as larger ones and comply with the law of least effort. This hypothesis predicts increasingly subtle visual communication with visual acuity, which appears to be valid if one compares the complexity and range of visual signals of primates with those of ungulates or rodents. A detailed study to test this hypothesis is still outstanding.

A second sign of "primitiveness" of dominance displays is the incorporation of functional elements into the display. We may thus find threats embedded in the broadside display, that is, the holding of a weapon system ready for use as is illustrated by the mountain goat (Figure 5-8). Here the head is averted as in most dominance displays, but in this case the "aversion" is the posture of head and horns just before the strike at the belly of an opponent. The Norway rat incorporates at least two functional elements into its broadside display. First, this posture permits the rat to effectively push the opponent away with the hip, and second, the rat raises her hind leg and pushes the opponent away. Thus, the display is here a preparatory posture to overt aggressive behavior.[1] In the varanid lizards, the broadside display (Rotter 1963) may be interpreted as a preparatory position for tail lashing. The dominance display of the Thomson gazelle (Walther 1964b) consists of a frontal approach with elevated head held ready for a downward strike. Granted the primitiveness of this gazelle (Geist 1974b), the iconic nature of this display may not be surprising. The root of dominance displays, therefore, may well be threats that in the course of evolution became exaggerated, complexified, and finally obscured.

The evolutionary development of dominance displays can thus be conceived as follows.

(1) They begin as an elaboration of weapon threats that exaggerate by various means the intention movement to strike; they are thus an elaboration on iconic communication mechanisms signaling overt aggression. As indicated above, this is no more than a hypothesis which remains or falls, depending on the investigations of dominance displays in animals of close relationship.

(2) Next, dominance displays incorporate elements signaling large body size. Thus we find the broadside orientation, the attention-guiding organs, and the behaviors that increase apparent body size. In addition, one finds mechanisms reducing iconic signals, and hence predictability, of actions by the displayer; arousal through unpredictability now becomes one aim of the dominance display.

We expect species at these levels of dominance display development to use such displays rarely, since they are closely tied to overt aggression. It is different in species in which dominance displays function also as attractants to the female, as we shall see later.

[1] This can be clearly seen in the film *Ratopolis* by the National Film Board, Montreal.

(3) At the next level, feats of vigor are included in dominance displays, as well as feats of weapon play. The pawing of cattle bulls and bison bulls in displays sending up clouds of dust (Schloeth 1961, McHugh 1972) are examples of the former; horning of the ground by these species and by cervids are examples of the latter.

(4) From a broadside display, the dominance display may move in the direction of an elaborate weapon display, as I argued for the caprids (Geist 1966b, 1971a) and for various other ungulate lineages (Geist 1971b, 1974b). In these displays the weapons are not shown in intent of use; there are no iconic signals. Thus weapons are held ready neither for attack nor for defence. Various attention-fixating mechanisms enhance the weapons.

(5) Dominance displays may also incorporate the displays of sexual organs. We see such in relatively primitive form in the Uganda kob (Buechner and Schloeth 1965), an antelope that often defends territories on a lech and in which females choose the mating partner. We find penis erection as part of the dominance displays in many of the Old World deer (personal observation); we find it frequently and most elaborately in the cercopithecine primates (Kummer 1968, Wickler 1967, Jolly 1972). In the latter, the size of sexual organs is a function of dominance; the penis, scrotum, and pubic hairs are diagnostically colored; the male sexual displays are used in typical dominance interactions, although they are also used under circumstances that suggest that the use of such displays is exceedingly complex (Kummer 1968). These phallic displays appear to have their root in the lessons subordinate males can learn in close proximity to a male guarding and breeding a female. The erected penis now becomes a symbol of the fear and pain they experienced that was inflicted upon them by the breeding male (see p. 90).

(6) A third direction is the incorporation of learned behaviors into dominance displays. This is best illustrated by the observations of Goodall (1971) on chimpanzees, in which one male—Goliath—learned to use empty petrol cans as part of his dominance displays and in so doing rose to the pinnacle of the male dominance hierarchy. Since chimpanzee males readily use broken branches in their dominance displays, the use of empty, shiny, and (in use) very noisy, petrol cans is maybe not very surprising. It is this trend, however—the use of objects and learned behavior in dominance displays— that is of more than passing concern to us, since in humans it is the essence of dominance displays and reaches such elaboration, complexity, or subtlety that it is next to impossible to recognize it as dominance displays. More of that later.

(7) A fourth direction taken by dominance displays is the advertisement of territories; territoriality is hence a dominance display (Geist 1966a). At the most primitive level the territory is advertised by making the individual conspicuous in external appearance, in voice, and in its choice of locality to display from. In the most advanced forms, the territory is advertised indirectly; in the case of mammals by extensive scent deposition, in the case of birds by the structuring and decorating of territorial symbols, as best illustrated in bower birds (Gilliard 1963). The territorial male in the most advanced bower birds is a drab, inconspicuous individual, but the bower he builds is large, ornate, and complex. Thus we find in mammals an olfactory extension of the individual in time, in birds we find a visual one. We also find

conspicuous, elaborate vocal signals in some territorial species, and consider these vocal extensions of the individual.

Dominance displays appear to change functions and complexify under specific ecological conditions. For instance, it is evident that visual and olfactory extensions of the individual permit it to signal territoriality while remaining inconspicuous. We can predict that such territorial advertisement will evolve where longevity is selected for, which is usual for species exploiting stable climax plant communities. Species with short individual life expectancies are likely to have territorial advertisements in which individuals draw attention to themselves directly.

In their most primitive form, in which they are elaborate threats, dominance displays probably function only to repulse conspecifics in agonistic interactions. We expect such displays in relatively solitary forms, since in gregarious forms dominance displays can automatically function as rank displays and attract females. If females can compare several males, then those males that display while terminating the displays of others are the dominant males, and these should be the females' choice as a mate (Orians 1969).

The same argument applies, of course, to territorial forms. Here, however, the female choosing a territory holder ensures not only competent genes for her children, but often also—depending on species—the resources to raise the children with. Thus, ecological conditions selecting for gregariousness and territoriality also select for a dual function of dominance displays: To repulse males and to attract females. Here dominance displays can enter into courtship.

In social forms, contact between individuals is frequent. Dominance displays containing strong signals of aggression are likely to lead to more overt aggression than displays with a minimum of content signaling overt aggression. Hence, in social species selection will favor arbitrary dominance displays increasingly. Granted that such species live on climax plant communities with scarce resources, it becomes adaptive to replace displays costing much energy, such as behavioral displays, with cheaper means; I suspect the morphological displays are such. We thus expect in extremely social forms strong external markings and patterning, but little behavioral display. This harmonizes with Collias's (1944) concepts that the more colorful birds and reptiles the less they fight, and the drabber their coats the more likely they will be to resort to combat. From the above, kin-selected species are excepted for reasons articulated below.

Dominance displays based on showiness of weapons appear to be correlated with specialized combat; in ungulates they are associated with large hornlike organs, and these are largely confined to northern forms and a few ecological specialists in subtropical areas for reasons explained elsewhere (Geist 1974a,b).

In species that form closed groups, such as wolves, baboons, many cercopithecine primates, and many colonial birds, in which species individuals know each other well over a long period of time, particularly in kin-selected species that compete intraspecifically not as individuals but as demes, dominance displays are expected to atrophy. In such species, in which cooperation between individuals maximizes their number compatible with resources, strife between individuals is expected to be rare, and so would be dominance displays. Moreover, the costs invested in dominance displays would detract from the resources required to maximize the number of competent adults in the deme to defend resources held in common. Therefore, dominance displays would not be adaptive, since they ultimately reduce deme size. Under kin selection, displays enhancing overt aggression would be common, since demes fight demes as groups (Brown 1974).

In contrary manner, in species in which strangers mix frequently in open societies, dominance displays are expected to be common, albeit economical. Such is the case with many social ungulates. Here dominance displays serve as strong memory joggers, reminding others of the individual's combat potential. Whether a society is open or closed is a function of strategies of resource exploitation; kin-selected species are found mainly in the tropics and productive areas of the subtropics (Brown 1974) for reasons I have stated elsewhere (p. 228−229). The ecological conditions governing the evolution of territoriality and its concomitant effects on sexual dimorphism I have detailed elsewhere (Geist 1974a,b).

The appearance of dominance displays in which learned elements are incorporated seems to be related to tool using. It requires considerable ability to learn and transfer behavior from one context to the next. Little can be said about it beyond this.

Human Dominance Displays: Biological Basis of Displays and Their Cultural Enhancement

The human body carries externally a large number of features comparable to display organs in primates. Guthrie (1970) pointed to this unmistakable parallel. In all higher primates, including man, display organs cluster on the head; in addition, our agile faces have display properties (Andrew 1963). This correlates with the arboreal habits of many, and the arboreal origin of all, primates; on a branch the movements of antagonists are, of necessity, restricted and they are most likely to face each other in agonistic encounters. In addition, the extremely high visual acuity of primates (Walls in Altmann 1966) makes a concentration of displays on the small head and facial area possible. The cephalization of display organs can thus be regarded as a consequence of arboreal adaptations, at least in primates.

The display features on the head appear to be the beard, fleshy and tinted lips, eyebrows, a prominent nose, eyes with a white sclera, a prominent chin, frontal baldness, gray hair, conspicuous ears, and a mop of hair. I must emphasize "appear to be." There is no experimental evidence to my knowledge confirming that these structures function in displays, conveying any advantage to the bearer. However, this should not discourage one from probing; the display structures could be vestigial.

Guthrie (1970a) correctly emphasizes that, in addition to display organs about the head, we have display features on our body such as the flat, broad chest in males, hair on the chest in some races, and exceptionally large sexual organs; Wickler (1967) points to the display function of the buttocks. In addition, one may mention the broad shoulders, and the vestigial hair patterns on the upper arms and shoulders that indicate that once large hair tracts covered the shoulders, probably to produce natural epaulets (Leyhausen in Eibl-Eibesfeldt 1970, p. 28).

If the features just described are not organs of display, functional or vestigial, then the following is totally incongruous: For every "display organ" mentioned we find a cultural enhancement. The enhancement may take the form of the following.

(1) *Decorative alterations* of the organ itself, such as combing, cutting, and braiding of scalp hair, beard and moustache, and slitting, stretching, and decorative scarring of ear, nose, lips, and penis.

(2) *Structural enhancements,* such as epaulets to broaden and enlarge the shoulders, tassels, lanyards, buttons, medals, shields, sashes, and embroidery to decorate the chest; penis sheaths and decorative stitching on pants to decorate the inguinal region; hats and various headdresses to enhance height and size of head, corsets and spreading skirts to ehance the waist and buttock regions.

(3) *Chromatic enhancements,* such as the coloring of lips, eyelashes, eyelids, and eyebrows; coloration of the face and the chest with war paint.

(4) *Auditory and olfactory enhancement,* the former with noise-generating decorations that produce characteristic sounds during a person's movements, the latter by the application of various artificial scents to the body. As in all structural, chromatic, or decorative enhancement, so there is considerable variability in the individual means of auditory and olfactory enhancement.

(5) *Behavioral enhancements* of natural or culturally enhanced organs, appearing to add to the attention-gathering qualities of the display organs. We can recognize on one hand behaviors that are largely biological, that is, hereditory, and on the other hand behaviors that are largely acquired through tradition. Behaviors that are probably innate in the ethological sense, that is, not acquired and learned through traditions, are postures we associate with pride and dignity: Erection of the body to the full height, expansion of the chest, jutting of the chin, slow, calm, conspicuously controlled movements, display of calmness and composure, and a pointed belittling of danger by acting nonchalantly, as if ignoring the danger or the opponent. Movements enhancing displays that are culturally formed are body motions enhancing the conspicuousness of headdresses and ornaments, as well as ceremonial movements.

In addition to the cultural enhancement of natural human display organs, we also find *cultural substitution*. This applies primarily to the substitution of culturally formed weapons for biologically formed weapons. Our biological weapons, teeth, fists, and muscle power for wrestling are by and large not enhanced in displays, or they are enhanced very little at best. Our culturally formed weapons, however, are enhanced visually by ornamentation and appropriate showy display behavior, as well as being steeped in rituals in which they play a central role, or being exhibited in mock battle. The weapon is not used in threat, but its virtues are sung, and it is exhibited in simulated combat or on appropriate *ersatz objects*. Here human dominance displays are strictly comparable to animal dominance displays in which weapons are displayed, such as are found in some horned and antlered ruminants. Here too, during displays, weapons are not shown in any posture reminiscent of threat, but they are shown to enhance their apparent size and mass, and they are exhibited in use by shredding sod or vegetation (Geist 1966a, 1971a, 1975a).

In addition to weapon exercise, dominance displays of humans and of large, social mammals may incorporate *acts of vigor and skill*. In humans they may take the form of vigorous male dances, ritualized combat, and sporting competition in which individual feats of arms are displayed (Eibl-Eibesfeldt 1967); in large mammals they are exemplified by the acrobatics of the gnu (*Connachaetes*) (Estes 1969), or the sod horning of cattle (McHugh 1958, 1972, Schloeth 1961) and red deer

(Heck 1956, Bützler 1974), the crashing displays of Rhesus monkeys (Altmann 1962), or the "dances" and charging displays of chimpanzees (Goodall 1971) and gorillas (Schaller 1963).

The biological basis for these displays appears to be the finding that strength and agility are correlated (Berg 1973). We also know, as will be discussed in the next chapter, that vigor is associated with high growth rates, an excellent diet, scarcity of disease and parasitism, in short, with exceptional health and physical fitness. Therefore, displays of vigor and skills become symbolic of strength, power, and superior capability in fighting.

We also note that dominance displays are largely, but not entirely, confined to males in humans—at least those connected with physical prowess. This, too, harmonizes with observations made on a large number of mammalian species in which dominance displays are largely a prerogative of the males. We have thus every indication that we perform dominance displays at the biological and cultural level that in principle are very similar to those of other large mammals. Before we go on to a discussion of the cultural expansion of dominance displays in humans, we must look at a unique form of apparently innate dominance display in man—laughter.

Humor, Laughter, Courtesy

That laughter has an element of malice and aggression has been recognized repeatedly, in fact, at least since Aristotle (Koestler 1964). That it is a dominance display *par excellence* in certain situations, and as such a uniquely human form of a universal vertebrate behavior, is less appreciated. Quite contrary to Koestler (1964), laughter is not a "luxury reflex" void of adaptive significance. It is a behavior of great value to individuals, one that enhances reproductive fitness, albeit one that cannot be understood outside the context of dominance displays, the human social feedback system (p. 333), and the adaptive strategies of our species in its terminal evolutionary phase, as I outline in detail later in this book.

A short review of the physiology of laughter and its psychology is useful at this point. As Koestler (1964) points out, it is a very complex reflex that involves some 15 facial muscles, spasmodic contraction of the thoracic basket, and periodic glottal stoppages. It can be explained physiologically as a signal triggered by successful "pattern matching," in which the two patterns, apparently incongruous, are suddenly perceived as congruous, provided it is also perceived that one is superior to the situation or person laughed about. As a signal, laughter communicates pleasure on the part of the laugher. It is the pleasure of self-assertion, recognition of self-worth, superiority, mastery, dominance, or success. Laughter can be punishing, therefore, to those who are the objects of laughter, since it exposes their inferiority, lack of mastery, unworthiness, or failure. Laughter is also infectious, so that individuals perceiving the small or large failing in others join in if one person is laughing; laughter lowers the threshold to perceptions called humorous, so further laughter easily results. As such, it becomes a mechanism of human social bonding, since laughter is an intensely pleasurable experience, a reliever of tensions built up in the body for various reasons. In this context, the act of laughing probably quickly restores the body to normal homeostatic functions, after arousal has prepared it for exertion; without laughter it might be that arousal would be difficult to diffuse physiologically.

The malice in humor, wit, and laughter lies in the detection of errors, mistakes, failures, personal weaknesses, transgressions of codes, etc., that the one who laughs can laugh at. As such, laughter signals superiority. For instance, it erupts at the recognition that the supposed danger in reality is not (e.g., a growl behind one's back, which on quickly turning we discover came not from a dog but a mischievous friend; a puppy barking and attacking our shoelaces makes us laugh since the signaled danger does not exist and we are clearly masters of the situation). Koestler (1964) emphasizes that laughter is produced as a consequence of tickling if tickling comes as mock aggression and is recognized as such; children laugh some 15 times more frequently if their mother tickles rather than a stranger.

The pain inflicted by laughter appears to have its origin in the common attribute of mammalian display, namely, denying the opponent any evidence of success in agonistic encounters. We noted earlier that in dominance displays the opponents acted as if ignoring each other, or the dominant in full display failed to respond to the threats of the subordinate. The dominant continues to act calmly, unexcitedly, as if supremely confident. Laughter is, of course, associated with relaxation, confidence, being the very opposite of arousal. Laughter on the part of the dominant thus goes beyond the acting of calm unexcited behaviors. Laughter in response to threat thus emphasizes the futility of the opponent's aggression; it deflates and belittles. Of course, laughter varies. There is at one extreme the arrogant insulting laughter, and on the other the gentle smile of superiority, or the baby's laugh upon recognizing his mother's face, signaling the simple mastery of pattern matching. Generically these smiles and laughters are the same.

When looking for humor in the animal world, we need not go further than our next relation, the chimpanzee. Hediger (1955, pp. 134−35) describes their practical jokes. Humor arises here from the volition of seeing another creature suddenly disadvantaged, or its body out of control, and its consequent struggles to regain its original position and composure. That is, the point of the game is to "startle" someone, put him at a disadvantage, and see him flounder or struggle. The act of the prankster becomes in this context a dominance display, since it demonstrates indirect mastery over the opposite's behavior. In principle, there ought to be "pleasure" at seeing conspecific competitors disadvantaged, as well as concerted efforts made at putting competitors in that position. Thus the chimpanzees' practical jokes are indeed a very basic primitive humor; practical jokes can be identified with some justification as simian humor.

In the foregoing context, the practical joke becomes a mild but nevertheless successful form of dominance display, and an instrument of social control. As I have discussed later (p. 333) under The Social Feedback System (Chapter 13), the primates have a rather refined social feedback compared to other mammals and the practical joke or prank arises quite naturally as one of the rather sophisticated social behavior mechanisms. Ecologically it is almost certainly related to kin selection, in which competitors must remain together owing to the nature of their strategy of resource exploitation (p. 333).

If we care to pursue humor further, we find its roots in play. Humor is a social play, an apparently peaceful way of self-assertion or even aggression. Play is often the exercise of competence; we gain great pleasure from demonstrating competence to ourselves and others, albeit we do not laugh as a consequence. We confirm our self-worth or superiority when, for instance, repeatedly hitting a target with an arrow. We do such an act over and over again for the sheer pleasure of it. That developing mastery is excellent for our health we learn from the studies of Seligman (1975).

It remains to be explained why in humans laughter became a uniquely human form of behavior, that is, why it was selected for. Granted the foregoing allusion to complex social feedback in primates, it appears logical that laughter as an extreme expression of superiority ought to be found in the primate species with the most sophisticated social feedback (p. 333). Yet how did humor and laughter rise from the simian level? The answer may be as follows. In Chapter 10, I argue that the gracile hominid *Homo africanus* must have had some strongly kin-selected ancestor. The chimpanzee shows much evidence of kin selection; here only a primitive or weak humor can exist, since alienation by strong humor could lead to strife and even dispersal. This would run counter to the very essence of kin selection, namely, maximizing the relative number of adults per unit of resource in order to maximize the defence of the deme's resources. Under conditions of scattered resources and the requirement for greater mobility, such as hypothesized for *Homo africanus*, selection would favor greater sexual jealousy on the part of the males with each attempt to maximize copulations. This would harden sexual competition between males and select for more and better dominance displays. On one hand, the troop's existence no longer depends on maximizing the number of adults in the troop, and it is therefore to the reproductive advantage of the dominant to keep other males subordinate and to expel those most likely to usurp him. On the other hand, any factor decreasing troop size reduces its efficiency as an adaptation against predation, and thus it is in the interest of the dominant to keep subordinate males in the troop. Since the troop does offer safety and social opportunity, it is very much in the subordinate's interest to remain, and as pointed out earlier one means to achieve this on the part of the subordinate is to fake femaleness. Yet there is another mechanism.

In the case of gregariousness as an antipredator strategy, enhanced sexual competition by males, and the requirement to keep a large troop size to minimize predation, humor and laughter serve the interests of both the dominant and the subordinate male. Laughter by the dominant reinforces the subordinate's perception of his inferiority as a consequence of some prank pulled by the dominant. Laughing as an audible signal would therefore be adaptive for the dominant. Clowning by the subordinate, a mimicry of physical ineptitude, causing the dominant to laugh, would be adaptive for the subordinate since it would be pleasurable to the dominant and reinforces his perception of superiority, thereby permitting the subordinate to remain in the troop. By clowning, the subordinate gains some measure of control over the dominant; it permits him to form bonds with strong individuals and exploit these bonds to his advantage.

We may note that *courtesy* is the antithesis of dominance displays; it elevates the social position of the social partner, since the one behaving courteously acts out submissive behavior. Hereby the social status of the social partner is openly recognized. As we shall see below, human dominance displays are elaborated by culture—and so are the acts of courtesy. The foregoing dwelt largely on the biological attributes of dominance displays that were at best culturally enhanced; we move next to the rules governing dominance displays at the cultural level.

Cultural Dominance Displays and Their Rules

At this point we enter a controversial and speculative area. It is not that cultural dominance displays of humans are not understood, or even studied, but the link of these apparently purely human actions to basic behavior systems of vertebrates is not likely to be met with favor. The gulf between ethology and human culture is bridged by at least one author, Otto Koenig (1970) with his work on "cultural ethology." In reading such books as Boren's (1972) *When in Doubt, Mumble* or Parkinson's books on his "laws," or Townsend's (1970) *Up the Organization*, or even Gordon's (1961) *Synectics*, one is repeatedly made aware of the intuitive knowledge of these authors of human dominance displays, ego manifestations, vanity, and the need for "ego polishing." This is, of course, explicitly recognized in textbooks on personnel management, such as that of Strauss and Sayles (1972). The following will be a brief description and classification of human dominance displays at the cultural level, and the rules they appear to follow; I discussed some of these in an earlier book (Geist 1975a).

(1) At all levels, dominance displays imply *superiority*. At the most primitive level, the level most closely recognized as biological and similar to the dominance displays of other vertebrates, is *implied superiority in aggression*. This is signaled at the personal level by the decoration of weapons and their conspicuous bearing, by evidence of past combat such as battle scars, and, in its cultural form, by medals and by behaviorally appropriate acts mentioned earlier that indicate serenity, calm, dignity, pride, etc. At the societal level, superiority in aggression is signaled by spectacular military parades, the display of advanced weapons, military maneuvers, publication of the size of the military budget, and financial and political encouragements of the military—industrial complex.

However, dominance displays encompass acts implying superiority in diverse endeavours, in sport, cultural activities, science, etc. Note, for instance, the competition among nations for gold medals in the Olympic Games and the concomitant listing of medals won by countries. Note the listing by village, city, province, and nation of the poets, scholars, inventors, scientists, adventurers, painters, composers, movie stars, musicians, etc., they produced. All human dominance displays imply superiority in one way or another.

(2) One *method* in which superiority is signaled we may call the *rule of alliance*. Here, an individual or nation allies itself with success or recognized symbols of success; we saw it above. We find this rule operating in complex primate societies, not only in humans, in which males align themselves with dominants and in their conduct take advantage of such alliances (Jolly 1972, p. 189). The rule of alliance is a far-reaching rule in human societies. The flags of regiments list their victories (never their failures); uniforms of soldiers reflect the cuts and symbols of militarily successful regiments of their own or different nations (Koenig 1970), so do the behavioral idiosyncrasies of regiments such as the British Rifle Regiment. In a form of verbal behavior called *"name dropping"* a person aligns himself with successful, highly placed acquaintants. A species of a genus with the same name is referred to in science as *"bandwaggoning."* Here the displayer reaches for

recognition by aligning himself with a topical, faddish, prestigious area of research. Another form of the same behavior I shall call *"harmonizing."* Here the displayer, vocally or in writing, supports a dominant individual in his contentions, and in the process the displayer hopes that all take notice and are duly impressed. Other common alliances with success are the use of ethnic dress or lab coat, accents, jargon, or argot—in short, a variety of symbols that permit stereotyping of a reference group, and in so doing influence any communication *a priori* in the direction intended by the displayer (Strauss and Sayles 1972, Chapter 10).

Somewhat more indirect alliances with success are those alliances with age, knowledge, prowess, mythical powers, the unknown, and wealth. The latter take the form of subtle and not-so-subtle displays with material possessions. They are comparable to territorial displays of vertebrates, but so are displays with and of knowledge. The latter are conceptual possessions, and persons tend to resent encroachment by others on their conceptual territories. This has been recognized repeatedly (e.g., Eibl-Eibesfeldt 1972, Greenbie 1973).

Displays of material wealth, or materialism, are a very costly form of human dominance displays. Unfortunately, they are also equated with "standard of living," implying that the greater the material wealth of individuals, cities, states, or nations the higher the necessary quality of life. Displays of wealth may be institutionalized, as they are in North American society, in which the size and age of a car; the size and location of the house; the exterior appearance of the house, lawn, fence, and backyard; the presence and absence of campers, power boats, or horses; all indicate the relative wealth of the owner: His possessions are displayed for all to see. Since this is so, there has been, and still is, competition among neighbors to come up to the level of material displays of the majority of the neighbors; it is called "keeping up with the Joneses." In ancient China, peasants practiced a reversal of this form of superiority display in the "cult of poverty" (Stover 1974). Since in the farming society of the Chinese "green circle" intense competition existed for arable land, and land holdings in excess of those required for subsistence led to jealousy and reprisal, peasants acted out poverty, hard toil, and ritual sacrifice to the village deity of gains above the bare necessities, in order to demonstrate externally the need for their piece of land, and to remain above reproach. Hence poor clothing, poor artifacts and dwellings, meticulously cultivated fields, long hours spent on cultivation, all signaled a moral superiority over those not of the ruling aristocracy that accumulated riches.

Displays of material wealth can take specialized form in conjunction with special prowess and skills. Thus there are displays by the wealthy with special cars, boats, yachts, big game trophies, art collections, or beautiful girls.

Alliance with age takes place when a suggestion, an idea, or whatever is promoted on the basis that it is sanctioned by the old and, by implication, wise. In promoting an idea sanctioned by the old and wise, the promoter hopes to gain recognition.

Alliance with mythical powers and the unknown invokes essentially religious sanction for an act, an idea, an argument, etc. However, such alliance can go beyond obvious religious sanctions; it can imply profound mythical knowledge. Thus persons may align themselves with concepts they have failed to grasp, but that they acknowledge and espouse in the hope of gaining recognition. A common present-day form of this alliance is the advocation of computer simulation and systems analysis. These tools have their legitimate place of course; the ignorant displayer is inevitably revealed if he advocates the use of these tools to solve problems they cannot solve.

(3) *The rule of the flip* is one that demonstrates a basic behavior of human dominance displays. Unless artificially frozen, displays increase in complexity over time until further complexity brings about no noticeable change in the display; then, at this point, the display reverts to its antithesis, simplicity. When this happens the display flips. Let me illustrate this by the visible Nazi hierarchy in the Third Reich. This group wore uniforms, as was the expressed wish of Hitler (Speer 1970). Dominance rank of military men is signaled by the insignia they carry, and also by their decorations and cut of uniform. Thus the more campaign ribbons, medals, lanyards, tassles, embroidery, etc., the higher the bearer's rank. The maximum load of such rank signals was carried by Herman Göring. It is quite inconceivable how one or two, or even a dozen more medals, could possibly have added greater distinction. Only a doubling or so of the number of decorations could have done that. However, such a load of medals would have dragged down the tunic and hampered movements—in short, it would have been ridiculous. That Göring was not quite sensitive to this was indicated by his use of cosmetics on his face. In short, it would have been quite impossible to outdo Herman by adding more decorations to one's array. How to shine in Herman's presence? Simple. Get rid of the decoration. This is what Hitler did, except for his Iron Cross. Koenig (1970) suggests that the dominant contrasts himself by being surrounded by decorated men while wearing simple attire.

The rule of the flip can be recognized in many instances; the extremely powerful and wealthy man who lives frugally and simply—as did Peter the Great or Lev Tolstoi; fashions that evolve in one direction and suddenly flip, as did miniskirts, which flipped to maxis; successful men who sought out and basked in the limelight until they suddenly flipped into extreme seclusion and secrecy, such as did Howard Hughes. But one of the least recognized of dominance displays, humility, is also due to the rule of the flip.

Humility may be a display of some great, renowned scholars. It is quite understandable that men who gain eminence display this through confident manners, and in some cases a touch of arrogance and haughtiness. A very great achiever can distinguish himself from the lesser great by an overt display of humility. Of course, there is no one to challenge him, and he knows it. In this form, humility can become the greatest form of arrogance.

It is self-evident that the rule of the flip is encountered most frequently where free competition prevails, so that the dominants are continually forced to enhance their displays. This is demanding of ingenuity and materials and threatens to destroy the meaning of displays. From this self-evident formulation it follows that where dominants have the power to do so they will stop, or at least slow down, the evolution of social dominance displays. This means in effect that a freeze is imposed on dominance and rank displays, and each display clearly signals the rank of its bearer. An individual may not assume the exterior trappings of a rank except by appointment, be it through heredity, achievement, exchange of goods or money, or conquest. The undue use of dominance displays should be punished by the dominants, much as it is in animal society (Geist 1971a).

A reverse movement of the rule of the flip is the decoration of a subordinate position with the trappings of power. For instance, doormen at the entrance of hotels. They are no less gaudy in their uniforms than highly decorated generals. To dress doorkeepers in correlation with their social rank would not reflect too well on the establishment; to dress them in ordinary clothing might not set them apart from customers; hence the flip to the extreme clothing style which fakes a greater courtesy and reverence for the customer than is intended or warranted.

(4) A peculiar variant of the rule of alliance is the *rule of transferral*. In this display the loser or subordinate emphasizes a particularly praiseworthy aspect of his conduct, or that of his group, so that his prestige may rise despite an obvious loss or defeat. This rule is exemplified by the manner in which the British treat the battle of Dunkirk during the second world war, or the Germans treat the sinking of the battleship "Bismark." In both instances a stunning military defeat is turned into a "moral victory," of sorts, in one case by emphasizing the efforts of a civilian population to save its army, and in the second by emphasizing the heroic nature and skills of the crew of the doomed battleship. In both cases what stands out is the praiseworthy, prestigious actions of the loser, not those of the winner. The rule of transferral is illustrated, of course, in the change of meaning of the cross in Christianity; the cross, a structure of shame, fear, and death akin to a gallows, guillotine, or garrotte, became the symbol for love, life, brotherhood, and the finest of human aspirations.

We encounter the rule of transferral at a more subtle level in everyday life. The worker who is dissatisfied with his monotonous job, but stays in it emphasizing that "the money is good," makes a display of the latter point, using the rule of transferral. It is a justification, a reinforcement of the behavior adopted by the displayer. One can multiply examples *ad infinitum*: He is a lousy shot, but a great hunting companion; in the restaurant the cooking is bad but the waitresses are pretty; the scenery is great, etc., etc.

(5) A further major variant of the rule of alliance is the *rule of antithesis*; it can easily be confused with the rule of the flip. This display is in effect when a minority group takes vocal issue with the majority, or with the establishment, and in demonstrating this opposition dresses and assumes mannerisms 180° deviating from, or antithetical to those of the majority. The hippies of the 1960's showed this behavior by growing beards and long hair; dressing with pointed sloppiness; choosing unisex clothing; espousing sexual licence, an antiwork ethos, and communal as against married life; sporting jewelry for males; nonaggressive opposition—all of which stand in antithetical relationship to the values, attitudes, and clothing worn by the "silent majority" at that time. Such external displays, of course, help greatly to cement relationships among members of the in-group, in this case the youth claiming allegiance to the hippie cult.

This rule operates wherever two groups are antagonistic to each other, or where an individual opposes a group. Note the assembly-line worker who wires the accelerator to the car horn or windshield wiper, or the glassblower who has been blowing glass rabbits with straight ears for many years and then starts to turn them out with drooped ears. Both examples illustrate self-assertion through opposition to the high and mighty (Strauss and Sayles 1972). At a more subtle level, we recognize it when an executive gets rid of his business suit after work, dons sloppy, comfortable clothing and goes bowling or beer drinking or works in the garden; it is a mild signal of opposition to his work.

The difference between the rule of the opposite and the rule of the flip is that the former describes a process by which displays change as they complexify; the latter is a conscious structuring of displays by one group or one person to oppose another group or a role that is disliked.

(6) A further class of dominance displays is *role playing*. It entails sticking strictly to the behavior expected from one's role, and emphasizing such behavior pointedly. A waiter who plays his role well, who takes pains in performing elegantly the little courtesies expected of waiters, gains our recognition and is called a "good waiter." Note, his actions match our expectations; our pleasure is the consequence of cognitive pattern matching. We have met this phenomenon a number of times. In the above example, the individual, by playing a role honestly and well, gained recognition; someone playing a role unbecoming to him is termed "modest" if he acts out a role of lower status than he actually occupies, and he is called a "fake"—or worse—if he acts as if he were superior to the actual role he is in.

(7) The next rule, like the rule of the flip, describes the behavior of dominance displays over time; I shall call it the *rule of the sink*. Symbols of privilege and status are progressively usurped by a broader segment of society, until these symbols characterize not a privileged sector of society but society itself. The salute, an example elaborated in great detail by Otto Koenig (1970), is one such example. It originated as a privileged behavior of the feared grenadiers and spread to all ranks of the armed forces. Many changes illustrating this rule are described by Barnes (1967) and Koenig (1970) for the military.

Any symbol that emphasizes the lowly rank of subordinates and conflicts with personal esteem is, in an open competitive society, fair game for the subordinates. Any external symbol of privilege will be contested and is likely to be usurped. At universities in North America this has been largely the case with separate elevators, bathrooms, and lounges for faculty. In industrial labor negotiations one apparently notices a continuous struggle between different unions, or even between skill categories, to usurp the privileges of others, while those with privileges try to ward off the opponents (Strauss and Sayles 1972).

(8) We may call the next rule the *rule of ritualization*. It states that symbols of privilege change over time in the direction of abstraction, until their original form and meaning are forgotten. The evolution of the eye to the miraboa as illustrated by Koenig (1970) is one such example; Koenig shows many examples of this rule. It predicts that as cultures change they carry in ritualized form many symbols and behaviors that we maintained for the sake of tradition rather than for functional utility. This rule also suggests that since symbols have a "life" of their own, they are quite resistant to extinction; this harmonizes with the idea of Davis (1965) that social structures are

quite resistant to change. We know from ethological studies that behavior patterns of intraspecies communication tend to be quite resistant to change (Krumbiegel 1954, Hofer 1972, Geist 1974a).

Innocent as the rule of ritualization may seem, it is not. It predicts the course that any form of human endeavor initially takes once it is subject to competition. Competitors outdo one another by adding "novelty," "rarity," and "uniqueness" to their work in order to differentiate it from that of others. Only the novel or rare attracts attention, so the theory of pattern matching predicts, hence complexification results from continuous attempts at making the subject matter "novel." To create a difference is to create "progress." In this process, the creation of novelty, lies also the root of abstraction, the removal of that which covers, obscures, or hides the essence. Abstraction may superficially look like simplification, but to understand it requires a great complexity of understanding. Both processes, complexification and abstraction beyond obvious recognition, are processes of ritualization.

Ritualization may serve either to satisfy "esteem," in Maslow's (1954) nomenclature, or "self-actualization," depending on whether it is the product of competition with others or with one's own self. In the latter, a person works primarily for self-satisfaction; he does not depend greatly on the positive response of others or the anguish of his competitors to continue striving. However, I doubt that self-actualization in its pure form can be found; in the final instance striving is dependent at least upon social approval and the recognition by others of the superiority of the work.

(9) The *rule of negation* is a common dominance display. It can be understood as follows: Not only is status raised by the superiority of one's own efforts, alliances, role playing, etc., but also by depreciating the efforts, alliances, roles, etc., of others. Hence, one can observe concerted efforts by dominants to extinguish the dominance displays of subordinates. This is shown quite clearly by various species of large mammals, and I have pointed this out in examples elsewhere (Geist 1971a) and in earlier parts of this chapter. This tendency to depreciate others is common at the cultural and conceptual level in humans. It is exemplified by "criticism." Someone who criticizes assumes that his understanding of the subject matter is superior to the one whom he criticizes. The rejection of someone's idea with a "No" or a "Yes, but. . . ." implies one's own superiority of understanding or, conversely, the inferiority of the person proposing the idea. The ready acceptance of a competitor's idea implies one's inferiority. Hence, a quick response of "No" to an idea is, first, an insistence on one's dominance; second, it is a conscious or unconscious blow or insult to whoever offered the idea; third, it forces the opponent to restate his idea, giving the one who rejected it time to marshall counterarguments or find means whereby the new idea can be at least temporarily defeated since otherwise there will be loss of face. The ensuing argument is verbal aggression *par excellence*.

The foregoing indicates that ideas are more readily accepted by friends, and that in interdisciplinary groups the cultivation of friendships is most important. It also explains why it is necessary in group processes to introduce formal rules such as lauding every new idea proposed before it can be criticized, lest the productivity of the participants declines owing to alienation (Gordon 1961). To be courteous to others is to identify their ideas, laud them, reiterate them to demonstrate that one has grasped them, and to make use of them. This, depending on the context, need not imply inferiority by the acceptor of ideas, but the converse; it can imply tolerance, openmindedness, fairness, couteousness, self-discipline, good judgement, magnanimity—in short, most laudable personal attributes.

(10) Another kind of dominance display is seen in *the rule of silence*. It was noted that animals acting as dominants gain by ignoring the subordinate and depriving him of a response to his threats of violence; this teaches the subordinate to be helpless (Seligman 1975), to lose the motivation to usurp dominance, and thereby become less of a threat to the dominant. We find, of course, the same behavior in humans. To thwart an opponent one only needs to keep him struggling in futility long enough, and he will give up. Or better yet, discourage any challenge by others by symbolic means or through the grapevine. The massive, almost brutal, architecture favored by the Third Reich (Speer 1970) is a plausible example of this. Something massive, colossal, cold, unresponsive, aloof, and silent is intimidating by virtue of being unresponsive, and teaches helplessness and the futility of resistance. Massive bureaucracies with their long response periods, errors, and intractability become teachers of helplessness, and in so doing ensure their survival.

The foregoing taxonomy and description of dominance displays illustrated a behavioral system common to higher vertebrates, including man, that has its roots in aggression. Even self-assertion is a form of aggression; here I agree entirely with Gilula and Daniels (1969). Human dominance displays, as a form of aggression, take on an incredible diversity, and yet they follow clear-cut rules. We have to learn not only to live with dominance displays, but to channel them into socially and ecologically relatively harmless channels; we cannot expect to get rid of them—and should never wish to. A person without self-assertion is a contradiction in terms. Moreover, dominance displays provide us with so very much that we can unstintingly enjoy—music, drama, plays, paintings, refined foods, clothing, better tools, etc. It appears to me that painstaking attention should be paid to channels of dominance assertion that are least damaging to society and our natural environments, and we should encourage these. Materialism as a dominance display clearly has to go, due to its environmental costs; increased involvement in decision making, craftsmanship, participatory sports, increased socialization, and problem solving in family life are all acceptable alternatives, as are many others.

Some characteristics of dominance displays are worth emphasizing: It exercises the process of pattern matching in which discrepancy between expected pattern and no pattern or a mismatching pattern generates tension, anxiety, or fear, and the subsequent matching apparently provides a parasympathetic response to the body in which the magnitude depends to some extent on the tension generated previously. We model our expectations on ideals; without such ideals no expectations can be generated; hence, behavior is likely to conflict with that of others which first raises anxiety, due to uncertainty in how to act, and also due to punitive responses by those affected. Reward, that is, internal parasympathetic stimulation, comes from the following sources: Stimulation of pleasure centers through pattern matching and by positive social stimuli (discussed in Chapter 13 under The Social Feedback System). This explains at once how we learn "to take pleasure from a job well done." We first receive reward from the social stimuli for a work, and by classical conditioning we then take pleasure from a well-made piece of work. We experience as pleasure, owing to pattern matching, the discomfiture of our rivals in response to our superior work, as well as the pleasure of friends, which is reinforced by socially positive stimuli. Both the product, and the process (behavior) count in dominance displays. The manifestation of "power" becomes explicable; individuals take pleasure from pattern matching, both from the compliance of others with their wishes, and from the discomfiture and pleasure they are able to elicit in others. Please note how small a step it is from the exercise of "power" to gambling: Gambling creates "expectations" and "tensions" and provides a match between expectations and reality on an unpredictable schedule. Since the tensions due to lack of match between expectation and reality are great, the parasympathetic reward is jolting when patterns do match, when the "expectations come true." Both in the search for power, and at the gambling table, a person exercises a physiological mechanism that creates a predictable environment, and that is what organisms are designed to strive for.

We can also recognize Maslow's (1954) hierarchy of needs as valid. Physiological homeostasis depends on the availability of materials for physiological functions, hence the priority of signals dealing with food intake, excretion, rest, and maintenance activities over others. For optimum maintenance of physiological functions we require the assistance of others during ontogeny, hence the social feedback system which in adult life permits bonding for various reasons, and in the search for it we manifest a "social need." However, the ancient rules of maintaining access to scarce resources, and doing it at the cheapest cost, automatically generates competition among individuals. In experiencing success in competition through the mechanism I described in this chapter, we generate a behavior recognizable as "self-esteem."

A superior performance in a given avenue of competition leads to self-competition, or striving for self-set goals. This is Maslow's (1954) "self-actualization." In short, the theory of dominance displays in conjunction with that on "social feedback" and pattern matching does explain Maslow's concepts of human needs. It also explains the notions of McClelland and Atkinson (Strauss and Sayles 1972, p. 18) of the positive needs of human beings, *achievement, power,* and *affiliation*, as well as the "negative need," fear of *failure*. Without going into detail, these are manifestations of pattern matching, dominance needs, and needs generated by the social feedback system.

Chapter 6

How Genes Communicate With the Environment—The Biology of Inequality

Introduction

In this chapter we shall examine an important, but poorly understood, attribute of genetic systems: their ability to communicate with the environment of individuals and to guide an individual's development in such a fashion as to enhance its reproductive fitness throughout its life. Little is known about how genes communicate, but communicate they do. From the examples to be discussed under the heading Population Quality one can readily deduce that such communication must entail negative feedback loops that control development in accordance with environmental dictates. Genes appear to have at their disposal alternative strategies of development which they switch on or suppress depending on the messages from the individual's environment. We may call these mechanisms epigenetic mechanisms. It is C. H. Waddington (1957, 1960, 1975) who had a deep insight into their significance to an understanding of evolution. His thoughts are greatly neglected, and invariably misunderstood when mentioned, in the polemics about human evolution; the inadequate neo-Darwinian paradigm is still king, as illustrated, for instance, in discussions by Wilson (1975), Trivers (1974), Alexander (1974), Durham (1976), or Ruyle et al (1977).

Waddington paid far more than lip service to the accepted notion that the phenotype is the unit of natural selection; he developed a theory explaining how environmental dictates are ultimately translated into genetic responses by species. He was concerned with how characters appear, become genetically enhanced, and finally, fixed in the developmental process of a species. My emphasis here is quite different. I am concerned with how genes in the developmental process adjust an individual to the environmental demands. Therefore, my emphasis is on communication between genes and environments rather than on canalization or fixation of a trait so that it develops despite the variability of environmental signals. Since the study of gene–environment communication is in its infancy, as one can readily detect when reading studies on differentiation, we cannot dwell on the mechanisms of that communication, not in detail at least. We can dwell on its logic and biological significance, however.

If genes were to express themselves unvaryingly in a fluctuating environment, then gene frequencies would fluctuate directly with environmental variations and a given gene would maintain itself only by a high reproductive rate of its carrier, the accident of some of its carriers finding themselves somewhere unaffected by the environmental variable, or by finding itself in the company of a dominant gene within a successful carrier. There may be other mechanisms, but these illustrate the point: Unvarying expression of genes can be permitted only for organisms with enormous reproductive rates and infinitesimally brief generation times, such as microorganisms. Yet even these vary phenotypic expression according to the dictates of the environment (Valentine and Campbell 1975). In long-lived organisms, genes that cannot vary their expression to fit the environment are weeded out by natural selection, sooner rather than later.

Let us for a moment take a gene's point of view. For it, evolution is bad. In fact, it is a calamity! Its first commandment must be to protect itself against evolution and permit it only under exceptional circumstances, and then only when it cannot be lost through the founder effect, genetic drift, phenodeviance, or inbreeding in small populations. We can, for the sake of illustration, compare genes to the board of directors in a large corporation which must protect itself against the vagaries of a free market. I am certain John Kenneth Galbraith could write a wonderful book on the survival strategies of genes; in fact, he might only have to change a few words here and there in such books as *The New Industrial State*. Genes, like managers or bureaucrats, must keep their organization competitive, they must not leave the future alone but actively shape it and ensure their organization's ability to deal with the future on its own terms (epigenetic mechanisms), they must buffer it against unexpected changes and thus not only keep a reasonably complete inventory of supplies but also maintain a pool of talent (genetic redundancy or canalization). In order to fit the organization to environmental demands, they must keep it from meager pastures lest they find themselves without a job; nor must they heedlessly drive into a new venture (dispersal) upon the mere promise of growth; they must diversify to maximize opportunity for income (heterozygosity); they must be aware of unprofitable mergers (hybridization); they must keep account of income and expenditures and invent ways to improve efficiency, and above they must ensure security by various alliances within and between corporations (pleiotropy, gene linkage) so as to compensate for each other's weaknesses and ensure access to raw materials, energy, and a safe milieu. They must tremble in fear of an environment akin to the free market in which the consumer (natural selection) is king.

In a variable environment, in species long lived relative to the fluctuations of environmental variables, a gene can survive only by virtue of directing the individual's development in such a fashion as to enhance the phenotype favored by selection—or drop out of the active scene altogether and become inactive and hence selectively neutral. Put differently, since selection acts on phenotypes, not genotypes, epigenetic mechanisms permitting genes to alter the development of individuals adaptively blunt the edge of natural selection or escape from it altogether. The corollary of this, of course, is that the better the epigenetic mechanisms, or other mechanisms of adjustment, are developed, the less likely it is that evolution will take place. The more perfect the *adjustment* the less likely is genetic adaptation. Once a gene has become widespread or fixed, it enters into competition with its own copies in different individuals on the basis of epigenetic mechanisms. That is, epigenetic mechanisms must be selected in a similar way to genes; here we know little, since we do not know what these mechanisms look like, except maybe in the case of controller genes and their evolution (Markert et al 1975, see Valentine and Campbell 1975). It is self-evident, however, that epigenetic mechanisms must be hierarchically organized so that mechanisms of the lowest order are buffered against natural selection by epigenetic mechanisms of a higher order, and so on. The foregoing emphasis on epigenetic effects notwithstanding, they are only part of the "insurance" developed by genomes of multicellular animals. As pointed out in the introductory chapter, there are others, such as gene linkage, polymorphism, pleiotropy, heterosis, and outbreeding.

Although "epigenetic mechanisms," strictly speaking, refers to physiological processes, they are simply one means of adjustment, and there is logically no reason the term could not have a meaning as broad as to include cultural mechanisms. Anything that changes individuals adaptively becomes, then, at least an expression or consequence of epigenetic mechanisms. However, for semantic reasons I shall retain a narrow meaning for the term "epigenetics," and let it stand strictly for the physiological responses, although I shall champion here—as did C. H. Waddington—the view that it is behavior that triggers these epigenetic responses, and that morphology is the consequence of epigenetic responses. We do not know much about the physiologic nature of epigenetic mechanisms. From the foregoing, however, we suspect that epigenetic mechanisms have certain analogies to genes: They adjust individuals while genes adapt species. We can validly speak of a K- and r-phenotype, as we shall see, that is, phenotypes of a species that act with very different reproductive strategies, as I discussed for mountain sheep and other large mammals (Geist 1971a,b). However, if epigenetic mechanisms and other means of adjustment adjust individuals to their environment, one does wonder when selection for genes themselves takes place.

Evolution, that is, genetic change, takes place in two situations—first, if energy and nutrients are available to maximize phenotypically an adaptive trait. Now the potential of each gene to develop that trait is fully exposed, the environmental variance of that trait is close to zero, and the genetic variance is near unity. With the genetic variance of the trait fully exposed, natural selection can proceed rapidly to remove genotypes not capable of maximizing that adaptive trait. We run into this phenomenon later when discussing evolution during dispersal and colonization by populations (p. 256). Under these conditions we expect the genome to produce the "dispersal phenotype," that is, individuals that are preadapted to confront and solve diverse problems. They are equipped to handle the unexpected, which so readily arises when dispersers move into habitats previously not occupied by their species. Such individuals ought to move about widely, to capitalize on the best forage sources, to minimize competition and to maximize experience, as well as to mate with dispersers from distant sources, thereby rapidly increasing heterosis and heterozygosity in the individuals of the colonizing population.

The second condition of gene selection occurs under conditions of resource scarcity. It is the same type of selection in principle as the one mentioned above. Here, however, organ systems compete for the scarce energy and nutrients available for growth, after the costs of maintenance are paid. Therefore, only those organs or attributes that aid reproductive fitness under conditions of resource scarcity are maximally developed, while less important attributes must be throttled in their development. Genes must work here in such a fashion as to create phenotype redundancy by whatever mechanism there may be. Waddington (1957, 1960) called this process canalization. In so doing, genes ensure their existence despite mutations deleterious to the development of a given phenotype; other genes cover up for the mutant. They conceivably also cover up the effects of inbreeding to some extent for, even if a nonfunctional gene becomes homozygous, others may still be able to cover up. Phenotype redundancy could be achieved through multiplication of advantageous genes within a karyotype, as by duplication of one or more chromosomes, linear gene replication, or unequal crossing over. However, processes of chromosome duplication also must call into force mechanisms silencing genes whose expression would detract from the development of adaptive phenotypes. The discussions of Markert et al (1975) are here most enlightening.

Selection for canalization would be expected whenever organisms or mechanisms are required to perform a constant function despite a fluctuating environment, or where a constant function or shape is highly adaptive. Thus, we expect phenotype redundancy for external characteristics in animals from such stable environments as the tropical rain forests or the deep benthic layers of oceans. We expect great phenotype redundancy for physiological processes controlling temperature in endotherms or the acid−base balance or the electrolyte content of blood. We can also express phenotype redundancy in the language of animal scientists: Tissues or body parts having high growth priority are likely to have great phenotype redundancy, since they develop during ontogeny almost without being influenced by the ontogenetic environment; the converse would hold true for tissues of low growth priority. We shall speak of this later in the chapter. These thoughts suggest some testable propositions, namely, that animals from the tropics can maintain lower effective populations sizes by virtue of phenotype redundancy than can animals from the variable temperate environments, without suffering reproductive wastage. Canalization of development in tropical forms would also explain the very long lag between the extinction of dinosaurs at the end of the Cretaceous and the appearance of the first adaptive radiation of mammals, in which we find forms vaguely resembling the former dinosaur adaptive forms. These propositions have significant implications for the conservation of mammals, since they indicate that we must not use uncritically concepts developed in the tropics when considering temperate zones and vice versa.

Under conditions of resource scarcity we then have selection for a phenotype adapted to cope with the consequences of a high density of conspecifics; we can term it the "maintenance" phenotype, or K-phenotype. Resources are reduced primarily by intraspecific competition; a high density of conspecifics favors the spread of parasites and pathogens; there may be more social partners to deal with; predators may pay more attention to the population. All these factors increase the cost of maintenance to the individual. Hence, the phenotype, under conditions of resource scarcity, can invest less in growth and reproduction, and must invest more in diverse mechanisms of individual maintenance. Therefore, it is smaller in body size, to reduce the cost of roaming and upkeep while maximizing food intake per unit of distance traveled, while also being able to use scattered resources, uneconomical for a large-bodied conspecific. It must be lethargic and must not readily enter into social interactions, in order to save resources. It must invest a relatively large amount of resources in antibody development to cope with parasites and pathogens. We expect it to stick very closely to the law of least effort. All in all, it is a phenotype that maximizes abilities to compete for resources needed for reproduction, while the dispersal phenotype—surrounded by an abundance of resources—maximizes abilities to deal with contingencies.

Because it was covered in Chapter 1, it need only be pointed out here that in maintenance populations there may be selection against heterosis. It may be too costly in energy and nutrients and subtract from the resources available for reproduction, without adding significant benefit to the individual. Therefore, in stagnant populations, we expect selection against choosing mates from distant genotypes, against roaming, and against tolerance of strangers. These predictions need to be verified.

It was also noted that dispersal phenotypes had to be able to control tastes and aversions rationally in order to increase their ability to adapt to new environments. In essence, they had to suppress the urges to do the formerly adaptive things and overcome aversions, in order to perform the actions now adaptive. Such are theoretical dictates. The reverse would apply to the maintenance phenotype; it ought to be less subject to rational control than the dispersal phenotype. In short, under conditions of resource shortage, which is the normal condition for populations, a phenotype quite different from the dispersal phenotype is adaptive.

The conditions under which genes are exposed to natural selection are also conditions in which genes can be lost accidentally. Thus, when populations decline owing to great resource scarcity, or when colonizing animals live temporarily in small populations, the rate of inbreeding is increased. This leads to the loss and fixation of genes resulting from the loss of phenodeviants, and to individuals with low heterosis and thus poor development (Wilson 1975, p. 72). Thus arises the founder effect—operating, of course, both in colonizing populations and in populations suffering decline—which may exclude from the affected population genes common to the parent population. One mechanism that prevents excessive gene loss is heterosis or hybrid vigor, as well as heterozygosity. Heterozygotes, for a number of reasons, have a higher reproductive fitness than more inbred individuals; they are less affected by environmental vagaries and usually have higher reproductive rates and more viable offspring (Mayr 1966). It is still mysterious why heterozygous individuals should grow and develop better than more homozygous individuals, that is, use energy and nutrients more efficiently and perform more work per unit of ingested energy. Yet clearly, heterosis is one mechanism maximizing genetic diversity in a population. It is a gene's way to make sure that it is not lost in small populations.

A second mechanism whereby genes are conserved, at least during colonization, is suggested by studies on malnutrition. Since malnutrition is associated with increases in congenital deformities, it appears that good nutrition may cover up genetic flaws (Williams 1971). Clearly, this would conserve genes in expanding populations faced with superabundant resources.

In a deteriorating environment, it is logical to expect animals to begin emigrating in search of better living conditions. Thus, deteriorating conditions ought to trigger emigration. As indicated earlier, individuals change into a new phenotypic dispersal form, nowhere better illustrated than in the insect world (Wilson 1975). In these instances, colonization would not be by subordinates (Christian 1970) but, on the contrary, by very specially prepared individuals. Genetic systems appear able to act on clues from the environment and alter phenotypic development adaptively. In this chapter, I have discussed rodent cycles from this point of view, since they appear to be phenomena ideally suited to maintaining rodents in environments with fluctuating habitat availability and local habitat extinction. Thus the development of dispersing phenotypes, during either environmental deterioration or increased availability, is ultimately a mechanism of gene conservation.

Once genes are fixed, they enhance their security by competing by way of epigenetic mechanisms, and it is then clearly adaptive to combine with others in order to enhance the range of responses of the epigenetic mechanisms. This should maximize the individual's plasticity so that it can respond to a greater range of environments. Once the organisms find themselves in environments within the range of adaptability by epigenetic mechanisms, selection on genes comes to a halt, so that evolution in effect falls asleep, not to wake until the population decreases in size or colonizes new habitat.

For years, private discussion among mammalogists hinged on the point that the differences observed between species within genera, tribes, or even families, could hardly be due to the direct action of genes, but must result from different degrees of expression of very similar genomes. Davis (1964), in his work on the greater panda, was brave enough to say that much, for he felt that rather small differences in the growth of different body parts distinguished the panda from the black bear. One gains the same impression looking at the vector distortions in different species, as illustrated by Thompson (1961). The paper by King and Wilson (1975) shows that genetically the genera *Homo* and *Pan* are apparently as close as sibling species. Clearly, the differences here lie less with genes than with epigenetic mechanisms.

Our understanding of the manner in which individuals come to differ in size, external appearance, and performance has been boosted recently by the work of Ooshima et al (1975). It was shown that during hypertension, a consequence of stress, the biosynthesis of collagen increases in the blood vessels, including the peripheral circulation. Collagen deposition on the walls of *constricted* blood vessels, of course, leads to the permanent reduction of the bore or lumen of the blood vessel, and, on top of it, may reduce the rate of passage of nutrient and gases between blood vessels and cells. Clearly, if the rate at which collagen is synthesized and incorporated into the blood vessel walls differs among organs, then we see here a mechanism by which the phenotypic expression of organs can be regulated. Given stress and resource shortages, organs of insignificance in such environments are reduced by virtue of lumen reduction in their blood vessels. Conversely, under optimum conditions vigorous activity keeps the lumen of blood vessels large, collagen synthesis is low, and the various organs can develop more fully. Tissues of low growth priority ought to be those that respond readily to stress with elevated collagen synthesis in their blood vessels, and vice versa for tissues with a high growth priority.

Under maintenance conditions, when resources for maintenance and reproduction are scarce, we expect a readjustment of growth priorities in tissues of high growth priority. Natural selection will thus make individuals increasingly more capable of maximizing the amount of resources they can spare from maintenance for reproduction. For instance, tooth and body size will be adjusted to be no larger than necessary for efficient reproduction. Individuals with organs of high growth priority that are larger than the size needed for efficient reproduction are selected against because such organs rob resources needed to maximize reproductive output. Thus, maintenance selection only refines the adaptive strategies evolved through the dispersal phenotype during colonization. No evolution of new adaptive strategies can, therefore, take place *in situ*; new "forms" arise during dispersal.

The foregoing discussions are not innocent in their implications. They suggest that in old stable environments we ought to find not just "living fossils," but the *actual ancestors* of extinct forms. For instance, the Sumatra rhino (*Dicerorhinus*), which is today apparently as primitive as it was some 30 million years ago, may be the living ancestor of the extinct dicerorhinid lineage that terminated in the woolly rhino (*Coelodonta*) of the ice ages. Similarly, the very primitive flat-headed cat (*Tetailurus*) may be ancestral to evolved felid lineages, while the cloudy leopard (*Neofelis*) may be the ancestor of the repeatedly evolved, now extinct, saber-toothed cat lineages. Unfortunately, one cannot prove or disprove this proposition at present.

One can, however, test another prediction, namely, that during climatic and geological revolutions on earth, there ought to have been waves of new species arising. It is then that environments change so much that epigenetic mechanisms no longer can buffer the genome against natural selection. The consequences are extinction of some forms and the evolution of others. Thus, major geologic and climatic revolutions are the end of some lineages and the beginning of others, and this is indeed what we find, nowhere better illustrated than in the evolution of the ice-age mammals, as I shall discuss in Chapters 8 and 9.

The concept of epigenetic mechanisms implies that in large long-lived organisms evolution is a rare event and that it tends to proceed rapidly when it does occur; evolution does not proceed continuously, but in steps, with long periods of genetic stability in between. Thus, the majority of variations between populations of a mammalian species ought to be phenotypic. Hence, much of our present-day mammalian taxonomy is called into doubt (Shackleton 1973).

We can now answer a question of long standing: What initiates selection for a given morphological feature? It is the behavior of individuals. Behavior is the cheapest means of adjustment. Specific activities result in physiological and morphological change of organs under heavy use, to phenotypic adjustment. If all individuals are subject to this adjustment due to a severe environmental contingency, then the differences between individuals in that adjustment are largely due to differences in genetics. Now the genetic variance is large. Natural selection rapidly acts on the genetic composition of the population, increasing the frequency of individuals genetically most capable of achieving the given adjustment. This much is Waddington's (1957, 1960, 1975) thinking. We add the following: New adjustments which lead to morphological change are expensive. They cannot, therefore, arise under maintenance conditions, but only during conditions of resource abundance. This condition is found during colonization when the dispersal phenotype exists. Therefore, new adaptations, diagnostic features, which differentiate the colonizing from the parental population, can arise only during dispersal.

The preceding discussions have some practical implications. The environment in which an individual develops, as well as the one in which it lives, leaves its mark on that individual's structure, physiology, and behavior. The environment manifests itself in inches of body length gained or lost, in cubic centimeters of brain, pounds of muscle, liters of tidal volume, days of illness, number of quarrels, milligrams of blood sugar, and so on. In short, it expresses itself in physically measurable form. Knowing how growth and development are affected by environmental variables, it is possible to form some judgement about the individual's environment by examining its structure and functions. Granted the validity of the diagnosis reached, it is possible to conceive and maybe implement alterations of the environment so as to promote optimum development.

Studies of genetic variation are the domain of the population geneticists; studies of phenotypic variation and its causes, however, are widely scattered through the disciplines of life science and have not been placed under one common roof. Animal scientists study it in order to maximize animal products per unit of feed invested, ecologists study it in order to understand the dynamics of populations, ethologists have begun to study it in order to understand behavioral variability of individuals and its consequent social manifestations, psychologists study it to develop psychological theory, and there are linguists, child psychologists, educational psychologists, epidemiologists, medical scientists, all of whom have an interest in, and perform studies on, the relationship between the environment and the individual or population. For this reason the study of phenotypic variability and its causes becomes an interdisciplinary effort, and does not yet have a theoretical underpinning of its own.

Phenotype Syndromes in Vertebrates

Individuals in populations vary. What is surprising is how severely rather small differences affect the resultant function of an individual. This is well illustrated by the study of Coulson (1968) on kittiwake (*Rissa tridactyla*), a small colony-forming cliff-nesting gull. Intense competition exists for nesting places in the center of the colony, with the less successful birds being pushed out to the colony's edges, while the unsuccessful ones form a large pool of nonbreeding adults. The gulls breeding in the center of the colony differ from those at the edge very slightly. They are a little heavier and have slightly longer wing feathers; otherwise there are apparently no noticeable external differences. Yet in their breeding biology and population dynamics, the gulls from the edge of the colony differ greatly from those in the center. They lay fewer eggs, their hatching success is lower, their clutches are laid later, they raise fewer young to fledgling age, they change mating partners frequently, and they suffer more than 50% higher mortality. Thus, small morphological differences appear to greatly affect the quality of individual kittiwakes, and a little development foregone greatly reduces the reproductive fitness of individuals. Coulson's study does not state how individual differences arise; it simply shows the consequences.

The studies of red grouse (*Lagopus lagopus*) illustrate both some causes and some effects of individual and population quality. The research on this bird was carried out in Scotland, and the following sketch is based mainly on the work of Watson, Jenkins, Moss, and Miller; a recent review of their studies is in Watson and Moss (1972), where further references are found.

The fate of an individual red grouse is largely determined by the maternal nutrition a few weeks prior to egg laying. It affects the quality of the eggs, which in turn affects the quality of the grouse raised from the eggs. Thus, poor nutrition is expressed in reduced clutch size, smaller eggs, lower hatching success, high mortality of chicks after hatching, reduced mothering behavior of the hen, and a low ratio of chicks to adults in the summer, while the surviving chicks grow up to be highly aggressive individuals, who capture large territories in the fall and have a lower mortality, and hence longer life expectancy, than individuals raised from high-quality eggs. The latter is counterintuitive and surprising, but we shall meet this peculiar relationship again in other vertebrate species. Low quality of individuals was associated with declining, high quality with expanding, populations.

The above syndrome of individual and population quality is eminently adaptive. The red grouse feeds primarily on the shoots of heather (*Calluna vulgaris*). On moors underlain by base-rich rock, the heather is nutritionally of high quality, which results in high-quality eggs and relatively large peaceful red grouse that hold down small territories and raise large broods. Conversely, on acid moors with poor quality heather, scrappy red grouse are raised that retain large territories which compensate in size for the low density of good forage on the moor. The phenotypic response of red grouse to the quality of the heather, therefore, apparently permits individuals to maximize their reproductive effort; the high-quality birds do not fight for territories in excess of that needed to raise a large brood, and the poor-quality birds enlarge territory size and thus enhance the chances of leaving at least a few offspring. Not only the geologic substrate of the moors, but also climatic and edaphic factors affected the quality of the heather and left their mark in turn on grouse populations. Moreover, there was a variability among the birds themselves in responding to environmental factors.

In the red grouse studies, we also have some inkling of the consequences of individual variations in quality. From the foregoing it is evident that low-quality birds that are territory holders maximize reproductive fitness by maximizing territory size and longevity, and high-quality birds by minimizing energy expenditure for activities other than reproduction; grouse unsuccessful in getting a territory, or losing one they held, die off rapidly in the following winter and spring.

The syndrome of individual and population quality was discovered independently, and considerably expanded in studies of mammals, primarily of domestic and wild sheep, goats, and deer. In Australia and Scotland, heavy postnatal mortality of lambs sparked a series of most important studies. Since I reviewed them elsewhere (Geist 1971a, pp. 284–87), I shall give only a sketch here. It was found that ewes on a low plane of nutrition bore lambs that were relatively small, and had relatively small and underdeveloped bodies compared to the size and development of the head; occasionally lumbar vertebrae were missing, the ossification of the bones was less advanced, they had a miniscule amount of fat at birth on which to survive the critical postpartum period, and their skin had relatively few follicles. These lambs were weak at birth; many failed to maintain body temperature and died of hypothermia. The ewes on the low-plane diet had underdeveloped udders; they were often exhausted at birth, and tended to desert their lambs. They had low milk production, and hence their surviving lambs had poor growth. Poor-quality lambs failed to catch up with high-quality lambs when put on a high-quality diet after weaning. They matured later sexually, and remained relatively small as adults.

This picture was reinforced by studies carried out on feral sheep, the Soay sheep of St. Kilda. These sheep live on a storm-swept island in the North Atlantic without husbandry, nor a predator to cull their ranks. The climate is uniformly cool with smaller temperature and precipitation fluctuations than in most terrestrial ecosystems. The Soay sheep live in a rather dense population at the verge of caloric bankruptcy. To the low quality syndrome we may add, in relation to studies on Soay sheep, exceedingly fragile skeletons, low hemoglobin and red blood cell counts, high rates of parasitism, low vitality and endurance, and heavy mortality of age classes subjected to energy expenditures such as during body growth, late pregnancy, lactation and after intense social activity (Grubb 1974, Benzie and Gill 1974, Cheyne et al 1974, Boyd and Jewell 1974).

While the studies in Australia and on St. Kilda were in progress, investigations on the biology of free-living mountain sheep in North America, and ibex in Switzerland, expanded the population quality syndrome still further (see Geist 1971a). My studies indicated that populations differed in quality and that low individual quality was reflected in small, listless lambs, which played little, were permitted to suckle for relatively short durations only, and although they tried suckling often, they were regularly discouraged. Such lambs fed on green vegetation very early in life, shed their lamb coats late, matured late sexually, and had relatively poor horn growth in their earlier years of life, but relatively good horn growth later in life, and enjoyed a longer life expectancy than high-quality sheep. The poor-quality rams remained low in the dominance hierarchy owing to their small horns, and therefore had a relatively slim chance of breeding an estrous female in any one year. In their social behavior the poor-quality sheep were relatively listless, resorted mainly to overt aggression, and performed fewer dominance displays. At the same time, when I published my finding that rams with a relatively poor growth of horns lived longer than rams with good growth of horns (Geist 1966d), Nievergelt (1966) published the same for ibex in Switzerland. He showed this phenomenon between populations; my study showed it within them. The sheep studies were picked up and expanded by Shackleton (1973) and Horejsi (1976) who went considerably further. Not only did Shackleton (1973) demonstrate beyond reasonable doubt the existence of the quality syndrome between two bighorn sheep populations, but he expanded on this by showing that the skull structure of sheep is subject to predictable variation owing to environmental quality. Thus, low-quality sheep have both a relatively and an absolutely shorter rostrum, while the cranial portions of the skull are less affected. This harmonizes, of course, with the growth priorities in the skull as seen by Hammond (1960). Shackleton (1973) also showed that high population quality was associated with an extended vegetation season and that, in essence, it was due to advanced maturation of the sheep. High-quality ewes had a higher reproductive rate and were better mothers, but apparently—and this requires closer examination—suffered a higher mortality (Geist 1971a). Horejsi (1972, 1976) showed that the quality syndrome outlined for lambs and ewes was distinctly recognizable between years in the same populations. In a very detailed study, Horejsi (1976) examined the behavior of wild bighorn females and their lambs. He showed beyond reasonable doubt an annual variation in the behavior both of the female and of the young, and tied this variability to the annual difference in plant phenology. Heimer and Smith (1975) showed that for Dall's rams from Alaska a good inverse correlation existed between the rate of horn growth and population density, a finding similar to Nievergelt's (1966) for ibex.

My colleague Petocz (1973) showed that bighorn rams in winter varied their social behavior with the severity of winter conditions. In a winter with high snow and low temperatures, the rams used overt aggression relatively often, but decreased overall social activity; in mild winters, or mild periods within winters when the snow blanket was absent or low and forage readily accessible, they reduced overt aggression but increased dominance behavior significantly. This study indicated, therefore, that general ontogenetic tendencies in social behavior, in response to a stressful environment, were similar to those of adult individuals subjected to stressful conditions.

Studies on the lungworms (*Protostrongylus stilesi*) of bighorn sheep by Uhazy et al (1973) suggest that the rate of lungworm infections is inversely related to population quality. This relationship is not firmly established, but can be inferred at present by combining the data by Uhazy et al (1973) with a knowledge of the populations sampled. Thus, bighorns from two populations of high quality (in Waterton Lakes National Park and on the Sheep River, see Geist 1971a) had low rates of larval infestations, while a stable poor-quality herd in Jasper National Park, which I judged to be of low quality on the basis of morphological criteria, had the highest infection rate. This harmonizes with the findings of Cheyne et al (1974) on poor-quality Soay sheep, as well as with the findings at the Behavior Farm Laboratory of Cornell University that stressed sheep and goats that developed poorly also carried a high parasite load (Moore 1968). The explanation for this observation can be found in Stein et al (1976). Psychological factors do affect the susceptibility of individuals to infections by parasites and pathogens, as well as affecting immune processes. It appears that stressful stimuli that increase the level of adrenocortical hormones lower the amount of circulating antibodies. Not surprisingly, dominant individuals or those raised in a relatively stress-free environment had the highest antibody titers. Stein et al (1976) also report a relationship between psychological factors and the development of cancers; stress reduced the survival time of individuals with some experimentally transplanted cancers, and it modified the incidence of occurrence of some carcinomas. Similar findings were reviewed by Seligman (1975) and discussed by Seifter et al (1976).

In addition to increased parasitism and pathologies, it appears that declining sheep populations, which would be low-quality ones, suffer also from high incidence of skeletal anomalies. Allred et al (1966) reported a high frequency of skull anomalies from desert bighorns living in a declining population (Hansen 1967) on the Desert Game Range in Nevada. These observations are supported by observations of my own on feral goats that were small in size, lived in stable populations, and had a high incidence of malformation and pathologies (Geist 1960). Detailed studies on pathologies of elephant populations by Sikes (1971) showed that elephants subjected to high density and poor nutrition in national parks developed extreme arteriosclerosis as compared to hunted low-density populations of elephants that were still able to exploit forest habitat. In addition, elephants from deteriorating habitats had noticeable alterations of the heart and its fatty mantle. Laws's (1974) studies of elephants extended the quality syndrome to elephants, since in his detailed studies summarized in that paper he showed that elephants on deteriorating habitat slowed their body growth, matured later, had reduced fecundity, and suffered greater calf mortality.

Studies of deer also support the contention that the quality of the environment is reflected in the phenotype and population characteristics of various deer. Body size, antler size, and reproductive rate can be controlled experimentally in *Odocoileus* and vary directly with the quality and quantity of forage (French et al 1955, Wood et al 1962). This was verified by D. R. Klein (1964) for field conditions, and applied also to feral reindeer introduced to an island in the Bering Sea (Klein 1968). In their pioneering studies of black-tailed deer, Dasmann and Taber (1956) noted that individuals on poor habitats showed less play behavior and were more skittish, as reflected in a greater flight distance. Earlier, Altmann (1953) had made similar observations on elk. Also, with increasing density black-tailed deer were observed to be more aggressive and more easily alarmed (Dasmann and Taber 1956). Riney (1954) found that starvation of red deer increased their irritability. These observations coincide rather closely with the independent finding on mountain sheep discussed earlier.

It has been known for a long time by reindeer managers that the quality of the environment was reflected in the structure, physiology, and reproductive performance of individuals. Thus, a favorable environment increased antler and body size, enlarged fat stores, and advanced the date of antler shedding and growth, as well as the shedding and growth of hair; it increased birth weight and rate of reproduction, advanced rut in males and estrous in females, and reduced the incidence of illness (Zhigunov 1961, Baskin 1970). We can also recognize the population quality syndrome in chamois (*Rupicapra*) upon careful reading of the monograph by Knaus and Schröder (1975) on this species.

In their studies of roe deer (*Capreolus*), Klein and Strandgaard (1972) noted an apparent reversal of the quality syndrome, in that roe deer from fertile soils were smaller and less well developed than those from poor, sandy soils. They were able to explain this anomaly by proposing that in the rather unique biology of roe deer, social strife in early summer—the period of maximum body growth—reduced the energy and nutrients available for growth, which in turn led to small body size in the deer. On the uncrowded poor soils, less strife existed during early summer, and most of the energy could be channeled into body growth. Although no detailed ecological studies have been made yet of low-quality and high-quality populations as such, one wonders whether the former may not be associated with an impoverished flora and aberrant food habits, including the intake of poisonous plants. Research is still outstanding on this topic.

From the large mammals it is instructive to go to small mammals and see how phenotypes change with expanding and declining populations. Much work has been done with microtines, field mice and their relatives, to explain their peculiar population dynamics. Much of the recent work has been reviewed by Krebs et al (1973). In part, at least, we find the familiar quality syndrome when examining vole populations: When populations expand, microtines of various species tend to be largebodied, disperse readily, and have a high reproductive rate. As the population density increases, so does the aggressiveness of individuals, while reproduction and dispersal decline. Simultaneously, a shift in the genetic constitution of the individuals remaining in the population takes place; how it is related to the phenotype of the individuals remains unclear. Left behind in the population (prior to and during the decline) are small-bodied, relatively unaggressive, individuals that disperse rarely and have a high rate of mortality. Thus the population collapses.

It is tempting to see in this population cycle a mechanism that exists by virtue of saturating all available habitat mosaics in an unstable environment if voles chance upon a patch of exceptional habitat. This is, in essence, the argument of Anderson (1970). The colonizers of a piece of rich habitat generate offspring that, because of heterosis and high phenotypic quality, disperse far and wide and colonize all marginal, as well as all good, areas of habitat. Granted the presence of grass fires, drought, or variable snowfall, the generation of high-quality individuals ensures, in essence, that they go far enough, and into diverse habitats from which they will recolonize other patches of vacant habitat. The fact that individuals in expanding vole populations have the largest adrenal glands (Chitty 1961), a condition associated with stress, may indicate only that the body is adjusted to high rates of energy expenditure. That this is so is indicated by the large body size, high reproductive rates, and great mobility of individuals in expanding populations. The individuals of low heterosis, on the other hand, suffer increased social contacts and produce poor-quality young, which is postulated to lead to a population crash.

The above mechanism is logically adaptive only in habitats with great fluctuations in food and cover, and having no conceivable advantage in biomes with great temporal and spatial stability. If this is indeed so, only time will tell. Note that it predicts microtine cycles in tundra and taiga subject to varying snowfall and frost regimes between years, to grasslands subject to fire, and its modern counterpart, the agricultural steppe with rotating crops. Where great habitat stability exists owing to a mosaic of small plots of diverse vegetation, where ecological successions maintain a steady mix of small plots of seral communities, microtine cycles are expected to be low or nonexistent.

Heterosis is another factor that appears to relate positively to the quality syndrome discussed, by virtue of enhancing reproductive fitness of individuals. It leads not only to increased body size, but in chickens it is also directly related to dominance (Craig and Baruth 1965). In reindeer, it leads to increased reproductive fitness owing to increased fecundity, birth weights, and calf survival (Preobrazhenskii 1961). In this experiment the reproductive success of stags exchanged among collective reindeer farms was compared with the success of stags raised by the herds that they bred. Heterozygous individuals are less sensitive to environmental change than inbred individuals (Mayr 1966).

The preceding discussion dealt mainly with studies under field conditions, which indicated the causes and effects of variations in individual quality. It indicated that quality and quantity of available food was largely, but not entirely, responsible for variations in individual quality. It also raised a puzzle, namely, that individuals of low quality at times exceed in longevity individuals of high quality. A little consideration of this concept shows that life expectancy must decline eventually as environmental quality deteriorates. Thus, life expectancy as a function of environmental quality must be an inverted U-shaped curve. In essence this was verified by the studies of Soay sheep, in which the poor-quality males had indeed a very short life expectancy (Grubb 1974). It is pertinent to note that the inverse relationship between feeding regime and life expectancy has been known in rats for a long time (McCay et al 1939, Brody 1945, p. 689, Ross and Bras 1975). Moreover, this relationship is valid even if the rats select their own diet. Thus, growth rates appear to be directly correlated with rates of senility. These puzzles aside, the quality syndrome appears to run as follows: poor-quality habitats are expressed by physically small young and adults, low reproductive rates, deferred maturation, skeletal deformities, specific shifts in adult body proportions and characteristics, increased incidence and intensity of parasitism and pathologies, poor mothering of offspring, reduced play and exploratory activity, low vitality, increased aggressiveness and use of overt aggression up to a point with a slump into apathy thereafter, and increased life expectancy—but a reduction in life expectancy if environmental quality continues to deteriorate. In addition, we must note the theoretical predictions—as yet unverified—that the dispersal phenotype, in order to maximize behavioral adaptability, must exercise "self-discipline." Moreover, heterosis or hybrid vigor ought to be adaptive only to the dispersal phenotype, not the maintenance phenotype. Therefore, we expect the dispersal—not the maintenance—phenotype to be tolerant of distantly related individuals and to roam widely. Furthermore, high mobility and dispersal is a necessary consequence of high reproductive rate lest individuals, by staying in the general place of birth, remove all food and the population crashes.

Social and Psychological Factors Affecting Population Quality

Although nutrition is a most important factor in the determination of adult characteristics during ontogeny of probably all mammals, it is not the only one. High-quality nutrition supplies the building blocks and the energy for an animal's growth and maintenance. However, a growing animal's blueprint, or genome, anticipates a rather specific environment in which it will perform rather specific activities. The "blueprint," interacting with the individual's milieu, provided the building blocks and energy are available, grows into the individual. Thus, the "proper" milieu, physical, social, and internal, must be there for "proper" development. Since the individual adjusts to some extent to compensate for deficiencies in its milieu, it may divert energy and nutrients from growth to pay the expenses of adjustment. The physiological mechanism, however, that permits the individual to adjust to unfavorable aspects of its environment by generating the necessary fuel for its metabolic machine, may severely damage the individual if exercised. This condition is, of course, known as stress. Thus, individual development may be affected by the absence of diagnostic features of its evolutionary environment (rendering its ontogenetic environment unfit, or at least suboptimal), by the deferment of energy and nutrients from growth to maintenance due to noxious stimuli, and by the deleterious effects of physiological mechanisms preparing the body to meet stress.

An examination of the relationship of social environment to ontogeny in various mammals shows that different species are affected differently. It appears that individuals from some species have an ontogeny governed closely by their genotype, while others depend to a great extent on environmental factors, in particular, an exacting social environment. Species requiring a "normal" social environment to develop normally to adulthood are invariably highly gregarious animals, while species that either live in solitude in adult life or have a very short life expectancy depend less on contact with conspecifics for normal development; cercopithecine primates, as studied by Harlow and his colleagues (Harlow et al 1971, see review by Hinde and Booth 1971, Jolly 1972, Hinde 1974), exemplify the former, and rats exemplify the latter (e.g., Eibl-Eibesfeldt 1955, or 1967, pp. 217−18).

There is by now much literature on developmental biology in mammals and other vertebrates. Another example of closed, as against open, genetic programing is the work on rabbits and cats exposing the mechanisms by which these species organize their perceptual fields in ontogeny. Kittens raised in a visual environment filled with dots but not stripes develop cortical neurons sensitive to spots but not stripes (Pettigrew and Freeman 1973); rabbits fail to develop such differential sensitivity (Mize and Murphy 1973), at least to the extent of cats, so that some gene−environment interaction occurs even here (Grobstein et al 1973). In humans, early visual experience can alter visual pathways permanently (Freeman and Thibos 1973). Different species respond somewhat differently to deprivation experiments; some—largely the social ones—whose behavior in adulthood depends to a lesser or greater extent on traditions transmitted by the social system, as well as on competence achieved in social interactions, suffer more than more solitary species do.

The effects of stress on the organism during ontogeny may be through deprivation of nutrients and energy, as well as through hormonal imbalances in the maternal body and in that of the offspring during gestation. The former effects may be comparable to the effects of malnutrition which in rats, and humans, leads to behavioral abnormalities (Levitsky and Barnes 1972, Scrimshaw and Gordon 1968). Other effects would be similar to those described for young mammals born to mothers on a low-plane diet (see above). These effects may be difficult to separate from those of excessive adrenocortical and adrenocorticotropic hormones, for instance, in the maternal circulation. In rats and mice it leads to greater emotionality and audiogenic seizures of offspring, and in rats to feminized behavior of males if they are applied to gestating mothers (Thomson 1957, Keeley 1962, Lieberman 1963, Weltman et al 1966, Ward 1972, Beck and Gavin 1976). Prenatal stress also decreases litter size, offspring viability, and the growth of offspring; alters brain potentials and leads to underdevelopment of the brain, as well as to biochemical changes in the brain, such as a reduction in the nucleic acids and protein fixation of brain tissues; increases the turnover of catecholamines in the brain; leads to abnormal adrenocortical activity in the young, as well as modifying their behavior (Petropoulos et al 1972). Neonatal stress increases emotionality in rats (Levitsky and Barnes 1972). Such rats also fought more, as was confirmed by the studies of Lamprecht et al (1972) and Henley et al (1973); they also explored less than the controls, a finding confirmed in primate studies (Hinde and Booth 1972). High noise levels experienced by mothers of human babies are apparently associated with abnormal development of the babies. They suffer retardation in learning to speak; some stutter (Junghans and Nitschkoff 1975). This speaks of prenatal interference in the growth of the brain. Hinde (1974) reviews the effects of stress on various vertebrates, primarily primates, as generated by isolating individuals early in ontogeny. Inevitably high levels of overt aggressiveness resulted. Normally raised individuals also responded with increases in agonistic behavior when isolated temporarily in various test situations, compared to animals kept in groups (Rowell and Hinde 1963). It may be that stressed individuals develop not only behavior and sight stimuli, but also odors that at once put others on guard, and predispose the stressed individual to high rates of attack. Thus, Morrison and Ludvigson (1970) found evidence in rats for an ''odor of frustration.'' Clearly, in stressed rats this could lead to enhanced hostility in encounters.

In addition to energy and nutrient deprivation, as well as toxic effects of hormonal imbalances, there are also the teratogenic effects of selective shortages of vitamins, amino acids, and minerals caused by the altered metabolism of the stressed maternal body (Timiras and Vernadakis 1972). Thus, shortages of these essential nutrients cause a great diversity of congenital deformities in neonates and increase the reproductive wastage of the individual and the population. A deficiency in zinc, for instance, during gestation of the offspring of rats leads to reduced brain size, impaired DNA synthesis, decreased RNA polymerase activity in liver and brain, and even teratogenic effects in the brain and behavioral abnormalities such as impaired avoidance learning; it also enhances aggression in the offspring (Halas et al 1975). The teratogenic effects are, of course, not easily separated from the effects of hormone imbalances in the maternal body or toxic concentrations of metabolic breakdown products.

In the social isolation experiments on cercopithecines, as reviewed by Hinde and Booth (1971), Jolly (1972), and Hinde (1974), many factors are varied simultaneously, so that it is quite impossible to ascribe the effects to specific stimulus deprivation, stimulus excesses due to the limiting environment, or to stress and its various secondary effects. Whatever the causes of ontogenetic maldevelopment incurred in isolation, the effects are long lasting. It produced Rhesus monkeys that played less, approached strange objects less, were hyperaggressive and generally hypersensitive, gave inappropriate social signals and developed such psychopathology that they were incapable of normal social life, at least until they were carefully "readjusted" over long time spans with the aid of other monkeys acting as "therapists." Chronic depressions also resulted (Akiskal and McKinney 1973). Chimpanzees, by comparison, were able to readjust socially a little better than Rhesus monkeys after longer experimental separation (Jolly 1972, p. 233). This primate work verified the earlier work of Liddell and his colleagues performed on sheep and goats (Liddell 1958, 1961, Moore 1968). It also verified the finding of Moore (Moore 1968) that the presence of the female protects the infant against stress. Stimuli that were upsetting, and in Moore's experiment fatal to lambs, produced little or no noticeable effects on infants in the presence of their mothers. The work carried out in Liddell's laboratory on the effects of experimental manipulation of mother−young relationships reads in part like a horror story (see the excellent review by Moore 1968) and leads to the inevitable conclusion that abnormal maternal care leads to abnormal offspring.

Another finding of Liddell and his colleagues (Moore 1968) is of the greatest significance: Females that were poorly mothered and developed into poor-quality adults, mothered their offspring poorly in turn, producing a low-quality adult. Subsequent work on monkeys and rats has confirmed this finding. The young of agonistic Rhesus mothers that mistreated their offspring raised similar youngsters (Mitchell 1968). We may assume that, if a normal female is subjected to a deteriorating environment, she will become a more irritated aggressive mother and pass on her behavioral traits culturally to her offspring, which passes it on in turn. increased aggression in response to increased population density appears to be widespread (Archer 1970); it also appears as a consequence of reduced availability of good nutrition, as I discussed earlier. In rats the effects of neonatal stress were passed down for two generations (Denenberg and Rosenberg 1967), as was the impairment of antibody formation in rats due to malnutrition (Chandra 1975). Thus, it is likely that once a population's quality begins to deteriorate, the process is not readily reversed. We may note that there is some evidence that in humans the same process operates, since parents with a history of child abuse tend to have had abusing parents (Parke and Collmer 1975).

An increase in population size up to carrying capacity and concomitantly heightened intraspecific competition, leads to a variety of effects on the individual. Most attention has been focused on the physiological and behavioral malfunctions (Christian 1963) so easily generated under experimental conditions in rodents. These studies are well reviewed by Archer (1970).

Since most concern has been with detrimental consequences of crowding, it has not been widely recognized that some stress generated by an increase in population density does have beneficial effects on individual development. Enhanced stimulation leads to large-bodied responsive rats; mild electroshock advances puberty in rats (Vernadakis et al 1967, Petropoulos et al 1968). Even malnourished or hypothyroid ones overcome to some extent their handicap if they receive enhanced stimulation, and can develop into normally behaving rats (Levitsky and Barnes 1972, Davenport et al 1976). Some stress during ontogeny may enhance the immune response of rats, thus preadapting them to fight at least some pathogens (see Stein et al 1976). Thus, the initial effects of depleted resources through intraspecific competition may be counterbalanced somewhat by enhanced stimulation in the increasing population. Thus, mild stress may be beneficial by keeping individual quality at a relatively high level; enhanced responsiveness should lead to more exploration, which in turn increases the individual's range of physical and intellectual activities, which in turn should lead to maximizing brain size and development, muscular strength, heart size and circulatory functions, renal clearance, etc. This concept harmonizes well with the notion that organisms require an optimum amount of arousal and that mild arousal is rewarding (Berlyne 1966); this may only be forthcoming above a minimum population density (Schultz, Binda in Archer 1970, p. 199). This leads to the conclusion that mild stress may be favorable for the developing, as well as for the adult, individual, a conclusion that Selye reached and popularized (Selye 1950, 1974).

Above an optimum density, however, social interactions enhance stress by various means and lead to severe deteriorations in growth, physiological functions, and behavioral competence (Archer 1970). In the natural setting, individuals probably leave, so that in some species we may consider that agonistic behavior is a normal mechanism of dispersal, which keeps the parent population within its carrying capacity while the dispersing individuals colonize the vacant patches of habitat (Healey 1967). This appears to be the normal mechanism of dispersal in r-selected species which exploit seral, rather than climax, communities of plants (Geist 1967b, 1971a).

Some of the effects of increased social interactions may be gross, others quite subtle. Among the latter, Archer (1970) lists the inhibition of behavior and thus reduction in ability to learn by subordinates faced with agonistic dominants. It leads to a reduction in exploration; consequently it must lead to morphological and physiological modification of brain and body, and various tertiary consequences on the individual's reproductive success. I listed earlier (p. 62−63) the costs suffered by subordinates as a consequence of social stress and interactions: chronic excitation, psychological castration, reduced size and competence of sexual organs, lowered resistance to infections, loss of weight, lessened physical development, lost nesting sites, mates, and access to scarce resources, raising inviable young, suffering more frequent body damage while healing such damage more slowly, suffering internal damage to organs through stress, and suffering early death.

Two Explanatory Hypotheses Based on Cortical Dominance, Growth Priority, and Blood Flow

The phenomenon of population and individual quality can be explained by a relatively simple set of hypotheses, which unify the diverse morphological, physiological, behavioral, and social syndromes with one basic explanation. It is this: As the quality of an environment deteriorates, so individuals suffer a reduction in the tissues of their bodies that have a relatively low growth priority. These tissues include sexual organs, secondary sexual organs, stores of energy, protein, vitamins, and minerals, and also—and most importantly—the cerebral cortex, and probably the peripheral nervous system. Myelination of the brain spreads from the fibers of the spinal cord to the medulla oblongata, pons, midbrain, diencephalon, and lastly, to the telencephalon or cortex. The cerebral cortex, in humans, is immature at birth. It exercises only limited control; many fibers are not yet myelinated. Clearly its tissues have a lower growth priority, and like all tissues with low growth priority they are grown late relative to related tissues with high growth priority (Timiras 1972). Hence, when environmental quality deteriorates, organs with low growth priorities, such as antlers, horns, strongly masculine and feminine features, the fat stores, skeleton, hide, and even muscles, as well as the cerebral cortex, are reduced in their development. In addition, the scarcity of vitamins and minerals can cause an increased frequency of teratogenic effects, leading to offspring of low viability and low competence as adults (Timiras and Vernadakis 1972).

The second, closely related, hypothesis is that during strong arousal the circulation to the cerebral cortex is restrained; if the animal suffers stress through chronic arousal then this condition becomes chronic, and the cerebral cortex is less perfused with blood than in the unstressed individual. This is what happens during hypertension when peripheral arteries, including those in the brain, contract and are then fused by collagen deposition so that their lumen remains smaller (Ooshima et al 1975). Its effect would be to reduce cerebral dominance over the subcortical behavioral programs, and to raise the threshold to incoming stimuli. This would lead in the strongly aroused individual to relatively "emotional," "irrational" behavior, with an emphasis on overt flight, aggression, and sexuality. The individual would show "behavioral regression," actions that would be considered more appropriate to juveniles than to adults (Wickler 1973). All this because the cerebral cortex would suffer reduced blood flow during strong arousal and thus would not be able to generate the forceful control programs to override the subcorticular centers. It may also be that the ability of the cortex to form control programs is a function of training, and rats deprived of such training responded quite emotionally compared to rats that had learned to exercise some control over their environment (Joffe et al 1973).

The hypothesis that during decreased corticular perfusion with blood the cortex has a higher threshold to incoming stimuli is based on the followings: Stimuli perceived by the sense organs cause local increases in the blood supply of the cortex. This is due to metabolic processes triggered by the stimulus in the cortex, leading to CO_2 release and a vasodilation, leading to an enhanced influx of blood (Guyton 1971). If the blood supply is throttled to the cortex, then obviously the overall metabolism of the cortex must decrease, leading to a reduction in the responsiveness of the cortex.

It is probable that reduction of blood flow to the cortex is a normal response of individuals to danger; it permits the "hard," hypothalamic programs to come forth and be used to aid the individual's survival or enhance his reproductive success. During stress, however, the individual probably develops hypertension, as is found in rodents in high-density populations, natural and artificial (Blain 1973), and as has been repeatedly shown in various mammals and humans as a response to noise stress (Junghans and Nitschkoff 1975). This chronically reduces cerebral perfusion with blood, leading ultimately to an insensitive, dull individual by virtue of higher cerebral thresholds to sensation, and an aggressive one, since signals mostly bypass the cortex.

The findings reported by Junghans and Nitschkoff (1975) that mammals subjected to noise stress suffer histologically detectable damage to tissues in the cerebral cortex, which may later spread to the subcorticular structures, is also in line with the hypothesis that blood flow to corticular regions may be throttled during stress. This is also supported by the work of Ooshima et al (1975), showing that cortical arteries during hypertension show an elevation in collagen synthesis and presumably deposition, leading to a reduction in size and permeability of the blood vessels. Freed of cerebral dominance due to hypertension, subcorticular neural circuits respond strongly. Thus, hypertension in old humans significantly reduces intellectual scores (Wilkie and Eisdorfer 1971); that reduced circulation in the brain leads to malfunction in older humans is also argued by Timiras and Vernadakis (1972).

Linking these two hypotheses we get the following picture. Environmental deterioration induces stress. Mild stress helps to adjust individuals, but as stress increases it leads to a reduction of tissues with a low growth priority, including the cerebral cortex. The mechanism by which this could be achieved is enhanced synthesis of collagen in peripheral arteries and arterioles, deposition of collagen, and a permanent reduction in the lumen and elasticity of peripheral blood vessels (Ooshima et al 1975). Growth priority could be controlled, of course, by the amount of collagen secreted in arteries of various tissues. Reduced blood flow to the cortex owing to the narrowing of arteries leads to a reduction in corticular dominance; hence, individuals respond less calmly, but usually emotionally and overtly, and are somewhat dull, lethargic, and insensitive. Increased overt responses of an aggressive nature, however, lead to further stress. This decreases the perfusion of the cerebral cortex with blood still further, reduces cerebral dominance even more, and so depresses the morphological and physiological development of the cortex even more, particularly in growing individuals. Stress leads to a reduction of nutrients and energy available for growth by diverting them to maintenance, reinforcing the underdevelopment of all tissues with low growth priorities. In addition, enhanced metabolic demands reduce the availability of vitamins, amino acids, and minerals for growth which, depending on the availability of these essential nutrients, leads to an increase in congenital deformities due to teratogenic effects. Since surpluses of energy and nutrients for reproduction and growth are curtailed, and the hormonal response to stress leads to a constriction of blood vessels, the developing young suffer underdevelopment, particularly in tissues with low growth priorities; they attempt to compensate by eating adult foods earlier and reducing unnecessary energy and nutrient expenditures. Hence, we see in bighorns the relatively low activity of young in low-quality populations, as well as their relatively frequent attempts to get milk from the mother, who often rejects them. In addition, owing to low cortical dominance and high thresholds to stimulation, the female is relatively insensitive to the demands of her young, but when she does respond it is often with overt aggression. In this way the female maintains some control over her limited resources (Trivers 1974). It leads to a reduction in the quality of mothering, as well as to the relative insensitivity of the female if her young are in danger, and her ready desertion of the young. This should lead to poor-quality offspring.

The offspring, in their turn, compete for maternal resources and dominance, and inflict upon one another some underdevelopment. Effective aggression at this age pays off with enhanced dominance during adulthood and consequent enhanced reproductive fitness. Effective aggression at this age may also increase the mortality of competitors and thus eliminate them as potential competitors in adulthood. It may also reduce the *intellectual* competence of the affected individuals, as argued by Chance (1969). Thus, investment in aggression may be adaptive to juveniles.

A converse argument, one advanced by Trivers (1971), also holds. If individuals grow up together and are likely to remain together throughout life, then acts that support an opponent may be returned at a future date. Thus, formation of cooperative bonds may be quite adaptive to individuals. This cooperation is termed "reciprocal altruism."

As already noted, the two hypotheses—that strong arousal impairs blood flow to the cortex, and that individuals suffer a reduction in tissues of low growth priorities as the environment deteriorates—both alone and in combination, explain and unify a set of diverse phenomena. They predict that actions typical of normal individuals under strong arousal and stress will be very similar to those of individuals raised in poor-quality environments. During ontogeny, the individual's basic survival mechanisms will be strengthened when exposed to a poor environment, and will permanently mark its adult behavior. They predict the impulsive, irrational, panicky behavior of frightened individuals, the "juvenile" behavior of stressed adults (see Wickler 1973), the hyperaggressiveness and hypersexuality of adults suffering regression due to an institutional environment (Kinsey et al 1948); these hypotheses predict that aged individuals with hardened arteries in the cortex should regress to childlike behavior, while their electroencephalograms should show great similarity to those of children, as is indeed found (Timiras and Vernadakis 1972). Strongly aroused individuals should not only act irrationally but also have great difficulty controlling their behavior; they should respond to fewer stimuli and be quite insensitive to subtle ones. The cortex of stressed individuals should be less developed physically, physiologically, and biochemically than that of unstressed individuals, because the former respond less to stimuli and hence exercise less physical and mental skills and so learn less, and this ought to lead to reduced growth and development of brain centers (Cummins et al 1973); the dendrites or cerebral neurons should be relatively shorter and unbranched, the neural cell bodies will be relatively small, the size and number of synapses will be reduced, there will be fewer glial cells and the RNA/DNA ratio of the cortex should be reduced (Greenough 1975), while the cholinesterase activity of the cortex ought to be reduced (Krech et al 1960). Since IQ correlates with the speed of signal transmission within the brain (Ertl and Schafer 1969), and learning enhances the conductivity of brain synapses (Deutsch 1971), and since both poor nutrition and stress would lead to a reduced cortical perfusion with blood and concomitant side effects, both poor nutrition and strong stress in humans ought to lead to a relatively low IQ in developing individuals. This is almost certainly found, although the effect of these factors independently is most difficult to evaluate (p. 369–70). The two hypotheses predict that up to a point the effects of malnutrition and stress will be identical in the neonate. This remains to be confirmed, but malnutrition and stress lead independently to behavioral abnormalities (Levitsky and Barnes 1972, Scrimshaw and Gordon 1968, Halas et al 1975).

The hypothesis in question can be applied variously: underdevelopment of the cortex is responsible for the appearance of basic overt behaviors (aggression, flight, and sex), while in individuals with a phenotypically highly-developed cortex these ought to be rare and complex, not iconic, but more abstract types of communicative behavior should be found. Hence, during maturation the latter behavior ought to dominate progressively; this is indeed found in mountain sheep (Geist 1968b, 1971a, Shackleton 1973); a delay in maturation is equivalent to a delay in the maturation of the complex behaviors mentioned. Illness ought to produce a temporary reduction in the blood flow to the cortex, and hence a reduction of corticular dominance, reducing intellectual, rational behavior in favor of an emotional one. Low levels of arousal, reducing blood pressure, should decrease blood flow to the cortex and reduce the sensitivity of an individual to external stimuli; strong aversive stimuli should lead more frequently to flight or attack in such individuals, owing to suppressed cortical dominance, than in mildly aroused—and therefore alert—individuals. Mildly aroused individuals should learn better than unaroused or strongly aroused ones, owing to enhanced cerebral perfusion. Individuals raised in environments with low or very high levels of stimulation ought to grow relatively poorly; again, available evidence discussed earlier supports this prediction. Individuals raised in environments with optimum stimulation should be relatively resistant to stress, compared to those with a relatively underdeveloped cerebral cortex, or those raised under conditions of malnutrition and/or stress. Again, the available evidence appears to support this prediction (Hinde 1974, pp. 231–34).

The foregoing allows us to add a number of criteria to the syndrome of, say, high individual and population quality. Such individuals will be rather alert, sensitive, somewhat restless, curious, and ready to explore; they will control their behavior well, rarely panic, and learn readily and well. We expect also that such individuals will have a higher rate of metabolism, higher threshold of pain, and a relatively higher rate of food consumption.

The effects of deteriorating environments on individuals and populations are now more predictable. During resource surpluses, in expanding populations, obviously the growth of low growth priority tissues is maximized, emphasizing precisely those features of individuals that are diagnostic of its subspecies, race, or species. Although variability may increase, phenotypic variability is reduced and selection proceeds rapidly because genetic variability is now exposed. It is evident that taxonomy in large mammals uses not only social organs (Geist 1971b), but also low-priority tissues, as a basis for classification. Therefore, deteriorating environments should produce individuals progressively less endowed with diagnostic features of their species or race. A reduction in body size means lower competence in agonistic interactions; a reduction in limb length means reduced running ability; a reduction in secondary sexual characteristics may mean reduced breeding success; reduced stores of resources means a greater possibility of caloric bankruptcy; while reduced corticular structures produce behavioral regression and lower parental competence.

Evolution and Adaptive Significance of Growth Priorities in Tissues: The Dispersal and Maintenance Phenotypes

How do these hypotheses harmonize with the discussion in the introduction to this chapter, that genes communicate with the environment during an individual's ontogeny and shape his development to fit the environment he lives in? The foregoing discussion at best only illustrates that the individual in unfavorable environments is a hopeless victim of circumstances, suffering underdevelopment, but certainly not adaptive changes. Such a discussion was necessary since it brings out the point that environments very unfavorable, such as those created purposely under laboratory conditions, are so far removed from the norms encountered by the respective experimental species that severe malfunctions are their response. Under field conditions individuals would probably have departed in search of new living places long before conditions deteriorated to the extent to which we can permit them to deteriorate in laboratories.

How then do genes control development adaptively? The answer appears to lie with different growth priorities among tissues, which also gives rise to the phenomenon of allometric growth. Growth priorities have long been identified in domestic animals and intensively worked with (Hammond 1960); in fact, they are the foundation of animal science. Neither their evolution nor their adaptive significance has been explained to date, however (Mayer 1966). I propose the following explanation: Tissues of *high growth priority* are essential to the organism's function under environmental conditions of low energy and nutrient availability and high intraspecific competition; they are highly canalized tissues, or tissues of K-selection. Tissues of *low growth priority* or low canalization are tissues highly functional under conditions of r-selection, such as during dispersal into vacant habitat. Tissues of high growth priority have been subject to selection for phenotype redundancy; their development is so essential under conditions of K-selection, that is, when individuals are competing for scarce resources at carrying capacity, that a multiplicity of diverse genes can be expected to contribute to the strongly canalized development of organs that are most adaptive under conditions of resource shortages. Growth priorities are evolved because the organs of an individual enter into competition for the body's limited resources of energy and nutrients. Organs least useful under conditions of resource shortage give up or pass up nutrients and energy. Such competition of organs within a body could be controlled by multiple genes, gene linkages, and pleiotropy of genes enhancing the growth of organs needed for intense resource competition by the individual. An alternative hypothesis is that the genetic control over growth priorities among organs is simple, and amounts to little more than altering the rate of collagen synthesis in the arteries of each organ, as is suggested by the work of Ooshima et al (1975). Organs of high growth priority respond little or not at all to stress by increasing the output of collagen in their arteries. Organs that respond sensitively to stress with enhanced collagen synthesis would, of course, show a low growth priority.

Let me illustrate this hypothesis by example: Under conditions of superabundance of forage, *normally encountered only by individuals colonizing a vacant or new habitat*, it is adaptive to prepare individuals for a relatively great diversity of situations as well as to make them opportunistic. To achieve this, individuals born to early colonizers ought to search out cautiously and explore new objects, situations, and social circumstances, and gain familiarity and competence in dealing with them. They must become generalists. They must store the newly acquired knowledge and motor patterns for future use. Therefore, during ontogeny such individuals should explore and play more than individuals born into stable populations of carrying capacity. A lot of physical exercise enhances learning very significantly, as was shown in rats by Hecht (1975) and in humans by Dru et al (1975). Physical exercise also raises the blood pressure, forcing open the lumen of blood vessels that lead to or are situated in organs of great use. This, in turn, permits increased nutrient and waste exchange along these blood vessels, and consequently an enlargement of the surrounding tissues. Clearly, for colonizing individuals for whom large body size is adaptive for competing socially, there ought to be plenty of exercise that should be embedded in play during ontogeny (Fagan and George 1977).

As a consequence of exploration, enhanced competence and concept formation, and greater memory storage, there is greater demand for neural mechanisms to accommodate them. Therefore, we would expect the brain to respond somehow to deal with these demands. Brain tissue enlarges in response to exercises of diverse nature, as shown experimentally (p. 31). The brain should grow largest in those areas that function to store memories, evaluate alternatives, handle sensory inflows, and control learned motor patterns. These regions of the brain are the lobes of the cortex. Thus we expect, under the above conditions which are founded on superabundance of resources, an enlargement of the cortex relative to the brain stem. This prediction fits nicely with Glickman and Sroges' (1966) empirical finding that large-brained mammals explore more than small-brained ones.

Because we expect a lower life expectancy in the well-grown dispersing individuals, we also expect increased competition by males for females. So investment in display organs is highly adaptive for males, since they face intense competition from males, and must attract the female. It is precisely under conditions of resource abundance that such investments by the male can be maximized; therefore, the amount of advertisement is a true genetic reflection of each male's degree of competence. Females ought to drive male competition to the utmost, which they can afford to do because of adequate resources for rutting. Extensive running by the estrous female notifies most potential suitors and makes them enter the competition. We therefore expect most vigorous activity in both sexes at rutting time in the expanding population with a superabundance of resources. This we tend to find in mountain sheep (Geist 1971a, Shackleton 1973).

In addition to preadapting the individual behaviorally to deal with diverse situations and develop solutions *de novo*, the dispersing individual also should be equipped physiologically to deal with diversity. In the colonizing population, this is achieved partly through heterozygosity, which develops due to the greater range over which the active dispersal-type individuals roam, thus breeding with individuals born and raised relatively far from each other's place of birth. Genetic heterogeneity may also be maintained by chromosome linkage in dispersing populations.

We noted that under conditions of superabundant resources maximum reproduction is possible and, of course, selected for. Individuals capable of flooding the newly colonized habitat with their genes by virtue of higher reproductive output increase their relative contribution of individuals in the succeeding generation. However, high reproductive rate carries a mandatory condition: The young must disperse. A phenotype of active, highly mobile, sensitive individuals is strictly obligatory. Individuals must be ready to move out as social interactions increase, rather than staying and fighting it out. If high reproduction were *not* linked with a phenotype of mobile sensitive individuals, then the population would rise and crash, just as did the reindeer introduced on a pristine St. Matthew's Island, as described by Klein (1968). Here the individuals could not remove themselves from each other due to the barrier of the ocean. Their population rose rapidly, the food was soon consumed, and the great die-off began. Maximizing reproductive rate, "r," can only be done by an obligatory link to a higher tendency to disperse on the part of individuals. So much for the "dispersal phenotype" under conditions of superabundance.

Let us now look at the reverse situation. Under conditions of intense competition for available resources, such as experienced in stable or declining populations at carrying capacity, individuals ought to intensify their competence to deal with *a few or only one* strategy of resource exploitation. They should reduce the diversity of behaviors they could use, and settle on only a few that have a high return in energy and nutrients relative to the amounts invested. Therefore, we expect a shift very early in life into activities that maximize individual *specialization.* We expect a reduction in play, and rather lethargic behavior that rarely deviates from activities essential to the ingestion and conservation of nutrients and energy. Instead of exploration and self-discovery, we expect imprinting and social learning to effectively transfer existing adaptive knowledge. A reduction in exploration and play, following the above argument, ought to lead to a phenotypic reduction in brain size. Note also that a reduction in activity should lead to a reduction in the development of the body and resultant reduced body size, so that only those tissues should be developed that maximize an individual's adaptive success in a population of intense competitors. This does include ability at overt aggression and sex, and reduction in dominance displays and courtship, since under these circumstances (K-selection) sexual selection does not reach the pitch and intensity it does in the vigorous short-lived individuals of the colonizing populations (see "dispersal theory," p. 256). We expect under these conditions of intense competition a reduced phenotypic development of the cerebral cortex. Thus, there is a relatively smaller poorly developed cortex, preadapted through reduced play and exploration to handle relatively fewer situations. Clearly, when unpredicted stimuli do appear, the animal with the poorly developed cortex is more likely to react with emotion, escape, and panic than the animal with the highly developed cortex, a cortex trained to handle new situations. The animal with the poorly developed cortex should have lower "cortical dominance" than the animal with the well developed cortex.

Different growth priorities in the brain are thus visualized as a means by which the genome adapts an individual to the uncommon event of colonization. Note that exactly the same argument can be made for antlers, organs of reproduction, and fat depots as for the cerebral cortex, all being tissues of low growth priority.

Now we ask again, how does a gene communicate with the individual's environment during ontogeny? The following hypothesis is apparent.

During ontogeny, sufficient quantities of high-quality food alter the blood chemistry of the gestating female. Through an unknown chemical messenger (probably a protein), the activity centers in the infant's central nervous system are altered in such a fashion that its "reward" and "punishment centers" have a high threshold, leading to development of an individual with high motor activity, "curiosity," self-discipline, social tolerance, and insensitivity to pain. That is probably enough to set the train in motion to produce the typical high "quality syndrome" discussed earlier, and is also the simplest testable hypothesis.

High activity entails enhanced muscular movement; enhanced muscular movement leads to enhanced body growth if the nutrition permits it, which in colonizing populations it does; enhanced activity leads the individual into contact with more stimuli and situations; greater excitability or sensitivity to stimuli leads to greater exploration and familiarity and greater brain growth, provided the neural mechanism rewarding *pattern matching* (p. 30) is highly sensitive, which we postulated above. Therefore, brain growth is enhanced. Rapid body growth accelerates the time at which reproductive size is reached, shortening ontogeny. Thus, enhanced activity ought to lead to enhanced body and brain development and enhanced maturation.

This hypothesis, that enhanced activity leads to enhanced development, can be tested experimentally if activity by an individual is artificially increased irrespective of the quality and quantity of food. Such experiments have, in fact, been done on rats, by applying electric shock or handling growing rats. The results are as follows: Malnourished rats subjected to stimulation by handling developed closer to normal compared with the malnourished unstimulated control rats (see Levitsky and Barnes 1972); adequately fed young rats developed better if subjected to mild electric shock (Vernadakis et al 1967, Petropoulos et al 1968). These experiments also explain why individual mice, raised in a communal nest made up of several litters, grow faster than individuals raised in a normal nest by a single mother (Sayler and Salmon 1969); the stimulation provided by the visits of several, rather than one mother, in addition to the stimuli of many moving nestmates, probably sets up a regime of mild stress leading to the same enhanced body growth that was apparent in the experiments cited above. The findings of Sayler and Salmon (1969) are thus compatible with the hypothesis that enhanced activity imposed by external factors leads to enhanced growth and development. A mechanism that translates intense activity into body growth may be hypoglycemia at the end of exercising; hypoglycemia stimulates the outpouring of growth hormones, an explanation favored by Schaefer (1970) in explaining secular growth in Eskimos.

The hypothesis that it is a factor in the food of gestating females that signals to genes to switch on behaviors conducive to great individual development is supported also by the following consideration: In northern ungulates, gestation is such that the developing fetus tends to benefit from the spring growth of vegetation. Poor vegetation periods, as caused by cool spring weather, would therefore affect the viability of young, as is indeed found (Hall 1973). Hence, if young are to be affected during gestation so as to develop the high quality syndrome, the message programing them to greater activity ought to be in the spring vegetation. We do know from our studies of mountain sheep (Geist 1971a) that play by adults, be they rams or barren ewes, takes place primarily in the spring and early summer, when the animals are feeding on sprouting vegetation, but not when the animals are on fattening fall vegetation. The same is true of moose which play exuberantly in the spring only (Geist 1963), and mountain goats which I saw play in the spring only, in two years of observation. This is confirmed by De Bock's (1970) observations of mountain goats. It must be noted that lambs decrease their play activity during the summer (Horejsi 1976), but increase such activity again, along with adults, during the following spring. This suggests that there may be factors in the sprouting vegetation which, if ingested in large enough quantities, enhance general activity, mobility, exploration, and play by individuals—both young and old. This factor could well be an amino acid or plant hormone that affects the sensitivity of the reward or punishment centres in the brain. A factor that raises the threshold of the reward and punishment centers would lead to vigorous, highly mobile, very inquisitive youngsters who would not easily be deterred by a painful stimulus. A recent study by Lytle et al (1975) shows that the amino acid tryptophane influences the seratonin levels in the brain of rats, which in turn decreases their responsiveness to electric shock (punishment). A diet low in tryptophane caused reduced brain serotonin levels and increased sensitivity to electric shock. The same may apply to other mammals. Enhanced general activity would, of course, lead to enhanced body growth, permitted by the abundance of resources. This proposition is, of course, testable experimentally. It also suggests how sudden sprouting vegetation could lead to enhanced reproductive activity in desert animals, or those exploiting the sudden superabundance of unpredictably occurring resources. That the availability of young sprouting vegetation may lead to play is suggested by Schaller's (1967) observations on chital following the first monsoon rains in India. It also predicts that play should be low, not only in animals from poor environments, as is indeed shown (Altmann 1953, Dasmann and Taber 1956, Geist 1971a, Horejsi 1972, 1976), but also in animals from rather constant tropical environments. I am not aware of data confirming or contradicting this hypothesis.

A paper by Zimmermann et al (1973), that belatedly came to my attention, fits closely with the theoretical model developed above. In a comparison of protein-deprived baby monkeys against controls on a diet with adequate protein, it could be shown that the protein-deprived individuals were socially less active, were fearful of strange objects (neophobic), had a low level of manipulative behavior, performed more poorly on some specific learning tasks, had greatly reduced sexual behavior, and exhibited a greater portion of all social behavior as overt aggression. One gets an image of sensitive, fearful, overreactive, withdrawn individuals as a consequence of protein malnutrition. Thus, under laboratory experimental conditions the laboratory monkeys in question showed much the same behavioral syndromes as found for other higher vertebrates under conditions of low individual quality.

Phenotype Plasticity and Man

Does the concept of environmentally directed phenotypic plasticity, which can be noted both in different populations of vertebrates and in insects (Wellington 1957, 1960), also apply to human beings? It almost certainly does. When reading such treatises as Tanner's (1962) on growth in adolescent humans, it is quite evident that environments during ontogeny affect human growth patterns. Psychological and nutritional factors are involved. Hostility of parents can even lead to a cessation of growth in children (Powell et al 1967a,b). Schaefer (1970) shows that secular growth changes in Eskimos are related to an increased consumption of disaccharides. He also marshals evidence to show that not "better" nutrition but increased sugar consumption may be responsible for secular growth changes in Eurasian and Asiatic countries. Excessive resources for reproduction and growth can lead to fat unhealthy bodies and organs and reduced life expectancy (Spain et al 1963).

The discipline of *constitutional medicine* (Damon 1970) appears to come closest to dealing with the subject of "population quality" as discussed here. Its aim, according to Damon, is to "develop a simple, clinically feasible and understandable, objective battery of external observations which will describe body form and which is correlated with body composition and is in turn related to behavior, physiology and disease" (Damon 1970, p. 185). The discipline is concerned with anthropometric measures, proportions of the body, body measurements analyzed by factor analysis, chemical body composition, and classifications of body form. I found in Damon's review paper no theoretical underpinning of the discipline, an underpinning such as would be derived from animal models subject to experimental verification. The discipline was primarily concerned with empirical studies of relationships between form and function; there was no theory to light the way. Consequently one found titillating bits of information but no conceptual cohesion [e.g.

he cites on p. 186 that Bjuruf found that the number of fat cells in men is mainly genetically determined and varies directly with the amount of chest hair, whereas the size of the fat cells is determined environmentally. On p. 188 we are confronted with the finding that stockier men leave more sons, and tall men have more children than short ones].

There is very real danger in pursuing empirical studies without a comprehension of how environment affects body form and function during ontogeny. In the absence of such a theoretical understanding, the worker does not have a basis for comparison or anticipation, and of necessity must rely on population averages of anthropomorphic measurements as norms. It is quite conceivable that in developing norms based on statistical averages, everybody measured could be ill and yet would be considered "normal." What is needed is a theory that describes how body form is affected environmentally, since then and only then is a worker able to identify those individuals grown closest to the ideal of a healthy competent individual with relatively long life expectancy. Unless one has a conception of body form as a consequence of an individual's ontogeny, statistical averages can be equated with normality and little guidance can be developed as to what environments people ought to experience during ontogeny. The theory of phenotype plasticity in relation to environment outlined is such a theory, albeit one requiring considerably more work to become an acceptable instrument.

The need for a general theory on how environment affects human biology is recognized by some authors, such as Dubos (1961), who laments that we shape the environment to pander to human comforts instead of training the body to accept stress and strain, and that we know too little about how to develop the power of individuals to adapt. Obviously, some 16 years after Dubos' lament we are still not much further. Hochstrasser and Trapp (1970) emphasize the need for anthropology and medicine to join forces in order to enhance understanding of the "biocultural nature" of man and the underlying processes. They argue for anthropologists to be integrated into the health sciences—a laudable point—but miss an even more fundamental one, namely that *all* professionals and decision makers ought to be familiar with the concepts of an optimum human environment so as not to contribute blindly to its deterioration.

Attempts at applying the concept of phenotypic plasticity to humans is thwarted because, among other things, no comprehensive account of human growth priorities is available. However, studies on the growth of the central nervous system indicate that the cerebral cortex, relative to other parts of the brain, has a low growth priority. One therefore anticipates that withholding nutrients as well as appropriate stimuli will lead to reduction of the cerebral cortex in man relative to the subcorticular structure. Data on this point are not available to my knowledge, but it is known that nutrition affects brain size, the number of neurons in the brain, the lipid content of the brain, learning ability, perception of geometric form, motor development, hearing and speech perception, problem solving, and IQ in children (p. 369–370, and Shneour 1974, Cravioto and De Licardie 1973). Studies on enhancing intellectual stimulation, such as those reviewed by Caldwell et al (1973), indeed show enhanced mental competence as a result. Whether intellectual stimulation also results in increased physical development, as theory predicts, was not investigated.

It must be stressed again that the human organism is plastic and can make up for reduced body growth through compensatory growth, once the debilitating environment is altered to a favorable one. This point is emphasized by Kagan and Klein

(1973) and Kagan (1976), who pointed out that even in primates severe retardation may be reversible with proper therapy, and that children grow toward "health" given the opportunity. Severe malnutrition in early childhood, however, leaves permanent scars (Shneour 1974). It is known that suboptimal ontogenetic environments do have a negative effect on man, and lead to morphological and behavioral abnormalities (Levitsky and Barnes 1972, Scrimshaw and Gordon 1968, Hinde 1974). Malnutrition in early childhood does have an effect on body size and IQ scores in late childhood, even if the ontogenetic environment is made favorable; whether these differences persist into adulthood is not yet established, according to Winick et al (1975). Children born prematurely apparently can catch up intellectually to better developed siblings by about age five (Osofsky and Rojan 1973). Undernutrition may not adversely affect ultimate adult body size, as Dreizen et al (1967) found, as undernourished individuals, given proper care in late adolescence, did catch up in growth. More importantly, these studies point out how much a good ontogenetic environment can compensate for early malnutrition, and thus support Kagan and Klein's (1973) contention. Malnutrition during gestation can lead to increased congenital deformity, a condition found even in an affluent society (Williams 1971); malnutrition also severely alters immunocompetence, and this in turn must compound the effects of malnutrition by increasing the drain on nutrients and energy imposed by higher rates of infection and a greater intensity of such infections (Chandra 1975). From Tanner's (1962) treatise on adolescent growth, it is evident that many factors accelerate or throttle growth. Those that accelerate it produce, in boys, individuals who tend to be assertive, self-confident, intellectually more mature and of a higher IQ, at least during the growth period. In later life these individuals were rated as better adjusted, had a great interest in conforming, were eager to make a good impression and, on average, had more prestigious jobs and better paid ones. Boys of poorer body growth tended to be aggressive by belittling others, a trait highly correlated with maladjustment. It is also known that body size and growth rate in boys is positively related to social class and success, and that taller girls on the average marry upward in social rank (Tanner 1962, Young 1971). Reduced body growth is associated with malnutrition, illness, increased family size, stress, psychological disturbance, poverty, and decreasing social status of family; it is a complex subject (Tanner 1962, North 1973). In women, short stature due to poor nutrition is strongly related to increased difficulties at childbirth and perinatal mortality (Thomson and Hytton 1973). Additional complexity is introduced by such studies as Teele's (1970) or Jaco's (1970), which indicate that a reduction in social relationships is associated in humans with various breakdowns in mental and social functions; distress during pregnancy may lead to abnormalities of childbirth, and is also associated with mental defectiveness of at least some children (Davids and Vault 1962, Stott 1969 in Joffe 1969). I mentioned earlier Schaefer's (1970) study of the effect of sugar on body growth. Hinde's (1974) discussions on the environmental conditions associated with aggression in humans are also most instructive.

How sensitive humans are to the psychological environment is indicated by several studies on the growth of school children during holidays as opposed to school time (Tanner 1962, p. 137). It was found that growth was throttled in most schools during the school term, but accelerated during holidays, independent of season. Our present theoretical understanding indicates a damning verdict of schools in which physical body growth is throttled. Reduced rates of body growth imply reduced rates of maturation, including mental maturation (Tanner 1962, pp.

207–22), and thus inequality of competence at the chronologic age in various national school examinations. It implies less than efficient teaching, since learning in neurophysiological terms *is* growth, and learning proceeds best when accompanied by adequate physical exercise (Hecht 1975, Dru et al 1975) and a high protein diet. Exercise should enhance psychological development, as it apparently does (Rarick 1973a, p. 220).

Other examples illustrating the sensitivity of human beings to environmental factors in their physical growth are studies on birth order, which show an average progressive decrease in body size with birth order (Tanner 1962), as well as such studies as those of Annis and Frost (1973), which show the effect of the visual environment on visual acuity in urban people as against those living in wilderness.

The sorting of data to support or contradict predictions derived from the concept of phenotypic plasticity is not made easy by the absence of a theory that predicts optimum ontogenetic development in humans. One aim of this book is to develop such a theory, even if it be imperfect, just as long as it stimulates discussion on this topic. In the meantime, it may be noted that some responses by human beings to their ontogenetic environment harmonize with those of other vertebrates, as discussed in the previous section.

In addition to environmental factors, heterosis is of major significance in explaining the increases in body size noted primarily in the industrialized nations in the last half-century (Tanner 1962). It appears that increased mobility of people has led to greater outbreeding; people more and more frequently marry a partner who was not raised in the same village, town, or even the same urban area. Coupled with improved nutrition, heterosis has apparently led to people of increasingly taller stature. This view, however, is challenged by Schaefer (1970) as telling less than the full story and ignoring nutritional changes.

It is important to note that *exogamy* systematically captures the advantages of heterosis. The heterosis referred to by Tanner is that above and beyond the normal heterosis generated by exogamy. The practice of exogamy, however, is recognizable as a cultural adaptation, to enhance individual quality by virtue of hybrid vigor. Heterosis is associated with enhanced fecundity, birth weight, and neonatal survival in reindeer (Preobrazhenskii 1961), and high social vigor expressed in terms of dominance rank in chickens (Craig and Baruth 1965). [Effects of inbreeding in humans are discussed by Cavalli-Sforza and Bodmer (1971) and Schull and Neel (1965); it is shown that it has a depressing effect on body growth and physical and mental competence.] Data of this nature make one suspect that exogamy is a cultural means of maximizing individual qualities, and, through these, population quality.

The foregoing sketch will suffice. The concept of human phenotypic plasticity will be controversial, since it raises the subject of inequality and challenges certain philosophical concepts about ourselves. Despite obvious dangers, I fail to see how this subject can be kept quiet, since it is of such paramount importance. It is crucial to our children. We are altering environments blindly, or—more correctly—are being driven into accepting increasingly unnatural environments owing to diverse economic and political dictates. It appears to me that to believe that humans are infinitely plastic, malleable, and adaptable is to be guilty of criminal negligence.

Chapter 7

Mammalian Systems

Introduction

The diagnostic features of man, features that differentiate us from all other mammalian species, can be considered to be the product of the last round of adaptation to a specific environment. At the other extreme are characteristics that we share with all mammals. They too are characteristics that evolved in response to rather specific environmental demands. They are also the adaptations on which all subsequent adaptations, including those distinguishing us as a species, have been based. That is, natural selection acts on established features when a species radiates into a new environment, so that every adaptation arising is based on a precursor or on an old adaptation from the parent species. Not only is the new adaptation based on an old one, but it is also a compromise between the old adaptation and an unachieved ideal, due to the resistance to change that has not been wiped out by natural selection.

As species pass through successive environments in their evolution, they form new layers of adaptations from existing ones and they obscure the old adaptations. We can consider the genetic adaptations of a species akin to an onion, in which the outer skin is the adaptation to the new environment, the next layer below is those of its parent species, and so on. One point that is false in this analogy is that in the onion one skin covers and hides the one below; in animals, however, we can still see old adaptations, at times merely vestigial, right beside the new ones. The oldest set of adaptations we shall deal with are those we share with all mammals, since they show well how evolution transforms the old into the new; it gives us pause to contemplate the implications of time in terms of major evolutionary changes, or the implications for eugenics; it shows the glories of the mammalian system, but also its liabilities. If we are a new form of life having a potential to survive and thrive a long time, then the implications of our mammalian features must be considered in our long-term plans for our future.

Our mammalian features distinguishable from reptilian ones are these: endothermy and homeothermy, the former meaning maintenance of a high body temperature by physiological means, the latter being control of body temperature despite environmental temperature fluctuations; a double system of blood circulation under very high pressure, in which the oxygenated arterial blood is kept separate from the oxygen-depleted venous blood by means of a four-chambered heart; unnucleated red blood cells of high density in the blood; a simplified circulatory system in which the renal portal vein is missing; breathing by a complicated mechanism of rib and diaphragm movements; a solid skull buried in the complex muscle mass which moves jaws, with a lower jaw composed of one bone element only, homologous to the reptilian dentary bone, and dentary – squamosal articulation, not quadrate – articular, as in reptiles; two occipital condyles which articulate the skull on the spinal column; three middle-ear bones (malleus, incus, stapes) instead of only one (stapes); ethmoturbinary or scroll bones in the nasal passages; a brain that fills the endocranium and floats, not in a large, but only in a small fluid space; the brain as the largest neural mass (in some dinosaurs this distinction was held by an enlargement of the spinal cord above the hind legs); high-voltage, slow-wave deep sleep; a secondary palate separating breathing and chewing chambers; teeth that are found only on the dentary, maxillary, and premaxillary bones; teeth diverse in size and shape, differentiated into incisors, canines, premolars, and molars; some teeth with two or more roots; teeth diphyodont or, rarely, monophyodont, that is, there are two (or sometimes only one) tooth generations, but never many generations of teeth, as in reptiles; no lumbar ribs, and cervical ribs fused to vertebrae; pelvic elements fused, not separated as in reptiles; a phalangeal formula of 2 - 3 - 3 - 3 - 3 rather than 2 - 3 - 4 - 5 - 3, as in reptiles; alveolar lungs with many-branched bronchioles; nitrogenous wastes excreted as urea instead of as uric acid (ureotelic as against uricotelic); skin glands differentiated into sweat glands, oil glands, pheromone-producing glands, and milk glands; hair coat present, scales rare or absent; viviparity (except the primitive monotremes, which are the only mammals to lay and incubate eggs); elaborate maternal care and nurture of young from the maternal body.

How is the constellation of characteristics, apparently unrelated to each other, to be explained? We must consider an ecological niche which, acting upon a typical reptilian system, will transform it into a mammalian one; that is, it will give rise to precisely the characteristics just elaborated.

What follows is an elaboration of a theory I first published in 1972. It is by no means an accepted theory, since there is no accepted theory of how mammalian features evolved. When I first became interested in the subject, it appeared to me that an ecological theory might unify these disparate features, which it did. It produced an internally coherent theory, in that it explains all mammalian traits listed plus some not considered by taxonomists, such as the behavioral features I shall discuss in this chapter; it draws logical links between environmental variables and the resulting adaptations, and is consistent with our present knowledge in a broad range of subjects.

The theory says that during the Permian, in cool but productive regions, mammal-like traits began to evolve in small-bodied reptiles which adapted to nocturnalism and exploited cold-numbed insects and small vertebrates for food while safe from any predation by larger-bodied diurnal reptiles. This is the ecological niche[1] or profession that led slowly to the perfection of the mammalian syndrome

---

[1] I am using the term "niche" here and throughout the book in Elton's, not Hutchinson's, sense, as the *role* or profession played by an organism; the *results* of an animal's "professional" activities is Hutchinson's *realised niche*. When discussing processes Elton's niche concept is the only viable one; Hutchinson's concept as an n-dimensional hyper-volume of physical factors is an attempt at making quantification possible.

and resulted, over the Mesozoic and Cenozoic eras, in successive waves of mammal-like reptiles and mammals, each wave being slightly more advanced in mammalian traits than the preceding one.

The ecological niche proposed led to selection for a raised body posture from the sprawling reptilian one, to a new muscle system which resulted in legs whose bones supported the weight of the animal, and to a new system of muscle tonus which, in addition to keeping the body erect on the legs, provided much of the heat needed for the animal to keep active. The increased energy demand selected for a ravenous appetite, for chewing with multicuspid teeth to fragment food for rapid digestion, for enlarged accessory digestive glands for enzyme production, for energy storage and for enhanced conversion of nitrogenous wastes to urea. The specific dynamic effect of its protein diet also increased the heat available to the animal. Increased energy demand selected for an increased oxygen supply through increased lung surfaces, which simultaneously permitted the large amount of carbon dioxide from metabolized food to escape. The increased energy demand also selected for a circulatory system that delivered the highest oxygen concentration to the areas of high metabolic activity and removed a maximum of metabolic wastes. This resulted in the separation of arterial from venous circulation, but only because blood pressures rose during peak activity and made the reptilian heart nonfunctional. Pressure in the circulatory system rose, since the heart had to pump against the enlarged capillary beds of the lung, the accessory digestive glands—in particular, the liver—and an enlarged kidney. The latter was required in response to the large amount of urea accumulating in the bloodstream because of the increased intake of food, particularly of protein for energy metabolism. Increased circulatory pressures selected for a heavily muscled large heart. Since a ravenous appetite led to a bulging of the gut, the gut—enclosed by abdominal ribs—encroached on the lung space and reduced its size. This reduced breathing efficiency. The problem was overcome by a selection for short abdominal ribs, which permitted the gut to bulge out laterally and ventrally, plus a septum which held the gut away from the lung cavity. The influx of muscles into the septum, combined with a sturdy rib basket surrounding the pleural space, led to the efficient diaphragm breathing of mammals, and increased oxygen intake and carbon dioxide expulsion. To retard the flow of heat from the body there was selection for a smaller surface-to-mass ratio of the body. This resulted in the shortening of the tail, and loss of the tail as a combat weapon. Its function was taken over by the mouth. To defend itself against the teeth and jaws of opponents, the mammal-reptile grasped the opponent's jaw in combat. This selected for strong jaws, strongly hinged jaws, sturdy skulls without prokinetic joints, a short neck, powerful neck muscles, and a large, strong occiput. There must also have been early selection for a body insulation, such as hair, to retard heat loss to the environment. Hairs probably evolved first as tactile organs and later functioned to markedly increase the apparent size of the animal. This display was adaptive in conspecific encounter since, if effective in intimidating opponents, it reduced the chance of combat. The erectile hair would be evolved to adequately enlarge the animal's silhouette in the twilight of dusk. Later hair evolved to function in thermoregulation. To be nocturnally active, the reptile required selection for improved olfactory, auditory, and tactile senses, as well as for an improved momory system for orientation and navigation. Simultaneous selection for improved chewing caused the head musculature to enlarge and creep over the skull, burying it within muscles. This produced the tender nose, lips, and fleshy face of mammals. The periodic exertions above the level practiced by reptiles selected for a quick release and redisposition of minerals from the bones, giving them a diagnostic histology. Such bursts of peak activity also selected for efficient mechanisms of heat dissipation, resulting in the evolution of sweat glands and effecting evaporative cooling by panting. Since increased metabolic rate shortened individual life expectancy, it also selected for intensified intraspecific competition for severe overt aggression and complex dominance displays. The requirement of the newly emerged young for food, which they could not procure for themselves in the cool night, selected for maternal behavior, that is, extension of parental care beyond hatching. Since neither the young nor the adults were protected on the surface of the land during the day, due to the activities of diurnal predators, there was selection for burrowing and nest building, which selected for strong appendages and a reduced phalange formula. This also permitted the adults in the insulated nest, even with imperfect homeothermy, to continue the digestive process by means of the waste heat from metabolism and lipogenesis. Philopatry, the stresses of nocturnalism, the difficult strategies for gathering food, and the high cost of metabolism and concomitant low surplus of nutrients for reproduction, selected for intense parental care. Lactation probably evolved as a means of minimizing excretion of feces and urine by the young and tainting the nest with food odors, so as to reduce the chance of predation. However, it also, on one hand, permitted a more efficient use of scarce nutrients, and on the other it allowed the female to supplement the nutrients in low abundance in the food from her body stores. The increased sensory inflow during nocturnal roaming, the requirement to orient and navigate by memory, and the demands of maternal care and increased social activity, as well as the requirements of ever-improving muscular coordination, selected for the begininings of the large complex mammalian type of brain. The shortened life expectancy of individuals, coupled with the loss of reproductive fitness due to loss of complex teeth, led to a selection for increased individual tooth life, and thus a reduction in the number of tooth generations; mammals reduced the number of tooth generations to two or even one. Mammalhood has its disadvantages: The mammalian system is up to six times more expensive to maintain than the reptilian one, is relatively poorly suited for deserts and hot climates, and leads to high intraspecific competition; it became competitively inferior to that of the allosaurus during the return of hot, dry climates during the Triassic.

Prelude to Mammalhood

If we are to understand our body system, we must go back a long time in vertebrate life, to the Permian period. This period began some 250 million years ago and lasted some 60 million years. Like our own age, it was an age of glaciations, of cool, moist climates, and of geologic revolutions (Robinson 1971, Kurtén 1968a). The continents of today were still united into supercontinents, Laurasia in the north and Gondwanaland in the south, but the big supercontinents were in the process of breaking up (Keast 1972). The geologic revolution brought to an end many plant and animal forms, as have geologic revolutions before and ever since (Cole 1966, Colbert 1955). But revolutions also signal the rise of new dynasties of life, and the Permian was no exception. There appears to be one link between the Permian glaciations, the cool, moist climates, and mammalian evolution, as suspected by various authors (Bakker 1971, Geist 1972, de Ricqlès 1972a-c, Feduccia 1973), and there appears to be another link between the insects that arose and radiated in the Permian and the mammals (Geist 1972). The insects, in particular, diversified into a gigantic subphylum to produce the most diverse and numerous group of metazoans on earth.

In the early Permian, vertebrate life was at a truly primitive level; advanced forms such as the amphibian labyrinthodonts and lepospondyli were becoming extinct. The existing reptiles were primitive. These reptiles—long before the dinosaurs appeared and began their march toward dominance—gave rise to a new dynasty of vertebrate life, the mammals. It was a major—and, insects notwithstanding, may have been *the* major—development of the Permian, for it ultimately led to creatures that differed far more from the pattern of animal life than do insects. These creatures are ourselves. Furthermore, during the Permian and the following Triassic periods, mammal-like creatures became the dominant life form on land. Eventually they lost out to the ruling reptiles during the late Triassic, when the climate became hot and dry on much of the land masses (Robinson 1971), the tropical forests replaced temperate ones, and the sea levels began to rise to their Jurassic condition. This need not detain us here.

The beginning of the mammalian system can be traced to some peculiar large-bodied, lizard-like reptiles of early Permian age (Colbert 1955, Romer 1966). These reptiles, the pelycosauria, were quite ectothermic (dependant on an external heat source to raise their body temperature to operating levels), as the evidence from the microstructure of their bones indicates (de Ricqlès 1972a,c, de Ricqlès et al 1972). They had large "sails" which were formed from spinal processes and covered by skin. These sails prompted the hypothesis, among others, that they were thermoregulatory organs (see Colbert 1955). The pelycosauria probably absorbed the sun's radiation with these sails and warmed their bodies with the warm blood that returned from the sails. Some calculations by Bramwell and Felgett (1973) give substance to this hypothesis. It is also a reasonable view granted the cool climates of the Permian. This, and the sails of the pelycosauria, suggest that gaining heat from the environment, as well as conserving it, must have been important to reptiles at that time, and that it is likely that various means to gain and conserve heat were evolved. The next step leads to the suggestion that maybe mammalian features are part of a syndrome of heat generation and conservation that arose as an adaptation to some aspect of the Permian environment. This simple hypothesis leads to surprisingly complex ramifications, and—as we shall see later—does indeed explain step by step the mammalian body system, as well as various attributes of mammalhood (Geist 1972). In addition, it illustrates how ecological variables are translated into morphological, physiological, and behavioral ones.

First we must examine the reptilian system. It depends for heat not only on the metabolism of the food it ingests, but also on direct solar radiation; it is ectothermic, as compared to a system that depends on metabolic heat, which is endothermic. Without solar heat or some other source of external heat, reptiles do not reach the optimum operating temperature of 30−38°C and remain sluggish, digest the ingested food slowly, and cannot sustain activity for very long. Even in the heat of the tropics, reptiles in the shade of a forest canopy do not quite reach the above-mentioned temperatures (Huey 1974). Nor are modern reptiles much good at learning at low body temperatures (Brattstrom 1974), and there is no reason to expect that ancient reptiles were any better off. All in all, reptiles require sunlight for activity, for they have inadequate physiological mechanisms to retard heat flow from their bodies; they are poikilothermic. They thermoregulate primarily behaviorally, by alternately increasing and reducing the duration and intensity of exposure to sunlight (Huey 1974). In cool climates this means that reptiles cannot be active at night. This is the clue to the evolution of the mammalian system.

A New Ecological Profession or Ecological Niche

Granted the cool, moist climates of the Permian and the adaptive radiation of insects, the opportunity then existed to feed on cold-numbed insects at night, as well as on cold-numbed reptiles (Rodbard 1953)—and their now unguarded nests— provided the nocturnal predator could retain enough body heat at night to maintain core temperature above 30°C and so be able to remain active. The depression of ambient temperatures at night provided an opportunity for a new ecological profession to arise, that of a forest-dwelling, small-bodied, nocturnal carnivore and insectivore—provided it could cope with the heat loss at night (that is, change from being poikilothermic to being homeothermic). Mammalian adaptations, as we shall see, clearly provide such capability.

It is pertinent to point out that several investigators independently hit on the idea that cold or severely fluctuating temperatures and nocturnalism were somehow associated with early mammalian evolution (Bakker 1971, Geist 1972, de Ricqlès 1972 a,b,c, Feduccia 1973, Kühne 1973). Heath (1968) felt that mammalian thermoregulation evolved in response to severe ambient temperature fluctuations. Feduccia (1973) suggested—correctly, I believe—that mammalhood found its origins in the cool climate of the Permian, but he failed to recognize the significance of nocturnalism in mammalian evolution and believed it to be a Mesozoic adaptation. Kühne (1973) also suggested a late appearance of nocturnal characteristics in the phylogenetic transition from reptile to mammal, contrary to Geist (1972) or the position I take here. Robinson (1971) stressed the ureotelic nature of mammals as a product of moist environments. The following is an expanded and more detailed account of a theory of mammalian evolution I published earlier (Geist 1972). The information I have become acquainted with since my earlier paper supports that theory. Note, for instance, Kühne's (1973) paper and the surprising similarity to my earlier work in the conclusions, although Kühne worked from an entirely different scientific base.

In addition to making accessible a rich food source in the form of cold-numbed insects and reptiles and their eggs, nocturnalism offered another major advantage. It offered safety from large diurnal predators. Not only are the capable reptilian predators, such as the sphenacodonts, chilled into inactivity at night, but they cannot have been expected to see in the darkness. It is probably not surprising that significant advances in adapting to this niche come about quite early, for the mammal-reptiles of the early Permian already show unmistakable evidence of endothermy in the histology of their bones (de Ricqlès 1972a).

One additional attempt to clarify the origin of mammalhood is significant. It is Olson's (1959) argument, based on death assemblages, that mammal-like reptiles evolved toward mammalian characteristics in aquatic habitats. Unfortunately he does not show how. One may ask toward what ecological profession did organisms with mammal-like characteristics evolve so that a steady improvement in mammal-like adaptations was the inevitable result? This Olson does not answer, and it is the very question that must be answered. The proposed hypothesis, about mammalian characteristics evolving in reptiles that were adapting to the nocturnal carnivore – insectivore niche in cool climates with productive plant communities, does fulfill the above requirement.

How was the reptilian system altered to meet the demands of nocturnal life?

If an external source of heat is not available, the animal can rely only on an internal heat source to promote activity. If a lizard-shaped reptile did so, it would lose heat quickly through its long body and resultant relatively large surface area. While a relatively large surface area is adaptive to reptiles in rapidly absorbing solar heat, it is a disadvantage in retarding heat loss. In a reptile, no external insulation in the form of hair, feathers, or a layer of fat below the skin would prevent the escape of heat. If heat conservation becomes critically important, as it does in low ambient temperatures, then additional mechanisms of heat conservation can be expected to evolve, above and beyond those typical of reptiles. Terrestrial reptiles may, for instance, retard heat loss physiologically through peripheral constriction of blood vessels (Bartholomew and Tucker 1963, 1964), as well as behaviorally by searching out appropriate microclimates. Reptiles may also maintain core temperature at close to 37°C by continual muscle contraction, as demonstrated in a brooding python (Hutchinson et al 1966). However, they have only limited means of retarding heat flow from their bodies.

A change to a nocturnal way of life in cool climates creates many problems for a reptile, of which heat conservation is only one. In order to be active at night, many changes of the reptilian system are necessary, changes that are not independant of each other, since an alteration in one system requires alterations in another, and since one change creates sequences of changes throughout the reptile's biology. In order to exploit the cold-numbed prey at night, the reptile would have to abandon the common reptilian way of life of spasmodic activity, which consists of lying in wait until something edible walks past and then darting forward and grabbing it. It would have to move about and search for its food, that is, exchange the irregular spasmodic activity for a sustained one. Selection for a body form and physiology that permitted roaming would have brought about the many mammal-like adaptations, as argued by Heath (1968). In order to move about at night, the animal requires an appropriate sensory apparatus. Vision, which most reptiles rely on, is not too useful; therefore, alternative sensory mechanisms such as smell, hearing, and touch must be improved in order to deflect prey and avoid dangerous situations. Moving about in search of prey is bioenergetically much more expensive than sitting in wait; Heath (1968) states that lizards at rest produce insignificant quantities of heat.

Sufficient energy to sustain a high level of activity must be liberated from the digestible nutrients in the gut. However, in reptiles digestion is slow. This might be good enough for the typical reptilian way of life but would be inadequate for the new demands. We can expect new mechanisms to evolve that would accelerate digestion. An increase in energy demand would lead to a quick exhaustion of available stores of energy, so a larger system for stored energy is required; this would be in the form of an enlarged liver and greater fat accumulation. Moreover, the metabolizable energy must be quickly distributed throughout the body, which would require some alteration of the reptilian circulatory system (Tucker 1966). Where more energy is burned, more oxygen is required, and metabolic wastes must be removed at a greater rate, requiring an enlargement of the respiratory and excretory systems. On top of all this, a reptile's muscular system simply is not built for roaming and is capable only of a short burst of high activity before giving out.

From the foregoing sketch it is evident that, in order to be active at night in cool temperatures, a reptile would have to undergo drastic alterations of its body system and behavior. Moreover, the whole system of linked adaptations would have to change almost simultaneously, which would drastically slow the evolution of the organism toward nocturnalism. In fact, it apparently took some 60–90 million years to accomplish it. What looks so simple at first glance, namely changing from a diurnal to a nocturnal existence, is really very difficult. Yet this is the change that best explains the evolution of the mammalian body system.

From Reptile to Mammal: Locomotion, Tonus, Heat Production and Control

Let us turn first to the problem of locomotion. Reptiles typically sprawl. The belly and rib cage support the body on the ground. The legs stick out sideways. For a short time only, a reptile can rise on all fours and scoot off. It cannot stand as we do for any length of time since its white muscles tire quickly; the bone structure of its legs does little to support the body weight economically. Moreover, reptilian skeletal muscles are largely innervated by two sets of nerves, one of which sets the muscles prior to contraction. This is time-consuming compared to the mammalian system, in which muscle tone is controlled by posture reflexes. Posture reflexes are controlled by the tension put on muscles, not by direct command from the brain. This is a simpler system than the doubly innervated one of amphibians and reptiles, but nevertheless it is superior and permits a somewhat better control of muscle tonus (Heath 1968). Thus, one of the first changes demanded by nocturnalism is an alteration of the skeletal and muscular systems to permit roaming about at a tolerable cost in energy. Necessarily, as Heath (1968) hypothesized, a shift in posture came first, including a reorganization of the muscle masses of the torso and appendages, followed by the evolution of a new tonus mechanism. It in turn created a new benefit—or liability, depending on the outlook: The new tonus mechanism generated much heat due to constant muscle activity and thus warmed the reptilian body, albeit at a significant cost in energy. The new tonus and contractile mechanism supplies somewhere between 30 and 60% of the heat required to keep the organism at 38°C. The muscles with the new tonus mechanism are the red muscles so typical of mammals (Jansky 1962 and Davis and Mayer 1955b in Heath 1968).

In addition, the development of a new thermogenetic mechanism, nonshivering thermogenesis, became adaptive. It depends on the calorigenic action of noradrenalin, and is produced primarily within muscle tisue, although the liver, heart, brain, and intestines add a small amount. It is particularly significant as a heat-generating mechanism for small mammals and less so for larger ones (Jansky 1973). Our hypothesis requires a small-bodied reptile evolving toward nocturnalism. For such an animal a heat-generating mechanism, in addition to the new topic mechanism, is clearly important.

Heath (1968) argued that not only a new tonus mechanism evolved but also a new central neural organization, to handle the increased sensory input and to produce finer muscle control. The cranial vault of therapsids increased, as is evident from endocranial casts of the brain. The olfactory bulbs enlarged. There must also have been the beginnings of neocortex differentiation, Heath argues, on the basis of parallelism in neural structures in monotremes, marsupials, and placentals, as well as the secondary motor system that characterizes mammals. The external manifestation in the latter is the pons which has been described in therapsids by Olson (1944 in Heath 1968).

As discussed earlier, we never escape economics when dealing with life. Since holding a muscle under tension costs far more calories than letting it relax, it is important to economize when standing. It is energetically expensive to stand on legs that sprawl out sideways. If the legs can be placed in pillarlike fashion directly underneath the body, then gravity relieves the animal of some of the cost of standing. The vertical alignment of bone elements supports the body weight, and the muscles act primarily in keeping the bone pillars aligned and vertical. This obviously costs less than forcing the body above ground with muscle tension. Thus, the leg structure of mammals reflects the law of least effort, granted the need for the animal to roam and maintain activity over long time periods.

I spoke in all-or-none terms when describing above why new muscle types and bone structures were necessary in the nocturnal niche for a reptile. This I did only for the sake of clarity. Even reptiles have the beginnings of postural reflexes; they are not devoid of them, albeit they are far less well developed and functional than in mammals.

In order to be continuously and highly active, the internal body temperature of a lizard, a turtle, a bird, a mammal, an insect or a tuna fish must be somewhere between 32 and 42°C (Carey and Teal 1966, Heath 1968, Green et al 1973, Carey 1973, Bartholomew and Casey 1977). If the creature is active in cool temperatures, then we expect to find various mechanisms to conserve heat, such as hairs, feathers, fat layers, or countercurrent systems in the blood circulation which shunt heat to the core of the body from the periphery.

Why the high operating temperature of 35−42°C? The answer appears to be that this is the temperature range within which the body's chemistry proceeds at an optimum rate; at lower temperatures the metabolic activity of the body is lower, and at higher temperatures the fragile proteins of the enzyme systems break down and chemical activity again declines. Although in the primitive protherian mammals such as the platypus (*Ornithorhynchus*) and the echidna (*Tachyglossus*), the body temperature varies between 26.5 and 36.0°C and is usually at 31°C, their enzyme systems still work best above 35°C (Aleksiuk and Baldwin 1973). Given the opportunity, the lizard *Diposaurus dorsalis* maintains a body temperature of 38.5−39.0°C through behavioral thermoregulation (Kluger et al 1975). Even the enzymes of fish, which normally have a low body temperature, work best at 37−40°C (Hoar 1966). Heat is thus required to speed up the chemical transfer of metabolizable energy to work energy, and the temperature range at which this takes place maximally is a function of protein chemistry. Organisms do not escape the limitations of chemistry and physics.

Another beneficial result of these high core temperatures appears to be a higher survival of the host after bacterial infection (Kluger et al 1975). Thus, a continuous high body temperature would serve as a protection against most microorganism whose enzyme systems probably would be deactivated in the "hot" interior of the endotherm. Even if microorganisms managed to infect the host, the work of Kluger et al (1975) on lizards indicates that at high body temperatures the host's survival would be better than at low temperatures; at body temperatures of 40−42°C, the survival of infected lizards was increased threefold over that at 36−38°C. Thus, the maintenance of continuous high body temperatures would be a protection against at least some bacterial diseases.

From Bolting to Chewing

In order to sustain a high metabolic rate and high body temperature, the organism must rapidly liberate digestible energy from the food it swallows. Even if a reptile swallows a lot of food it still has no way to digest it quickly. It bolts food in large chunks. This of course exposes only a relatively small surface area of the food to the action of digestive juices. If food were shredded into small pieces or masticated, the surface-to-mass ratio of the food would increase and the rate of digestion would increase. It would also permit increased transfer of energy and nutrients across the gut wall, provided the absorptive surface of the gut is increased, along with the size of the accessory digestive organs and their rate of enzyme production.

The first small step in the direction of increasing the surface area of the ingested food is to "bite" the prey often and put many holes into it. This permits nutritious blood and body juices to flow out and digestive enzymes to flow in. The next step is to bite or tear the prey into small pieces and swallow these, rather than to bolt the prey in one piece. The next step is to improve the efficiency of chewing (repeated biting) by strengthening the jaws and the muscles that move it as well as the skull, plus differentiating teeth into holding and cutting or crushing teeth (Hopson 1969, Barghusen and Hopson 1970). The teeth within the mouth of a primitive reptile are, after all, much the same shape and size.

The paper most significant to an understanding of the evolution of mammalian teeth is the work of Kühne (1973). He goes into considerable detail and explains the functional significance of tooth replacement, differentiation of teeth into canines, the appearance of recurvature, cutting edges, serration on tooth margins, thegosis (the nonmasticatory sharpening of teeth), drift (the meeting of teeth for mutual support, undershearing, and mastication), roots, tongue and groove (interlocking of teeth to counteract tongue during mastication), anisognathism (unilateral function of tooth rows in scissorlike cutting of food), and the reduction of tooth generations. I shall not go into detail here, but refer those interested to Kühne's (1973) paper.

We should note here one advantage of the protein diet to our evolving nocturnal reptile: Protein produces the highest specific dynamic effect of all foods. It releases more heat in being metabolized (Brody 1945, Kleiber 1961) than do other foods.

Alterations of Digestive Glands and Ureotelism

An increased flow of metabolizable energy and nutrients across the gut wall is a mixed blessing. It creates problems for the reptilian system. For instance, the liver must be enlarged so that it may handle the increase in metabolizable energy and nitrogenous wastes; this change also permits increased heat production from the larger liver. Even if an increased flow of energy permits the reptile greater activity, an increase in the metabolism also increases the rate of metabolic waste formation. The body is flooded with nitrogenous wastes, as well as by carbon dioxide. If our hypothesis is valid that mammalian characteristics evolved in response to nocturnalism in temperate climates in a relatively small, forest-dwelling reptile, then the most probable mode of nitrogen excretion is in the form of urea. Ammonia is poisonous and requires large amounts of water for its excretion; it is thus found mainly in reptiles closely associated with water, such as marine or freshwater turtles and crocodilians (Bellairs 1968). Uric acid is at the other extreme. It is a compound excreted by reptiles that are adapted to a desert environment and that must conserve every drop of water. Even the best water resorption mechanism in mammals cannot compare with the water excretion economy of uric acid excretion; it takes at least ten times as much water to excrete urea as it takes to excrete uric acid (Dawson and Bartholomew 1968). Urea excretion is apparently associated with terrestrialism in reptiles, but oly where water is reasonably abundant (Bellairs 1968). Since an evolutionary impetus for nocturnalism only exists where there is a large food resource of cold-numbed insects and reptiles available, and such a large food resource presupposes high biological productivity, and it in turn demands water, reptiles adapting to a forest environment can be expected to excrete urea and hence be ureotelic. Thus, the ureotelism of mammals supports the hypothesis examined here that the mammalian system evolved in a forest environment. This supports the thesis of Robinson (1971) that therapsids compared to sauropsids lived under mesic conditions.

Bigger Kidney and Control Glands

In the hypothesis examined, the nocturnal, forest-dwelling reptile is a carnivore and insectivore. Most of its food energy comes from protein. After the protein is burned in the metabolic machinery for fuel, there is plenty of nitrogenous waste floating in the bloodstream. This nitrogen would be converted to urea by the liver; however, the liver must increase in size to handle the waste problem. Conversion of ammonia and other nitrogenous wastes into urea solves the problem of toxicity, but it creates another one, osmotic pressure. Osmotic pressure is created in the bloodstream by an increase in urea, and this would cause a flow of water from the intercellular spaces into the blood. If this were to continue, the circulatory system would burst. Thus it is advantageous to remove urea as rapidly as possible. Since the water content of the body is reasonably constant, an increase in urea excretion is achieved by increasing the rate of body-fluid cycling through the kidneys, as well as evolving larger kidneys and a superior mechanism of resorbing water that the kidney has filtered. The mammalian type of kidney with its filtration apparatus (glomerulus) and its long resorption loop (loop of Henle) appear to be the evolutionary answer.

If body fluids circulate at a higher rate in mammals than reptiles owing to accelerated metabolism and its consequences, then it is evident that mammals require larger regulating glands to synthesize the necessary regulatory hormones. We would expect in the mammals larger ductless glands which are associated with metabolic activity as are larger thyroids, adrenals, and pituitaries.

Larger Lungs, Bigger Heart, Simpler Circulation

A high rate of activity demands an increase in the absorption of oxygen from the air. This requires a larger lung area, an increased ventilation rate for the lungs, an increase in the efficiency of clearing the lungs of used air, and an increase in the capillary beds of the lungs. The internal surface area of the lung can be increased by convolution. A larger lung area, of course, at once permits a greater escape of carbon dioxide from the blood into the air. This solves one waste disposal problem. Later we shall tackle how the increased ventilation rate and lung clearance was achieved.

A larger lung to facilitate respiratory gas exchange, a larger kidney to facilitate excretion of urea and resorption of water, a larger liver to handle the increased flow of metabolizable energy and nitrogenous wastes, larger accessory digestive glands, all increase the size of the capillary beds and thus increase the resistance against which the heart must pump. Hence, the organism requires a larger heart with heavier, more powerful muscles and larger, thicker-walled arteries to handle the increased blood pressure. Simultaneously it becomes adaptive to decrease the work load of the heart as much as possible.

The work load of the heart may be decreased by decreasing friction within the circulatory system. This can be done, first, by getting rid of redundant blood vessels—and reptiles do have some redundant blood vessels (Romer 1962). Second, the lumen or bore of the blood vessel leading to the capillary beds should be increased, and third, the length of the tube connecting the heart and capillary beds should be shortened. This reduces the surface-to-volume ratio of the blood vessel as well as the absolute area with which the blood makes contact, and thus friction should be reduced to a minimum. If these steps are taken, the complex circulatory system is changed into the simplified, elegant circulatory system of the mammal.

The heart deserves some special attention. The reptilian heart is not only too small for the tasks indicated above, but also badly constructed. It is three chambered, and there is some mixing in the heart of the arterial and venous blood since the ventricle is only partially divided. It is to be noted that the purpose of increased lung surfaces, ventilation rates, and lung clearance is to provide more oxygen to the metabolic burners of the animals, and remove metabolic wastes faster. Hence, a mixing of arterial and venous bloodstreams due to an incomplete separation of the two tends to defeat the purpose of the earlier adaptations. Thus, blood from the lungs carrying an oxygen supply enters the left atrium, passes to the left part of the ventricle, and is pumped to the systemic circulation via the aorta. Blood returns from the body via the vena cava to the right auricle, and is pumped to the right part of the ventricle, from where it proceeds to the lungs via the pulmonary artery, rids itself of CO_2, and loads up with O_2. Since mixing occurs in the incompletely divided ventricle, it is here that an alteration of the system may be expected.

Even reptiles and amphibians with three-chambered hearts have some means of decreasing the mixing of venous and arterial blood (Romer 1962, White 1968, Johansen and Hanson 1968, Bellairs 1968). They evolved a wall (septum) within the ventricle, and even two walls in some forms. The venous blood is shunted mainly to the lungs and the less acid arterial blood to the body, by an appropriate bending of the ventricular septum. Yet some mixing of bloodstreams still is possible. Granted a reptilian heart with a partially divided ventricle, the heart must fail under heavy exercise. As the demand for oxygen is increased under heavy exercise, the pressure within the circulatory system rises as more blood is pumped against the resistance of the capillary beds. Under high pressure, the septum must give way to the onrushing bloodstream and fail to separate bloodstreams. The heart must become larger and heavier, with more and more of the opening between the left and the right parts of the ventricle closed off. When the septum is complete and the opening between the ventricular halves are closed, we have the typical mammalian heart.

Adjusting Surface-to-Mass Ratio

The previous alterations would be for naught were the animal to allow heat to escape unrestrained. Calories are hard to come by, and waste cannot be tolerated since it reduces reproductive fitness. In order to reduce the flow of heat to the exterior, the animal can reduce its surface area relative to its mass. It should become more round. A reptile can start in this direction by reducing tail size. Let us take a cylinder 10 cm long and 4 cm in diameter, a cone 3 cm long of the same diameter, and a second cone 10 cm long. We make an "animal" of these with a tail 10 cm long. It will be 23 cm long and have a surface area of 213.3 cm. It will weigh 179.8 g, given a specific weight of one. Let us keep it the same weight, but reduce its tail length by half. The new tail will be 5 cm long, tapering to a cone in the last 3 cm. Our animal is now 18.3 cm long and has a surface area of 200.1 cm, a reduction in surface area of 6.2% (Geist 1972). This is nothing to sneer at. Indeed, one finds that mammal-reptiles reduced tail size greatly (Romer 1966, Kuhn 1970).

A Consequence of Tail Reduction: New Weapons and Defenses

However, reducing the tail causes problems for our reptile. In general, lizardlike reptiles do not bite in fighting, but lash out at one another with the tail (Rotter 1963, Bellairs 1968, Eibl-Eibesfeldt 1967). No tail, no lashing. What weapons to use? Biting is the next logical thing to do. In fact there is a little evidence indicating that a lizard deprived of its tail will bite (Eibl-Eibesfeldt 1967). The most likely defence against biting, or against any weapon system, is to grab hold of and hang on to the opponent's weapons (Geist 1966b, 1971a). This disarms the opponent as long as his weapons are held and he is not free to use them. For the reptile, this means grasping the opponent's jaw before he can bite and holding on for dear life. However, reptiles have thin, tubular jaws composed of several bone elements rather than one, and they are thus structurally weak. Furthermore, they are weakly hinged. It would not be difficult—given a powerful trunk and legs—to break such jaws and rip them out, or at least damage the weak musculature that operates reptilian jaws. Thus, reptiles defending themselves by grasping the opponent's jaw place themselves in mortal danger. If the reptile is to bite in combat, it requires a heavy, well-hinged jaw and strong muscles, as well as a solid skull and a strong joint between skull and body supported by strong neck muscles.

Indeed, the paleontologic evidence indicates that the early mammal-reptiles did move in the direction of heavy, massive jaws of great structural strength, stronger jaw hinges, a very sturdy skull, and a strong occiput on a relatively short, stout neck (Kuhn 1970). Some oddities like the massive anterior end of the lower jaw in some gorgonopsians, e.g., *Lycaenops*, appear to be explained as an adaptation to jaw grasping in combat. In addition, heavy jaws with large muscles permit stronger teeth, more precise chewing, and easier tearing of chunks of meat from the body of a prey animal. These adaptations, as indicated earlier, would aid in the rapid digestion of the prey so necessary for the sustained activity of our hypothetical nocturnal reptile. A shift to the anterior part of the animals with combat adaptations is also indicated by the evolution of small hornlike organs in herbivorous mammal-reptiles (Geist 1972); this view has found support in the studies of Barghusen (1975) on therapsids.

Toward Homeothermy

A second way to reduce heat loss is to grow larger; this also reduces the surface-to-mass ratio. It has the drawback that, although a larger animal eats less food relatively, it eats more absolutely. Thus, it must roam further than a small one in search of food. It is also more conspicuous and so more vulnerable to predation, it leaves more signs and scent, it can hide in fewer places, and would have to work more to dig its own hiding places; in short, it can be found and eaten more readily by diurnal carnivores. The only place where getting big is of real advantage to the evolving mammal-reptile is in temperate to cool climates. Here their ability to control their internal temperature to some extent would permit them to be active when large ectothermic reptiles could not. However, such an advantage would vanish with the return of warm or hot climates (Robinson 1971, Geist 1972). Indeed, during the Mesozoic the ruling reptiles and mammals would be separated by the day/night change as Kühne (1973) suggested; the reptiles would be active during the day, the mammals at night.

Evolution of Hair

Heat loss can be reduced by developing an external insulation, be it of fur, feathers, or fat. We have no way of knowing whether the early mammal-reptiles did evolve external insulation, but it is likely that they did. Their descendants, the mammals, obviously do have hair. We know that other warm-bodied creatures of relatively small size tend to have external insulation, e.g., moths and birds. There is evidence that the flying reptiles, the pterosaurs, had hair coats, for the imprint of hairs is found on fossilized pieces of skin (Romer 1966). We noted earlier that mammal-reptiles may have had sensory hairs on the face. If so, they almost certainly had hair. External insulation is a far more effective means of reducing heat loss than is a reduction of the surface-to-mass ratio (Geist 1972); it is almost certain that it was evolved rather early.

The evolution of hair is problematic. In order to obtain even a probable answer as to how hair evolved, we must proceed in a roundabout way. Fighting, like any other activity that requires exertion and high energy expenditure, is a great luxury to reptiles. Their white muscles tire fast; their system is soon saturated with lactic acid, a product of anaerobic respiration, and it takes them relatively long to repay their oxygen debt (Brattstrom 1974). It is hence adaptive for reptiles to avoid exertion as much as possible, and replace it with displays that have the same effect. It is also adaptive to restrict the use of these displays, again because they are bioenergetically expensive to the reptilian system. In correlation, we find that reptiles submit easily and quickly. As Evans (in Brattstrom 1974) pointed out, hierarchy in captive reptiles was achieved without fighting, on the basis of body size alone, since subordinates quickly assumed submissive behavior in front of dominants. This indicates that reptiles are very sensitive to body size differences of conspecifics, and respond to larger ones by becoming submissive (see Brattstrom 1974). Moreover, granted their relatively long life expectancy, they can permit themselves the luxury of delaying their ascent to breeding status and dominance. This luxury, as indicated later, is not available to mammals owing to their short life expectancy. We can assume that as mammal-reptiles evolved, they were at least as sensitive as normal reptiles to size differences of conspecifics and initially submittd just as easily to a larger-bodied one. With shortening life expectancy, however, increased self-assertion became more adaptive. Of course, this also made it more adaptive to feign larger than actual size, the effect of which would be to make a cocky rival submissive. In the day-active monitor lizards, we find various means of body enlargement, such as erecting the body high on the legs and depressing the ribcage so as to increase apparent body size (Rotter 1963). In the dusky world of the early mammal-reptiles, however, something more drastic was called for in the poor light conditions; a simple broadside display would no longer do. I propose this more drastic measure was relatively long erectile display hairs. When erected in the twilight, they would greatly enlarge the apparent size of the animal; they originally evolved as tactile organs, and later became insulating hair. This concept also explains the great frequency of body displays that increase apparent size in mammals, as well as the common erector pili effect during agonistic encounters. In essence, the thermoregulatory function of hair is secondary; one original function, when few hairs were present and could not significantly retard heat loss, was to enlarge the apparent size of the animal in agonistic encounters.

To achieve homeothermy, the animal must have not only a favorable surface-to-mass ratio, external insulation, and countercurrent systems, but also various means of expelling excess heat or generating excess heat on demand. The latter is achieved in part by shivering thermogenesis and in part by an increased heat production from an enlarged liver. If we grant mammal-reptiles the ability to suddenly increase metabolic output above the level common for reptiles but below that of mammals, that is to a level more than 6 times but less than 40 times basal (Brody 1945, Gordon et al 1968), then they are in need of quickly dissipating heat. Since metabolizable energy is transformed to work energy with an efficiency of only some 25%, some 75% of the degraded energy will escape as heat (Brody 1945). This requirement explains why mammals have far superior cooling mechanisms than reptiles; they possess sweat glands and can also dissipate considerable heat through panting (Hoar 1966, Bellairs 1968).

In addition to physiological and morphological means of reducing heat loss, the animal can also use behavioral means; it can choose to go where it will lose relatively little heat. Heat is lost from the body through radiation, convection, and conduction. About half the heat is lost through long-wave radiation (Joyce et al 1966, Hoar 1966). An animal can conserve a considerable amount of heat if it moves to a habitat in which the body radiation is reflected back instead of being lost to the open night sky. Forests and dense bush provide an excellent radiation shield (Moen 1968); moreover, forests trap incoming radiation, as well as reradiating back the heat they absorbed during the day to create a greenhouse effect (Clarke and Brander 1973). A small, warm-bodied creature in a cool climate would therefore do well to choose a dense forest as a habitat. This would also reduce its heat loss through convection and conduction, since wind, which increases convective heat loss, is reduced in forests, and a forest floor covered with litter and small plants acts like an insulation layer if dry, reducing conductive heat loss. For a creature for which heat conservation is critical, the forest is therefore an ideal habitat. This is one reason why it is postulated as the evolutionary habitat of the mammalian syndrome.

A New Breathing System

We must again return to physiological mechanisms that permit our evolving nocturnal reptile a high level of metabolism over long time periods. In order to fuel the highly active, but poorly insulated body, the animal must eat a lot of food. If it increases its rate of feeding significantly, its gut will soon bulge. In the reptilian body, however, the gut can bulge forward. If the animal feeds to capacity and its gut is filled to distension, then the gut encroaches on the lungs, leaving very little lung space for breathing. The animal will have a hard hime-breathing and will have to compensate for the reduced lung space by taking quick, shallow breaths. Hence, the more it eats, the less it can breath and the less oxygen is available to fuel its metabolism. Obviously, this is an impossible situation. A way must be found to let the gut bulge away from the lungs. This is, however, not difficult initially. All that is required is a shortening of the reptilian abdominal ribs. Then the gut can bulge downward. This works fine as long as the animal stands. If it lies down, the bulging gut again rolls forward into the lung space. Something is needed to hold the gut away from the lung space and clearly separate gut space from lung space. A tough, fibrous wall, a septum separating lungs from gut, will do. Next, the strength of the rib basket must be increased lest pressure on the fibrous septum draws the ribs inward and reduces lung space. Moreover, a strong rib basket over the lungs will not permit the dead weight of the body to reduce lung space significantly when the animal is resting. The septum between gut and lungs can cause another problem. If it is rigid it will, of course, restrict the movement of the rib basket, and thus reduce the animal's capacity to breath. If it is a *loose* septum, however, it would automatically straighten out when the rib basket expands during inhalation, and in so doing, it would push back the gut. What an opportunity! If muscles were introduced into the septum, the dividing wall between gut and lungs, the septum, could push back the gut actively during inhalation and thus increase the volume of air inhaled. This would make the septum the familiar *diaphragm* of mammals. Thus, primitive diaphragm breathing is an active pushing backward of the gut mass while the rib basket is expanded. When the rib and diaphragm musculature relax, the rib basket collapses, the gut rolls forward reducing the lung space, and the air is forced out. Note what diaphragm breathing depends on: a freely suspended gut mass, unconfined by abdominal ribs, a very strong rib basket, and a muscular dividing wall separating peritoneal from pleural space.

The above is an explanation of how the simple reptilian breathing mechanism could be altered into the complex mammalian one, a subject that has previously not had a satisfactory explanation, although Brink (1956) came close to it. Note that it is a logical consequence of the ravenous appetite of the mammal-reptile evolving toward its new profession as a nocturnal carnivore and insectivore in a cool climate. The ravenous appetite is still found in the most primitive of mammals with poor mechanisms of heat conservation, such as the Australian platypus (Johansen 1962 in Hoar 1966) which requires daily a prodigious amount of food (Walker 1964). A ravenous appetite apparently compensates for poor thermoregulation.

More Erythrocytes

Granted the mechanisms of the hypothetical reptile to sustain a high level of metabolism, it would be odd if its blood retained the low level of red blood corpuscles typical of reptiles. The more red blood corpuscles, the more O_2 can be carried from the lungs to the body, and the more CO_2 can be carried from the body to the lungs. The mammalian condition of a high count of red blood corpuscles should therefore not be surprising (Hoar 1966).

Olfaction, Scent Glands, Touch, and Hearing

We shall now turn to the sensory, neural, and psychological adaptations that permitted nocturnal life. Vision obviously is of little use at night, but a keen sense of smell, touch, and hearing are most advantageous. We find in the early mammal-reptiles—as expected from our hypothesis—evidence for increased use of smell in the form of enlarged olfactory lobes in the brain (Edinger 1964 in Heath 1968), as well as the beginning of mammal-like turbinary bones in the nose which serve for the attachment of the olfactory sensory epithelium (Brink 1956, Moulton 1967). In addition, the turbinary bones with their cover of capillaries serve to warm the incoming cool air, and may also serve to absorb some heat from the expelled lung gases.

We find deep depressions in front of the eyes in some skulls of the early mammal-reptiles, as if these contained a preorbital scent gland such as is found in some mammals today. Clearly, odoriferous glands would be a great advantage, since they would allow unambiguous identification of individuals in the dark, or they could be used for territorial marking. We have no way of knowing whether the early mammal reptiles did indeed possess odoriferous glands; but we do know that odoriferous glands, used in intraspecific communication, are by no means rare in reptiles (Blair 1968). There is little doubt as to the usefulness of such glands in nocturnal life, and that mammals carry an almost incredible assortment of largely unexplained odoriferous glands. Mammals may also cover their bodies with various excretions and secretions; we assume that they live in a strong olfactory Umwelt.

One can gain an appreciation of the complexity of the mammalian olfactory environment from the studies of Schultze-Westrum (1965) and, in particular, the work of Müller-Schwarze and his colleagues (e.g., Müller-Schwarze and Müller-Schwarze 1971, Quay and Müller-Schwarze 1970, 1971, Quay 1959) and Mykytowycz and his colleagues (Mykytowycz 1965, 1966a,b,c, Mykytowycz and Dudzinski 1966).

In the skulls of the early mammal-reptiles we also note numerous openings (foramina) that permit nerves and blood vessels to pass to the exterior of the skull. This speaks for an increasing amount of muscular and sensory tissue external to the skull. In part it is caused by muscle masses climbing onto the skull roof and attaching themselves externally in order to move with precision and power a strong jaw element with diversified teeth. In part it may be caused by an increase of sensitive skin on the skull, which would increase the rate of tactile information delivered to the animal. The foramina and nares openings suggest that a soft, fleshy nose and lips were present in mammal-reptiles (Kuhn 1970). In the nasal area of the upper jaw one finds peculiar deep pits which Brink (1956) interpreted as evidence for vibrissae or sensory hairs (see also Kuhn 1970). Although van Valen (1960) rightly cautions against too hasty an acceptance of this interpretation, it—together with the foramina in the skull—is in line with the view that an increased inflow of tactile information would be adaptive in nocturnal life. Note that blind cats deprived of vibrissae bump into every obstacle, but not those with vibrissae (Altmann 1966); obviously, vibrissae would be most adaptive in a nocturnal setting.

There is no evidence from the middle ear of the mammal-like reptiles that their hearing was any more sensitive than that of the present-day reptiles (Hopson 1966). Both have a similar middle ear structure. However, the complex middle ear of the mammal evolved twice, apparently, once in the therian lineage to which all except two genera belong, and once in the protherians which contain the remaining two genera (Hopson 1966). The advantage of the complex middle ear, derived from the former elements of the reptilian jaw, as can be traced both paleontologically and embryologically (Portmann 1959, Romer 1962), lies in its sensitivity to a greater range of sound frequencies (Grinnell 1968). Again this fits well with the view that increased auditory activity would be adaptive in the nocturnal forest environment, as do the findings of auditory activity in nocturnal as compared to diurnal reptiles (Blair 1968); the complex middle ear, however, arises very late in the evolution of the mammalian syndrome.

"Mnemotaxis"

Orientation in the darkness can be achieved over short range simply by memory. If the animal can remember how far and in which direction it is, say, from its hideout, it can quickly dart into cover as needed. This means the animal would have to learn its location over a short period and store the pertinent information in its brain. Learning is greatly facilitated in reptiles by keeping them at an optimum body temperature (Brattstrom 1974, Krekorian et al 1968). Orientation by learning and memory could, therefore, become a most functional attribute in our hypothetical nocturnal reptile. The better its learning and memory capacity, the better its organs of touch and smell, and the quicker it could dart back to its hiding place should danger threaten. Here we find the first ecological requirement for reorganizing the "conceptual" parts of the reptilian brain in accordance with the needs of nocturnal life to generate a refined ability of hypothesis formation. The mammal-reptiles did indeed begin to evolve away from the small brain that typifies reptiles toward a larger and more complex brain in the sensory, motor, and integrative regions, as is well discussed by Heath (1968). The ability to orient by means of memory has, incidentally, been wrongly called "mnemotaxis" by Frankel and Gunn (1940); it has nothing to do with a taxis.

An additional benefit to the evolving mammal-reptile of the ability to remember locations in space is the ability to remember where food was abundant, thus permitting it to return to these favorable localities. To a generalized reptile, on the other hand, this is not nearly as important. It finds a place where food comes by regularly and waits there. Moreover, it needs less food than a mammal. For an active roaming animal, however, "navigation" is important, since it opens the possibility of exploiting food sources only temporarily available and returning promptly to the more predictable food sources available over long stretches of time.

For long-range navigation, scent and scent trails are important in the dark, such as we find in the slow loris (*Nycticebus coucang*) or the kinkajou (*Potos flavus*), small forest-dwelling mammals of modern times (Seitz 1969, Poglayen-Newall 1966). Since scent evaporates as the animal deposits and moves away from it, the animal can return home by smelling the scent where it is weakest. The intensity of scent can be equated to time. The ability to translate scent data into time data also probably requires an improved neural apparatus, that is, a better brain. From the foregoing it is quite evident that the requirements of nocturnal life would select increasingly for an improvement in learning and memory capacity, and for a better neural mechanism to handle the increased flow of sensory data from olfactory, tactile, and kinesthetic senses. Increased roaming would, of course, increase the flow of sensory data, which the animal would process by noting deviations from the expected sensory data. This, as has been discussed earlier, is apparently how vertebrates process data. It would lead to increased "intelligence" of the type I defined earlier.

In comparison to fish and turtles tested in psychological laboratories, mammals and birds appear to have a number of differences in their modes of learning that are relevant to this discussion. An excellent review of comparative analysis of learning is found in Bitterman (1975). Fish and reptiles show a strong bond between stimulus and response, with the consequence that a large reward leads to a proportionately slower extinction of responses to the signaling stimulus; it leads to random probability matching if confronted by randomly reinforced stimuli, and it leads in fish to a slow extinction of a response to a consistently reinforced stimulus. Mammals, on the other hand, readily shift in their responses in a predictable fashion: they maximize energy and nutrient intake by diverse means and reduce the energy costs of behavior to a minimum. Thus, extinction of a response to a consistently reinforced stimulus is quicker than to an intermittently reinforced one. It ought to be. The rat has apparently a "hypothesis"-forming mechanism which generates "anticipation" on the basis of past experience. Clearly, if consistent response is followed by consistent reinforcement and then switched to nonreinforcement the animal ought to perceive the inconsistency due to its expectations, and quit since it is wasting energy. Similarly, in random probability matching experiments, monkeys maximize, as do rats, or at least follow a specific strategy. They ought to. This is how they maximize gain per unit of time. Similarly, in the "dimension transfer experiments," it is shown that mammals and birds can generalize and improve their performance in acquiring food in the experiment by generalizing from previous experience; fish do not act in a comparable fashion.

The above experiments indicate quite clearly that *per unit of time* mammals and birds can gain more resources by having a hypothesis-formation mechanism linked to the mechanism generating the stimulus−response (S-R) behavior. This is clearly a necessity for warm-blooded homoiotherms owing to the very high caloric cost of living entailed in keeping the body temperature high and constant. In short, the advanced learning systems of the bird and mammal can be regarded as a consequence of the selection for homoiothermy and endothermy. Poikilotherms are "permitted" to form strong S-R links and have the luxury of slow extinction of responses, since their life processes per unit of time are slow, and a large amount of food ingested at reasonably consistent intervals is not necessary for their existence. The refined ability of mammals to form hypotheses and match the actual against the expected, finds its roots in nocturnalism, probably as described above under "mnemotaxis."

The Nest for the Young and for Digestion, Energy Stores, and Digging Paws

In order to be nocturnally active, our hypothetical reptile has to undergo further fundamental changes. Note that it must provide a constant heat source to incubate its eggs in order to raise young. In the forest there is little sunlight for the incubation of eggs, except along watercourses and clearings maybe, and even there the eggs would cool off at night. Our nocturnal reptile is too small to do as crocodiles do and build a big nest of rotting leaves. The rotting vegetation gives off heat and incubates the eggs. Somehow the reptile must incubate the eggs either by sitting on them or carrying them about internally until they hatch. We do not know how this problem was solved initially but, granted that the small animal did carry its eggs in a pouch, then it is evident how heat could be supplied to the eggs during the night.

We noted that the animal ate a lot and had a bulging gut. It is also evident that our hypothetical small-bodied nocturnal reptile would hide during the day to avoid detection by carnivorous diurnal reptiles. This, however, would remove it from an external heat source, the sun, so essential to digestion. We know that after bolting food reptiles seek out an external heat source that promotes digestive activity (Rotter 1963). Our hypothetical reptile still has poor physiological thermoregulatory mechanisms and can maintain a high ambient temperature in its hideaway only by actively structuring an appropriate microenvironment. It would have to build a nest to retard heat loss from its body. This puts some new demands on its reptilian way of life; since the animal must collect nesting material and shape it appropriately, it requires new neural mechanisms to perform the appropriate behavior of selecting, transporting, and shaping nesting material.

As a consequence of digesting food, the animal must store the digested energy. Granted its ravenous appetite, it will have to store quite a bit of energy in a limited body capacity. The evolutionary answer is to expand the glycogen stores by enlarging the liver and enlarging fat depots. One gram of fat stores 9.5 cal of energy, while a gram of protein or carbohydrate stores only about 4.5 cal (Kleiber 1961, Brody 1945). Fat storage, however, is calorically expensive, since for every 1 cal stored as fat, at least 1 cal is lost as heat (Blaxter 1960). Thus about 50% of the animal's energy in excess of maintenance during the diurnal rest period is given off as heat. However, in the nest the heat is not wasted. It ensures, first, that the animal is kept warm and its digestion can proceed if its thermoregulation is less than optimum and second, if eggs are in the animal's pouch, it keeps the eggs at much the same incubation temperature during the day as it does at night.

When the young hatch, they are far too small to go out and forage during the cool night, since their unfavorable surface-to-mass ratio would quickly drain heat from their bodies. They cannot go out during the day into the warm sunshine, for in so doing they expose themselves to the predation pressure of diurnal reptiles. Moreover, their sensory apparatus would likely be more appropriate for night than for daylight activity. The way out of this dilemma is to keep the young in a nest with an appropriate microclimate, while the female or the male goes out to collect food for them. This is another new requirement placed on the reptilian system. The female must satisfy not only her own demands, but those of the young as well. Maternal behavior is a rarity in reptiles, and none feeds its young (Bellairs 1968). Again this new requirement must have selected for new neural mechanisms, for looking after and feeding the young is no simple matter. Since the female, and possibly the male, must exert effort to provide in excess of their own needs, one can anticipate further selection to improve the morphological, physiological, and behavioral mechanisms described earlier.

The adult and young required protection during the day. This they obtained probably in localities hard for diurnal predators to reach, such as burrows or cool caves. Digging burrows frequently would require sturdy phalanges, and it is likely that it is this selection pressure that simplified the complex reptilian phalange formula by selecting for sturdy phalanges. One way to get sturdy phalanges is to reduce the phalange formula from 2-3-4-5-3 to 2-3-3-3-3. The necessity for keeping the young warm would reinforce the necessity to keep hidden but warm during the day, and both would select for burrowing and nest building. Kühne (1973) independantly reached similar conclusions; he indicated that the limbs of *Oligokyphus* support the notion of fossorial adaptations.

Evolution of Parental Care and Lactation

Parental care has received considerable attention from students of animal behavior, and was reviewed and placed in evolutionary context by Wilson (1975). The theory he elaborated, in combination with the theory of mammalian evolution presented here and in Geist (1972), predicts the evolution of viviparity and intense maternal care so well that granted the validity of Wilson's theoretical constructs it serves to verify the ecological theory of mammalian evolution, or vice versa. In combination, the two theories predict that mammals ought to evolve parental care.

Wilson (1975) lists four features that force organisms through natural selection into parental care. They are philopatry, K-selection, unusually stressful environments, and resources that are scarce, or that require relatively great effort to procure. Let us begin with K-selection: when great resource scarcity exists, adults can spare relatively little of their resources for reproduction. Exactly the same condition is achieved if the organism lives off a very abundant resource base, but its cost of maintenance is so high that it can spare only a small amount for reproduction. That is essentially the condition of the evolving mammal-reptile. Although food is very abundant in its nocturnal environment compared to that available to diurnal reptiles, its cost of existence is many times that of diurnal reptiles, and with its inefficient mechanisms for heat conservation it always is hard up for food.

Because of its high body temperature, we expect a short life expectancy in our evolving mammal-reptile, which in turn would select for high reproduction. Increased birth rate, however, is out of the question, owing to perpetual resource shortages. Increased net reproduction can only be achieved by increased juvenile survival. Therefore selection pushes for a very few highly developed and capable young per reproductive pair. This, in turn, is maximized by a high parental investment (Trivers 1974) per offspring.

In reptiles under conditions of low resource availability for reproduction, we find not only a reduction in the overt reproductive effort, as measured by the relative clutch weight, the number of clutches, the simplification of courtship, and sexual dimorphism, but also by an enhancement of parental care and an increase in viviparity (Tinkle 1969). We therefore have reason to believe that mammal-reptiles subjected to shortages of resources for reproduction would evolve in the very same direction.

Philopatry, that is, the formation of a nest and concomitant loyalty to a piece of ground, is another factor selecting for parental care, as elaborated by Wilson (1975). We noted that mammal-reptiles had to escape diurnal predators, as well as maintaining a high body temperature during their diurnal rest and hiding phase, lest they be unable to digest the food they had gathered the previous night. If they were immobile, they generated little heat, and in their cool hideout could not digest properly; if they had been mobile, they could have been detected by diurnal predators. The solution to this dilemma was a nest with thick insulation, so that heat flow from the body would be retarded, and the low heat output of normal physiological processes, plus that generated by lipogenesis, would maintain the animal at a high ambient temperature. Philopatry was thus mandatory for the evolution of mammal-reptiles, and it preadapted them to incubation of eggs and parental care.

An unusually stressful environment is another factor selecting for parental care. For the adult, but even more for the offspring endowed with reptilian adaptations, the cool night represented the stressful environment. I elaborated in the preceding subchapter why the young could not leave the nest after hatching, but could do so only when their thermoregulatory ability and their body size were equivalent to that of their parents. Hence they had to be fed until they were large enough to leave the nest and not fall victim to hypothermia shortly afterward.

Another condition favoring extended parental care is scarcity of foods, or difficulty in procuring them. To the evolving mammal-reptile, food was not scarce and probably not particularly difficult to procure, but the great demand for metabolic fuel made it essential that strategies for obtaining food were maximized. We noted that unlike reptiles, mammal-reptiles had to wander about and search for food—obviously a new behavioral dimension that vastly complicated life. This required well-developed young which learned quickly and well the diverse ways of acquiring food.

Since fouling of the nest increases the chance of the nest being found by predators, selection would quickly work in such a manner as to reduce food consumption at the nest site to a minimum. Hence, the bringing and digesting of food at the nest site by parent and young would be selected against. It would be advantageous to separate the inescapable end-products of digestion—feces and urine—as far as possible from the nest. This could be done by the adults ingesting the urine and feces of the offspring. However, as they grew larger, this would present a serious problem, since it would reduce the capacity of parents to ingest food, and yet they have to maximize food uptake to pay their own metabolic costs and the cost of raising young. Removal of feces and urine is therefore not an alternative that is adaptive with large young. The way out appears to be lactation. It allows the parent to digest the food for the young and deposit the concomitant feces and urine far from the nest, reducing the risk of the nest being detected by predators. Secondly, it ensures a minimum amount of food per unit of body growth, owing to the balance of energy, protein, minerals, and vitamins for optimum growth. This can be inferred from the correlation between the fat and protein content of milk and the postnatal growth rates of various mammalian species (Krumbiegel 1954). It can also be predicted that milk produces a relatively low amount of feces and urine. The provision of a balanced diet to the young is also to the advantage of the female, since she can use for her own needs the amino acids, fatty acids, and carbohydrate molecules in excess of those needed for milk production. Milk production, there-fore, makes for relatively more efficient use of rare nutrients by the mother, com-pared to the bringing of food to the young. It also permits the female to supplement the food from her own body stores of rare nutrients, thus promoting the growth and development of her young when the direct feeding of the young would not. It permits the female to harvest and store resources for reproduction well before lactation, and to make the scarce resources still available for her reproductive effort. It increases the flexibility of habitat exploitation in mammals. However, lactation could only have evolved gradually from a precursor, such as the direct feeding of young, and we shall learn the evolution of lactation only in general terms (Long 1969). It is self-evident that little evolution toward lactation could take place prior to the evolving mammal-reptile's ability to use its bones as a mineral store, and rapidly dissolve and replace bones through complex Haversian systems.

In the early phases of therapsid evolution, viviparity, particularly of large, well-developed young from small isolecithal eggs, was out of the question. Such eggs develop into tiny embryos that which depend on the mother for nutrient and waste exchange. Such an exchange demands that the female be able to gather food to pay the cost of maintenance for her and the embryo, and the larger the fetus the higher such costs. Such costs in turn demand well-developed organs of excretion, oxygen uptake, CO_2 excretion, circulation, digestion, locomotion, and better sensory locomotion. Clearly, the birth of large young is only compatible with superior mechanisms to handle metabolites, and egg laying is compatible with, and indica-tive of, poorer thermoregulatory, excretory, and other capabilities. In view of the small embryos of marsupials and egg laying in monotremes, while only placentals give birth to large young, and in view of the differences in thermoregulatory capac-ity and in metabolic rates even between marsupials and placentals, we can conclude that therapsids were largely egg laying. However, since viviparity shortens the "nesting" period, it obviously frees individuals increasingly for a roaming exis-tence in which temporary food sources can be exploited. Yet the flexibility selected for in the mammal-reptiles made it inevitable that selection favored increasing viviparity in *variable* environments. It is here that we must look for the origin of placentals.

Diphyodonty

We saw that for our hypothetical reptile there will be a strong selection pressure for maternal care and for feeding of the young. The latter can be accomplished by means other than lactation, such as the regurgitation of food by adults. In this case we expect the young to hatch or be born with fully functional dentition; we also expect a multiple replacement of teeth as the young grow to adulthood. This view supports the findings of Hopson (1971), who reevaluated the available material on tooth replacement in mammal-reptiles (see also Kühne 1973). He also stressed that the single replacement of teeth in mammals (diphyodonty) is probably an indication of mammal-like maternal care with lactation. Diphyodonty cannot be traced back beyond the late Triassic, a time when mammal-reptiles were more mammal than reptile (Hopson 1971).

Tooth replacement in mammals differs drastically from that of reptiles, in that reptiles continuously replace teeth, and a crocodile may replace teeth some 50 times in its lifetime (Osborn 1973), while mammals typically have only one replacement (Krumbiegel 1954). What were the reasons for the sharp reduction in tooth replacement? Granted the increase in metabolism in mammal-reptiles, we would expect a reduction in life expectancy of the individuals. Life expectancy tends to be inversely related to rates of activity and the expenditure of energy, and it is a good rule of thumb that each unit of protoplasm has roughly the same number of calories to burn before it expires (see Brody 1945). When the life expectancy of individuals was decreasing, mammal-reptiles evolved increasingly more complex dentition, in particular, large, complex, multirooted teeth. The size and complexity of these teeth suggests, first, a relatively large energy expenditure in their growth and, second, that they had great significance to the individual during daily life. A complex tooth lost meant not only a large amount of energy required to restore it, but also a possible reduction in the efficiency of mastication, and thus in the liberation of metabolizable energy from the food. The loss of a complex tooth became a disadvantage. Therefore it became adaptive to grow harder—that is, slower-wearing—solidly rooted teeth that remained functional a long time. Hence, while the life expectancy of individuals decreased, the life expectancy of each tooth generation increased. The ultimate outcome would, of course, be a single tooth generation rather than two tooth generations. Indeed, such large single-tooth generations are typical of marsupial mammals, while some short-lived placentals like shrews and bats resorb the milk dentition prior to eruption, and have only one tooth generation (Krumbiegel 1954).

Kühne (1973) linked the reduction of tooth generations also to lactation; that is, lactation permits the reduction of tooth generations since the young, in ingesting milk, are not dependant on teeth for survival and can apply the energy from tooth growth to other parts of the body. He also made an important point: Granted a milk dentition or absence of dentition, then tetrapods experienced for the first time an ontogenetic period, a period of "youth," in which they were dependant on adults. It can be added that for ontogeny, maternal behavior is mandatory.

It is evident from the foregoing that a great load is placed on the remaining tooth generation, which is expressed in increased tooth consumption (thegosis, see Kühne 1973). In fact there is apparently a fourfold increase in the consumption of tooth substance compared to the reptilian condition (Kühne 1973). This increase correlates roughly with the increase in the cost of maintenance, which is some two- to sixfold.

New Bone Microstructure

The high rate of activity of our hypothetical mammal-reptile would require a high rate of replacement of its enzyme systems. This requires not only the ready availability of proteins and molecules, easily degraded to energy, but also a large store of readily available minerals. During peak activity, there must be a high rate of flow of minerals from the blood plasma to the enzyme systems. The blood in turn draws minerals from the gut, but if there were temporarily no food in the gut the animal would be hard put to obtain minerals, unless it possessed an exceedingly well-developed system of resorbing minerals in the kidney. A high level of these minerals in the blood is compatible with homeostasis only as long as the activity of the animal is high. Once the activity is dropped, the valuable minerals must be excreted unless they can somehow be stored. The mineral store in mammals is their bones. By virtue of the complex Haversian systems, the bone matrix is readily absorbed into the bloodstream and readily redeposited (McLean and Urist 1968). Since reptiles have the capacity to raise their peak metabolic output to only some 6 times that of their resting metabolism (Gordon et al 1968), while mammals have a capacity of some 40 times (Brody 1945), it is evident that homeostasis may be difficult to maintain in the absence of a mineral store. Reptilian bone is largely avascular, while mammals have densely vascularized laminar bone. This gives us a tool to investigate the extent of endothermy, or at least the high activity in early land vertebrates. Vascularized laminar bone is found in the mammal-reptiles, in the flying archosaurs, the pterosaurs, the dinosaurs, and even in the ancestors of dinosaurs, the thecodonts (de Ricqlès 1972a,b,c, de Ricqlès et al 1972, Bakker 1972). These findings thus support the contention that the early mammal-reptiles were rather active creatures probably with a high body temperature, as is argued here primarily on ecological bases.

Sleep and Its Early Evolution

The biological significance of sleep is still surrounded by controversy, as is amply illustrated in a recent review of that subject by Meddis (1975). Complex sleep has evolved at least twice in vertebrates, once in the placental mammals and once in birds, while a less complex sleep is found in marsupial and monotreme mammals. It is probably more than a coincidence that placentals and birds happen to have the highest body temperatures and the closest physiological control over thermoregulation. In contrast to mammals, reptiles do not show deep sleep as manifest by deep-voltage, regular-wave electric activity in the brain; they do show periods of inactivity, however.

Activity, and even alertness on the part of an animal are associated with increased metabolism, that is, a catabolic activity during which tissue is broken down and burned for energy. As long as an animal is active or alert, metabolism is likely to be such that anabolism, that is, tissue growth, is retarded if not brought to a halt. In particular, tissues of low growth priority, such as neural tissue or fat, ought to be most affected by the absence of periods of anabolism. Clearly, periods of quiescence would be needed for growth of tisues.

In mammals, the tonic mechanism supplies a large percentage of the heat of the resting animal. In favorable ambient temperatures, our evolving mammal-reptile could easily dispense with the tonic mechanism and save the fuel. This appears to be accomplished by deep sleep in which the body suffers a great loss of tonus. It is also adaptive to enter such a low state of activity in order to permit lipogenesis to convert energy in excess of need into fat. I emphasized earlier that only through lipogenesis could food taken in during the night be converted into a compact form for use during the following night, while digestion cleared the alimentary tract for filling when the animal awakened.

The foregoing hypothesis predicts that sleep will increase during stages of intense ontogenetic growth, after periods of intense catabolism, while sleep deprivation interferes with growth processes. The foregoing hypothesis has nothing to say about complex sleep; it may be that it evolved to keep the animal—whose body temperature is very high, and who thus might wake easily—from awakening, as is suggested by some of the work reviewed by Meddis (1975).

Mammalian Liabilities: High-Cost, Short Life, High Aggression, Poor Performance in Heat and Desert

The mammalian system, magnificent as it is, does have its liabilities. The mammal may be unconscious due to sleep for a large part of its life, depending on its species. Its cost of maintenance is a staggering 2.5 to 6 (or more) times that required to maintain a reptilian system of similar dimensions (Templeton 1970, Robinson 1971, Spector 1956 in Hoar 1966). Simply to keep alive, a mammal must consume somewhere between 2 and 6 times the amount of food a reptile of similar size would need. Expressed in another way, there can be up to 6 times as many reptiles as mammals for the same land area. Moreover, the mammalian system has a clear advantage over the reptilian system only in cool moist climates. In climates with warm days and nights, reptiles can be just as active as mammals day or night, and that at considerably less cost. In tropical climates and forests with warm nights, perfectly good nocturnal reptiles can evolve and remain reptiles, since here the selection for endothermy and homoiothermy cannot be expected. Moreover, reptiles are likely to have an advantage over mammals in hot, dry climates, if such reptiles excrete uric acid rather than urea and are thus capable of greater conservation of water than are mammals. This latter argument is developed by Robinson (1971).

The short life expectancy of mammals automatically selects for individuals with increasingly effective mechanisms to gain access to the prerequisites for reproduction in short supply—primarily food. This predicts that intraspecific competition should be more severe in mammals than in reptiles. Short life expectancy selects for aggression (Geist 1974 a). An indication of the validity of this hypothesis would be not only a higher incidence and greater severity of agonistic interactions in mammals, but also a greater elaboration of visual, vocal, and olfactory display mechanisms which, as indicated earlier, achieve much the same effect as aggression but at a considerably lower cost to the animal. Our understanding of vertebrate behavior, particularly reptilian behavior, is not yet at a level where meaningful comparisons can be made; the above hypothesis still remains to be verified, but the apparently harmless fights of reptiles and their apparently rather simple visual and vocal displays are in line with the expectations (Bellairs 1968, Rotter 1963, Modha 1967, Eibl-Eibesfeldt 1967, Blair 1968, Brattstrom 1974).

During the Triassic, warm, dry climates began to spread (Robinson 1971). In tropical and hot desert conditions, the mammalian system offers no particular advantage over the reptilian one, since it is expensive to maintain energetically (Geist 1972) and requires more water compared to uricotilic reptiles (Robinson 1971). It is for this reason not too surprising that the mammal-reptiles declined, and, save for small nocturnal forest-dwelling forms, were replaced by the ruling reptiles. In those climates perhaps the only intrinsic advantage mammalhood offers over the reptilian condition is its superior intellectual capacities, which may well have prevented reptiles from dominating in today's tropics and hot deserts.

Chapter 8

Life Forms And Extinction

Introduction

Our time is becoming an age of planning. The future, so Galbraith (1967) argues, is not only contemplated, but shaped so as to leave as little as possible to the mischief of chance. Yet our planning has not yet turned to consideration of a very basic question: How are we to perpetuate our species ad infinitum, how will we replace one generation of individuals by another of equal or superior competence? If we are to address this question, be it with the wishful notion that we are a new dynasty of life about to fill the earth with glorious deeds of our kind, as Robinson (1972 a) suggests, or more humbly, only wishing to contemplate our survival, a brief concern with the life forms of the past is sobering.

In planning our future as a species, we should have to deal with time dimensions that are totally beyond our present conventional wisdom to handle, and yet these are the time dimensions in which the history of past life is measured. It deals in hundreds of thousands and millions of years. Some day our planning horizons will be extended that far. In fact, the day is here, forced upon us by the advent of nuclear power. Nuclear wastes require safeguarding for up to 200,000 years, lest the poisons spread, to the detriment of all living things.

We also note from the following review that history issues us a stern warning. We cannot take comfort from the history of unique life forms or of giants of their respective lineages. We are both a unique life form and a grotesque giant among the primates. If history were to repeat itself, one might predict that our stay on earth would be a short one. Primitives survive, not highly evolved forms. Of course, by being unique our history is not comparable to that of other species, but the warning must stand just the same. It does not preclude our becoming a new dynasty of life, and a long-lived one at that, but we cannot take it for granted that this will be our destiny.

A review of the life forms of the past also makes evident the enormous power of the environment over the genetic material that falls into its hands to mold; the expression of this power is convergent evolution. That, too, is a warning. We cannot take our humanity for granted outside an environment that reinforces its very characteristics. Unless we pay attention to the environment that nurtures us as that unique life form, we shall, over time, lose the very genetic base that makes us humans. Therefore, our concern with guarding our genome ought not to be expressed so much in eugenics as in environmental design.

The history of life forms from the past is, of course, also interesting for its own sake. Some cherished notions, such as that condemning dinosaurs as unsuccessful forms, take a drubbing. We shall do very well indeed if our life form rises to become a dynasty akin to that of dinosaurs. We note the ever-increasing rate of biological change and may contemplate its consequences. We note how climatic and geologic revolutions produce waves of extinction and evolution of new forms, and we have reason to ponder the criteria for biological success and the lessons of the review for the conservation of animals and plants. We may also note that, being a unique life form, the chances of something similar to ourselves ever arising again on earth if we should destroy ourselves is practically nil. This conclusion becomes evident when we discuss the circumstances of our evolution in later chapters. We have reason to suspect that the talk about the existence of intelligent life in the universe is rather glib talk. The evolution of "intelligent life" as we know it is so unlikely a phenomenon that no justification exists even for its search. We are alone, a unique and precarious life form that has yet to prove its viability. Glib talk about extraterrestrial intelligent life detracts from facing up to the consequences of this realization and can create only a false security that, somehow, we are "not alone." We are. Nobody will step in to save us. Nobody stepped in to save dinosaurs either!

The history of vertebrate life shows the continued rise and fall of dynasties, the evolution of dominant giants which inevitably perished, and the accelerating pace of evolution of new forms, and it points to the precarious existence of the mammalian system. It shows that the mammalian system is barely, if at all, superior to that of the reptilian system. Mammals had no apparent hand in the extinction of the ruling reptiles but almost perished themselves at the end of the Cretaceous period; they were subordinate to some extent during the early Tertiary period and they became subordinate to birds on some islands. It took the mammals some ten million years after the fall of the ruling reptiles to develop a significant adaptive radiation.

The history of mammals followed closely that of climatic changes. During the first half of the Tertiary, they evolved an imposing diversity of life based on warm, forested land. Then ice ages began to appear on the southern continents and the first large adaptive radiation succumbed to a rather modern fauna which evolved together with a new plant form, the grasses. When the first cold tremors of the ice ages in the northern hemispheres gripped the land, the mammals declined in number and diversity. However, the ice ages heralded a new and glorious age in the history of mammals, an age of rapid evolutionary turmoil, the appearance of bizarre unique forms of life, and the ascent of man.

Dynasties of the Past

Since man has risen to become the most dominant species on earth, and has the potential to control not only his own destiny on earth but that of other species, it may be instructive to take a short look at the fate of life forms from the past. We cannot here treat life in its full complexity, its myriads of acellular organisms, invertebrates, plants of simple and complex organization, its various terrestrial, marine, and arboreal forms; we can only look at our closer relatives, the vertebrates, and even here confine ourselves to some highlights of the past 250 million years or so.

There is no need to treat life from the past by describing the various evolutionary lineages, which has been so well done by Romer (1966) or Colbert (1955); rather, one can concentrate on the conclusions derived from extinct forms, as well as of primitive and advanced life forms on earth today. Moreover, the viewpoint with which we shall pursue the discussion will be ecological, for an understanding of the relationship of organisms to their environment is the ultimate goal of this book.

We noted earlier that beginning with the Permian, some 250 million years ago, the mammal-reptiles rose in dominance and were the most widespread and diverse terrestrial vertebrates until the end of the Triassic some 90 million years later, when the allosaurs or ruling reptiles succeeded them. The allosaurs ruled to the end of the Cretaceous period, a time span exceeding 120 million years, which is approximately twice as long as the Age of Mammals in which we now live, and half as long as the evolutionary history of the allosaurs.

Before their demise, the mammal-reptiles had evolved a minimum of 350 genera (Kuhn 1970) in 8 suborders, as described by Colbert (1955), and had given rise to at least 5 mammalian orders (Multituberculata, Symetrodonta, Triconodonta, Docodonta, Eupantotheria); they had diversified into small and large, even gigantic, forms, into herbivores and carnivores, and into aquatic forms; they failed to evolve arboreal forms. They exceeded the genera evolved by the alloaurs by one-third (Kuhn 1970). Yet, while they flourished, so did the early lineages of the allosaurus which were ultimately to out-compete them in the warm, dry climates of the late Triassic (Robinson 1971).

The ruling reptiles evolved, in total, 12 orders (Colbert 1955). They extended their realm beyond that of the mammal-reptiles by evolving arboreal as well as marine forms, and they too gave rise to a new class of vertebrates, the birds. In fact, they conquered the air twice, once in the form of flying reptiles, the pterodactyls, and once in the form of birds. Whereas the mammals continued as small-bodied forms practicing in the ecological niche of insectivores and small carnivores, ex-cepting the rodentlike multituberculates, the ruling reptiles began to fill the world with giants. In warm climates the productivity of the land is very high, hence the vegetation grows very dense wherever there is a contact of soil and water, hence the ever-present energy supply for growing huge herbivores, and hence the evolutionary impetus to evolve huge carnivores. The parallel evolution of large carnivores with large herbivores had occurred previously in the mammal-reptiles, and it was to occur again in the Tertiary and the ice ages with mammals.

We can see another parallel evolution in the Jurassic and Cretaceous: the parallel evolution between plants and allosaurs (Ostrom 1961, 1964, 1966, Clemens 1968). There is noticeable, almost abrupt, shift to new allosaur forms with the evolution of flowering plants, the angiosperms, in Cretaceous times. This is the time of duck-billed dinosaurs, the hadrosaurs, the horned dinosaurs or ceratopsians, and the largest of terrestrial carnivores, the tyrannosaurs. Moreover, it was also a time when mammals became increasingly abundant and diverse, although they maintained their small body size, and a new invertebrate fauna evolved (Clemens 1968).

By mid-Cretaceous times, some 20 to 30 million years before the dinosaurs vanished, the mammals evolved the marsupial and placental system. We can trace the divergence backward from the evidence of fossil teeth, the most durable structure of the mammalian body. The structure of teeth suggests that, with the appearance of modern plants, mammals began increasingly to avail themselves of fruits (Clemens 1968), which is not a difficult shift to make for primitive carnivores. The placentals remained a rather insignificant group in the late Cretaceous, but the marsupials flourished. They were very much at home in areas with moist tropical climates. Their home was North America, not Australia where they happen to be abundant today (Clemens 1968).

The end of the Cretaceous, some 60 to 70 million years ago, is signaled by the extinction of the ruling reptiles a subject that has fascinated scientists and laymen alike up to the present day. What few realize is that when the dinosaurs became extinct, the mammals also declined and almost vanished; in particular, the marsupials, so abundant in the late Cretaceous, became almost extinct (Clemens 1968). Moreover, it took the mammals a staggeringly long time, some 10 to 15 million years, to recover from the waves of extinction that swept the terrestrial and marine faunas at the end of the Cretaceous.

Various hypotheses have been advanced to account for the extinction of dinosaurs (Colbert 1955, Axelrod and Baily 1968, Bellairs 1968, Steel 1970, Bakker 1971, Feduccia 1973), almost all of them ignoring the fact that many mammalian groups also succumbed, while not all dinosaurs became extinct. The last of the allosaurs, the crocodiles are very much with us yet, as were, for a while during the Age of Mammals the ancestors of the dinosaurs, the eosuchians (Colbert 1955). Extinction was highly selective, just as it has been at the close of other geologic periods. There have been extinctions at the end of the Jurassic, Triassic, Permian, and Tertiary. Nor was the demise of the dinosaurs a sudden event, but a reasonably gradual one (Steel 1970, van Valen and Sloan 1977); so that hypotheses based on catastrophic events that caused sudden extinction simply do not stand up to scrutiny. Stegosaurs became extinct before the ichthyosaurs and pterosaurs, and these in turn were survived by ceratopsians (Colbert 1955). Whatever did in the allosaurs came slowly; it destroyed the terrestrial and marine giants as well as the remaining pterosaurs, but left the marsh-dwelling crocodilians to live on, as well as the lizards, snakes, and turtles. Even the eosuchians, the ancestors of dinosaurs, survived the Cretaceous at least for a time (Colbert 1955).

The hypothesis that, with angiosperm evolution dinosaurs experienced a shift in food habits to which their bodies could not adapt and hence died of constipation, is amusing in its innocence, as is the view expressed by Schatz (in Steel 1970, p. 92) that dinosaurs "burned out" due to the higher O_2 content of the air resulting from the increased abundance of angiosperm plants. As indicated earlier, dinosaurs, in the form of hadrosaurs and ceratopsians, and modern plants lived side by side for some 30 million years (Clemens 1968, Ostrom 1961, 1964). The hypothesis that mammals increased in number and ate up the eggs of dinosaurs also suffers from innocence, since alligators and crocodiles and numerous reptiles which lay eggs and build nests are still with us, despite egg predation by mammals. Moreover, this hypothesis fails to explain why mammals suffered drastic extinctions at the end of the Cretaceous.

The hypothesis that mammals outcompeted dinosaurs does not appear tenable. In what is today Montana, the displacement of ecosystems containing dinosaurs by ecosystems that do not can be followed in the paleontological record. Dinosaurs were part of a warmth-loving fauna that was displaced at the end of the Cretaceous by what van Valen and Sloan (1977) call the *Protungulum* community, which contains no dinosaurs and lives in plant communities indicating temperate climates. Dinosaur and *Protungulum* communities existed contemporaneously, and it is the latter that contained mainly placentals, while the former contained, besides dinosaurs and lizards, many marsupials and multituberculates. Moreover, mammals failed to evolve ecological equivalents to dinosaurs for 10 to 15 million years after the demise of the big reptiles (Bellairs 1968). Nor do we find a concurrent decline of allosaurs and an increase of mammals in the same biotic community as one would expect, while such a relationship can be demonstrated betwen allosaurs and therapsids when the latter declined during the Triassic (Robinson 1971). The competitiveness of the mammalian system, despite its high cost of maintenance, or because of it, is not much superior to that of reptiles or birds, for in the beginning of the Age of Mammals following the Cretaceous, birds came close to filling the role of the dinosaurs, as we shall see below. Although mammals became abundant during the early Cenozoic they were only beginning to develop ecological professions vacated by the death of allosaurs.

The explanation for dinosaur extinction must lie somewhere other than in easy answers, as indeed it appears to do. Close scrutiny of allosaur remains, a careful comparison of their abundance in time, the relative distribution of herbivores and carnivores, and a better understanding of ecology as a whole, have brought us closer to what is probably the correct explanation. A comparative study of bone histology indicates that dinosaurs were endotherms (Bakker 1972, de Ricqlès 1972a,b,c), and probably rather active creatures that were probably as costly to maintain as mammals of comparable size. This view is supported by the finding that dinosaur carnivores were just about as numerous, in relation to available prey, as are mammalian carnivores (Bakker 1972). If the carnosaurs were ectothermic, we would expect more predators relative to prey. However, it is not at all certain yet that Bakker's finding is valid. A third line of reasoning advanced by Bakker is that a reptilian type of metabolism would hardly be enough to permit sustained activity at high level for dinosaurs. Thus, the huge *Tyrannosaurus* would move at a maximum rate of about 3 miles an hour if he had a reptilian metabolism. It is reasonable to assume that its maximum rate of movement was somewhat higher than that. Most of the other evidence Bakker marshalled to support endothermism in dinosaurs is irrelevant or false (see Bennett and Dalzell 1973, Feduccia 1973), but the evidence from bone histology, community structure, and rates of movement can be accepted as reasonable.

Granted that dinosaurs were endothermic although they probably operated at a lower body temperature than mammals do, they had nevertheless a deficiency in their adaptation to control heat loss. They had no hair or feathers; they were naked. We know this from pieces of fossil skin (Romer 1966, Colbert 1955, 1961). Thus, they must have been sensitive to heat loss, particularly as small-bodied youngsters, as is also suggested by their large surface area relative to their body size. Note, for instance, the long tails and necks. Given the bulk of their large forms, they would probably have had difficulty getting rid of heat under conditions of prolonged exercise in warm weather; and their large surface-to-mass ratio may well have been adaptive in getting rid of excess heat. Conversely, the large surface-to-mass ratio would have been maladaptive during cool temperatures. Great variations in diurnal and seasonal temperature could have been the cause of dinosaur extinction, as argued by Axelrod and Baily (1968), and with reference to ecology in the detailed paper by van Valen and Sloan (1977).

Dinosaur and Cretaceous mammal extinctions coincide with a gradual reduction in the world temperature at the end of the Cretaceous (Worsley 1971, van Valen and Sloan 1977), the shrinking of tropical belts, the rise of mountains, and thus more extreme weather conditions. We note that dinosaurs decline in species diversity and that primitive species last the longest (an old rule, as we shall see), so that the eosuchians, probably less homeothermic than the specialized dinosaurs, survive their descendants into the early Age of Mammals (Colbert 1955, 1961). Van Valen and Sloan (1977) describe the faunal changes during dinosaur extinction in Western North America in great and admirable detail. Extinction coincided with climatic and ecological changes and was quite gradual. A decline in world temperature and tropical zones would also explain the decline of the warmth-loving marsupial mammals and the ascent of the placentals, which are generally better adapted to cool environments by virtue of higher body temperatures and better mechanisms of thermoregulation (Johansen 1962 in Hoar 1966, Dawson and Hulbert 1969).

The belief that the reptilian system is vastly inferior to the mammalian one is not well supported by the history of vertebrate life. Even today in the Age of Mammals we still have very ancient reptiles with us, such as the Rhynchocephalia, the primitive tuatara of New Zealand, as well as the crocodilians, the snakes and lizards, and the turtles. There are also more species of reptiles on earth today than of mammals, some 6000 compared with about 4800 mammal species. Obviously, reptiles have not done badly since their ruling days and are quite abundant in tropical environments.

Following the extinction of allosaurs, birds began to fill the niches vacated by large reptiles, both in marine and terrestrial environments. Giant birds are concentrated largely in the Paleocene, Eocene, and Oligocene. Thus, we find giant penguins in the early Tertiary which were up to six feet tall. They disappeared with the ascent of the seals and sea lions in Miocene times. We also find in the early Tertiary huge carnivorous birds the size of ostriches, such as the Eocene diatrymiformes of Europe, the gastornithese of Europe and North America, and the huge cariamas from Miocene South America (Kuhn 1971).

Furthermore, on islands with no mammals or primitive mammals such as New Zealand, Madagascar, Mauritus, and New Guinea, large birds occupied the large herbivore role. Thus, until its occupation by man, we find the elephant-birds (aepyornithiformes) on Madagascar, the moas (dinorniformes), huge geese (*Cnemiornis*) and rails (e.g., *Aptornis*) on New Zealand, and the Raphidae, better known as dodos, on Mauritus. On new Guinea the large and—to man—dangerous cassowaries persist to the present (see Kuhn 1971, Martin 1967, Battistini and Vérin 1967).

The Age of Mammals has three recognizable periods. The evolution of mammalian orders and archaic mammal faunas in the early Tertiary, the evolution of the modern fauna from the mid-Tertiary to the ice ages, and the evolutionary turmoil of the ice ages. Within 10 million years of the close of the Cretaceous, the majority of mammalian orders had evolved. We can discern an early wave of giants that lasted to the middle Tertiary and produced the largest terrestrial mammals ever, the indricothers. These were huge, giraffelike rhinos. Following the Cretaceous, a cooling of the climate set in, (Cook 1972, Sparks and West 1972), apparently correlated with the appearance of glaciation at the South Pole (Dunbar 1970, Markov 1969, Bandy et al 1969, Denton et al 1971) and a period of mountain building. There was a reversal to warmer climates during the Miocene followed by cooling (Addicott 1969). A dry climate probably prevailed in the northern hemisphere during the early Miocene followed by the rapid evolution and differentiation of grasses and grazers which replaced the extensive forests and browsers of the early Tertiary. Simultaneously, a wave of extinction swept the archaic mammals, while relatively modern forms began to make their appearance. In South America the spread of grasses apparently preceded that in the northern hemisphere—as is understandable, owing to the earlier glaciation of the southern hemisphere—for we find diverse herbivores evolving complex hypsodont dentition by the early Tertiary, much earlier than in the northern hemisphere (Patterson and Pascaul 1972).

The mammal fauna from the mid-Tertiary, despite some strange-looking creatures like the protoceratids and entelodonts, would be less strange than in the preceding epoch. We would find horses, antelope, camels, giraffes, rhinos, elephants, pigs, and an avifauna not strikingly different from those of our own day. In the seas we would see whales, seals, penguins, and fish very similar to some present-day forms. This fauna lasted through to the end of the Tertiary. However, the climate was gradually becoming cooler even in the northern hemisphere, and when the first cold tremors took hold some 3 million years ago and heralded the coming of the northern ice ages, more and more forms vanished. By the Villafranchian, the long period of minor cold tremors prior to the full glaciations, about one-third of the mammalian orders and suborders evolved had disappeared. The impoverishment continued into the ice ages or Pleistocene, but the evolution of new forms during this unique age soon increased mammalian diversity. We will treat it a little later, for it is the age not only of the evolution of some of the most bizarre animals the world had ever seen, it is also the age of the ascent of man. For more detailed accounts of the history of mammals during the Tertiary, I refer to the volume edited by Keast et al (1972) in which further references can be found, as well as to such works as Scott (1937), Thenius and Hofer (1960), Colbert (1955), Martin and Wright (1967) and Kurtén (1968a).

Extinctions

We must pause a little. We noted earlier that the ruling reptiles out-evolved the therapsids, while the mammals and birds out-evolved the reptiles, not in number of species but in orders and suborders. Thus, according to Colbert (1955), we find:

Mammal-reptiles	8 Suborders or orders	Time required, 100 million years
Reptiles	34 Suborders or orders	Time required, 280 million years
Mammals	57 Suborders or orders	Time required, 180 million years
Birds	33 Suborders or orders	Time required, 180 million years

Obviously, the endothermic classes diversified more than their reptilian relatives. Although the figures for orders and suborders are approximate and may change with taxonomic taste (Kuhn 1970), the message implied in the figures may be valid. The obvious uncertainty is that modern mammals and birds are well known compared to the extinct forms of reptiles from the Mesozoic; there may have been more orders of reptiles during the reign of the allosaurs that we have failed to discover.

A second point arising from the study of succession of life forms is that extinction and evolutionary rates are dependant on climatic changes and geologic revolutions. Hence it is understandable that the warm/cold oscillations of the Pleistocene led to massive extinctions and an incredible turmoil in the evolution of mammals, as we shall see later. Also, man emerges as a totally new force that, with the advance of technology, broadens its niche to the detriment of the megafauna.

Thirdly, extinctions were most selective. Some forms survived and still survive. These are generally relatively small-bodied animals. On closer inspection we note that the survivors are often primitive forms of their respective lineages and are ecological generalists. This is an old recognition going back to Cope (Stanley 1973. Note, for instance, that the eosuchians survived their descendants, the dinosaurs, and were still found in the early Tertiary (Colbert 1955).

The same phenomenon is exhibited by the Hyracoidea. They evolved giants in the Oligocene which died out, while relatively primitive small forms survive to the present (Hahn 1959, Cook 1972). The same is shown by the South American caviomorph rodents (Patterson and Pascaul 1972); by Old World deer (Geist 1971b); to some extent by the American Xenarthra among whom, besides some small generalists, a few small specialists have also survived (Patterson and Pascaul 1972); by the rhinos of the *Dicerorhinus* lineage, by the beavers (Castoridae); by the bears, whose most specialized form, the cave bear *(Ursus spelaeus),* failed to survive although the primitive and advanced generalists did; by the pigs, whose most evolved giants disappeared in the late Pleistocene of Africa (Cook 1972); by the bison *(Bison)* (Geist 1971b); by the Alcelaphinae, a tribe of African bovids (Cook 1972); by the elephants, whose most advanced forms in the genus *Mammutus* died out at the end of the Pleistocene; by the felids, which have lost all but the lion among their highly evolved forms; by the tragulids (see Thenius and Hofer 1960); and in part by the giraffids, since a generalist and an old-established specialist survive.

The vulnerability of specialized and large-bodied members of a tribe or family to extinction is not difficult to explain. The adaptations of a species are related to a specific environment, and they limit the evolutionary response of that species when it encounters new environments. Large-bodied species, in particular, are in difficulty. Some of their adaptations are strictly alterations of their body system to permit them to exist as relatively large-bodied forms (Stanley 1973, the problem of similitude). The more specialized or the larger bodied they are, the fewer are the organ systems that are readily changeable to fit the new environment, and the more stationary the organism becomes in its change with time. Conversely, the more specialized organ system can do a few tasks very well but has lost the ability to do others, and with this has lost or reduced the potential for moving in a new adaptive direction. A species with a generalized organ system can still do this. The evolutionary potentials are greatest for a small-bodied mammal whose organ systems are capable of doing many tasks, and whose body is still unencumbered by secondary adaptations that support large body size.

A second type of survivor is the specialist whose generalized relatives have either been out-competed by another animal group, or who has succeeded its parent species of eons and whose niche has narrowed in time. The horse may be one such example. The small, primitive forest horses of the Eocene are replaced by "duiker-like" bovids, cervids, and rodents, while the early savannah and steppe radiation of horses, the hipparions, were succeeded by the single-toed horses; only a small number of species survive to the present. Another example is the specialized eden-tates, such as the tree sloths and anteaters, which survived the demise that overcame the diverse edentates at the end of the Pleistocene.

Furthermore, survival is associated with old, stable, productive habitats such as the tropics and subtropics, or with oceanic islands isolated from contact with conti-nental forms where, to his day, we find a large collection of archaic forms or "living fossils" (Thenius and Hofer 1960). In contrast to the terrestrial environments, the marine environments are not characterized by the same extent of extinctions. In general, small-bodied ecological generalists from warm, stable, productive habitats survive best, while large, bizarre, ecological specialists from unstable habitats survive least well. This has clear implications for today's conservation practices. Since large body size and ecological specialization go hand in hand, it is the large forms that are most likely to become extinct if ecological changes come about.

Life Forms

We note that ecological equivalents, which often—but not always—have surpris-ingly similar body form, replace each other in geologic time. Thus we find the following equivalents among reptiles, mammals and birds:

Extinct reptiles	*Equivalent mammals*	*Equivalent birds*
Ichtyosauria	Dolphins	—
Pterosauria	Bats	All flying birds
Ankylosauria	Glyptodonts	—
Ceratopsia	Rhinos, brontotheres	—
Sauropterygia	—	Penguins
Struthiomimus	—	Ratites
and *Ornitholestes*		

Conversely, we find some life forms that were never evolved by reptiles, such as Mysticeti, the baleen whales; Rodentia and multituberculates, gnawing herbivores; and all cursorial (fleet-footed) quadrupeds. Some life forms were only evolved by reptiles: sauropods, the bipedal, striding forms with supportive tails (kangaroos do not fit this adaptive syndrome); Sauropteriygia, the long-necked, heavy, quad-rupedal dinosaurs; Crocodilia, the long-tailed, aquatic, sprawling, quadrupeds with carnivorous habits.

Nevertheless, many life forms that came into being in the Mesozoic are still with us today. Time moves on, orders vanish, but the same life forms reappear, filled by reptilian, mammalian or avian systems. Some life forms, like the fish, have enorm-ous duration, having been filled not only by divergent orders of fish, but also by amphibious reptiles and mammals. We note that large-bodied diverse life forms are short lived, that specialization brings diversity, and that generalists are constrained to a few life forms, albeit they are long-lasting ones. In this respect the mammalian life forms are most instructive.

If one examines mammalian diversity, it is noticeable that most mammals fall into clearly recognizable life forms, regardless of their ancestral origin. For instance, the wolf *(Canis)* as we know it is a placental mammal, but it had a counterpart in the marsupial wolf, the now extinct *Thylacinus*. These animals have surprisingly similar skulls and even body shapes. We assume that similar ecological professions produced here similar adaptations. Also, the hyena *(Crocuta)* which is, in appearance, ecology, and behavior a dog or wolflike animal, has no relation to wolves whatsoever, but is an offshoot of the viverids, a diverse group of tropical carnivores. Bourlière (1973a) compares the larger mammals of the African and American tropics and shows a striking resemblance in size, shape, coloration, and food habits to mammals which, despite different phylogenetic origins, have adopted the same ecological niches (see also Dubost 1968).

We find different lineages evolving incredibly similar life forms, e.g., mice *(Muridae)*, which evolved into "squirrels," as did the cloudy rats *(Crateromys)* of the Philippine Islands (Walker 1964, p. 959); shovel-toothed mastodons, which evolved in three lineages independently, as illustrated by *Gnathabelodon, Amebelodon,* and *Torynobeldon* (see Scott 1937, pp. 291,292,299); kangaroos, which evolved into "rats" down to the scaly tail (see Walker 1964, p. 102); the "horse," which evolved three times—in the true horses, in the rhinos (Hyracodontidae), and in the South American Litopternans *(Diadiaphorus, Thoatherium,* see Scott 1937, pp. 562,565,566); the "rhino," which evolved five or six times independently, at least three times in the rhinos proper (rhinocerotids, acerathers, dicerathers), in the form of brontothers, and in South American toxodonts (Scott 1937, Thenius and Hofer 1960). How often the form "mouse" or "rat" evolved in the mammalian lineages nobody will ever know, but it is the most popular life form of mammals.

Below is a list of common mammalian adaptive syndromes:

Adaptive syndrome	*Genera and family*
"Mouse"	Many rodents, insectivores, and many metatherians in the Phalangeridae, Dasyuridae, Caenolestidae
"Rat"	Many rodents, a few insectivores: *Echinosorex,* a primitive hedgehog; *Microgale,* a tenrec; *Solenodon,* the kangaroo rats, in particular *Hypsiprimodon, Potorus, Caloprymnus, Aepyprymnus,* and *Bettongia;* many Paramelidae; Phalangeridae; and Didelphidae.
"Wolf"	The Canidae in general, the Hyaenidae, and *Thylacinus* among the Dasyuridae
"Saber-toothed tigers"	*Smilodon,* and all Machairodontidae, *Thylacosmilus,* Borhyaenidae
"Cat"	Felidae in general, *Cryptoprocta* among the viverids, as well as the civet cats *(Poiana, Genetta, Viverricaula, Viverra, Civettictis, Prionodon);* among metatheria, *Dasyurus* (Dasyuridae)

Adaptive syndrome (cont.)	*Genera and family (cont.)*
"Hedgehog"	All advanced Erinaceidae, many Tenrecida *(Setifer, Hemicentetes, Dasogale, Echinops)*, and the prototherian Tachyglossidae
"Mole"	All Talpidae, Chrysochloridae, and among metatheria, the Notoryctidae
"Rhino"	Arose twice among the Rhinocerotidae (once in the acerathers), Brontotheridae, in embrythopods *(Arsinoitherium)*, in Dinocerata, and among Toxodonta (in *Trigodon,* see Scott 1939)
"Anteater"	Myrmecophagidae, and in metatheria the Myrmecobiidae
"Monkey"	The cerocopithecine and platyrhine primates
"Horse"	Equidae, among the Rhinocerotidae, *Hyracodon* and *Triplopus;* among the Litopterna, the family Prototheride and members resembling three-toed horses *(Protarotherium, Licaphrion,* and *Diadiaphorus)* and one-toed horses *(Thoatherium)* (latter were more monodactyl than any horse)
"Pig"	All Suidae, Tayassuidae, and extinct Entelodonta
"Marten"	Most Mustelidae, many Viveridae and *Dasyurops* among the Dasyuridae
"Squirrel"	The Sciuridae, but also the *Tupaidae* (insectivores)—showing a close resemblance to the palm squirrels; among the Muridae, the cloudy rats, in particular *Crateromys;* and among the metatheria, the Phalangeridae of the genera *Gymnobelideus, Petaururs,* and *Dactylopsila*
"Flying squirrel"	Among Sciuridae, many species *(Petaurista, Aeromys, Eupetaurus, Pteromys, Glaucomys, Hylopetes, Petinomys, Aretetes, Trogopterus, Belomys, Pteromyscus, Petaurilus, Iomys)*. All Anomaluridae (scaly-tailed squirrels *(Anomalurus, Anomalurops, Idiurus)*
"Duiker"	All Cephalopinae, also the unrelated *Neotragus, Nesotragus, Tetraceros, Noemorhoedus, Capricornis;* among the cervids, *Muntjacus, Elaphodus, Mazama, Pudu;* also *Tragulus* and *Hyemoschus* compared to *Cuniculus,* a dasyproctid rodent from South America
"Otter"	Several mustelid genera *(Lutra, Pteronura, Amblonyx, Aonyx, Paraonyx, Enhydra)*; among insectivores, the Potamogalidae *(Potamogale)*, and among metatherians, the dedelphid genus *Lutreolina*
"Bipedal rodent"	The Heteromyidae; in the murids, *Notomys;* in the cricetitids, *Hypogeomys* of Madagascar; the Dipodidae; and *Pedites* (spring hare) of the Pedetidae

Adaptive syndrome (cont.)	*Genera and family (cont.)*
"Horned ungulate"	The Bovidae, Cervidae, Antilocapridae, Giraffidae, Merycodontidae, Protoceratidae
"Chalicotheres"	The extinct Chalicotheriidae and some Homalodotheriidae among the South American Notoungulata
"Shovel-tuskers"	Platybelodontidae, Amebelodontidae Tritilodontial ("spoonbills")
"Water pig"	*Choeropsis* and the rodent *Hydrochoerus*
"Rabbit"	Most Lagamorpha; *Dolichohis,* the mara among the South American rodents

One can get a quantitative picture of the frequency with which mammals exist in various life forms (Table 8-1). The life forms are arranged in order of frequency of occurrence. We note at once that almost half of all mammalian species are found as

Table 8-1.   Life Forms and Species Numbers

Life Form	Number of Species
Mouse (below 200 mm body length)	1232
Rat (exceeding 200 mm body length)	1009
Bat	785
Squirrel	271
Mole	143
Streamlined aquatic species	127
Horned quadruped	118
Monkey	97
Duiker (including rodents)	93
Marten	76
Spine-covered or hedge hoys	67
Ground squirrel/marten	61
Duiker (excluding rodents)	60
Rabbit	55
Cat	52
Flying squirrel	51
Lemuroid	48
Dog	46
Kangaroo	39
Water mammal, primitive	37
Water mammal, advanced	90
Otter	23
Armored skin mammals	15
Hornless ungulate	14
Lorisoid	13
Bear	11
Badgerlike	10
Brachiator	6
Anteater	5
Rhinos	5
Elephant	2
Hominoid	1

one-fifth, squirrels for about 6%, moles for about 3%, horned quadrupeds for a little less than that, while remaining life forms trail by an increasingly wider margin. Even without the widespread megafaunal extinctions of post-Pleistocene times, this picture would not be altered appreciably. One also notes Zipf's law mirrored in these figures, for it appears that the larger and more complex a life form, the more it deviates from ancestral forms the less frequently it is encountered. Obviously "mouse" and "rat" are the most successful life forms and are the most likely to reevolve in future geologic periods. In contrast to the common life forms which are clearly recognizable evolutionary successes are the unique life forms, of which we are one. These are life forms whose success is suspect, since they fill a unique ecological niche and, given past history, are not likely to last long. Successful life forms reappear in evolution, unique life forms can be expected to be short-lived experiments of nature. Since we are a unique life form, this has implications for our strategy of survival; from the history of life forms on earth, we cannot take comfort, but only a warning.

It may be argued that we are a unique life form about to embark on populating the earth with a new dynasty of vertebrate life, which is as far removed in its ultimate system from mammals as are mammals from reptiles. This view is doubtful, for we are by zoological standards an advanced hominid form and a giant among the primates. Giantism within a lineage is associated with specialization. However, if we grant that we are the eohomos who potentially could be the ancestors to the allohomos, the new class in vertebrate life, then we would have to plan our survival for a minimum of 125 million years to do only half as well as dinosaurs did.

Chapter 9

The Ice Ages

Introduction

The change from the later Tertiary to the Pleistocence or major ice ages in the northern hemisphere was not sudden. The average mean temperature, at least that of the temperate regions, began tc decline as well as to oscillate. The oscillations increased in frequency and amplitude to culminate ultimately in the major glaciations. Each major glaciation was characterized by several stadials during which the ice front advanced, and several interstadials during which it retreated. Climates comparable to today's characterized the interglacials, while during the glacial maxima climates colder and more variable than today's ranged over the Earth. The foregoing is valid for the northern hemisphere only; in the southern hemisphere the ice ages had begun already in the early Tertiary and there were, as far as can be determined, no stadials or interstadials comparable to those found later in the north (Markov 1969, Dunbar 1970, Margolis and Kennet 1970, Denton et al 1971, Sparks and West 1972). After a temporary upswing during the middle of the Miocene, temperatures continued to decline through the Pliocene into the Pleistocene (Addicott 1969).

The causes of glaciations have been much debated (Sparks and West 1972, Chapell 1973, McCrea 1975, Hays et al 1976). At present, the theory gaining ground is that ice ages may be caused by reasonably regular variations in the Earth's orbit, a theory associated with Milankovich's name. For the later discussion and verification of this theory I point to Hays et al (1976). However, there appear to be some modifying factors. Denton and Karlén (1973) found a periodicity of about 2500 years for minor glaciations that correlated with a regular variation in the solar corpuscular activity (Bray 1971); about 900 years coincide with the expansion of glaciers and 1600 with their retreat. In recent years, weather patterns have changed as the mean earth temperature has begun to decline, so that there is an increase in the snow cover of the northern hemisphere (Kukla and Kukla 1974), and an increase in snowfall and cold temperatures over the old epicenters of glaciation such as the

Rocky Mountains, Hudson Bay, Labrador, Scandinavia, and the Alps, as well as an increase in aridity in the major deserts, which may indicate a return to glacial conditions. However, increased spread of deserts may even more likely be a result of human mismanagement of the land, a view most plausibly argued by Kai Curry-Lindahl (1972a,b, 1974). But an increase in deserts is typical of glacial periods (CLIMAP 1976). The causes of recent weather changes appear complex at best, and are very controversial (Hobbs et al 1974, Bryson 1974); however, man's activities on Earth, particularly the liberation of excessive quantities of dust, may be one causative factor. It appears now that significant weather changes occur very fast and bring on ice age climates within centuries (Bryson 1974).

Preceding the major glaciations which began about 1.2 to 1.8 million years ago with the Nebraskan glaciation in America and the earlier Donau and Beiber advances of the Alpine sequence in Europe (Birdsell 1972, Cooke 1973), there was a long prelude of at least 2 million years before the major glaciations. It is called the Villafranchian period in Europe and the Blancan in America. It was characterized by climatic fluctuations of lower amplitude than those of the major glaciations, as well as by smaller, mainly montane glaciations. The exact dating of the Pleistocene—which includes the Villafranchian in the view of some authors but not others—is not well agreed upon (Glass et al 1967, Klein 1973); but this matters little for our present purpose. Whereas we know little about the extent and frequency of Villafranchian glaciations, we know that there were probably eight major glacial advances during the 1.2 to 1.8 million years of the major glaciations (Lozek 1971). During a glacial maximum or stadial, the total snow and glacier fields are thought to have covered some 60 to 70×10^6 km^2 in the northern hemisphere, which is about 30×10^6 km^2 more than is found today (Kukla and Kukla 1974).

The most commonly adopted scheme for naming glacial events is the Alpine sequence in Europe, which I have adopted from Kurtén (1968a) and altered in accordance with the findings of Sancetta et al (1973) of the time scale of glacial events as based on deep sea cores, and Cooke's (1973) correlation of American and European glacial events (Table 9-1). I am adopting this scheme despite the fact that the timing, as well as the absolute time scale, of glacial events has been disputed and will be disputed (Glass et al 1967, Emiliani 1970, Lozek 1971) because the Alpine scheme appears to be the best compromise available.

During glacial advances, the distribution of ice was localized. Major centers of glaciation were in Europe, the Alps, and northern Scandinavia (Frenzel 1968), while in Asia major glaciers were found on the Taymyr peninsula (Giterma and Golubeva 1967) and the major mountain chains such as the Tien Shan, Altai, Himalayas, and the lower mountain ranges of eastern Siberia, the Djugjur, Stanavoi, Kolyma ranges, and others. In North America there were several epicenters: one in the Keewatin area, one in Labrador, and in the Rocky Mountains, the Coast Ranges, St. Elias Range, and Brooks Range. From here the ice spread and coalesced to follow the pull of gravity and the push of accumulating firn on the glacial epicenters.

The ice-free areas on the continents, as well as a few areas at the glacial margins which were not overridden by ice, became the refugia for the flora and fauna, where it not only survived but flourished and prospered, and from which it dispersed in postglacial times. The small ice-free areas protruding above glacial ice are called nunataks and serve as refuges for plants and animals adapted to cold and dryness.

Since glaciers only withdraw and do not disappear entirely in mountainous

Table 9-1.   Pleistocene Chronology (An Approximation)[a]

Epoch	European Sequence	American Sequence	Dates (B. P.)
Holocene	Postglacial	Postglacial	0– 11,000[b]
Pleistocene	WÜRM II STADIAL	WISCONSIN	11– 20,000[b]
	interstadial		40– 50,000[b]
	WÜRM I STADIAL		67– 73,000[b]
(major	Eem interglacial		73–127,000[b]
glaciations)	RISS II STADIAL	CARIBOU HILLS	127–200,000
	Ilford interglacial		
	RISS I STADIAL		
	Holstein interglacial	Sangamon	200–350,000
	MINDEL STADIAL	ILLINOIAN	350–550,000
	Cromer interglacial	Yarmouth	550–700,000
	GÜNZ II STADIAL	KANSAN	700–900,000
	Waalian interglacial		
	GÜNZ I STADIAL		
Villafranchian	Teglen interglacial		900,000–1M[c]
(lower			
Villafranchian	DONAU II STADIAL		1–1.1 M
begins = 4.0M)	interglacial ?	Aftonian	1.1–1.35 M
	DONAU I STADIAL	NEBRASKAN	1.35–1.40M
	interglacial ?		
	BIBER STADIAL	EARLY NEBRASKAN?	1.6–1.9 M
	interglacial ?		
	PRAETIGLIAN		
	COLD PHASE	SIERRAN	2.3–2.5 M

[a] After Kurten 1968a, Cooke 1973

[b] After Sancetta et al 1973

[c] M = millions of years

epicenters, even during the interglacial periods, while large areas are subject to cold climates, such as the present tundra regions, there is a continuous impetus for the evolution of plants and animals adapted to periglacial and arctic conditions. Although the periglacial ecosystem evolved early in the Pleistocene, the alpine and tundra ecosystems appear to have evolved not earlier than the middle of the major glaciations and spread into North America only in late Pleistocene times (Hoffman and Taber 1967). Thus, the animals we tend to associate with the tundra are of relatively recent origin.

During glacial maxima, sea levels dropped as water became locked up in huge continental glaciers. Deserts increased in size and shifted. Land bridges rose between continents and led to major faunal exchanges, faunal mixing, and extinctions. During glacial retreats and separation of glaciers, corridors opened between glacial refugia, and the flora and fauna flooded into new habitats vacated by the ice and met and mingled in the glacial refugia. The retreat and advance of glaciers thus not only caused a repeated contraction and expansion of the ranges of plant and animal species, but also created faunal instability owing to relatively frequent meetings of related and also unrelated faunas, such as those of North and South America. In addition, there were great differences imposed on the ecology of the land in glacial and interglacial periods which greatly affected the quality of habitats and probably led to the oscillations in body size observable in large mammals (Kurtén 1968a).

They also probably led to an increase in faunal turnover rates above those of Tertiary times (Webb 1969). Although the glacial advances and retreats appear to be far apart on a human time scale, the last glacial retreat occurring some 12—13,000 years ago, they are very close together on a geologic time scale. We live in an interglacial today and a new ice age will almost certainly return in a few millennia at the most. It becomes understandable from the foregoing why the ice ages were a time of unprecedented evolutionary turmoil in which some of the most bizarre and unusual creatures evolved and colonized the Earth, man included. As we shall see later, the home of the man we can consider human is not Africa with its warm climates but the rich, diverse, and demanding periglacial environments at the fringe of glaciers in Eurasia.

In the following chapter a first reconstruction of the periglacial ecosystem, in particular that on the sunny lee side of glaciers, will be made. It shows that this ecosystem was a pulse-stabilized one, depending for its existence on loess and water liberated by the glaciers. On the lee side of glaciers a dry sunny climate, characterized by temperature extremes, existed which, in conjunction with the fertile loess and highly localized water sources, created highly productive mosaiclike plant communities and a diverse, rich fauna. In addition, the heavy load of silt in stream channels draining glaciers produced fertile floodplains. Some of the puzzles of Pleistocene zoogeography are explained by the theory; it also explains megafaunal extinction as a function of the ecosystem collapse that occurred during deglaciation. The periglacial ecosystem approached some African ecosystems in diversity, and was apparently far more productive than the climax ecosystems at comparable latitudes and altitudes. Its ecological peculiarities are reflected in the adaptations of Pleistocene mammals, and—as we shall see later—in man. This view goes against the present views expressed, for instance, in Sparks and West (1972, p. 20), who considered that little life existed in the periglacial zones during major glaciations.

Tertiary and Pleistocene Giants and the Dispersal Theory

To understand the significance of the ice ages to mammalian evolution and ultimately the evolution of man, we must take another look at the Tertiary mammal fauna.

In mammals, the brain changes. The neocortex is smooth in archaic mammals but begins to show folding in carnivores and ungulates, and establishes relatively specific patterns of folding by mid-Tertiary times; moreover, folding evolves independently in diverse mammalian lineages, apparently owing to a relative increase in the size of the neocortex (Tyszka 1966, Oboussier 1971). In general, carnivores exceed herbivores in relative brain size (Count 1947, Jerison 1969).

There is an increase in the brain size of mammals from the early Tertiary to the Pleistocene (Edinger, in Thenius and Hofer 1960, p. 208, Jerison 1961), and it accelerated even more in animals adapting to the periglacial and arctic ecosystems. This is illustrated not only by the evolution of man, which we shall deal with later, but also in the ruminants in which the northern forms have, on the whole, a larger brain than forms belonging to older faunas in warm climates. Thus the Pleistocene cervids have larger brains than the older bovids from Africa (see Oboussier and Schliemann 1966, Kruska 1970b) while a comparison of brain weight of warm-climate and cold-climate deer or pigs indicates that the cold-adapted forms have the larger brains. The differences in relative brain size may be greater than indicated by the data of the authors cited, since African antelope compared to northern forms have less fat, thinner hides, and less hair and horns, as indicated by the high dress-out percentage of meat (McCulloch and Talbot 1965). This indicates that, per unit weight of active protoplasmic mass, northern forms have more brain. These points, however, require further data, although they are in line with the general trend toward larger brains in the more recently evolved forms. These evolved largely in response to glaciations. Moreover, convergent evolution should produce much the same adaptations in the brain as it does in body form in species adapting to similar habitats with similar strategies, as has indeed been found to be the case in some arboreal mammals (Diamond and Hall 1969), and in the brain morphology revealed through endocranial casts in pterodactyls and birds (Hofer, in Thenius and Hofer 1960, p. 40; see also Tyszka 1966, Hofer 1972). We can infer from the changes in relative brain size that this organ became relatively more important to successful life during the Tertiary for most mammals and even more so during the ice ages.

Although the Tertiary produced some very large-bodied mammals, the Pleistocene produced on the whole many more giants, although none reached the dimensions of the peculiar hornless browsing rhinos, the indricotheres of the Oligocene. The mid-Tertiary and early Tertiary saw the evolution and extinction of a wave of very large-bodied creatures such as the rhinolike brontotheres, dinotheres, and arsinotheres, the giant hyrax (*Titanohyrax, Megalohyrax*), the large forest horses (*Hypohippus*), a variety of rhino lineages, anthracotheres, astrapotheres, pyrotheres, and entelodonts. These large forms appear to have been adapted to forests or rich savannah habitats in warm climates and disappeared with the spread of grassland in Miocene and Pliocene times, apparently as a consequence of glaciations at the South Pole and a cooling and drying of climates.

During the late Tertiary some giantism occurred, such as that of caviomorph rodents in South America (Patterson and Pascual 1972) or of camels in North America (Scott 1937) as well as of the divers mastodonts (Scott 1937), and is apparently related to ecological specialization, which is most striking in the grotesque shovel-tusked mastodonts of the upper Pliocene of North America. This relationship between giantism and ecological specialization is also indicated in at least some of the earlier giants, such as the tree-browsing indricotheres, the highly hysodont *Arsinoitherium,* the apparently aquatic astrapotheres, and the selenolophodont giant hyrax. Furthermore, when examining the giants of the ice ages we find it indicated again; thus, the horses increased in size with ecological specialization, so did the elephants compared to the mastodonts, so did the New World deer and Old World deer, which is noticeable if one compares the older tropical and newer subarctic forms, and so did the hominids, as we shall see later. We thus have good reason to suspect that giantism in the Tertiary is associated with ecological specialization, as is the case for giantism in tropical climates. I have explained the theoretical justification for this elsewhere (Geist 1971b), and it need not concern us yet. An important point is that ecological specialization need not lead to giantism, although giantism appears to be associated with ecological specialization.

The Tertiary giants were noticeably different in one respect; very few of them evolved the large, elaborate if not grotesque, horns, antlers, and tusks that are so commonly found in ice age ungulates. *Arsinoitherium* of the Oligocene was one such exception (Cooke 1972) as was *Prolibytherium*, a giraffid from the lower Miocene (Hamilton 1973). There is nothing in the Tertiary record to equal the antlers of the giant deer (*Megaloceros*), moose (*Alces*), reindeer (*Rangifer*), red deer or elk (*Cervus elaphus*), or the tusks of Pleistocene African suids, or the tusks of the last of the mastodons, or the mammoths, or the rhinos from the cold zones (*Elasmotherium, Coelodonta*), although here *Arsinotherium* comes close. There are many huge-horned forms in the Pleistocene record or among living mammals, such as the true goat and sheep or Asian and African buffaloes. Since hornlike organs are not only weapons and defensive organs, but also display organs (Geist 1966a, 1971a,b) and are products of social evolution, fair warning is served that something happened in the Pleistocene that did not happen to the same extent in the Tertiary. Ice age evolution apparently produces more emphasis on combat, aggression, and dominance displays, which is curiously harmonizing with human evolution, as its product is the most destructive aggressive mammal yet evolved.

Moreover, there is now some evidence that the cold-adapted ungulates, which are evolutionarily more recent than the warmth-adapted ungulates, are more damaging in combat than the African or Asian ungulates (Geist 1971a, 1974a,d). This point requires confirmation, but there is some theoretical justification for it considering the short rutting season in the temperate climates and the relatively low predation pressure compared to subtropical areas.

There is an explanation for this phenomenon which I have elaborated on elsewhere (Geist 1966a, 1971a,b and 1974a,d). It runs as follows: During dispersal into uninhabited terrain such as becomes available after glacial withdrawal, selection favors increased body size as well as a specialization of combat and defensive adaptation during the population's dispersal phase. This happens because the population is faced for some time with a surplus of forage and individuals grow very large and act very vigorous. I have elaborated on this in Chapter 12 in relation to the evolution of early man, and we touch on it when discussing population quality. The important thing is that the more frequently a species is thrown into the expanding population phase the more selection there is for improved weapons, defenses, and display organs due to enhanced social competition. Clearly, the more stable a habitat the less likely there will be evolution toward improved weapons, etc., while instability promotes it. Since the Pleistocene is characterized by unstable climates, expanding and contracting habitats and the establishment of connections between refugia and continents, we except an acceleration of evolution toward giantism and improved combat and display organs. Thus environmental instability accelerates social evolution. It does not, however, lead to an increase in ecological specialization in cold or periglacial ecosystems (Geist 1971b). In tropical or pre-Pleistocene habitats, dispersal was for theoretical reasons *preceded by ecological specialization* in allopatric populations; I follow here Mayr (1942, 1966). Hence, ecological and social adaptations evolved in parallel. Thus we can often expect giants to be ecological specialists and have rather oddly shaped grotesque skulls and horns which, due to social selection during dispersal, achieved their diagnostic form. The grotesque ice age species achieved this simply because of a more frequent dispersal, due to greater instability of their ranges compared to Tertiary forms as well as the greate differences in seasonal carrying—capacity of habitats during stadials; the more advanced species remained ecologically much the same as their ancestors. It is also evident from the foregoing that stable ecosystems, such as the tropical forests and savannahs, ''preserve'' species, reduce evolutionary rates, and hence contain many ''living fossils.''

Another reason for excessive hornlike organs is also provided by the instability of the periglacial and cold zones. The population sizes are set by the availability of winter ranges. In summer there is a superabundance of forage per individual which the latter converts to the best of its ability into a new hair coat, more bone in the skeleton, fat in the body, and vitamins in liver and fat. There is no shortage of food, and thus it costs little to grow large, heavy antlers, horns, or tusks. As I have discussed elsewhere (Geist 1974d), tropical forms cannot afford to waste energy in fat storage or the growth of huge hornlike organs. Hence the appearance of the bizarre giants in the ice ages and cold, unstable terrestrial ecosystems (see also Geist 1977 & Chapter 5).

The increased bizarreness of large-bodied forms is, in part, a function of the increase in body mass itself, for such an increase automatically creates structural problems for a large animal which a small one does not have (Thompson 1961). Thus, Stanley (1973) points out that for small mammals such as shrews and mice unspecialized limbs and claws are sufficient for climbing, while large mammals such as leopards, tree sloths, and primates require specialized claws or hands or prehensile tails to do it. In a similar vein combat adaptations change. As I have pointed out elsewhere (Geist 1966b, 1978a) large-bodied mammals cannot fight in the manner of small mammals without wrenching legs, breaking bones, and rupturing internal organs; they must specialize to deliver, as well as catch and absorb, blows of exceedingly great force (Geist 1966b, 1971a). To do it, they evolved combat organs in many forms and specialized combat behavior in all. In addition, one finds structural alternations on the skulls, neck vertebrae, and tendons which serve to absorb the heavy impact of large fast-moving bodies. We can also recognize alterations in body plan, such as the reduction in hindquarter size compared to the front quarters as one means of increasing manoeuverability during combat, an alteration typical of the highly evolved bisons, mammoths, and deer. Any acceleration in the selection for efficient combat organs and defences will at once create body alterations that strike us as bizarre in their uniqueness.

Increased body size brings with it a reduction in metabolism per unit weight, the opportunity to live on forage of low digestibility, an increase in longevity, increased selection against overt fighting, increased structural demands on limbs and combat organs, increased susceptibility to being spotted by predators, and also increased ability to ward off predators and increased opportunity to emancipate information transfer from the genetic to the social (traditional) realm (see Geist 1974 a,c).

One important additional consideration explains giantism. We noted that under conditions of resource surplus, selection for large body size was favored. Resource surpluses are a function of high productivity or an immature ecosystem, following Margalef's (1963) argument. The more mature an ecosystem, the lower the energy flux per unit of biomass, that is, in a mature ecosystem a relatively larger amount of energy is fixed as biomass. Thus, a larger seasonal energy pulse is available for growth and reproduction in an immature ecosystem. The size of the seasonal energy and nutrient pulse is determined by the difference between the highest and the lowest seasonal carrying capacity for a given species in an ecosystem. In a mature ecosystem such a pulse is small of necessity. Hence, we expect in a mature ecosystem relatively small-bodied individuals of a given species and in an immature one relatively large ones. Northern and cold temperate ecosystems are immature compared to southern or warm-climate ecosystems according to Margalef (1963). Clearly, we expect the largest-bodied forms of a species in the north, and the smallest-bodied forms in the south, or the largest form during stadials and the smallest during interglacials as shown by cave bears (Kurtén 1976). The same argument can be made for montane versus lowland ecosystems; the former are less mature and should have the larger-bodied races. *Note that the foregoing explains Bergman's rule that northern and montane forms are larger bodied compared to southern and flatland races.*

Clearly, giantism was selected for during repeated episodes of dispersal, and it was maintained if species settled in relatively immature ecosystems. As we shall see, the Pleistocene created some highly productive, immature ecosystems in the vicinity of continental glaciers. Furthermore, we have a clue suggesting that in warm climate mammals, giantism will be associated with the exploitation of food resources such that a great seasonal difference in maximum and minimum carrying capacity is found. Clearly, many giants will be desert forms, or inhabitants of the floodplains of flooding rivers.

In viewing the external appearance and behavior of ice age mammals, and considering the syndrome by which population quality may be evaluated, the following is striking: Relative to mammals from unglaciated, subtropical, or tropical areas, ice age mammals are characterized by exceptional development of tissues that normally have low growth priority, such as antlers, horns, fat, hair, dermis, and cerebral cortex. Note also the increase in body size, an indicator of enhanced population quality. All this speaks of selection under conditions of resource surpluses, at least during the season of intense body growth.

The elements are now at hand to explain much of what happened during the Pleistocene. First, there is a high turnover rate of species (Webb 1969). Second, many typical ice age creatures are giants of their respective families (Geist 1971a,b 1977). This includes not only the bears, canids, beavers, marmots, elephants, horses, deer, bison, sheep, pigs, hippopotamus, glyptodonts, and the enormous edentates of ice age vintage, but it also includes man, who is the third largest primate among some 163 species alive today.

We note in successive glaciations the appearance of giant forms whose body size, however, decreases with time, such as the large bison and elephants in North America (Guthrie 1970b) or the aurochs (*Bison primigenius*), cave lion (*Panthera spelaea*), and hunting dog (*Lycaon*), which appeared in Europe in mid-Pleistocene times and shrank in size thereafter (see Kurtén 1968a). The strong fluctuations in body size seen in various ice age creatures (Kurtén 1968a), such as mammoths, bears, reindeer, and sheep, are a function of the ecology of periglacial ecosystems, as I will explain below when describing periglacial ecology during glacial advances and deglaciations.

We also have species of ice age creatures which remain relatively primitive but have rather more advanced relatives, such as the forest bison (*Bison bonasus*) and the larger and long-horned steppe wisent (*B. priscus*). Here, as in tropical forms (Geist 1971b), the form that is older is forest-adapted, while the ecologically and socially evolved form utilizes the steppe. Bears also have evolved in a progression of forms of increasing size from tropical forests, to temperate forest, to alpine and arctic tundra, to arctic seas, as represented by the very primitive sun bear (*Helarctos malayanus*), the Asiatic and American black bears (*Ursus thibetanus, U. americanus*), the brown bears (*U. arctos*), and the polar bear *(U. maritimus)* (Herrero 1972). Simultaneously they gave off side branches such as the more herbivorous cave bear (*U. speleaus*) and the carnivorous giant short-faced bear (*Tremarctotherium simum*) who must have preyed on the edentates, mastodonts, and elephants of America. Thus we have elements of the large mammal fauna, which have adapted progressively to different habitats, which developed during the ice ages, and to different niches within the same habitat while increasing in body size.

It is now also evident why mammals that disperse across continents increase in body size and reach their largest size at the end point of their long journey. This is well illustrated by the saber-toothed tiger (*Smilodon*) who reached its largest size in Patagonia, and was of more modest size in Eurasia, as the parent genus (*Megantereon*) was (Kurtén 1968a); *Magantereon* was about the size of a panther, while *Smilodon neogaeus* from South America was larger than a lion. Mountain sheep show the same phenomenon (Geist 1971a), as do Old World deer (Geist 1971b). In addition, if one follows one genus through time, there is a tendency in some genera either to increase in size with time, such as *Canis* (see Zeuner 1967), or to increase in such specialization as hypsodonty or increasing molarization and reduction of premolars. Examples of the latter are various African suids such as *Mesocheorus, Metridiocherus*, the African elephant *Elephas recki* (Maglio 1972), or *Megantereon* (see Kurtén 1968a), or *Homo*, as we shall see later. Such gradual increases in body size and/or efficiency of tooth structure appear to be largely restricted to the areas of warm climates unaffected by major glaciations; species adapted to the periglacial or cool zones, however, appear to fluctuate in body size. We can follow the increase in body size very well in some families and subfamilies of Pleistocene mammals. Thus, the marmots began with primitive, small, solitary forest marmots in America, which may have been similar to if not identical to the woodchuck (*Marmota monax*). They spread then into the Cordilleran region where, in the less mature ecosystems, they increased in size and became more social while adapting to the cold climates (Barash 1974). These are represented by *Marmota flaviventris* and *M. caligata* in North America and by *M. camtschatica* in Eastern Siberia. These primarily cold-adapted mountain forms gave rise to an adaptive radiation of mountain and tundra marmots in Asia (*M. himalayana, M. menzbieri, M. caudata*), as well as to the steppe marmot (*M. bobac*), which is more specialized still, both ecologically and socially. Marmot evolution culminated in the region furthest removed from North America, in the Alps, with the alpine marmot (*M. marmota*). Marmot specializations in the morphology of skulls, teeth, and front legs, as well as shortening of the tail and increased socialization, can be clearly traced from forest, to alpine tundra, to steppe (Bibikov 1968).

The same is probably true for the group of ruminants encompassed by the rupicaprids, caprids, oviboni, and eucerathers, that is, the goat, antelope, sheep, goats, musk oxen, and extinct shrub oxen, although this is less well documented. However, we do find the most primitive forms in the subtropical forest biome; these are serows (*Capricornis*). Next came the diverse mountain forms, the goral (*Neomorhaedus*), takin (*Budorcas*), chamoise (*Rupicapra*), mountain goat (*Oreamnos*), and the secondary radiation of true goat and sheep starting with the thar (*Hemitragus*), leading on to the audad (*Ammotragus*) and true goats (*Capra*) and sheep (*Ovis*). The size and ecological and social specialization increase generally with increasing distance from the Himalayan region (Geist 1971a). The real giants with greatly enlarged horns are also most distant, both in ecological specialization and geography. They are the musk oxen adapted to taiga (*Symbos, Botherium, Praeovibos?*) and the high arctic or dry, cold tundra (*Ovibos*), as well as the eucerathers (*Euceratherium preptoceras*) (Hibbard 1955) of the mountains of North America, where they became extinct in late Pleistocene times. Again ecological and social specialization is a function of ecological specialization, geographic dispersal, and ecosystem maturity.

Here we find again the peculiar phenomenon that during the evolution of these mammals the tail shortens; in ungulates with hornlike organs this leads to the nonsense correlation or inverse relationship in size between "horns" and tail, a relationship found in Old World deer, caprids, bison, elephants, and rhinos (Geist 1971b). If we add to the foregoing the changes in fauna when continents joined, the major extensions, such as that of the old South American fauna with its great diversity of peculiar ungulates and marsupial predators, or the major megafaunal extinctions that occurred during the late Pleistocene and around which controversy continues to rage (Martin and Wright 1967, Klein 1972, Martin 1973), the development of the totally new ecosystems, the alpine and arctic tundra (Hoffman and Taber 1967), and the periglacial ecosystem, and the evolution of the most bizarre mammal yet, the human, then we can appreciate that the ice ages brought on a period for terrestrial fauna that at present appears unparalleled in the earth's history. This is a new age.

The Tundra Ecosystem

The ice ages created two new types of ecosystem, which differ appreciably from each other, the tundra—be it alpine or arctic—and the periglacial ecosystems. During glaciations, the periglacial ecosystem predominates, while during deglaciations the tundra ecosystem does. During the interglacial period, the periglacial ecosystems become so inconspicuous to us, due to their small size and great diversity, as well as—with few exceptions—their inability to hold the complex trophic structure of large mammals that apparently characterize the periglacial ecosystem during the glacial periods, that only few have realized its great importance to an understanding of historic ecology and the evolution of man (Geist 1971a, 1975a, Klein 1973). Man first adapted to the rich periglacial ecosystem, and here reached his advanced form as Cro-Magnon and Aurignacean man, probably some 40,000 years ago, as we shall see later; he adapted to the tundra ecosystem and arctic marine ecosystem much later.

During the interglacial, the tundra becomes large and dominant. The essence of the tundra ecosystem is permafrost (Péwé 1966, Washburn 1973) with a thin, active layer of low biologic productivity; relatively undeveloped soils; a highly seasonal vegetation period of short duration; low rates of decomposition; a vegetation that is low in fiber and high in protein, carbohydrates, and minerals and which—like alpine vegetation (Johnson et al 1968, Klein 1970a,b)—is highly digestible and nutritionally superior to the coarse-fibered vegetation of the tropics; highly localized areas of high productivity in the vicinity of running water; the predominance of sedges, bryophytes, lichens, herbs, and prostrate shrubs in the vegetation—which reproduce not only by seeds but also, more commonly, by vegetative means in response to unpredictable temperature and snow conditions (Beschel 1970); a sparse, highly mobile, large-mammal fauna with strongly seasonal food and reproductive habits; a meager small-mammal fauna that usually hibernates; a great influx of birds in summer; highly mobile fish populations; a seasonal insect activity; and a harsh climatic regime with severe cold throughout much of the year and frequent blizzards. Moreover, there are short-term and long-term climatic fluctuations which cause great fluctuations in animal populations (Uspenskii 1970, Vibe 1970) in the high arctic, and extremes in daily hours of sunshine from winter to summer. In the conception of Margalef (1963), it is a relatively simple, immature ecosystem even at its climax.

The cold-adapted large mammals of the tundra that do not hibernate are characterized by being relatively large, which reduces heat loss through a reduction in the surface-to-mass ratio, and having thick hair coats with rather high insulation value (Schoelander et al 1950); they store fat from the superabundance of summer for use during the rutting season or for fetal growth and adult survival in winter; they also probably all store vitamins in their liver and fur, as well as metabolic salts in their skeleton which are depleted during winter, as caribou do (Zhignunov 1961); they produce a rich milk that allows rapid growth rates of the young during the short vegetation season (Klein 1970b); they tend to produce single young rather than twins, probably in response to the need to grow the young as large as possible during the short vegetation season, although moose and deer (*Odocoileus, Capreolus*) form exceptions. As indicated earlier, they tend to be specialized in their social organs and behavior, and tend to fight more damagingly than relatives from lower latitudes (Geist 1971a, 1974a,c,d), and they are highly seasonal in their activities. Some, like the wolf, caribou, sheep, saiga, and even moose, tend to roam far, even migrate, in the course of a year (Mech 1960, Kelsall 1968, Geist 1971a, Bannikov et al 1961, Edwards and Ritcey 1956); others, like mountain goat and musk oxen (Tener 1965), are reasonably sedentary. They also appear to be adapted for long individual life spans and low reproductive effort per individual. Reflecting the low plant productivity the ungulate biomass on the tundra appears to be very low, probably less than 100 lb standing crop per square mile in most areas (Petrides and Swank 1965 and Table 9-2), which is some 30 to 100 times less than on equatorial African savannah (Watt 1973, Bourlière and Hadley 1970).

Some of the generalizations discussed above may or may not fit locally, since tundra types vary greatly (Fuller and Kevan 1970, Zhigunov 1961). One should include with this biome also the norhern fringe of the boreal forest, for it is used by most tundra creatures as a refuge in winter. There are now many good books on the tundra and taiga and it need not detain us longer. It is the extreme terrestrial habitat

Table 9-2. Large-Mammal Biomass Within the Periglacial Ecosystems of Kluane National Park, Yukon Territory, Canada[a]

Species	Estimated Population Size [b]	Weight of Individuals (kg)	Biomass (kg)
Dall's sheep	4400	80	352,000
Mountain goat	850	80	68,000
Moose	350	370	129,500
Caribou	50	100	5,000
Mule deer	20	80	1,600
Feral horse	50	370	18,500
Grizzly bear	250	150	37,500
Black bear	100	100	10,000
Wolf	50	30	1,500
Total biomass			623,600

[a] After Hoefs 1973; weights taken from Guthrie 1968.

[b] Area of census 7160 km^2 (2800 mile^2); thus, standing crop is estimated at 89 kg/km^2 (499 lb/mile^2).

that man adapted to late in his history, and is of lesser importance than the periglacial ecosystem to an understanding of man's evolution.

The Periglacial Ecosystem: The St. Elias Range

We can get a reasonable picture of the periglacial ecosystems at lower latitudes as they must have existed during glacial periods, or even in interstadials when glaciers withdrew somewhat, by examining the evidence from archeological excavations of Pleistocene human habitations, such as the species of mammals and birds that are found in kitchen middens, the pollen profiles and the fossil soils, as well as by studying periglacial ecosystems of today (see Washburn 1973). Here, however, we must exercise extreme caution in our selection. Large glaciers exceeding 10,000 square miles in size are found today in the Antarctic, in Greenland, on some high arctic islands such as Baffin Island and Ellesmere Island, and in North America in the St. Elias Range on the Alaska–Yukon border. The huge antarctic glaciers unfortunately are not relevant to the understanding of the periglacial ecosystems of Eurasia and America owing to the entirely different faunas. The Greenland glaciers are somewhat more relevant. Humans do live in close proximity to glacial ice, but their food base is largely the sea, not the land. Still, such forms as caribou, musk oxen, and wolf do live here in close proximity to glaciers as they did in the Eurasian periglacial ecosystems. The Greenland and high arctic glaciers do not maintain the diverse terrestrial life that Eurasian Pleistocene glaciers did. They are also at far too high a latitude; the rich periglacial ecosystems developed at lower latitudes. This leaves only one large glacial system, which is in the right faunal zone, is an old epicenter or Cordilleran glaciation (Denton and Stuiver 1967), and is reasonably far south (just north of the 60th parallel) in the Northern Hemisphere, for consideration; this is the St. Elias Range in North America.

The glaciers and ice fields of the St. Elias Range exceed 12,000 square miles in size; this does not include the ice fields and glaciers of the adjacent Wrangall Mountains in Alaska or the glaciers of the Canadian and Alaskan Coast Ranges to the south. The Seward-Malaspina glacier alone in the glacier system of the St. Elias Range covers some 1650 square miles of area and contains about 420 cubic miles of ice (Sharp 1960). Even if the other glacial systems are less massive, there are probably still in excess of 3000 cubic miles of ice in the St. Elias Range.

The snowfields that feed the glaciers of the range are formed when the moisture-laden Pacific air is forced to rise off the icefield ranges that rise sharply from the ocean to become the tallest mountain chain in North America. The snow falls mostly on the oceanic side of the range. The snow condenses into ice which fills the valleys of the ranges and escapes to the oceanic and continental side through cuts in the mountain chain. On the oceanic side huge piedmont glaciers are formed, or the ice moves into fjords and then falls into the sea. The cool temperatures and high snowfalls on the Pacific side keep the rate of glacial ablation lower than the sunny, warm, and dry climate on the continental side. It is this side we are interested in.

The glacial system of the St. Elias Range teaches first the following: The windward, moist, cloudy side of the glacial system is quite different in its ecology from the leeward, dry, and sunny one. Second, even on the lee side there are very great differences in the ecology of periglacial ecosystems depending on a number of factors. The ecosystem of concern to us is the one generating the loess steppe.

Before going on to it, it is pertinent to point out a number of features of the St. Elias Range that make it the most representative of all glacial systems for studying periglacial ecology if we are interested in periglacial large mammal faunas and human evolution. It has a remarkable diversity of mammals; we find here today 34 species of small mammals (including fox, coyote, wolverine, marmots, and beavers, but excluding bats) and 13 species of large mammals, including horses, which went feral in the range. About 15 species of large mammals became extinct in the adjacent Alaska–Yukon refugium during megafaunal extinction in late Pleistocene times, including the horse. Present today are 3 species of cervids of 4 subspecies (caribou, moose, mule deer, and blacktailed deer), 2 species of bovids (Dall's sheep and mountain goats), 2 species of bears of 3 subspecies (grizzly, Alaska brown bear, and black bear), 1 large and 2 small canids (wolf, coyote, and red fox), and probably 2 felids (lynx and—less certain—mountain lion). At the end of the Pleistocene there were musk oxen, modern bison, horses, and elk; we also have reason to believe that other large mammals found in the Alaska–Yukon area were present, although not necessarily all at one time, such as lion (*Panthera atrox*), saber-toothed tiger, short-faced bear (*Tremarctitherium*), forest musk oxen (*Symbos*), *Cervalces*, camels (*Camelops*), long-horned bison, yak (*Bos*), mammoth, mastodont, ground sloths (*Megalonyx*), and probably saiga antelope (from Martin and Wright 1967, Scott 1937, Guthrie 1970b, 1968, 1966, Goldthwait 1966, MacNeish 1964). Hence, it is probable that in late Pleistocene times a fauna existed close to or in the St. Elias Range which had almost as many species of large mammals as small mammals (29 to 34), a situation we find today only in some African ecosystems which also have about the same number of species of small and large mammals (Guilday 1967). This is a most important point: The northern periglacial continent ecosystems had a trophic structure resembling that of productive African ecosystems, and that is for very good reasons, as we shall see later. Even today the ratio of large herbivores to small herbivores (7 to 23) in the St. Elias Range is similar to that of the Great Plains prior to European settlement (5 to 17), despite the better productivity of the Great Plains, and is similar to that of the Eurasian Steppes (7 to 27) (Guilday 1967). This does not include the elk (*Cervus elaphus*), which have been reintroduced close to the St. Elias Range, and which have formed a population of doubtful viability that has not reached the St. Elias Range, nor the bison (*Bison bison bison*), which were introduced in the adjacent Rubi Range but have not fared well. On the continental side of the St. Elias Range, the overall ratio of large mammals to small mammals is 13 to 34, for a total of 47 species of mammals which live at the glacier's edge, within some 2000 square miles, a diversity found in few terrestrial localities and none at a latitude above the 60th parallel.

The second point about the fauna of the periglacial zones of the St. Elias Range, besides species diversity and relatively large numbers of large mammals, is its close affinity with the "mammoth" fauna of the last glaciation in Eurasia (Vereshchagin 1967). The recent as well as the late Pleistocene fauna of this area overlaps strongly with that of Eastern Siberia, and was remarkably similar to the late Pleistocene fauna of the Ukraine and southern Russia (Klein 1973). Some Eurasian genera found in eastern Siberia, such as the woolly rhino (*Coelodonto*), cave hyena (*Crocuta spelaea*), sable (*Martes zibellina*), blue hare (*Lepus timidus*), musk deer (*Moschus moschiferus*), roe deer (*Caprolus pygargus*), and the antelope (*Spirocerus*), failed to get into North America, but they are compensated for in the Pleistocene by American forms such as *Cervalces,* the mountain goat, American deer (*Odocoileus*), mastodont, ground sloth, short-faced bear, black bear, coyote, saber-toothed cat, porcupine, and snowshoe hare (*Lepus americanus*). The old and recent faunas had in common mammoth reindeer, horse, camel, bison yak, wapiti (*Cervus elaphus*), musk ox, saiga antelope, mountain sheep, moose, lion, wolf, red fox, arctic fox, brown bear, lynx, wolverine, hare, marmot, and ground squirrel, not to mention similar smaller mammals of the same or closely related species.

The third reason today's periglacial ecosystem in the St. Elias Range is relevant to an understanding of the larger continental periglacial ecosystems of Eurasia and America is that the mammals of the St. Elias Range, in a few species, appear to attain the maximum phenotypic body size for their species. Unfortunately, the data are still very "soft" on this point, and one hopes that future studies will throw some light on it. It is important to note, however, that during the Pleistocene of Europe the bear (*Ursus arctos*) reached its largest size during stadials (Kurtén 1968a). The largest living bears are the Alaska brown bears (*U. arctos middendorfi*) which live beside, or maybe even at times on, the glaciers on the oceanic side of the St. Elias Range. These bears reach the incredible density of one bear per 0.6 square miles, with local densities of up to ten bears per square mile. Moreover, their reproductive rate is exceedingly high, with litter size ranging from 2.19 to 2.36 cubs (Troyer and Hensel 1964, Pearson 1972). Furthermore, Pleistocene caribou are reported to be of very large size (Banfield 1961). We do not know for certain how large caribou from the St. Elias Range get, since only one has been measured and weighed to my knowledge (they are rather large caribou to judge from the sightings I have made). It was a bull and weighed 508 lb (A. M. Pearson, personal communication). The largest-bodied caribou in America are associated with snowfields and alpine glaciations. The reasons for this will become apparent later. Pleistocene mountain sheep and wapiti (Stokes and Condie 1961, Guthrie 1966) are very large compared to modern sizes. Some Dall's sheep in the St. Elias Range appear to be exceptionally large, and the few weights taken suggest that they exceed Dall's sheep elsewhere by almost one-third in body weight and are just as large as Alberta bighorn sheep (M. Hoefs, personal communication). Moreover, the Alaska moose (*Alces alces gigas*) found in the Range is the largest subspecies of moose, but it is not known whether the St. Elias moose exceed other moose in size, since no data are available for comparison. On the continental, dry side, however, the grizzly is not exceptional in size, and has low reproductive rates and densities compared to the brown bear on the Pacific side, probably owing to the lower productivity of this area and the reduced salmon runs (Pearson 1972). Hence, not everything is exceptionally large in the St. Elias Range, but the trend toward large phenotypic body size appears to be there.

In addition to the mammalian characteristics of the periglacial areas of the St. Elias Range, which suggest a close affinity to Eurasian periglacial zones during the late Pleistocene, the flora shows the same relationship. This can be seen by comparing the pollen data, as described by Russian workers for the Würm glaciations of eastern Europe (Klein 1973), or those of Frenzel (1968) for northern Eurasia and Bonatti (1966) for the Mediterranean, with the plants presently found in the St. Elias Range, as shown in the work by Hultén (1968). Of particular interest here are plants of the loess steppe of which a small amount still exist in the St. Elias Range. Genera of interest are *Eurotia*, which is only found in the St. Elias Range in northern Canada and Alaska, *Artemisia*, of which nine species are found, and *Choenopodium*. In addition, the St. Elias Range is rich floristically owing to elements of the coastal Pacific flora, the Beringian—American montane, American prairie, and high arctic flora mingling in this range.

For all the above reasons the periglacial ecology of the St. Elias Range appears to be pertinent to an interpretation and understanding of the dry periglacial ecosystem of Würm and Wisconsin times.

The Elements of the Periglacial Ecosystem

There is no shortage of attempts to reconstruct the Pleistocene environments (e.g., Cornwall 1970, Washburn 1973). I am making use of these; however, I look at the periglacial features with the eye of an ecologist and so view them somewhat differently from the geologists.

The periglacial ecosystem is a *pulse-stabilized ecosystem* which is maintained in the early stages of plant succession by climatic and edaphic factors, and is *highly productive*. It is both immature and complex compared to the tundra and taiga at comparable elevations. The major factors are: the finely ground *fertile glacial debris in the form of loess,* silt, and sand with its complement of diverse nutrients which the glaciers spew out annually; the great physical instability of the fine rock debris and its enormous abundance, which caused it to form a wall in front of the glacier and forced the rivers to cut deep channels through it; the *coarser glacial debris,* which is spewed forth by the glaciers and pushed about by ice movements to create an unstable land surface, resulting in braided shifting glacial streams, moraines, and fluctuating lakes, and which is ultimately responsible for much of the discontinuous mosaiclike distribution of plant communities, as well as for complex distribution of fish faunas (e.g., Lindsay 1975); the continuous presence of *water*; its strong *seasonal* abundance, which causes a continuous outward movement and redeposition of fine and coarse glacial debris; the creation of silt flats on flood plains subject to desiccation and wind erosion when stream levels drop or lakes drain; the formation of *aufeis*, frost disturbance, and ice cores in the ground, which move rocks upward and damage plants; the maintenance of green vegetation from beginning to end of the vegetative season along water courses; the formation of snow and loess drifts in winter, which protect vegetation and have many other consequences; the *extremes of temperature*, both on a daily and a seasonal basis, which cause deep permafrost *aufeis*, ice lenses, ice veins and desiccation, and which accelerate or halt glacial ablation, control the build-up of snow blankets as well as their rate of depletion, affect the flow of water from the glacier, create up-currents and down-drafts along the glacial fronts, and indirectly contribute to cloud cover and duration of sunshine; the great frequency of winds, in part katapatic winds generated by the glacier (Washburn 1973); and the *extremes in wind velocity*, which also vary on a daily and seasonal basis and more silt away from the glacier, distributing it over a wide area, bury ice veins and *aufeis*, preserving them for millenia, bury plant and animal remains (Vereshchagin 1974), move snow and organic debris, pack snow into hard layers, desiccate and sandblast plants, rob animals of heat, and cause mosaiclike patterns of microclimates. There are also the interactions between loess, wind, temperature, and water which increase the diversity of ecological conditions around glaciers; as well as such phenomena as differential melting and the consequently unstable debris surfaces on glaciers, which may have a perpetually changing plant cover; the *albedo effect of glaciers,* which may locally increase plant production owing to increased solar radiation; the *sorting action* at the glacial termini which produce mineral licks so very important to cold-adapted large mammals; the *clear skies on the lee side of glaciers* which at night act as a heat sink, robbing the periglacial area of heat, while during the day permitting intense solar radiation, a rise in temperature, desiccation, and thermal currents. The permafrost may be deep, but so will be the active layer. In total these factors produce, on the continental or lee side of glaciers, an environment of extremes, a mosaic of biotic communities due to great microclimatic and fertility variations, a perpetually young ecosystem with a great diversity of plant and animal species, as well as great productivity despite inclement climatic factors.

The foregoing are conclusions based on personal experience (Geist 1971a, 1975b), on the ecological studies of plant successions in the vicinity of glaciers, such as that of Viereck (1966), on the recent experiences and studies of colleagues in the St. Elias Range (Pearson 1972, Lindsay 1975 Hoefs 1973, 1974, 1975, Hoefs et al 1975, Hoefs and Thomson 1972), as well as various papers on biological, agricultural, climatic, geological, and glacial studies (Kerr 1933, Bostock 1948, 1952, 1957, Porsild 1966, Sharp 1947, 1949, 1951a,b, Drury 1953, Tsukamoto 1963, Krinsley 1965, Denton and Stuiver 1967, Borns and Goldthwait 1966, Terasmac 1967, Muller 1967, Hughes et al 1969, Rampton 1970, Cornwall 1970, Johnson et al 1968, Denton and Karlén 1973, Washburn 1973), and on recent studies in Siberia (Vereshchagin 1974). Above all, they are based on an integration of this information as guided by observations during my stays in the St. Elias Range between 1965 and 1972.

Glaciers are not hostile to life. Far from it. One quickly forgets such fairy tales if one is exposed to the throbbing life in sight of ice, although perceptive anthropologists such as Klein (1973) suspected that life was rich and diverse in the vicinity of glaciers after evaluating the archeological and paleontological record. Some of the highest densities of mountain sheep are found just adjacent to glaciers, while large bands of sheep move deep into the glacier fields along nunataks. Mountain goats, grizzly, lynx, ground squirrels, pikas, and weasels cross glaciers from nunatak to nunatak (Murray and Murray 1969). One must have stood waist-deep in lush herbaceous vegetation, inhaled the scent of *Angelica*, and seen massive black moose feed beside a glacier while listening to the perpetual murmur, creaking rockfall, and gurgling waters of the glacial ice. Mountain goats scramble along the glacier's lateral moraine to mineral licks at its terminus; golden-crowned sparrow (*Zonotrichia atricapilla*) nest within a dozen paces of blue glacial ice; grizzly bears strip the heavily laden branches of the soapberry (*Sheperdia canadensis*) on glacial outwashes at the terminus. Above all, though, one's ideas change about periglacial life when one contemplates that moose spend the winter *on* a glacier, where deciduous forest has found root in a thin layer of volcanic debris on top of glacial ice. This can be seen on the terminus of the Klutlan glacier in the St. Elias Range.

One of the most spectacular phenomena are the loess storms. The glacier acts like a giant mill that grinds rock on the bedrock it traverses, as well as within its innards, and spits out the finely ground silt with the water flow from it. During summer, when much water moves from the glacier, it distributes the silt and sand on the floodplains of the braided glacial streams. When water subsides, or when streams change channels, or glacial lakes drain, the silt comes to the surface and dries into fine dust. It is a rich dust, rich in nutrients for plants, and the bigger the glacier the more rock it grinds, and the richer and more complete the loess is in nutrients. During morning and evening, strong winds howl down from the glacier. They pick up the loess, whirl it high into the air forming dust clouds thousands of feet high, and blow the loess far downwind from the glacier. Here it accumulates on obstacles in the wind's path, such as hills, ridges and trees, and falls into hollows behind obstacles or on flat plains, or is blown onto water bodies that carry it away. It is a fertilizer that is blown down the wind's path, continuously picked up, and continuously replaced by the glacier's action.

The constant deposition of loess on the plains, coupled with the action of sudden floods pouring from the glacier in season, would continuously erode the river banks in such a fashion that very steep crumbly "loess cliffs" would flank much of the river. It is these loess cliffs that collapsed under the weight of mammoth to cover and bury these giants, and ultimately permit their carcasses to freeze into permafrost. It is quite likely that these cliffs permitted the "jumps" of large ungulates at a much greater frequency than do our present river banks with their eroded rounded contours.

Now we can answer some of the puzzles of the periglacial paleontologic record from the Würm glaciation of Europe. It appears that some faunal elements typical of semiarid plains, the tundra and boreal forest, as well as the broadleaf forest, occurred together. Thus Klein (1973), in reviewing the archeological evidence from the Ukraine of late Pleistocene hunters, reports the snow lemming (*Dicrostonyx*) and the great jerboa (*Allactaga*) on the same site. Vereshchagin (1967) lists as occurring in the same site reindeer, a tundra and alpine species; saiga and horse, creatures of the dry dusty steppe; together with red deer, *Bison*, and *Bos*, which indicate the presence of some deciduous forest. At another site, Vereshchagin (1967) lists forest wisent (*Bison bonasus*), aurochs (*Bos primigenius*), saiga, moose, red deer, wild boar, roe deer, onagers, horses, and beavers. Moreover, the summary lists of this author imply that between the Urals and the Altai in the Pleistocene early man could have had as mixed a bag as camel, moose, reindeer, yak, and roe deer. On the face of it, one would claim such a faunal mix as impossible, since camel, saiga, horses, steppe marmots, and jerboas belong to one fauna; reindeer, snow lemming, arctic foxes, and musk oxen to another; and forest wisent, aurochs, wild pig, red deer, and roe deer to a third. These three faunas tend to be very far apart toady, although some mixing between tundra and taiga faunas occur in Siberia where reindeer, moose, wapiti, and roe deer may be found in the same valley in Yakutia (Egorov 1967). The most puzzling aspect is finding inhabitants from the cold, dry steppe with inhabitants of the permafrost region. The common explanation offered is that the "periglacial steppe" was something quite different from any habitat we have today (Garutt 1964, Klein 1973) and that conditions prevailed such as cannot be experienced today. To this I take exception, for one glance at the relict, but still actively maintained, loess steppes in the St. Elias Range unravels the puzzle.

We can now understand why ungulates from the periglacial steppes tended to have large, often high-crowned or even overgrown, cheek teeth and, where comparison has been made, their tooth rows were longer than those of their modern counterparts (e.g., Ruttern et al 1972, Banfield 1961). Windblown silt deposited on the vegetation appears to be responsible for this. It imposes wear additional to that from the plant fibers on the teeth. Thus Dall's sheep living on the silt-dusted slopes of Sheep Mountain in the St. Elias Range have a short life expectancy compared to sheep from other populations (calculated from Hoefs 1975). They also have a high frequency of diseased jaws, which can be attributed to the tooth damage imposed by silt, and which in turn reduces the ability of these animals to masticate efficiently and hence increases their mortality owing to caloric bankruptcy. One can explain in a similar vein the rather short life expectancy of Asiatic urial and argali sheep compared to ibex or American-type sheep (Geist 1971a). The former are inhabitants of rolling desert and steppe terrain in which fine, abrasive, windblown dust is common; the latter inhabit the rugged rocky slopes and cliffs in which dust is comparatively uncommon. The difference must have been particularly acute during glacial times when argalis must have been living at the edge of the silt-generating glaciers. Similarly, the dentition of horses, mammoth, and bison may have been enlarged and hardened to withstand silt abrasion.

Owing to the work of Hanson and Jones (1976) on the significance of sulphur in soil, water, and mineral licks to wildlife, periglacial ecosystems on the dry lee side of glaciers take on an additional significance. It is likely that, due to the flood and recession cycle of water over the silt and loess of glacial streams, as well as intense sunlight, a large amount of sulfates of calcium, magnesium, and sodium would come to the surface and enrich the surface loess, with actual encrustments of these compounds in some places. Such deposits may account for the lenses of mineral depositions so sought after along eroding loess banks beside rivers by various ungulates today. The presence of these metallic sulfates would greatly enhance the production of ungulates for reasons so well espoused by Hanson and Jones (1976), since sulfates not only increase the protein content of forages, especially with sulfur bearing essential amino acids, but ruminants in particular may synthesize such amino acids in their rumen directly from inorganic sulfur.

Habitat Mosaics as a Function of Periglacial Climatic and Edaphic Factors

Let us visualize the conditions on the lee side of the huge Scandinavian or Taimyr ice masses in south central Russia, western Europe, and central Siberia. The weather systems would have unloaded their snow load somewhere over the epi-center of the ice mass, so that skies would usually be clear on the lee (south) side of the glacier. The topography would be undulating due to the moraines deposited by advances and retreats of the glacier, as well as to sand dunes shaped by the strong—even violent—winds coming daily from the glacier, and the many deep cuts left by the ever-changing river channels. The cold winds would blow from the glacier to the south, while the clear skies would ensure strong solar radiation on the south-facing slopes and a severe heat loss from the north-facing slopes to the heat sink of the open sky at night. Drifts of fine loess and snow would form on the lee side of hills and ridges above the sites receiving intense solar radiation. These would usually be free of snow, owing to the low snowfalls expected on the lee side of the glacier and the evaporation of snow by the sun. Snow and loess would also accumu-late in the hollows. Here, protected from the abrasive action of windblown sand and cold desiccation, we would find dwarf birches and clumps of spruce, as well as sedge meadows and willow flats, depending on the availability of water.

In winter the strong winds would blow layers of loess over the frozen river channels and *aufeis* and bury the ice; since river channels would shift, some frozen rivers remained cut off and became more deeply covered by loess with each windy day. This may be one origin of the many ice veins with their columns of frozen loess described by Vereshchagin (1974) in the Taimyr region of Siberia.

A belt of wet tundra changing to dry tundra would protrude on the north slopes of ridges beyond the range of spruce trees, to terminate in relatively bald areas caused by the desiccation and sandblasting of strong winds. The lowlands would be deeply filled with windborne loess, silt, and sand deposits on which loess steppe would form, while the protruding undulating uplands would be covered with taiga-tundra on the north slopes, rich grasslands on the south slopes, and dwarf birch flats, willow flats, wet sedge meadows, and cotton sedge meadows on flat areas. Thus, there would be sharp vegetation zones. Permafrost zones with tundra vegetation would abut rich grassland on one hand and *Artemisia*-covered loess steppe—continuously formed from windblown loess—on the other.

The presence of steppe marmot indicates a loose soil layer deep enough to permit hibernation, and the absence of permafrost in the immediate vicinity of its burrow. This large social rodent would probably hibernate in deep loess deposits on south-facing slopes close to large, slowly melting, snowbanks which would extend the season of sprouting vegetation for the marmots (Bibikov 1968).

From the edge of the glacier, across the moraine fields, and on the high uplands some way from the glacial terminus, one would find predominantly tundra and taiga vegetation. The broad river channels cutting through the moraine fields would bring most of the loess and fine debris together with water well beyond the moraine margins to an area of meandering braided streams. It is here that the loess would usually become airborne and be dumped on the land beyond to accumulate on flat areas and in hollows. The further from the glacier, the more frequently the land would be covered by loess during the daily windstorms. Hence tundra and taiga would give way to loess steppe and grass steppe. The colder the climate, the longer the ground would stay frozen, the less loess would be blown about, and the wider the taiga-tundra belt would be around the glacier's edge. As climate becomes warmer, the ground is less frozen, there is more desiccation, and more loess is blown about; loess is an unstable substrate, does not hold water well, and since it is constantly reworked by wind while new loess is deposited and old loess and sand are blown away, it can be colonized only by plants resistant to extremes of temperature and to snowblasting; it does not grow trees which could permit snow to accumulate and permafrost to form (except in deep pockets protected from wind where snow accumulates). Hence we can and do have situations where permafrost areas abut loess steppe that is free of permafrost. This should explain the first puzzle, how snow lemming, reindeer, and arctic fox can be in sight of jerboas, steppe marmots, saiga antelope, and horses.

From attempts to reconstruct the Pleistocene climates of Europe (Washburn 1973), it is evident that warm, moist air would be drawn from the Mediterranean and Black Sea region toward the glaciers. Such air could move up the river valleys and be trapped by the cold air blowing from the glaciers in the river canyons. This would probably allow floristic and faunistic elements typical of warm climates to move deep into cold periglacial steppe along the water courses. Hence the peculiar mingling of bones from arctic, steppe, and forest creatures in the middens of Paleolithic hunters.

Biomass of Large Mammals

Another aspect that is puzzling is the impression left by the fossil record, not only of a great diversity of mammals, but also of an abundance of these animals. Of course, the fossil record does not provide any measure of abundance, except relative abundance, yet it does stimulate one to see whether productivity in the periglacial ecosystem could have been higher than in the tundra. The theoretical consideration that the periglacial ecosystem is pulse-stabilized and young suggests that it is most productive. Fortunately, there are some data to answer the question of whether the periglacial zone is indeed more productive than the tundra, and even to estimate the upper limit of large mammal biomass which could have existed in the Alaska−Yukon refugium. Hoefs (1972) has provided data, which I have put into Table 9-1, on the abundance of ungulates in Kluane National Park, which now covers much of the St. Elias Range on the Canadian side. The total area of potential habitat enclosed in the park is about 2800 square miles, or some 7160 square kilometers. The park boundary excludes the most productive lowland area within and just outside the range, so the estimate of biomass will be low. There are now nine large mammal species in the Park, including feral horses; I have estimated the number of horses most conservatively in Table 9-2. Even so, the standing crop is about 500 lb/mile² (89 kg/km²), almost five times that cited for tundras (Petrides and Swank 1965). If the more productive lowland sites were included, and horses, elk, musk oxen, and bison were reintroduced and allowed to stabilize their population, the standing crop would probably be close to 1000 lb/mile² in a climatic regime similar to that found today.

It is worth noting that agricultural productivity adjacent to the St. Elias Range, despite the cold climate which—depending on locality—allows only for a 15-to 50-day frost-free period, still comes close to that of the Alberta prairie more than 1000 miles to the south and east (Tsukamoto 1963). The point is that the periglacial ecosystem appears more productive by far than the tundra at comparable latitudes and altitudes with similar weather, and could have supported a much more complex trophic structure of large mammals than does the tundra today.

The mosaiclike plant communities formed through sharp microclimatic soil and moisture differences as described earlier would extend the sprouting season of vegetation and make it available for feed as well as for a varied fall vegetation. The spring vegetation is rich in protein, vitamins, and minerals; the fall vegetation is rich in seedheads of plants with their high protein and fat content, as well as the frozen herbaceous vegetation so favored by northern ungulates (Geist 1971a). These factors would increase the quality of available food throughout spring, summer, and fall. The physical structure of the plant communities was largely such that ungulates and rodents had almost total access to the photosynthetic layer. Grasslands, tundra, dwarf birch, willow flat, sedge meadows, and loess steppe all have plants of low height; the photosynthetic layer is compressed compared to that of forests. There were few trees whose crowns were beyond the range of herbivores, as is typical during interglacials. Moreover, the abundant sunlight on the leeside of glaciers would ensure the maximum photosynthesis that edaphic factors permitted. Thus not only quality, but quantity, of plant food available to ungulates would be very great. From Table 9-2 we also note that the ratio of wolves to ungulates is about 1:112, a ratio close to the 1:130 that Guthrie (1968) found in the Alaska Pleistocene fossil record.

Large Mammal Habitats in the Periglacial Zone

We have not dealt with the river valleys. These are rather deep cuts in the surrounding loess topography with several terraces as well as a floodplain; we expect to find many steep loess cliffs along the streams. The floodplains would be protected from much of the strong wind that blew down from the glaciers. On the north-facing ravines and slopes, we would expect a forest of spruce and birch, and rich grasslands on south-facing slopes and plateaus surrounded by south-facing slopes. During very cold periods with short vegetation seasons, we would expect larches to replace the spruce. There would be sedge meadows in the floodplains, willow flats along the gravel bars, pine groves on dry terraces, poplar groves on alluvial deposits, and dwarf birch flats on wet terraces surrounded by north-facing slopes. The further south the river went the more elements of warm deciduous forest would creep into the vegetation surrounded by south-facing slopes, particularly on sites that would receive a shot of flood water in spring and remain boggy all the year round. The albedo effect would create a microclimate quite favorable to elements of a warm-country flora and fauna; furthermore, we expect warm, moist oceanic air to move up the river valleys, if the reconstructions of European Pleistocene weather maps are trustworthy (see Washburn 1973).

On the fringes of the river system closer to the glacier, where there would be a mixture of boreal forest and steppe along the river course, we can expect roe deer, moose, and even red deer to thrive, just as they do in Siberia today, and just as in central Yakutia we would expect the moose and roe deer to go into colder zones than the red deer (see Egorov 1967). The roe deer, which is a small animal and thus very much affected by heat loss from convection (wind) and radiation, occupies the river valleys in Yakutia which have a rich mixture of steppe and taiga. In winter they stay on steep south-facing slopes wherever there is adequate browse from willows and a thin snow blanket. We expect this habitat to have been available in the periglacial zone. On the south-facing slopes, roe deer can absorb some sunlight, which is important, as we know from Webster's (1971) work; he found that cattle would save about one-third of a megacalorie of fat from oxidation for every megacalorie of sunlight absorbed. Roe deer also prefer dwarf birch flats. During the night roe deer presumably move under a heavy canopy of conifers, since such habitat is essential to them. This is also not surprising according to Moen's (1968) work on white-tailed deer. They escape significant radiation heat loss on clear nights by going into cedar thickets, which act as a radiation shield against the clear sky. Furthermore, the steep river valleys would permit roe deer to escape from the strong, cold glacial winds; the heat balance of a small animal is much more affected by wind than that of a large animal. In addition, the snow blanket would remain soft, being protected from the wind by canopies of trees, and would permit roe deer to paw for forage if the need arose. Furthermore, an area with soft, thin snow and patches of snow-free ground, traversed by an occasional cornice of hard, windblown snow, offers very good footing to a fleeing roe deer. This little deer would probably have done quite well along the rivers granted plenty of sunshine, thin soft snowblankets, and a mosaic of meadows, steppe, dwarf birch, willow flats, and taiga. It would have been restricted to rivers, however, and relative to the occupants of the steppe or tundra it would have been rare. Moreover, it would not have been very profitable to hunt, granted its solitary and secretive nature; thus, its scarcity in Pleistocene bone heaps left by early man is not surprising. It would have become more common further downriver in increasingly warmer areas. Moreover, it is capable of significant migrations and could have moved upstream in summer and downstream in winter, optimizing its seasonal habitat requirements (Egorov 1967, Heptner et al 1961).

Moose would be a relatively rare animal in the periglacial zone, since their winter habitat would be confined to the willow thickets on the floodplains and gravel bars of glacial rivers and, exceptionally, to some willow flats surrounding moraine lakes close to the glaciers. They too would be relatively difficult to hunt granted the low snowfalls and the light, fluffy snow in areas protected from wind, which we should expect in the periglacial zone. Moreover, this is a dangerous animal, which even Siberian hunters in the past century rated more dangerous than bear (Heptner and Nasimovich 1968). We do not expect it to be common in Pleistocene kitchen middens, as indeed it is not (Klein 1973).

Red deer would become increasingly common with extensive floodplain sedge meadows and rich grasslands, but they would penetrate deep into the periglacial zones along river valleys. They would also utilize dwarf birch flats, willow groves, and the rich grasslands on steep south-facing slopes and on terraces opposite south-facing slopes during winter. In summer they would quite likely move onto the moraine country in the periglacial zone to feed on the tundra vegetation, and sedge meadows within the boreal forest zone.

Caribou would find ideal conditions in the periglacial zone for a number of reasons. The moraine country would provide good feeding grounds in early winter in the dwarf birch flats, which would have on drier sites a good cover of favorite forage lichens of the genera *Cladonia, Steriocaulon*, and *Certraria*. In wetter dwarf birch flats the lichen component would be reduced, but it would be compensated for by *Festuca*, a grass that grows under the snowblanket and is important for winter green forage, which is essential for reindeer, as well as short sedges and various horsetails (*Equisetum*). There would be sedge meadows around moraine lakes and marshes, supplying caribou with sedges in winter and the roots of forest marsh plants in spring.

Some of the more open areas of the spruce forest on the moraines would supply lichens and the tundra would produce some evergreens, such as *Empetrum*. Since the constant flow of water keeps vegetation green along the water courses throughout the vegetation period, frosts and the first snowfalls would preserve the green vegetation which can act as a most important forage throughout the winter. In particular, marshy hollows would preserve plenty of winter green. As the snow blanket is expected to stay soft and thin, much of the vegetation would remain available to the caribou throughout winter. In particular, marshy hollows would preserve plenty of winter green. Since the snow blanket is expected to stay soft and thin, much of the vegetation would remain available to the caribou throughout winter. On windblown sites along gravel bars of water courses, caribou would find such plants as *Dechampsia*, an important winter food. During summer the sedge meadows, cotton sedge meadows on drier sites, tundra vegetation, and dwarf birch/willow flats would furnish the forage for lactation, fattening, and antler and hair growth, as well as for rebuilding the skeleton made thin, fragile, and porous if metabolic salts are not available in winter in the forage—an unlikely event in the rich, well-fertilized periglacial habitats—and the storing of vitamins A, D, and E in liver and body fat for the following winter.

Caribou could escape the insect pest by heading onto the ablation moraine or moving far onto the glacier on glacial debris or into the glacial zone on a ridge not overridden by ice. Here the cool winds blowing off the glaciers would provide relief from biting insects. Simultaneously the larvae of warble flies (*Oedemagena tarandi* L.) would tend to fall onto glacial debris or wet ground, rather than the dry dusty ground they require for pupation (Savalev 1961, Kelsall 1968). A similar fate would befall the nasal warble fly (*Cephenemya trompe* L.).

Excellent winter green forage, the reduced need for emigration, and an abundance of diverse vegetation from wet plant communities such as wet tundra, wet meadows, marshes, and riparian communities, as well as a reduction in expenditure of nutrients and energy to avoid insect pests and nourish warble fly infestations, predict that reindeer would have been phenotypically large animals. The fertilizing effect of loess predicts that the forage would be rich in all essential nutrients; hence caribou would have large antlers. The presence of some windblown loess on their forage would predict accelerated tooth wear and possession of relatively large tooth rows. All this is indeed found in Pleistocene reindeer (Banfield 1961).

The moraine hills with their sedge meadows, windblown ridges, and loose, dusty snow would also have been favorable for musk oxen. They do thrive in the periglacial zones of Greenland and the Canadian High Arctic islands, and became extinct relatively recently in the St. Elias Range itself. They would not likely be a common element in the kitchen middens of Pleistocene man wintering along river courses, since they would remain close to the glaciers in winter, and if hunted and killed would be boned out for easier transport.

Climatic Factors and Megafaunal Extinctions

One additional point needs to be made: In winter the cold air streaming off the glacier would ensure a wide cold zone around the glacier's margin. If warm air moved up from the south, it would inevitably melt snow which would freeze and crust at night, making it difficult for ungulates to paw through snow and reach their forage in areas with relatively deep snow. Apparently, such warm, moist oceanic air could be expected to move toward the glaciers, granted the present reconstructions of European Pleistocene climates (Washburn 1973). In areas with low snowfalls, warm chinook or foehn winds would clear forage and make it more available. The cold air zone and cold winds from the glacier would ensure that the snow would remain soft and powdery in the periglacial zone, permitting ungulates to forage successfully. If the influence of warm air were common to the south of the ice shield, there would be a zone somewhere south of the glacier of frequent snowfalls, winter rains, and alternate freezing and thawing, creating a zone quite unsuitable for wintering ungulates. Hence, we can expect a dense wintering population of ungulates, hares as well as carnivores, within the periglacial zone—including the chinook zone—in winter, and only a smattering of wintering ungulates to the south of it, at best. The condition described here is found in Greenland, where the most important wintering ranges of caribou, arctic hares, and musk oxen are adjacent to the glaciers, while warm-air storms with winter rains make much of the land some distance from glaciers unsuitable for them owing to periodic icing (Vibe 1970, and personal communication). Hence, Greenland's herbivores fluctuate with the weather cycles; they are abundant when dry, soft snowblankets predominate or chinooks clear the snow from their wintering areas, and scarce when heavy snowfalls, winter rains, and icing conditions reduce their winter range to the immediate vicinity of glaciers, as well as causing wholesale starvation. The records by Nasimovich (1955) of starvation caused in ungulates by icing conditions are also most pertinent.

It should be evident from the foregoing that during periods of cold, dry climates the tundra and taiga would expand, as would the cold steppe on the loess deposits. The ground would stay frozen for up to ten months in the taiga belt, as is indicated by the presence of larch and birch and the reduction of spruce and pine. Trees would be greatly reduced in the immediate vicinity of the ice. During warmer periods, much of the loess steppe would change to grasslands and much of the tundra would turn to boreal forest. The vegetation season would be extended and productivity would be increased, while foehn winds in winter would keep the winter forage of ungulates free of snow. However, there would come a point where snowfalls and icing, due to alternating warm and cold spells to rains, would begin eliminating many large grazers from the periglacial zone. We would expect this to occur during deglaciation, when reindeer and musk oxen would be confined increasingly to the immediate vicinity of glaciers, while wide areas of grassland and meadows would become inaccessible to bison, but not to horses, since the latter are not only very efficient pawers and easily remove snow but also prefer the loess steppe where winds continue to remove snow. A gradual increase in snowfalls and icing during winter would lead to increased hardship and lower reproduction by the large grazers (see also Vereshchagin 1967). We should find that their population quality would deteriorate, as would be indicated by reduced body size of adults and an increase in malformations and abnormalities in the skull, as has indeed been found for mammoth (Heintz and Garutt 1965 in Kurtén 1968a, Vereshchagin 1970), but which remains to be investigated for other ungulates, such as horses and bison. This should not be confused with the general reduction in body size of many mammals in postglacial times, as discussed, for instance, by Kurtén (1968a) or more speculatively by Edwards (1967). With warmer unstable weather during deglaciation, the skies would be cloudy more often, more snow would fall, the snow would often be harsh, and foraging would be more and more difficult. There would, however, be an increase in the productivity of the land. Forests would spread as climatic fluctuations decreased in amplitude; the loess steppe would subside to become a rich grass plain into which elements of southern faunas such as gazelles and onagers would spread. Red deer and forest wisent would become more numerous in the floodplain forests, which would eventually spread to the surrounding benchlands. In the old periglacial zone, however, icing conditions would tend to lead to wholesale starvation of bison, horses, reindeer, hares, and mammoths, while the ice age hunters, finding themselves with reduced game herds, increased hunting pressure on the survivors and diversified. Musk oxen would retreat to the tundra to the east of the Scandinavian ice sheet, while reindeer would continue to exist in decreasing numbers in the periglacial zone and expand populations in the adjacent tundras. In essence, I am adopting for the European scene much of Hester's (1967) view of megafaunal extinction.

The mammoth and woolly rhino are of some interest here. It is known that both were predominantly grazers, both from their tooth structure, the anatomy of soft parts where preserved, and from the remains of plant material found in the mouths of frozen carcasses (Garutt 1964, Vereshchagin 1967). Pollen analysis indicates that these forms were at times far from treed landscapes. If grasses and sedges continued to be their main forage in winter—supplemented of course by a significant fat layer they deposited for winter use during the superabundance of summer—then it could only have been tall sedges and grasses such as grow in marshes in combination with low dusty snow. Heavy snowfalls in winter will simply flatten the vegetation to the ground, making it almost unavailable for ready grasping. Mammoths probably did clear snow with their tusks, as is indicated by the wear surfaces on tusks. Their trunk had a tip that could have grasped bunches of grass and pulled them out (Garutt 1964). Soft, powdery snow would have greatly facilitated feeding, providing the herbiage continued to stand reasonably upright in the dry state. This suggests that mammoth and woolly rhino probably utilized much of the coarse grasses, sedges, rushes, and reeds growing in marshes and along water courses which are not eaten by any large herbivore today. This hypothesis is in agreement with Guthrie's (1968) findings that fossil mammoths were found closer to main drainage systems than were bison, for instance. It also agrees with the distribution of mammoth reported by Vereshchagin (1974) in Siberia. This hypothesis is also supported by the theoretical concept that large-bodied ungulates can feed on relatively low-protein, coarse-fiber forage, while small-bodied ungulates must select increasingly finer and more nutritious forage; in addition, the digestive system of the mammoth would have permitted rapid passage of forage, and the animal could have increased energy uptake by a rapid passage of forage in which only the most digestible parts were digested (Bell 1971). This would explain why mammoth and woolly rhino would be found with such arctic forms as white foxes (*Alopex lagopus*) and caribou, as well as with steppe animals such as saiga antelope, steppe wisent and horses, and taiga animals such as moose, roe deer, beaver, and wolverine, the latter indicative of taiga along rivers (Garutt 1964, Vereshchagin 1974). The mammoth would probably move about considerably, utilizing relatively rich, tall grasslike vegetation, be it in the green or dry state. If this is so, it would have been most susceptible to deep snow and icing, as well as to extremely cold, dry climates which reduced such vegetation in favor of dry tundra with a low herb layer. Hence, it would reach its highest population density and largest phenotypic size in interstadials with long vegetation periods, probably low snowfalls and frequent chinook winds. This view is most compatible with the presently available data on the body size variations and malformations (Vereshchagin 1970) and times of appearance and extinction in various areas (Kurtén 1968a, Garutt 1964, Vereshchagin 1967, 1970, 1974, Klein 1973).

The woolly rhino was probably less a tundra creature than the mammoth—although in the summer both would probably be found on the tundra, for their remains are associated with tundra vegetation (Garutt 1964)—and more an occupant of somewhat dry, productive grasslands and meadows on benches and floodplains of rivers. With the reduction of such grasslands in favor of either tundra or deciduous shrubs and forests, it would decline, no doubt hastened in its demise by Pleistocene hunters. I am not aware of studies indicating a decline in population quality of the last woolly rhinos. It, as well as the steppe wisent, may have survived the woolly mammoth by a few millenia, as recent findings have indicated (Klein 1973).

The steppe wisent's fate throughout Eurasia and in Alaska may have been much the same as that of rhino, mammoth, and horse. It could have declined with icing in winter during glacial melt-off. It was associated more with grasses on uplands, as indicated by Guthrie's (1968) findings in Alaska; horses and mammoths here preferred the river valleys. The steppe wisent did not survive the deglaciation in Eurasia, although forest bison did, and probably spread to North America to become the modern bison (Geist 1971b, Geist and Karsten 1977). Horses also became extinct over all but the dry parts of central Asia, where they are still found with onagers, camels, and saiga antelope. If milder climates during deglaciation caused much of the loess steppe to change to grasslands, then increased snowfalls in combination with icing should have caused massive die-offs of horses in Siberia and Alaska andsled to their local extinctions. The large wapiti, an inhabitant of the grassy areas and diverse communities of the river valleys, also became extinct in Beringia (Guthrie 1966), but survived in the mountains of central Siberia and in the forested areas of Manchuria and Transbaikal. In short, large-mammal faunas associated with northern sedge meadows and grasslands became extinct. The wapiti has been reintroduced in the Yukon, as has the prairie bison; neither introduction has fared well. In Alaska bison have, however, been successfully established (Guthrie 1968).

Megafaunal extinctions are a striking part of the Pleistocene faunal history (Martin and Wright 1967, Webb 1969, Kurtén 1976). Some of these can be accounted for by climatic and ecological changes that must have taken place (Hester 1967, 1970, Kurtén 1976), others by the activity of specialized hunters (Martin 1967, 1973, Bryan 1973), while others may have been through direct competition by evolving man. These extinctions left vacant niches; there has been no replacement by ecological equivalents, although the diversity of small-mammal herbivores and carnivores increased during megafaunal extinctions (Webb 1969).

It is commonly assumed that the interglacials were "better" periods for human habitation of the glacial refugia than the glacial periods. I believe this assumption to be in error. Although the interglacial climate is milder, it brings about tree growth which mitigates against ungulates developing as high a biomass as on the steppe. Trees act as snow traps, and reduce wind at ground level. Hence deep, soft snow hinders movement. To master it one requires snow shoes, and this presupposes a relatively highly developed technology. From the standpoint of hunters, interglacial periods are a disaster—and the evidence does bear this out (see Chapter 14).

One last point: Glacial times must have been times of low abundance of waterfowl and migratory birds, due to the large size of the ice masses covering the northern hemisphere and expansion of the deserts. The face of the Earth has been greatly altered from Pleistocene times by an interglacial climate, human dominance, and the consequent alterations of the floral and faunal elements, plus the rearrangement of some for the service of man. Today, we can barely grasp the significance of the ice ages, and yet they are supremely significant, for in our last major step in evolution we adapted to the rich hunting grounds of the periglacial zone and were transformed into modern man. We must now look at our past history.

Chapter 10

Prehumans

Introduction

In approaching an analysis of our own evolution, we may pause to look at that of other mammals. The fate of lineages of large mammals is of particular interest, for here the fossil record is better than with small mammals, and many of these lineages spread over large areas of the globe much as hominids did.

We encounter the same phenomenon repeatedly: In a given lineage, species judged primitive are found in tropical and subtropical forests; such species appear earlier in the fossil record than advanced forms; more highly evolved species occupy grasslands, deserts, and cold climates. Complexification of social organs and increased gregariousness run parallel with ecological specialization, except in the most northern forms. The lineages apparently evolved from mesic to xeric, aseasonal to seasonal, stable to unstable, warm to cold, low elevation to high elevation—in short, from rather benign physical environments to increasingly inhospitable ones,

The species judged primitive tend to be small bodied with relatively simple display organs and an adaptive strategy based on resource defence, be it in pair territories or in demes defending a common resource base. The advanced species are often—but not always—grotesque giants, huge in body with showy, odd, social organs, illustrated nowhere better than by the parade of advanced cervids and bovids from cold environments and the recent Pleistocene record. I have described this phenomenon in somewhat greater detail in earlier publications (Geist 1971a,b, 1974 a,c,d, 1977 Geist and Karsten 1977). The evolution of horses from diminutive browsers of the early Tertiary to fleet-footed medium-sized grazers of the late Tertiary to giants of the Pleistocene plains has been described by Simpson (1951). The African antelope are seen to have evolved from tropical forest browsers to grazers (Estes 1974) while some, like the gazelles, gave off large-bodied, highly gregarious, offshoots in northern cold climates (Geist 1974d). The same basic process of evolution from forest to grassland has been postulated for the macropods (Kaufman

1974b), and the process of evolution just described also appears in some birds, particularly the gallinaceous birds, as I noted recently (Geist 1977).

In birds, however, we have an additional phenomenon, only weakly discernible in large mammals. As evolution proceeds from forest to savannah and onto open plains, so sexual dimorphism tends to change from monomorphic to dimorphic and again to monomorphic. In ungulates we find the same (Geist 1974a). In gallinaceous birds and ungulates, monomorphism in primitive forms tends to be through male mimicry, resulting in well-armed, often colorful, aggressive females. In highly social species of ungulates, such as the American buffalo, caribou, oryx, and eland antelopes, the monomorphism or reduced sexual dimorphism is also based on male mimicry resulting in male-shaped females. This tendency is expressed in few gallinaceous birds, such as the eared pheasants. The general trend is, in advanced forms, toward *female* mimicry, resulting in adults of female or juvenile plumage, great tolerance, and poorly developed weapons. A similar procession can also be seen in Australian parrots (Brereton 1971).

Before beginning an analysis of human evolution, one has, therefore, to ask whether our genus evolved much in the same fashion as did other lineages of geographically widely dispersed large mammals and birds. Did we evolve progressively from tropical forest dwellers into increasingly more challenging physical environments, into savannah, grasslands, deserts, temperate zones, and finally the periglacial and arctic zones? Did we begin as small-bodied late Pliocene forms and evolve into grotesque giants of our family? Is the fossil record compatiole with the hypothesis just advanced? Are our characteristics derivable from the adaptive syndromes of species successively adapting to more inclement environments? I suggest that these are rhetorical questions and that our genus evolved much as did other lineages of large Pleistocene mammals.

The Early Hominids

The early history of our kind is shrouded in the mist of antiquity and in the mysteries of crushed, deformed, incomplete, and scarce fossils. It is bedevilled by differences in interpretation by experts and by a very incomplete understanding of the relation between ecology and morphology, behavior, and culture. Little is gained by discussing the early primate history of our ancestry, and this can be pursued in various serious or popular syntheses, such as Cartmill (1974), Kurtén (1972), and Young (1971) or, for the layperson, in Pfeiffer (1972). We begin our enquiry by taking a hint from other large mammals and looking for an ancestor from tropical forest from which hominids evolved.

We require in this ancestor a K-selected species, one with a low reproductive rate, single births and long ontogeny, as this is quite typical of our species as well as of the anthropoid apes, our closest ancestors. It must be close zoogeographically to the australopithecines, the next adaptive step in our ancestral lineage. Moreover—and this I shall elaborate on later—it ought to be a kin-selected species, for only such a species could ultimately give rise to a good many human features. Suspicion focuses on *pan*, the chimpanzee, or a form very similar to *Pan* as an ideal model for the origin of human evolution. This is supported by the very close genetic relationship of *Homo* and *Pan* discovered by King and Wilson (1975), a relationship no greater than that of two sibling species, as well as very similar chromosome morphology in the chimpanzee, gorilla and man (Miller 1977). It is supported by the chimp's primitive tool-forming ability, by its social signals (Goodall 1968, 1971), by the recent linguistic experiments with chimps (Premack 1976) and by data that do suggest that chimpanzees are a kin-selected species. In kin selection, natural selection maximizes the number of adults as defenders of a common resource per unit of defendable resource. This results in adults most capable of inflicting serious harm with weapons evolved to maximize surface damage and therefore pain (Geist 1978a), as well as in relatively low sexual dimorphism. *Pan* seems to be under moderate kin selection at least. Note the following characters we expect in a kin-selected species: single births, low reproductive rates, long ontogeny, great tolerance of youngsters by males, adoption of orphans, food sharing by "distance calling" which attracts chimps to newly found food trees, a very complex system of signals which I have termed "social feedback system" and discussed in greater detail in Chapter 13, a rather low sexual dimorphism, sharp weapons in the form of long canines, free splitting and reuniting of social units among chimps familiar with each other, no conspicuous display organs solely for the purpose of dominance displays among group members, and a very low level of sexual jealousy (Goodall 1968, 1971). Unfortunately, Goodall worked with chimps that knew each other quite well and she apparently observed no encounter between different demes. In such encounters, we expect demes to confront each other, display, and attack with resultant severe wounding or even death. We expect the members of one deme not only to drive out, but to destroy a hapless member of another deme, opportunity permitting. Some of Goodall's observations, such as the eruption of dominance displays between male groups that met, the loud vocalizations that could function not only to attract members of a deme to a source of food but also to warn members from other demes to stay away, hint at kin selection. So do Bygott's (1972) reports that chimpanzee males were observed killing and eating young chimps from groups other than their own. So do the "carnivals" described by Reynolds and Reynolds (1965), which these authors suspectd were encounters between relatively unacquainted groups. Precisely because chimps have long, sharp canines, the theory of aggression (Chapter 5) dictates that meetings between strangers ought to lead to intense and long-lasting sessions of dominance displays and occasionally to very damaging fights. In these fights individuals may be considered to sacrifice themselves on behalf of the genomes of their deme. Only in that context do the terrible canines of chimpanzees have a logical explanation.

It appears that *Pan* is so close to the origin of the hominids that it is a form frozen in its evolutionary development since the middle Tertiary, and that it is quite comparable to other primitive ancestors or near ancestors of other mammalian lineages. *Pan* appears to be to the hominids what *Helarctos* is to the ursids, *Capricornis* is to the rupicaprids and caprids, *Dicerorhinus* is to the now-extinct woolly rhinos of ice-age fame, or *Muntjacus* is to the old-world deer. All these relics are so similar to good ancestral forms that if they were not alive their existence would have to be postulated.

The next step away from the forest-adapted *Pan*-like ancestor is expected to be in the direction of grasslands, beginning with the ecotone between forest and steppe, namely the savannah and riparian vegetation communities. The most probable adaptation is from forest to mesic savannah and, in particular, the highly productive riparian communities. The widely accepted notion expressed by Montague (1960) and others, that during the Miocene and Pliocene a drying trend forced arboreal apes to the ground to assume a terrestrial life on the savannah, is simply not tenable. It ignores the great opportunities of the forest−steppe ecotone. Such a sweeping statement ignores the availability of river valleys and highly productive floodplains with marsh and plains vegetation studded with well-spaced trees or clumps of shrubbery. It also ignores the fact that it would take a specialist adapted to life in the savannah, let alone in the dry steppe. In order to adapt to dry steppe, or even dry savannah, a lineage must first adapt to the wet savannah or the forest-grassland ecotone. The great diversity of "unspecialized" apes which arose in Miocene and Pliocene times (Young 1971) suggests that these primates adapted to forest−grassland ecotones. It would have made them highly adaptable species with a broad range of social signals, food habits, and antipredator adaptations, adapted to live in the forest and in the open (Geist 1974a), which set the stage for the hominid line to move into the moist savannah and riparian grassland communities. It is here that in the literature we encounter the polemics surrounding the australopithecines.

In the late Tertiary, well before the minor ice ages, two lineages of hominids arose and were already recognizable as distinct some four or even five to six million years ago (Eckhardt 1972, Tobias 1973). One form is large bodied and robust, apparently with a brain like an ape's, barely larger than that of a chimpanzee (Hofer 1972), and the other a smaller, lighter, and more gracile one with a brain of a size virtually identical to the robust one. Brace (1973) champions the argument that these two forms are simply male and female of one species. Others disagree. The differences in dentition between the two forms are allometric ones and of the order found between gibbon and siamang and are not due to major differences in adaptive strategy (Pilbeam and Gould 1974). Although the teeth overlap in size, Robinson (1972a) dismisses the overlap as irrelevant, arguing that it is the sharp increase in tooth size from incisor to molass that separates the robust from the gracile form. On balance, the evidence appears to suggest two different species of australopithecines (Howells 1973a, Howell and Coppens 1976), a conclusion closely supported by the detailed study of tooth structure based on adequate samples by Read (1975). Although the majority of students of hominid evolution accept the two-species concept, there is disagreement about how to name them. Robinson (1972a) suggests *Paranthropus* for the robust and *Homo* for the gracile form. Others recoil from this proposition and put both forms into the same genus, *Australopithecus,* arguing that forms so closely related ought not to be generically separated (Le Gros Clark 1964, Young 1971, Tobias 1973, Oxnard 1975).

The robust form, *Australopithecus boisei,* is commonly represented in the late Pliocene−early Pleistocene strata of Africa in the Lake Rudolf deposits (Maglio 1972, Leakey 1976a,b, Day 1976, Walker 1976), in the Olduvai Gorge, in Ethiopia (Howell and Coppens 1976), and in South Africa (Robinson 1972a, Tobias 1973). On the whole, although specimens from different localities differ somewhat, not unexpectedly if phenotypes are shaped to adapt to environmental vagaries, the robust form changed little over geologic time. It had a long existence from probably more than four million years ago to at least the beginning of the major glaciations, a span exceeding two, if not three, million years. It may even have given rise to an Asian giant, *Meganthropus* (Robinson 1972a, Howells 1973a).

The antiquity of the gracile form is probably as great as that of the robust one (Leakey 1973b, Tobias 1973, Howell and Coppens 1976). It is found considerably less frequently than the robust one (Maglio 1972, Leakey 1976b). It apparently occurs as a rather diminutive form and gives rise to a larger one, still an australopithecine (Howell and Coppens 1976), and it is generally fingered as an excellent ancestor for the hominid line (Tattersall and Eldredge 1977, Howells 1973a).

It is our expectation that from a forest dweller at the origin of the hominid line would arise forms exploiting the steppe–forest ecotone or savannah. According to Robinson (1972a) and strictly on the basis of fossil assemblages, the robust form occupied a moist environment and the gracile form a drier one. This can be equated with mesic and dry savannah. Robinson's suggestion thus fits squarely with our expectation derived from analogy with other lineages of large mammals. Again, by analogy with large mammals, we expect the robust form to be the more primitive and closer to the ancestral form than the gracile one. One finds that this is suggested by Robinson's (1972a) as well as Howells's (1973a) descriptions. Yet the similarities of robust and gracile forms are so close that we have to follow Oxnard (1975) and accept them as of one basic adaptive syndrome. Let us now attempt to structure internally consistent adaptive syndromes for the robust and gracile australopithecines.

Adaptations to the Wet Savannah

What are the characteristics of moist savannah and floodplain and riparian plant communities?

Marsh communities show a great productivity gradient (Verner and Willson 1966); they have a great diversity in plant and animal life, and they are less subject to oscillations in the availability of food than are terrestrial communities, due to the continuous presence of water. Furthermore, the photosynthetic layer is greatly compressed compared to the typical forest ecosystem, and consequently the forage resource is compressed and concentrated. This makes the production of the photosynthetic layer almost totally available to an ape-size primate, and permits him to reach higher population densities than in the forest ecosystem whose photosynthetic layer it can only partially exploit.

Let us assume an arboreal omnivorous primate begins to adapt to the marsh and savannah communities along water courses. He is faced by the following: a seasonally fluctuating forage area, the size of which varies with the flooding of rivers and the annual rain cycle; a tall grass layer which impairs visibility and hides both predators and prey; a highly productive foraging area along the water course where growing plant matter is available, particularly the highly nutritious meristematic parts, as well as nutritious roots and tubers buried in soft soil that can be had during the dry season; a diversity of seeds, berries, shoots, and fruit; diverse birdlife, which can be beneficial in warning when predators approach, and which offers food in the form of eggs and young birds; diverse insect and invertebrate life, which also offers protein food; diverse large mammals—far more diverse than the present fauna (see p. 254)—which often must be avoided, who pound useful trails into the tall grass, and many of which hide their small defenseless young close to the water course, thus making them occasionally available as food; a diverse fish fauna which becomes accessible during the dry season as shallow ponds—filled by flood waters during the rainy season—begin to dry up; an occasional grass fire and a magnificent opportunity to feed on disabled insects, reptiles, fledgling birds, etc. (Komarek 1965, 1967).

Thus it is evident that the food supply is concentrated geographically. This is important. Our primate does not need to roam far in search for food, not with the productivity as great as that of tropical mesic savannahs (Bourlière and Hadley 1970). We therefore do not expect morphological adaptations indicative of sustained economical roaming or of high-speed running. Such adaptations, according to Robinson (1972a), are not shown. We do, however, expect some evidence for bipedalism, because it is this that will permit a hominid of the size of australopithecines to exploit a roughly six-foot layer of savannah vegetation. Kortland and van Zon (1969) noted that chimpanzees living in savannah use a bipedal gait more frequently than do chimpanzees living in forest. According to Jenkins (1972), Howells (1973a), and Oxnard (1975), the australopithecines must have been bipedal, albeit a bipedalism quite different from our own.

Bipedalism could not have evolved as a means of minimizing the cost of locomotion, as running on either two or four legs is equally expensive (Taylor and Rowntree 1973), while Schmidt-Nielsen (1972b) suggests bipedalism to be nearly twice as costly as quadrupedal locomotion. Nor can one make a case for bipedalism being continuously used, forced upon the australopithecines by the savannah. The individuals can be expected to spend much time crouched down close to the ground, pulling out growing vegetation or buried edible plant parts. Yet, at times the individuals would be forced to stand for long periods of time to reach food overhead, such as seed heads, fruits, flowers, birds' nests, shoots, insects, fledgling birds, etc. It would be more economical to stand and make little steps laterally in exploiting a food source overhead than to continuously return to a quadrupedal poition for every shift in location. Thus, primitive bipedalism is probably a necessary adaptation to exploit the 6 to 8 foot photosynthetic layer of the savannah or marsh communities, because it would minimize energy expenditure in foraging and reduce muscle fatigue. It would, of course, also permit predators to be spotted, the common claim for the evolution of bipedality (Robinson 1972a, Kortland 1967). However, predator spotting would not require continuous bipedality; it would be required for foraging among relatively tall, fragile herb and grass stalks and within the thin, often thorny branches of low shrubs.

Since the australopithecines have been postulated to lie phylogenetically between the apes and the human, it is "natural" to think that they ought to show intermediate characteristics. One must resist this temptation. Adaptations are shaped by natural selection for functional efficiency, not by phylogenetic determinism of some sort. The expectation that australopithecines were knuckle walkers, similar to the great apes, has been one such "natural" expectation. Tuttle's (1969) investigations of australopithecines and humans revealed no evidence for knuckle walking. Since australopithecines had long strong arms (Leakey 1973b), it is still being suggested that they used them similarly to apes (Kay 1973, Henry 1973). They probably did, though not for moving on the ground. The arms were used on the ground for other purposes, namely, for rather sophisticated means of obtaining vegetable food.

If we grant that the diversity of food would be reduced during the dry season, then it follows that the robust hominid would concentrate on the plants growing along the water's edge, or where the groundwater came close to the surface. Plants are the only reliable food supply. Animal foods such as a few stranded fish might be welcome, but would not be a reliable food supply if only small areas were exploited. Reliance on plant food would select for the large grinding molars found in robust hominids, and we can expect to find in the fossil record a pelvis which will suggest a relatively large gut mass. Herbivores, compared to carnivores, require a larger and more complex digestive system. Moreover, if the robust form did reach 150 to 200 lb (70 to 90 kg) in weight, as has been variously estimated (Robinson 1972a), then it would require about 4 to 8 lb (1.8 to 3.6 kg) of wet plant forage per day, provided it digested plants as efficiently as a ruminant.

It has been postulated by Clifford Jolly that hominids evolved as seed eaters and that "small-object feeding" played a crucial part in their evolution. Comparisons are made to gelada baboons (*Theropithecus gelada*) whose similarities to hominids are not insignificant, both having small incisors and large molars, fatty sitting pads, and, in the female, a great similarity between the perianal and breast displays. I am also following the view that hominid evolution revolved about foraging for small pieces of relatively concentrated, easily digested, food, but I must differ from Jolly (1970) in refusing to confine these "small objects" to seeds. Seeds, almost certainly, formed an important part of the diet of hominids, but as seeds constitute only a small fraction of the plant biomass they would force a primate—particularly a large one—to forage very widely, hold down a very large home range, and be built as an agile roaming form. The fossil evidence contradicts this view for the robust form (Robinson 1972a). Large body size, as found in the robust form, would favor the selection for feeding of less digestible plant matter, for the same reason that large body size in ungulates and rodents favors the utilization of relatively coarse fodders (Bell 1971, Rensberger 1973, Geist 1974a). This would amount to niche expansion, and hence to the exploitation of a larger share of the vegetation, culminating in a higher population density, greater geographical distribution, and greater survivability of the robust hominid form. If the robust form was indeed a seed eater, then we would expect a relatively small gut mass; future analysis of the pelvis of the robust hominid may determine whether the gut of this form was large or not. Furthermore, if the robust form was indeed a specialized seed eater, and the gracile form an omnivore with a larger niche breadth, then the gracile form should outnumber the robust form in the fossil record where they are found together. This is apparently not the case (Maglio 1972). Thus, it appears most likely that the robust form from the moist, tall-grass savannah and floodplain communities would exploit more than seeds. Note that the seed-feeding hypothesis assumes continuous seed production, a most unlikely assumption for a savannah with distinct seasons, or for riparian communities with distinct floods. It is therefore probable that the robust australopithecine had to feed on plant matter that was both abundant and available over the greater part of the year. I suggest this plant food was the meristematic parts of grasses.

Had these robust australopithecines relied only on fruit, leaves, berries, grubs, or seed heads, their hands would not have evolved away from the shape typical of chimpanzees. Yet the hands of australopithecines and even of their dryopithecine ancestors are, according to Napier (1962), closer to those of modern man than to apes, except for a shorter thumb. What selected in australopithecines for a more efficient power, and ultimately precision grip, above and beyond that of the chimpanzee? Clearly, the australopithecines and dryopithecines had to do something that enhanced the reproductive fitness of individuals with their hands that required very precise control over muscular contractions and was so important that it selected against knuckle walking. That something, I propose, was feeding on grass meristems and on tubers buried in soft, but firm, soil.

Up to the present, it has not been recognized that in grazing, ruminants are not simply biting off the top of grasses, but are pulling out grasses so as to pull free the sweet, tender, and highly digestible growing parts of the grass or sedge blades. I first became aware of this when studying mountain sheep, when I became curious about the noise of grass blades gently rubbing against their upper palate during grazing, just before the grass blades snapped or pulled free (Geist 1971a). Such grazing, of course, maximizes the digestibility and nutritional quality of the forage, an essential aim of foraging strategies in ruminants whose gut is cleared only as fast as the food is digested (see Blaxter et al 1961); horses, which can clear the gut voluntarily, can increase their energy uptake by ingesting more coarse forage and, after absorbing the digestible parts, quickly pass it out (Bell 1971). In ruminants the replacement of upper incisors by a soft palate may, in part, be an adaptation for pulling meristematic parts of grasses and sedges from the plant while grazing. Similarly, the robust hominid form could have concentrated on pulling the meristem of growing grasses from the blades. This, however, requires a controlled, careful pull, as anyone who has tried pulling free the meristematic parts can attest; too much pull and the blades or stems snap. The pull has to be controlled. The same applies to pulling roots and bulbs from the soil. If too much pull is applied the vegetative portions of the plant simply break, whereas a smooth, steady pull will draw the root from the ground. Therefore there would be an evolutionary impetus for evolving hands with relatively fine motor control and a good power grip (but not precision grip), since such adaptations would pay off with more energy and nutrients from the meristems of grasses and sedges and the roots and bulbs of various plants. Selection for dexterity and a fine motor control over arm movements presumably would be even greater in the gracile form of hominids, because in the drier short-grass savannah many more plants have evolved fleshy roots and bulbs, which permit them to stay dormant until the rainy season arrives. This hypothesis deviates considerably from the views that bipedalism arose because of the need to free hands for carrying weapons, tools, or food (Wolpoff 1968); it suggests that manual dexterity had adaptive value *prior* to tool use.

The dentition of the australopithecines gives additional support to the notion that, in the change from a chimpanzeelike ancestor to the hominid condition, grinding of food had become important. The clue is the reduction of canines and the ability to use rotary grinding movements of the jaw (Campbell 1974, p. 230). In juveniles, the deciduous molars may be worn off so much that the cusp pattern becomes obscure (Wood 1976). As discussed under the adaptations of the gracile form, small-bodied animals must, by and large, select food of high digestibility, for they require more energy and nutrients per unit of body weight per unit of time than do large-bodied forms. Highly digestible foods are usually soft and require little crushing. Hence, simple up-and-down chewing movements will suffice, which allows for the existence of large canines, of course. These would interfere with rotating chewing movements of the jaw. If a species adapts to foods that require a lot of chewing, and effective chewing, then canines are in the way, and selection will either reduce them or lead to their loss as other organs take over the weapon function. Adaptation of hominids to riparian or savannah communities would probably impose just such selection because chewing, even of meristems, fibrous roots, and seeds, would select for rotary chewing movements and therefore greater grinding ability. Consequently, only small-bodied forms, specialized in feeding on plants, ought to have long canines, while in large-bodied forms canines are absent or vestigial and other weapon systems compensate. This hypothesis is extremely well supported by data from ungulates.

Granted the geographically concentrated food source in the riparian tall-grass savannah, then the food resource may become defendable (Brown 1964); it is concentrated in an area the animal can somehow control and defend. This means that by defending an area, the animal can hoard food for its own and its children's benefit. Hence, we find in birds and mammals that exploit marsh communities with strong productivity gradients, a strong evolution toward territoriality, polygyny, and sexual dimorphism (Verner and Willson 1966, Orians 1969). The female maximizes her reproduction by staying close to a male who successfully chases off others and thus either removes competitors from the available food, to the female's benefit and that of her unborn or suckling offspring, or reduces the harassment of females by young males. Harassment is expensive bioenergetically (Graham 1962, Webster and Blaxter 1966, Geist 1974a). In both cases sexual dimorphism will be selected for (Geist 1974a). Therefore, the expected social system would be based either on territories occupied by a male plus one or a few females, or by a one-male group with peripheral groups of males.

We can exclude the first alternative in regions where the foraging area fluctuated owing to annual river flooding or strong seasonal rainfalls (Geist 1974a). Since Pliocene Africa was drier than it is today (Clark 1970, p. 55), we can expect strong wet−dry seasons and consequently a strong regular flooding of rivers. This condition would extend the annual difference between the minimum and maximum carrying capacities of the riparian habitats and result in a relatively large energy and nutrient pulse per individual. If this account is valid, it explains the relatively large body size of the robust form, following the arguments I made (p. 191) from Margalef's (1963) reasoning. The fluctuating habitat would select against territoriality and for one-male groups with females. This second alternative develops because females can stay close to a large male and escape costly harassment by smaller males; it would lead to sexual dimorphism, which is apparently found in the robust hominid form (Robinson 1972b). We can expect the males to have been quite aggressive so that the bands would be well spaced, much as is found in most forms in which we find a strong sexual dimorphism (Etkin 1954).

We must now turn to the strategies of predator avoidance in the tall-grass savannah. Daylight hours would presumably have been safe, since the abundant avifauna and large mammals along the river courses would soon warn of predators. Predators such as lions, leopards, hyenas, and hunting dogs hunted mainly at dusk and during the night (Kruuk 1972, Schaller 1972). Therefore, these were the dangerous times, and it was critical for the hominids to be prepared. Trees were an obvious place to escape the primarily ground-hunting predators; in tall grass the robust hominids probably did not venture far from trees due to their inability to readily detect distant predators. It would have been safest to stay close to trees, granted the tall-grass savannah they are postulated to have inhabited. If the robust form of hominids lived in relatively warm, mesic climates, spending the night in trees would have been no great problem from the point of view of heat loss for so large an animal. At relatively high constant ambient temperatures, trees are good places to stay at night; cliffs are better where night temperatures are low, for reasons discussed later when considering the adaptations of the gracile form.

How hominids escaped leopards is not clear, unless leopards refrained from making kills in trees due to the danger of falling off and dislocating or breaking bones. Moreover, these early hominids were exposed to a whole series of predators now extinct (Kurtén 1972), such as the scimitar cats (*Homotherium, Therailurus*) or the dirk-tooth cat (*Megantereon*) (Maglio 1972), of whose habits we know nothing.

We can, however, hypothesize that escape from predators by climbing trees was an obvious and most readily available means in the tall-grass savannah. And the scanty morphological evidence of the extremities of the robust and gracile hominid forms supports this view (Robinson 1972a, Oxnard 1975). The scapula and clavicle, as well as the phalanges and talus of *Australopithecus* are similar to that of the orangutan, according to Oxnard (1975), a good indication of rather acrobatic climbing and—so Oxnard suggests—arboreal quadrupedality. The adaptations of the arms and hands must also be viewed in the light of antipredator adaptations, not only in the light of locomotion on the ground and tool usage, as is usually done. The strong arms of the australopithecines are thus seen as arising from their antipredator adaptation, the arboreal habits they had to assume at night when predators were most active.

Much has been made of male primates acting in unison to attack, threaten, or distract predators (Jolly 1972, Ardrey 1963). The studies of Kortland (1967) have been most instructive here, in that they show that savannah-dwelling chimpanzees are more likely to gang together and advance on a predator than chimpanzees from forests. We can expect at least as much from the robust hominids. It is likely that their most effective defence was not so much their use of branches and sticks in such antipredator displays, but their mimicry of vocal threat signals of predators, as I shall discuss later in this chapter.

The following image of the robust australopithecine now emerges: It moves about largely bipedally, but in a manner distinctly different from our own. It is capable neither of great speed nor of silent stalking, nor of economic movement over long distances, as we are. It is never far from trees into which it escapes from predators and where it stays during the nights. Therefore, it has great arboreal ability, as reflected in its anatomy (Oxnard 1975). It can afford to be near trees as it exploits only small home ranges, and this is possible because it exploits highly productive riverine savannah or moist tropical savannah. It lives in one-male groups and groups of peripheral males feeding on growing vegetation but supplementing its diet with various animal foods gathered opportunistically. Australopithecines would be heavily haired, permitting infants to hold onto the hair coat of the mother when she climbed into the trees at nightfall. The males would not only be larger but probably also somewhat distinctly colored at puberty, maybe by a graying of the head hair, just as mountain gorilla males "turn gray" at maturity (Schaller 1963). In the tall grasses of the savannah the head would be the most visible part of the animal, and we can expect display signals to have been concentrated in the head and shoulder regions. This could include, besides the gray head hair, a beard and long erectile hairs on shoulders and neck (Leyhausen in Eibl-Eibesfeldt 1970). We can expect the robust hominids to have been noisy—a good way of keeping together in tall grass, and a good way of communicating when a predator is spotted. However, we can expect this vocalization to have been in the low-frequency range, which does not carry far and is thus unlikely to attract predators. The complexity of their vocal communication was probably on a par with that of other social apes, which evoke specific sounds in specific situations, and even have different sounds for different predators, as do vervet monkeys (Struhsaker 1967). Noisy mobbing is probably also a good defence against predators, if an opportunity for mobbing exists, since the predator whose hunt is spoiled regularly in a given area is not likely to revisit that area. The robust hominid would be rather abundant and living in relatively high density owing to four factors: It would be a herbivore with a relatively low energy expenditure in procuring food, and a high conversion rate of energy and nutrients into tissues; it would exploit the most productive habitat available—the tall grass, riverine savannah, and adjacent marshes; it would exploit a rather stable habitat that favors a high biomass per unit area; the photosynthetic layer is compressed and almost totally accessible to a man-sized primate. In short, granted an understanding of how ecological parameters are translated into adaptive strategies, one can interpret the paleontological evidence in a broader way than has ben done to date.

Robinson (1972a) suggested that the robust hominid form gave rise to the gracile one. There is merit in this hypothesis, since the evolutionary progression in various mammalian lineages, be they horses, deer, pigs, rhinos, bears, or elephants, has been from the forest to the savannah and then to the steppe, from the warm and wet habitats to the hot and dry or cold and dry ones, as indicated earlier. The hominids, following Robinson's hypothesis, would have taken the same course. The small body size of the gracile form would be a consequence of adaptation to a more mature ecosystem than that inhabited by the robust form; it would be a contradiction of Cope's Law, but little significance should be attached to this. The gracile form would be a scaled-down version of the robust one, differing in food habits, and differing in morphology, in part for reasons of allometry, a secondary consequence of smaller body size (Pilbeam and Gould 1974).

Tobias (1973) suggests that the gracile form gave rise to the robust one; Robinson (1972a) argues for the converse. It is most likely that once both forms existed contemporaneously they were both specialized and somewhat different from a common ancestor. This may indeed have been a small-bodied form but similar to the robust one, differing from it by reason of allometry. It is likely that the robust form gave rise to a large Asian form, *Meganthropus;* there is, however, reason to doubt that the robust form was derived from it, as Robinson (1972a) suggests. More of this later.

Adaptive Syndrome of the Gracile Hominids

The adaptive strategy of the gracile form can be shown to be somewhat different from that of the robust form, but it cannot be equated with that of human beings, as Robinson (1972a) has suggested. I favor the conservative interpretation that the gracile form is at best a variation on the robust one, as it in no way precludes the notion that it was indeed the adaptive syndrome of the gracile form that gave rise to the human one. The gracile form appears to have been adapted to feed on small packages of food of high caloric and nutrient density (Jolly 1972) in the dry—rather than the moist—savannah (Robinson 1972a). Primitive predators and small-bodied herbivores typically feed on small packages of easily digestible matter whether it be plant or animal matter. This assumption is central to the adaptive syndrome I shall develop for the gracile form.

Since bits of food of high caloric and nutrient density, such as seeds, fruit, flowers, tender shoots, grubs, and bird and reptile eggs are widely scattered geographically, particularly in relatively unproductive ecosystems, an individual must roam far and spend considerable time in collecting such food. However, owing to the high digestibility of such food, only small amounts are required to fulfil daily food requirements compared to fibrous plant food. Because the body size of the gracile form was very small, about half that of the robust form (Robinson 1972a), individuals of the gracile species required an absolutely smaller amount of food per day. Thus, the food habits of the gracile form imposed roaming on individuals, and it has been argued by Napier (1967), Robinson (1972a), and others that the body form of the gracile species is an adaptation to speedy economical roaming. Its pelvis is relatively shorter than that of the robust form and thus a little more like that of human beings. It has a rather human foot and leg (Howells 1973a), although the first metatarsals are not nearly as large as those of the later hominids (Young 1971), nor did they have to be, given the light weight of the gracile species.

Although Napier (1967) and others suggested that the human body form can be explained, in part, as an adaptation to reduce energy expenditure during foraging, he did not go far enough, nor into sufficient detail in his explanation, nor did he confront the apparent contradiction that the cost of bipedal locomotion is nearly twice that of quadrupedal locomotion (Schmidt-Nielsen 1972b). Clearly, some advantage is gained for the hominid adaptive syndrome by assuming a bipedal posture which more than offsets the increase in locomotory cost. In the robust form, bipedalism functioned in harvesting food from a six to eight foot vegetation layer. In the gracile form, it took on a new function; it became an adaptation maximizing energy and nutrient intake through *passive competition* (also termed "scramble," see Wilson 1975). The argument runs as follows.

(1) In open country where visibility is unimpeded by vegetation, such as in a short-grass savannah, it is to the advantage of an individual to associate with conspecifics. In the open, an individual is visible for a great distance, and conversely can detect predators at a distance. To protect itself, it must scan the country. A herbivore, and even an omnivore—quite unlike a carnivore—are pressed for *time* to acquire enough food. Plant food is of low digestibility compared to animal foods; the animal must feed selectively, the more so the smaller it is in body, and so requires food of increasingly greater digestibility (Bell 1971). It therefore requires time to scan the area for food, and time to move about in search of food.

A carnivore, by contrast, can ingest a relatively large amount of highly digestible food in one sitting, and is then free to search for more. Whereas a herbivore must keep its senses focused toward the ground in search of food most of the time, a carnivore can scan the country for herbivores. Thus, a carnivore has an average or better chance to detect a herbivore first, rather than vice versa, and can conceal itself from the herbivore and proceed to stalk undetected, so a herbivore, out in the open plain, must somehow maximize the chances of detecting a predator at a distance. If alone, it can do so only by frequent disruption of feeding—which does not permit it to maximize the intake of food for reproduction and growth. Hence, by associating with others, a herbivore can take advantage of the scanning by other individuals and thus ingest a maximum amount of food while still being relatively safe.

This tendency to clump, however, requires as a mandatory adaptation a signal given by all individuals in a group that all is well, a "Stimmfühlungslaut," as I discussed under Communication (p. 45). The moment this sound stops, all recognize the signal of alarm.

The foregoing predicts, of course, that individuals in groups spend less time scanning the open country than do individuals by themselves (e.g., Murton 1968), that the efficiency of detecting predators is greater for the group than for the individual (e.g., Carl 1971), and that individuals of gregarious species ought to be much more ''nervous'' and unapproachable when alone than when in a group, an observation repeatedly made for caribou (Kelsall 1968, p. 44). At present, the available evidence is limited, but entirely in line with the above predictions.

We may note that the same line of argument, applied to forest-dwelling species, correctly predicts dominant elements of their social systems. Thus, where individuals are obscured from sight so that predators cannot readily detect them, nor can the prey readily detect the predator, the individual reduces its chances of being killed by a predator by reducing chances of detection. It is adaptive to be solitary, silent, and cryptically colored, and to rely heavily on the senses of hearing and smell and on vocal warning calls (Geist 1974a).

Another protection from predators enjoyed by the individual when in a group is that an individual on the periphery of the group is more likely to be taken by a predator than an individual in the center of the group. This concept is well recognized (Wilson 1975, p. 38). Thus, if an individual can join others, *and* place itself at the center of the group, it reduces predation on itself substantially.

(2) However, the clumping increases the energy cost of foraging and reduces the time available for foraging by each individual. Each new number added to the group adds to the cost of living of the others. There comes a point at which an increase in group size is uneconomical for all involved and the group should stop growing in size. If clumping is highly adaptive, and the resource is undefendable—and scattered bits of plant and animal matter of high caloric or nutrient density can be considered undefendable—then selection enhances mechanisms of passive competition. Active competition, that is, removing competition from forage by means of overt aggression or the threat of it, becomes too costly in energy and nutrient relative to the gain in energy and nutrient, and it may reduce group size with a concomitant increase in the risk of predation by the threatening individual. Not only will the group size drop as individuals leave the obnoxious conspecific, but the aggressive individual may find itself increasingly deserted, and so more often at the *edge* of the group or even alone, and thus a prime target for predators. Moreover, dispersed bits of food may separate individuals so far that threat becoms ineffective.

(3) Granted a dry short-grass savannah of relatively low productivity and an adaptation to foraging on undefendable diverse forage of small size found widely scattered, then individuals compete by enhancing their ability to spot the maximum amount of edible matter, minimizing the time required to reach it, maximizing the area that they can cover during daily foraging, and minimizing the cost of doing this. This selects for diversity and opportunism in food habits, for broadening the food base of the species, and for economy in searching for food.

The first prediction fits well with human food habits; they are extremely diverse. From the foregoing, we have reason to assume that this diversity is a very old adaptation indeed, forced upon hominids by selection for passive competition.

The second prediction runs into difficulty: Bipedal locomotion is almost twice as costly as quadrupedal locomotion (Schmidt-Nielsen 1972b) and is thus not a means of minimizing the cost of locomotion, despite the fact that the light, almost fragile build of the human body compared to anthropoid apes (Hediger 1965) does suggest economy in locomotion (Napier 1967, Robinson 1972a). Nor can one argue that the light bipedal body is an adaptation for speedy locomotion, as this assumes that bipeds run faster than quadrupeds. This is not so, as a comparison of running in mammals shows (Krumbiegel 1954, p. 80).

The upright posture, however, *elevates the eyes* almost to the maximum possible height and thus permits a much larger area to be scanned than is possible from a quadrupedal posture. The upright position of the torso, plus the elongated hind legs and neck, thus maximize the elevation of the eyes and minimize the time required for spotting edible matter and reaching it. The scanning is made more efficient by the excellent visual acuity of higher primates (Walls 1942 in Altmann 1966), by binocular vision, and by the agile neck, which provides quick, frequent, wide, and economical scanning in all directions.

In this context, a *light* body build is logical, not only because it is more economical than a heavy body, but also because it does not have to withstand, in this adaptive syndrome, the forces of frequent overt aggression. Passive, not active competition is selected for. Had it been otherwise, powerful bones, joints, heavy muscles, fascia, and tendons, plus an assortment of morphological defensive adaptations (Geist 1966a, 1971a, 1977), would have been selected for.

The long legs could also help reduce the time between spotting the food and ingesting it, although it would not reduce the cost of locomotion. The cost of locomotion does not vary over a wide range of speeds in animals (Brody 1945, Schmidt-Nielsen 1972b). In essence the human body form does what the body form of the secretary bird (*Sagitarius serpentarius*) does in the African plains or, for that matter, the form of the heron and stork.

Note this important point: Grouping in the open is a function of herbivor omnivory, not carnivory. Only if hominids were feeding on plant matter as gatherers would there be any advantage in congregating in the open country, feeding in competition. It argues that the typical human body form evolved under conditions of gathering plant food, not hunting animals, as Washburn has argued (Napier 1967).

In addition to the body form just discussed, hominids would also benefit from enhanced manual dexterity. It would be most useful in quick, efficient plucking of food and accurate picking up of small food particles, and it would be economical in repetitive tasks such as turning stones or leaves over to get at grubs or accurate picking of ripe fruit from bunches comprising fruit in various stages of ripening. Great manual dexterity would be the greatest asset in selecting quickly small pieces of highly digestible matter, regardless of origin.

Diversification of food habits demands diversification of physical skills, leading to the unique human ability to do a great many tasks very well, an ability called "lack of specialization" and referred to by many students of man. Coupled with this ability, as a logical necessity, is great curiosity and a drive to explore. Clearly, the foregoing shows that the adaptations of the gracile form would be strongly moulded by passive competition in the direction of human characteristics. In contrast, the robust hominid probably relied to a large extent on active competition for its food, as reflected in the sexual dimorphism and as expected from the defendable nature of its food resources. The wide dispersion of food in the dry savannah would have forced the gracile form to compete for food passively.

Since the highly digestible parts of plants (fruits, seeds, shoots, flowers) account for only a small fraction of the plant biomass, they can support only a small biomass of herbivores (see Jarman 1973). Thus, among primates, those concentrating on small packages of highly digestible food have a relatively low biomass density (Hladik and Hladik 1969 in Jolly 1972); the same applies to African antelope (Jarman 1973). This argues that the gracile form was rare compared to the robust form which, according to hypothesis, fed extensively on fibrous vegetation. In addition, the upland savannah, the probable habitat of the gracile species, has a lower plant productivity than the moist savannah occupied by the robust species. Furthermore, food items widely scattered in small packages cost more energy to procure; hence, the gracile form would be able to put relatively less of the ingested energy and nutrients into tissue; it would perform similarly to a carnivore (Watt 1973). These four factors all predict that the gracile form was less abundant than the robust form, a prediction supported by the fossil record (Maglio 1972).

It is likely that individuals of the gracile form experienced less seasonal fluctuations in the carrying capacity of their habitats than did the robust form. The latter, according to the hypothesis followed here, lived along flooding rivers or wet savannah with strong wet and dry seasons. Moreover, the roaming ability of the gracile form would tend to "equalize" the daily food intake in quality and quantity throughout the seasons. This means that the annual pulse of surplus energy and nutrient was very small or nonexistent for individuals of the gracile form. Therefore the individuals of the gracile form would be smaller in body size than those of the robust form following Margalef's (1963) arguments.

The home range of the gracile form is expected to be small compared to that of the later hominids. Since endurance at maximum sustained muscular effort varies as the cube root of body mass, it can be calculated that a 50-lb gracile hominid could sustain maximum effort only about 0.7 times as long as a 150-lb hominid, although the gracile hominid would probably be just as fast as the larger one (Hill in Thompson 1961, p. 31). We therefore have reason to suspect that endurance was not a strong point of the gracile form. Thus, the gracile hominids were probably not long-distance roamers; therefore, they could hardly have been hunters, and probably had not developed the ability to stalk which is required in systematic hunting, for this requires a striding gait. In short, they were agile, speedy short-distance roamers of no great endurance. Endurance is reflected in home-range size, so that the home ranges of carnivores, because of their greater roaming, are larger than those of herbivores of comparable size (Watt 1973). As a rule of thumb, home-range size varies as the three-quarter power of body weight (McNab 1963).

Meat Diet: For and Against

The relative scarcity of highly digestible plant matter during the dry season would select for catholic food habits and therefore for individuals capable of exploiting animal food sources. Much has been made by Ardrey (1963) of Dart's evidence for carnivorism and cannibalism in the gracile hominid form. Brain (1972) recently pointed out that much of the damage to skulls examined by Dart could be due to damage caused during the fossilization process which Dart was apparently unaware of. When Dart (1949, 1953, 1957) published his works, they were largely ignored and ridiculed, although subsequent events have supported Dart's contention in part. The field work on wild chimpanzees by Goodall (1971), Bygott (1972), and Teleki (1973) has shown conclusively that even this subhuman primate has a great fondness for meat, that it is a capable, though opportunistic, hunter of small monkeys, young antelope, and adolescent baboons, and that on rare occasions it kills and eats its own kind (Bygott 1972). Baboons will also kill and eat cape hares, antelope, fawns, bushbabies (*Galago*), and vervet monkeys (Altmann and Altmann 1970). There is no reason left to suspect that the gracile hominid was less fond of meat and would not kill prey given the opportunity.

The generalized dentition of the gracile hominid form has been interpreted as an adaptation to meat eating (e.g., see Young 1971, Howells 1973a). I disagree. The "generalized" dentition of the pig (*Sus scrofa*) does not indicate carnivorous food habits. The australopithecine dentition indicates no more than an adaptation to chewing of relatively nonabrasive food. Moreover, the lack of a massive occiput, large mastoid processes, massive neck, back and shoulder girdle, the lack of massive facial muscles, as well as the small body size and consequent lack of great muscle power, argue against the gracile hominid worrying off bite-sized pieces of meat, as Neanderthal man probably did (see p. 290). It is a most difficult matter to chew mouth-sized pieces out of raw meat, as anybody can demonstrate to satisfy himself by performing the experiment, starting with, say, a large piece of fresh flank meat from a deer, one of the more tender parts of that animal.

Dart's work also suggested strongly that the gracile hominids had a tool culture based on bone, and the findings of primitive artifacts in association with their remains or in layers containing their fossils suggest that they were tool users (Le Gros Clark 1964, Leakey 1970, Isaac et al 1971). Again, the studies of Kortland (1967) and Goodall (1968, 1971) showed that even chimpanzees were able to fashion primitive but functional tools. The artifacts attributed to *Australopithecus* are of a kind needing little elaboration of the skills beyond those of chimpanzees.

What has been overlooked is that such evidence, even if it were not questionable, as is indicated by Sutcliffe (1970), Brain (1972), and Pilbeam and Gould (1974), does not constitute a case for hunting nor for hunting being important as a means of obtaining food for the gracile form. Granted the small size of the gracile form and the validity of Went's (1968) calculations of the forces which men of different body sizes could generate when striking, then gracile forms equipped with clubs could do very little damage. Thus Went calculated that the strength of a blow of a three-foot creature would only be 1/25 that of an ordinary man; that of a five-foot person only 1/8 that of a seven-foot giant. A gracile hominid would be virtually incapable of killing big game by means of clubs. This does not argue well for the hypothesis that bipedalism evolved in response to a need to free the hands for weapon use (Wolpoff 1968). Nor can a case be made from the ecological conditions that hunting was essential to these early hominids and anything but opportunistic in nature. Furthermore, there has been little appreciation of the difficulty of adapting from a largely herbivorous diet such as that of the chimpanzee, to a carnivorous one. Moreover, even in present-day hunter−gatherer cultures from Africa, meat plays a most subordinate role in a diet which is largely vegetarian and obtained by gathering (Woodburn 1968a, Lee 1968). The case for hunting by the early hominids has been greatly overplayed, and the idea that mid-Pleistocene men hunted cooperatively in groups appears to me a case of wishful thinking (Binford 1970, Campbell 1974); it ignores the very real differences in cognitive structures required by hunters as opposed to those of herbivores. Nor is there any sign of recognition that cooperation is a function of a specific kind of social system based on complex food sharing. Such a system could not have evolved prior to the evolution of a strong male−female bond.

This did not come about until *Homo* became a systematic hunter (see p. 251). That anthropologists have an inadequate appreciation of what hunting entails is, in my opinion, amply demonstrated in their writing about it (e.g., Clark 1967, Washburn and Lancaster 1968). There are exceptions, like Laughlin (1968), yet even he accepts that early men (here defined as *Australopithecus africanus* and *Homo erectus*) were skilled big game hunters. I shall discuss this later. For the present, I suggest that *A. africanus* was only a surface gatherer of food, and in intense passive competition with its conspecifics for the scarce resources as these were essentially undefendable; hence the emphasis on speed, roaming, and light body weight in order to quickly gather the required food, rather than on organs of aggression which would aid in displacing conspecifics.

A meat diet carries several inherent dangers. There are nearly a hundred diseases which can be transmitted nowadays from animal to man by handling meat and hides, which presumably could have been transmitted to the early hominids also (Cowles 1963, Hull 1955, Lieberman 1958). Hominids catching and handling small mammals, birds, and reptiles would often run the risk of parasitic, bacterial, and viral infections. They would mainly catch the young, the old, and the diseased because these are the easiest to catch (see, for instance, the experiences of Schaller and Lowther, 1969). Eating carrion would add more hazards, such as death from botulism and other toxins. In this context, it is most noteworthy that baboons and chimpanzees kill and eat mainly *young* animals, which because of their age alone are unlikely to carry a heavy load of parasites and pathogens (Goodall 1968, 1971, Strum 1975). Chimps also catch the arboreal red colobus and blue monkeys which, because of their arboreal habits, are likely to be free of parasites and disease. Chimpanzees do not eat carrion and could not be enticed to take meat (Goodall 1971). Clearly, such selectivity in prey and food would greatly reduce the risk of chimps contracting pathogens carried by the prey animals.

It may also be noted that the intestines of young animals are less likely to contain a diverse fauna of parasites, and that eating such intestines would not be as harmful as eating the tripe of adult prey. For a description of the eating habits of some present-day hunters in Africa, and their total disregard for the parasites of their prey, I refer to Cowles (1963). He also notes that total extinction of groups has occurred due to disease, and this despite the fact that today's humans are well-adapted flesh eaters compared to the evolving gracile hominids we are discussing.

From the foregoing it is evident that the best strategy for initiating the carnivorous habit is to kill and eat young animals rather than disease-prone old prey, and it is better to concentrate on large rather than on small game since this would reduce the frequency of contact with prey animals and therefore the chance of contracting disases. It is inconceivable that so small-bodied a species as the gracile hominid form would be able to kill anything but young prey consistently, since confrontations with larger animals would in a short time lead to injury and—directly or indirectly—to the death of the small, ineffectual hunter. The findings of Dart that adult baboons had died a violent death possibly at the hand of gracile hominids, even if valid, does not contradict this view; these baboons need not be victims of hunting, but rather the occasional victims of a dispute over entry into the caves.

The present fossil and tool finds, as well as ecological considerations and the evidence from present-day hunter–gatherers, does not argue in favor of the view that the gracile hominids—or their decendants for millions of years—were hunters. The evidence suggests that they were gatherers, primarily of plant food found above the soil surface; occasionally they killed opportunistically, although they did so probably more frequently and consistently than chimpanzees do, or baboons, for that matter (Strum 1975). There was no ecological necessity for systematic hunting, given year-round availability of plant food. Although the evidence for these creatures using weapons and killing prey, as well as their own kind, is less than adequate, it is possible and is not shocking in view of what we know of cannibalism of wild chimpanzees today (Bygott 1972) or about intraspecific killings in other large mammals (Geist 1966b, 1971a, Schenkel 1966b, Schaller 1972, Pearson 1972, Miller 1972, Payne 1972, Russell 1972, Larsen 1972).

One piece of evidence in favor of the plant-eating habits of the australopithecines is the large opening of the false pelvis, which implies a large and bulging gut (Robinson 1972a). This makes sense only if these hominids ate a considerable amount of plant matter, since it requires a large and complex gut to properly digest such food. The arguments by Leopold and Ardrey (1972) against plant food on the basis of toxic substances is quite irrelevant, in view of the fact that plant foods are basic to modern gatherers (Dornstreich 1973), are the dominant food of primates, and were almost certainly the dominant food of early hominids, as implied by their relatively large dentition, particularly that of the robust form.

Social System

The social system of a species is not only the product of its adaptive strategies by which it extracts its share of energy from the ecosystem but also a consequence of ancestral adaptations. It cannot be predicted in total by considering only ecological factors such as resource distribution, abundance, dispersion, seasonality, productivity, or cover type (Geist 1974a,c,d, Altmann 1974). As indicated earlier, there is reason to suspect that the gracile hominids' social system was derived from a parental type that was kin selected.

In kin selection, a deme, that is a group of genetically closely related individuals, maximize the number of adults per unit of defendable resource. Only in so doing can the deme guard itself against the encroachment of neighboring demes. From this, it follows that intrademe aggression would be reduced to a minimum, as would dominance displays, while interdeme aggression, deme display, and intolerance of strangers would be very high. Owing to the small amount of resources available for reproduction, a long ontogenetic development and a low birthrate would be selected for. Deme members can be expected to give aid, where resources are very limited, to the reproductive effort of other deme members (Brown 1974). The level of sexual jealousy would be rather low among males, as is seen in chimpanzees (Goodall 1971) and lions (Schaller 1972, Bertram 1975). Under kin selection, adults can be expected to show low sexual dimorphism (Brown 1974) for several reasons. Since every adult is required for the defence of resources, there would be an advantage to the aggressor if the sex of smallest body size and lowest combat competence could be selectively attacked. If sufficient injuries could be inflicted on members of the sex of lowest combat competence, the attacked deme is likely to lose members. It is also adaptive for the attacking deme to select all those members of the opposite deme that leave the group or are somehow conspicuous. Therefore, there is selection against individuals that move away from other members of their deme and act or appear to be different. Kin selection thus reduces sexual dimorphism, producing a society of look-alikes.

However, the moment a kin-selected hominid moves into an ecosystem in which the resources it depends on can no longer be defended, the whole social system must change. This happened at the latest in the gracile form which lived in the African savannah, under conditions of continuous plant productivity throughout the year but discontinuous plant productivity areally. This forced the surface feeding *Australopithecus africanus* to scatter widely, move appreciable distances in search of forage daily and seasonally, and alter its home range seasonally. Under these circumstances the resources are too scattered in time and space to be defendable and kin selection comes to an end.

In the open country, large mobile groups form in response to predation pressure. Strangers can enter the group readily. Therefore, more genetic exchange between neighboring groups takes place, and each group becomes genetically more heterozygous. It becomes adaptive for males to mate with a distantly related female, since heterosis could enhance their reproductive fitness (Mayr 1966, Preobrahenskii 1961), particularly that of males competing with others for estrous females. Dispersal becomes adaptive. This increases the chance for genetically diverse, unrelated males to cluster about estrous females. Male competition is increased, as are dominance displays. The female enhances her reproductive fitness by making estrus conspicuous and making males compete over her. However, since she is likely to suffer considerable harassment during estrus, this period ought to be short. Under kin selection, estrus also should be conspicuous, but it ought to be long, because this permits the femal to bind the largest possible number of males to the deme and thus ensure the largest possible number of defenders for her and her offspring. Once kin selection is relaxed and group fights decrease in frequency, so do group displays, coordination of activities of deme members, and the adaptive significance of canines. Large canines would have been a hindrance to the austrlopithecines because they interfere with the efficient grinding of food by restricting the lateral movements of the jaw (Campbell 1974, p. 230). Secondly, long, sharp canines, adaptive to a genotype whose members may sacrifice themselves in group combat, become unadaptive out of the context of kin selection. Large canines inflict painful wounds which in turn are likely to trigger retaliation in kind; this is incompatible with gregariousness as an anti-predator adaptation. (see p. 79). Since competition is high among males for females, selection enhances dominance displays and sexual intolerance. One of the display organs enhanced under selection during sexual competition is the phallus of the male, just as we find in cercopithecines with intense sexual competition by males. This would explain the vestigial phallic display in humans (p. 350). Other display enhancements were likely to have been the beard, frontal baldness, gray hair on the head, and erectile hair on the shoulders, the vestiges of which are traceable on our body to this day (Leyhausen in Eibl-Eibesfeldt 1970). Laughter and humor as dominance displays and bonding mechanisms would also be selected for.

Sexual dimorphism is likely to be selected against. If gracile hominids were forced to disperse widely while foraging, then it would have been adaptive for females to mimic males in order to reduce harassment by courting males. In the large groups postulated, males would have very great difficulty controlling the activities of subordinate males, so that females would have to protect themselves against harassment from small males (Geist 1974a). This hypothesis explains well

the reduced sexual dimorphism of highly gregarious ungulates and my be applicable
to primates. Gracile hominids apparently showed a low sexual dimorphism.

Once kin selection diminishes, it is no longer in the reproductive interest of
individuals to support the reproductive effort of others. Therefore, child rearing
becomes a burden charged entirely to the female. We can safely assume that under
savannah conditions, the males of *Australopithecus africanus* would have had a
very small role in child raising, if any. As child raising falls squarely on the single
female, there would be intense selection for child care by the female. The male
maximizes his reproductive fitness by maximizing the number of inseminations.
Pair bonding does not take place. We do not find it in gregarious savannah-dwelling
primates. The productivity of the landscape is adequate and continuous enough for
the female to carry the burden of nursing and caring for the offspring. Pair bonding
takes place in nonhuman primates only in the gibbon (Carpenter 1964) for reasons
explained in the following subchapter (Territoriality). It also must have been present
in the next species, *Homo erectus,* for reasons to be discussed later.

Given the formation of groups as an antipredator strategy, there would be certain
characteristics selected for in the social behavior of *A. africanus*. There would be
selection against lone and peripheral individuals. This would eliminate excessively
aggressive individuals shunned by others, as they would find themselves relatively
often alone or at the edge of groups, and thus subject to predation. There would be
selection for highly effective submissive behavior in males that would allow them,
particularly as young individuals, to stay close to large, dominant adult males, and
thus stay with the group. Since it is most unlikely that in the absence of kin selection
or pair bonding, males would care for children, the most promising avenue for
submissive behavior to follow is *female mimicry*. Thus, in submission or appeasement, young males most likely mimicked females in estrus; they would present
sexually to dominants. Thus, males would live in a homosexual society, akin to that
of baboons. This would explain the very high prevalence of homosexuality among
males of *Homo sapiens,* particularly when they are under duress (Kinsey et al 1948,
Mileski and Black 1972). Had there been kin selection, submission and appeasement would have taken the form of *juvenile mimicry*.

Since safety from predators still lay in numbers, there would be selection against
disruptions that would cause groups to break up. Therefore it would be adaptive for
dominant males to break up the fights of subordinates, just as is found in savannah-
dwelling cercopithecines (Jolly 1972, Wilson 1975). Furthermore, intensely aggressive individuals would select against themselves, not only because others might
shun them, but also because they would contribute to a decline in the size of their
group and thus increase the risk of predation on themselves. On the whole, we see
the society of *Australopithecus africanus* emerging as a typical Old World primate
society fitted for life on the arid savannah with seasonal shrinkage and dispersion of
resources. Theirs would not yet have been a human society but rather an age-
grade-male social system, in the terminology of Eisenberg et al (1969).

The presence of predators would have had additional effects on the social behavior and habitat preference of the gracile hominid to those just described. There
would have been selection for noisy individuals within the large bands. The cessation of noise would now function as a danger signal. They could have been noisy, as
are other savannah-dwelling primates (Jolly 1972), because in the open they are so
conspicuous that noisiness does not add to their conspicuousness (Geist 1974a).
They probably also had a predator warning system specific to predator species or

groups of predator species; cercopithecines can do as much (Altmann 1967). Once predators were sighted, however, individuals of the gracile hominid form would likely ascend trees or cliffs. Their apelike scapula (Robinson 1972a, Oxnard 1975) argues for good climbing ability, which even the larger and heavier *Homo sapiens* displays when in danger.

Because the small-bodied fragile *Australopithecus africanus* had hardly the strength to defend itself against predators, it sought safety in being active during daylight hours when predators could be spotted flocking into areas with good visibility, congregating in rapidly moving troops, and seeking safety in cliffs or trees during the night. Our diurnal habits and our fear of the dark may have their roots here, as has probably another characteristic, one we share with apes, namely, our ability to snore (Schultz 1961). This speaks for relatively great safety from predators at night.

We may safely assume that gracile hominids preferred cliffs to trees wherever nights were cool. In cliffs, individuals might huddle and thus collectively reduce their surface-to-mass ratio, and thereby reduce their rate of cooling. Cercopithecines also huddle in cool surroundings (Schultz 1961). Cliffs give better protection from convective heat loss than do trees, no small matter to a small-bodied hominid in an open landscap where air currents move freely. Cliffs reradiate the heat stored during a sunny day. Infants can be readily taken within a group of adults during the night when they are huddled on a cliff platform or cave. The infant can be shielded less effectively in a tree. Thus, under conditions of warm days and cold nights—a feature of dry savannah and steppe even at low latitudes and altitudes—gracile hominids probably selected cliffs over trees to spend the night in because of the microclimatic and spatial advantages of cliffs and the safety they granted from at least some predators. Cliff climbing does not require a grasping foot; the latter is probably essential for tree climbing and living in trees.

The foregoing has shown that the three postulates about *Australopithecus africanus*—that he lived in arid savannah; that he lived off widely scattered, small, highly digestible packages of food that was high in calories and nutrients; and that he arose from a kin-selected ancestor similar to *Pan*—are compatible with the morphology of the gracile hominid, its dentition, and its low sexual dimorphism. These are revealed by the paleontological record. The above three postulates are also fundamental to the evolution of a large number of features typical of humans, such as sexual jealousy, exogamous mating, lack of anatomical organs specialized for combat, the existence of humor and laughter, the residual phallic displays and homosexuality, the development of display organs far beyond the level reached by *Pan*, great manual dexterity leading to the use of tools, catholic food habits, relict estrus in women, diurnal habits and fear of the dark, iconic vocal communication, and a great tendency to explore and play.

Territoriality

In view of the public attention the concept of territoriality has received owing to the writings of Ardrey (1963) and others, and in view of the concern that architects and urban designers have shown for territoriality (Lyman and Scott 1967, Hall 1959, Sommer 1969, Newman 1972), it is essential that this concept be discussed in relation to human biology. Let me state from the outset: There is no theoretical reason to believe that there was classical territoriality in *Australopithecus* or *Homo* or biological selection for characteristics that would promote territory formation.

A territory, a defended piece of land, is formed under rather specific ecological circumstances (Geist 1974a). It must have all the requisites for life throughout the year; the resources must be continuously distributed and continually produced, and it must be so small an area that an individual can oversee it, spot intruders, and prevent their entry and use of resources defended in the territory. The territory thus protects rare, defendable resources (Brown 1964). It does not occur in species exploiting shifting resources, unless, during the breeding season, a stable resource can be exploited for the benefit of the young. Where seasonal events make movement necessary, only a breeding territory may develop in a species; where daily climatic occurrences such as snowfalls make individuals shift to different localities, territoriality cannot evolve either. Nor can it evolve where resources are so concentrated that masses of conspecifics flock in, and defence is costly and useless (e.g., a carcass for hyenas, Kruuk 1972). Where resources cannot be defended, such as when they are highly concentrated or widely dispersed, not active but *passive* means of competition are selected for. This includes a greater ability to bolt food, as in the hyena (Kruuk 1972), and better roaming and food-finding ability, such as is reflected in light body weight, upright stance, and appendages evolved for roaming in the genus *Homo,* for instance. For a territory to arise, the resources must be defendable. Thus, the year-round territory can be found only in relatively specific ecological situations, as in subtropical and tropical plant communities of moderate productivity, such as tropical rain forests.

If a species develops a strategy of exploiting a habitat as a year-round territory, a very specific sexual selection follows as a consequence: The female is faced with a severe burden in defending the territory. Under this territorial system, the possession of a territory not only permits reproduction, but also survival, since all usable habitat is parceled out. Reproduction depends here on continuous year-round productivity, not on a seasonal energy and nutrient pulse, as in species that are seasonal territory holders. Thus, to be without territory is to be condemned to habitats that do not conform to the species' niche and, therefore, to having a short life expectancy, or, alternatively, to be subject to persecution by territory holders and have a short life expectancy due to the costs of harassment and wounding. Clearly, under these circumstances, there will be selection for intense aggression, and thus very determined attempts by young adults to gain a territory by evicting territory holders. Note that such selection would evolve organs of attack and defence, as well as "warning coloration," notably lacking in the genus *Homo*.

Under the above circumstances, a female alone on a territory would be constantly faced by highly aggressive, determined intruders. Yet, she must carry the reproductive burdens in addition to those of defence. Hence, it becomes adaptive for the female to acquire a territory defender, a male. Now an intruder is faced by two territory defenders, namely by the monogamous pair. Thus, monogamy is a means to defend a scarce resource, a concept well recognized in socioethology (Wilson 1975, p. 330). It is adaptive for the intruder to concentrate attacks on the territory holder of its own sex. By removing that territory holder, the intruder may acquire territorial status, and live. Thus, it is to the advantage of the territory-holding female if she resembles the male in external appearance, behavior, vocalization, and smell. By retaining the male, she first saves on the cost of excitement and harassment that go with each change of partners; second, she maximizes her reproductive fitness by retaining the male who bred her. This follows because it is to the advantage of an intruding male to destroy the young of the previous male in order to free resources for its own reproduction (note that some langurs and lions do just that, see Sugiyama 1965, Hrdy 1974, Schaller 1972, Bertram 1975). Conversely, it is adaptive for the territory-holding male not to jeopardize its reproductive investment by readily re-placing the female that carries and nurses his child. We therefore expect only gradual pair formation and rather long periods of courtship in the year-round terri-tory holder.

Male mimicry by the female, and hence a reduction in sexual dimorphism, carries considerable benefit for the couple occupying a classical year-round territory. It increases the *ambiguity* to the intruder: His chances of attacking the partner of his own sex are now reduced from 100 to about 50. The *redundancy* of the message, that the territory is occupied, is doubled, however. This should reduce the risk of attack on the territory-holding couple (Geist 1974a).

If the female defended a territory, it would also select against behavioral estrus. I discuss this in detail later (p. 332). In essence, estrus attracts strange males, increasing territorial defence; it maximizes the chances for the male's defeat, since the sexes can now be identified owing to the female's estrous condition, thus reducing the female's reproductive fitness. Clearly, under conditions favoring pair territories, the partners will be look-alikes, will smell alike, and will act alike. Reduced sexual dimorphism implies equality of functions.

Granted the limited resources of the territory, it is in the female's interest to select a male as *small* as possible, but still capable of being a good territory defender. Hence we do find females in some territorial forms that are slightly larger than males, e.g., in the gibbon (Schultz in Campbell 1974, p. 301). It would also be in the female's interest, and in the male's, too, if the male could utilize resources other than those required by the female for gestation and lactation. This hypothesis predicts a difference in tastes and food preferences between sexes, as well as different food habits. I am not aware of research pertaining to this hypothesis in primates; it has been validated, however, for ptarmigan (*Lagopus mutus*, Gar-darsson and Moss 1970).

Could not gracile hominids have lived in such pair or family territories? Probably not. The parent species with the proper social system for this type of territory would not lead to the evolution of characteristics typical of hominids, such as close social bonding of many individuals or homosexual male submission, nor to the sophisti-cated social feedback system of our genus (see p. 333), nor to females who still secrete estrous pheromones to which, however, males are insensitive (see p. 332). Nor is it conceivable how a new type of bonding, based on the social feedback system and extended sexual receptivity, could arise from it, since in the territorial form male−female bonding based on cooperation in aggression does exist, while sexual relations are truly minimal. We can see this in the gibbon, the only primate with pair bonding in the context of year-round territorial life (Carpenter 1964). Human pair bonding is vastly different from that of the gibbon. Therefore, one cannot bridge the pair−territory adaptations and those of *Homo sapiens*; but one can bridge those described earlier for *Australopithecus* and those of our species. More of this later.

However, there are other territorial systems, group territories held by demes of closely related individuals (Brown 1974), male-defended territories with several females, and pure mating territories, such as those of many African bovids (Buechner 1961, Walther 1965, Estes 1974, Wilson 1975). Territory inhabited by a pair year round is quite uncommon. Is it not possible that early human social systems were based on these, rather than on the pair—territory? Again, probably not. All these territorial types, excepting the group territory held by a deme, select for *strong* sexual dimorphism. Such is not found in the evolution of *Homo*.

Again, we must digress for a moment and discuss how sexual dimorphism arises in the polygynous social system, based on individual males defending territories or guarding female groups. The argument runs as follows: Under conditions of dense or highly concentrated resources, it becomes adaptive for a female to persist in attacking to gain entrance to a territory since the resource abundance, by definition, is greater than can be maximally exploited by a single female. It is to the disadvantage of the defending female to continue to resist the second female, since she only incurs cost but gains no reproductive advantage. It is also to the disadvantage of the defending male to resist the second female, since she can turn surplus resources into offspring that have half the genome of the male. It is not in the male's interest to look like a female, but to maximize his abilities as a defender of his territory against other males. It is also to his reproductive advantage to inhibit territorial defence against other females by his own females. So males will differ strongly in external appearance under the above conditions, become almost exclusively defenders of the resources available for reproduction, and will reduce aggression among *females* under their control. This latter has been demonstrated for domestic chickens (Craig and Bhagwat 1974). The foregoing argument was originally developed by Verner and Willson (1966) and Orians (1969), and elaborated on in Geist (1974a). Since in *Homo* we find no strong sexual dimorphism, it is quite unlikely that we ever occupied multifemale territories.

This leaves only the possibility that the gracile form could have lived in a group territory defended by the deme as a whole. The possibility for that kind of territoriality is quite slim, however, in a dry savannah. Territoriality in humans, therefore, has at best a cultural base, but not a biological one akin to that in territorial mammals. Thus, territory is not "in our genes," at least not in those we acquired since the *Australopithecus* phase, and we cannot invoke nature to justify territoriality in man.

The adaptive syndrome of the australopithecines was distinct from anything we know today and not intermediate in an arithmetic sense between apes and humans, as Oxnard (1975) emphasizes. It was a successful form of life as the long history of australopithecines in the geologic record attests. Contemporaneous with them, but apparently replacing the gracile form, are remains of true humans. These are subsumed under the title of *Homo habilis*. Tobias (1973) and Howells (1973a) derive it from the gracile form. There are opinions to the contrary, claiming that australopithecines are not in the human line of descent because humans are contemporaneous with australopithecines. However, this is like arguing that the Old World deer could not have been derived from the Muntiacinae, nor the caprids from the rupicaprids, because the primitive members of the respective lineages are contemporaneous with the advanced ones.

Homo habilis is larger in body and much larger in brain size than the gracile form (Leakey 1973a,b) and has a hand apparently capable of a true power grip. We find it as early as 2.9 million years ago and, in some layers containing its fossils, we also find the first crude stone tools, camping places, and remains of what appear to be butchered large mammals. The tools date back as much as 2.6 million years (Fitch and Miller 1970, Isaac et al 1971). We cannot be certain yet that it is *Homo habilis* rather than contemporary australopithecines that made the tools. The tools are pointed pebbles, choppers, and polyhedrons found among a large amount of waste flakes (Isaac et al 1976, Isaac 1976, Chavaillon 1976). Even some bone tools may have been present, according to Chavaillon (1976). One may recall here that Dart (1957) championed the concept of a tool culture among australopithecines based on bones and teeth, an osteodontokeratic culture. He argues with merit that long before the first stone tools were crafted, hominids must have used natural objects as tools. Bone tools among *Homo habilis* ought not to be surprising from Dart's perspective.

Significance of Tools

Why make tools? One cannot make a case for crafted tools being essential to the gatherer's way of life as proposed for the gracile australopithecines. We must ask whether the ecological conditions dictated the use of such carefully made tools. Tools such as branches can be assumed to have been plentiful and would be good enough to dig a root from loose, shallow soil. A stick is good enough to smack a snake over the head with, but the small-bodied gracile hominids could hardly have been in a position to club down big animals. If the daily food intake can be satisfied by gathering, then there is little impetus for tool usage, except for cracking nuts and tough seeds. For this purpose, however, no crafted tools are required; stones will do.

It takes several attributes before tool using is possible: firstly, the ability to conceive three-dimensionally, that is, to visualize or imagine objects with well-defined properties of size, proportion, texture, weight, and color; secondly, the ability to make precise value judgments on what to select and what to reject; thirdly, to store diverse learned experience and alter value judgments accordingly. Fourthly, it appears that self-recognition is also an essential attribute, but since chimpanzees and orangutans can be demonstrated to have self-recognition (Gallup 1970, Lethmate and Dücker 1973), we are safe in assuming that early hominids had it as well. Fifthly, it requires that the tool-user "bond" with the tool and carry it about, particularly if it is a tool hard to find or make, but is required often for a successful life.

Almost all animals that use tools must have these attributes, albeit their conceptual abilities may be very selective and they often cannot generalize. The above intellectual abilities may have different neural mechanisms and development in different species (Alcock 1972). To make these perceptual qualities operational requires manual skills. Primates have great manipulative skills, though none reaches the proficiency seen in man. It is interesting to note that ground-dwelling savannah-adapted cercopithecines, such as baboons and macaques, have the greatest manual dexterity (Jolly 1972). Gracile hominids from the savannah thus probably evolved great manual dexterity prior to tool use. From the structure of the gracile hominid skeletons, however, there is little reason to believe that such skills were as well developed as ours.

In order to *choose* a tool, let alone make one, one must first produce a mental image of it. Without such an image, no choice is possible between a tool and a useless object. The choice of objects according to their appropriateness for a task may be shaped by experience, but it has to be preceded by a mental criterion for making a choice, a mental template, as it were.

We know from the work of Wolfgang Köhler (1927), Birch (1945), and Schiller (1957) that even chimpanzees can form quite an adequate concept of what is required in given situations, and can form primitive tools toward that end. The early gracile hominids could be expected to be at least as competent. We can claim an increase in their skill above that of chimpanzees only if we can demonstrate an increased ecological need for using tools. However, we must first return to the principles of tool use itself.

The ability to distinguish proper shape, weight, size, etc., for tool use implies the necessity of *value judgment*. *Any* choice depends on value judgment. The value judgment may be quite rudimentary, but it must be there just the same. Animals can do as much. Note the oyster catcher, preferring six eggs beside his own nest to the three eggs in it, or regularly choosing the larger egg over the three smaller ones laid (Tinbergen 1951). Regardless of how we camouflage the oyster catcher's choice in scientific jargon, the bird still exercized a choice depending on the crierion that "larger is better" or "more is better."

A precursor to value, or an indivisible aspect of it, appears to be our sense of "beauty." At least, so I understand Santayana (1896), the father of scientific aesthetics, to say, and it makes sense to me. The sensation of "beauty" would be triggered physiologically as mild arousal after perceiving a rare, contrasting stimulus object or event that fits a conceptual pattern. For instance, a flower is "beautiful"—that is, so attractive that we look at it again and again—only if it is relatively rare, contrasts strongly by virtue of color and size with its background and, of course, if we already have formed the concept "flower." Such an aesthetic experience, however, probably has its roots in a far more functional one, namely, in experiences felt internally that are guides to values. A strikingly beautiful ax is to me a highly functional one, one that fits my conception of what an ax "ought to look and feel like." I propose that the sensation of beauty originally was a guide to the valuable, and the valuable must have been originally in the image of the functional. A tool is not really valuable until it is "beautiful.". It follows from the foregoing that the sense of beauty would be expanded and made more sensitive along with the sense of value during our evolution as tool users and tool makers. The origins of the sense of beauty appear to lie in perceptual mechanisms common to higher vertebrates, as is implied by the example of the oyster catcher noted above. Other examples could be added; they all show that animals chose according to the criterion that "larger or more is better." Values are thus not "inexplicable reaction of vital impulses," as suggested by Santayana (1896, p. 24), but messages of a physiological nature of considerable survival value in our evolutionary history.

We can come a step closer to understanding the sense of beauty by using the principle findings of gestalt psychology that visual patterns are organized spontaneously and involuntarily in such a way that the simplest available visual structure results (Arnheim 1971, p. 3). An analysis of what is perceived as beautiful suggests that the beautiful fits with such basic perceptual patterns. The removal of "disorder" and "confusion" appears beautiful. It appears that we are dealing here with some basic inherited neural organization that we are largely slave to, and from which we can deviate only after intensive training. We are dealing with a deep structure of our psyche, very much as discussed by Stent (1975). At the same time one can see that, given that findings of gestalt psychology, perception is organized to give spontaneously the simplest structure, that we are very much subject to Platonic ideals. It is clear at once why Plato could argue that the only reality is the ideal into which we squeeze the world.

After visualizing, recognizing, and choosing a tool, the animal must also see its connection with a given objective; the tool becomes a part of a mental "gestalt" in which not only physical objects but also actions and processes are contained. This is the prerequisite to *using* tools. The animal must have an image of how to use the tool toward a desired end. Without such a "gestalt" no trial and error learning, and consequent improvement, is possible. Again, chimpanzees can do as much, as shown by their goal-directed tool use and their improvements in tool use, be it in the laboratory or in the field.

The same argument as for choosing a tool applies to shaping one. You must have a rather good notion of what a finished tool will look like before beginning to build or shape the raw material. We must also possess a sense that tells us how close our tool is to the desired shape. It is here that the sense of beauty, enhanced to exhilaration, enters and tells us when the tool has reached a shape close to that of our mental image. Tool building thus depends on one's ability to imagine and one's ability to apply past experience. Clearly, the better the imagination and the greater the experience, the better the tool.

The next step in tool use is improved value judgment. It may be important not only for recognizing, choosing, and using the tool, but also for keeping it. The tool user must "bond" with the tool! Forming an attachment to a tool is inconsequential only if similar tools are abundant and can be procured at will. Otherwise the tool maker who discards a rare tool imposes on himself the hardships that lack of an appropriate tool at the appropriate time bring about. The tool, if it is important for daily survival and hard to get, should be kept. This, however, requires *foresight* of future events and some matching of past experience with the anticipation of repeating them. This means that spatial visualization and process visualization are required, in which the tool, the individual, the object, and the process form a "gestalt." Please note that in order to be a specialized tool user, the animal develops increasingly the notion of "what is mine" and "what is thine."

The mechanism by which a tool user "bonds," or forms an attachment to a tool, an inanimate object, remains something of a mystery. There is one ray of light which may be the correct explanation of how bonding to a tool takes place. We know that, for instance, swords were given names by knights, while rifles were named by American woodsmen. Siegfried's sword Mimung, or King Arthur's sword Excalibur, are examples of the former; Davy Crockett's rifle Betsy is an example of the latter. It could be that in use a tool becomes "personified." If that inference is correct, then no consistent tool use and tool making was possible until the genus *Homo* had acquired the ability to form personal bonds. Rocks carried over long distances from their origin in the form of tools would therefore be evidence for bond formation by *Homo,* and the beginning of a human rather than primate social system.

The foregoing did not illustrate how tool use evolved. It merely analyzed the prerequisites for tool use and making, and showed that some of those prerequisites are already found in chimpanzees. Therefore we have no reason to suspect that they would not be present in the australopithecines. In addition, one ought to mention that the ability to generalize is also a prerequisite for plastic, diverse, adaptable tool use. It is also a prerequisite for evaluating diverse alternative strategies. The adaptive syndrome of the australopithecines would not demand as careful planning of diverse complex strategies as that of humans from the late Pleistocene. However, the environment of the gracile hominids did impose some requirements for plotting strategy, albeit not for complex strategies. The wide spatial and temporal distribution of food items, owing to seasonal flowering, growing, and ripening of forage items in the dry savannah, would require some notion of where to go at what time. It would require that the early hominids were good observes and able to move in response to clues. There would be an economic return for those who possessed the ability to move to the proper place at the right time; many mammals have evolved such an ability. Compared to their forest-adapted ancestors, the gracile hominids would meet a greater diversity of herbivores and predators in the dry savannah, and would require abilities to evade them, or to take advantage of their presence. A tree-dwelling primate can, to some extent, ignore what is going on below him on the ground; not so the ground dweller—he is forced to make adjustments. There is also a greater range of competitors for the same resources concentrated into the thin photosynthetic layer of the savannah. In particular, the great range of predators on the savannah probably forced diverse strategies of antipredator behavior on the gracile hominids. Like other plains game, they would have to recognize when predators are likely to be dangerous and when not, in the interest of conserving resources (Schaller 1972, Kruuk 1972).

The dry season would promote such behaviors as digging holes in creek beds to obtain water, but baboons can do this (Kummer 1968). Chimps can dig purposefully with sticks (Köhler 1927). From here it is a small step to excavating tubers from dry, loose soil or pounding holes into termite hills to "fish" for termites frequently and regularly. The transition to tool use does not appear great provided, say, tubers and termites formed an important part of the diet. An occasional surprise by predators in the open steppe would promote the gathering of rocks and branches, as well as a banding together to smash their weapons on the ground, brandish them, and frighten off the predator, much as savannah-dwelling chimpanzees are apt to do (Kortland 1967, Jolly 1972).

The roots of tool use are found in primates in such activities by baboons and howler monkeys as dislodging rocks and branches against predators, in manipulative ability and curiosity of so many primates which come in handy when poking and searching for concealed insects and larvae, and in the primitive tools made and used by chimpanzees (see Alcock 1972 for a review). These creatures are clearly capable of discrimination and, to a lesser extent than we are, of choosing the right object for a task, but their value judgment is clearly wanting, and even more wanting are the judgment and ability to bond that make a creature keep and carry a tool. It is important to note that no case can be made, either from ecological considerations, from the paleontological record, or from anthropological studies of primitive cultures, that sophisticated tool use in the advanced stage is essential to sustaining a gathering economy in the productive savannah. Much exploration work in this regard could be done by experimental archeology. The improvement in stone tools was advancing at a negligible pace but, given some three million years, it did improve; the improvement was on a scale similar to the improvement of teeth in the suids and elephants that are found in Africa along with hominid remains (Maglio 1972). It argues for a very stable way of life.

Evolution of Conscience, Imitation, and Faith

We assume that tool making enhanced the reproductive fitness of individuals practicing it. This, in conjunction with the ability of juveniles to learn by observation and imitation, sets the stage for acquired behaviors to influence natural selection. We call this element in the evolution of our kind "tradition." A number of noted authorities on evolution agree that it is a dominant element in human evolution (Waddington 1960). Our purpose here is to delineate the manner in which tradition acts in evolution and the biological system on which it depends.

The advantage of tradition is that accumulated knowledge can be passed on from one generation to the next, giving the receiving generation maximum time to improve already proven knowledge and skills rather than suffer the consequences of inefficiency by developing such knowledge. If tradition is to maximize a favorable trait under selection, then it must be transmitted accurately from generation to generation. This requires, first, that the juvenile should be able and willing to mimic the adult and that the adult should reinforce the young in its attempts at imitating. Second, it requires that individuals deviating from traditional knowledge and skills experience a punitive emotion; it requires a "conscience." Consequently, individuals will tend to stick to the acquired proven behaviors. Third, the acquired traditions must be supported by a powerful positive emotion which we may equate with faith or conviction, so that traditional skills are practiced despite obstacles. Here are the roots of mimicry, conscience, and faith. They can be selected for and enhanced genetically only if traditions enhance the reproductive fitness of the receiver; they are a *mandatory* part of a system of transmitting adaptive acquired knowledge.

Let us briefly examine the implication of the above concept. Clearly, the more a species depends on traditions to exploit its environment, the longer an ontogeny its individuals must have to acquire and practice adaptive traditions, and the more emotional the adults of that species will be, since they must compare more and more the acquired knowledge with the strategy demanded by the environment. Thus an increase in brain size, equivalent to an increase in diverse intellectual and motor skills, should be associated with increased emotionality. This has long been known to psychologists (Hebb 1966). The more a species depends on traditions, clearly, the better its individuals become at imitating, as well as at self-discipline. The latter is a consequence of "conscience" punishing a deviation from tradition despite the temptations of a given situation, plus the conviction of faith in traditional knowledge and skills rather than in opportunistic insights.

Note also that the juvenile must be less emotional during ontogeny than the adult in order to receive what is new to him. Again, this is well known for anthropoids and humans (Hebb 1966). Conversely, the adult must be very sceptical about new knowledge, since ready acceptance of new knowledge could destroy adaptive traditional knowledge. For the adult it pays to accept new knowledge only after very careful examination and testing, incorporating it only if it enhances the adult's efficiency at making a living. The juvenile must have an appetite for being indoctrinated. This means not only that he must learn readily, but also that he must be responsive to social reinforcers from adults. He need not doubt the knowledge and skills transmitted by the adult as long as these enhance reproductive fitness; on the contrary, he must accept them and practice them with great faith. The adult could not be a deceiver since his life depends on the very traditions he transmits.

Note also the following. Under normal maintenance conditions, in which the individuals of the maintenance-phenotype are readily discouraged owing to sensitivity to punishment (discussed in Chapter 6), a tradition will be transmitted faithfully from generation to generation virtually without change. Even slight change ought to be emotionally upsetting to such individuals. Early humans can be expected to have pioneered new traditions only during their dispersal phases and frozen these traditions at once at carrying capacity when the K-selected phenotype appeared. Since dispersal lasts only a few generations at best, only small changes can be expected in the cultures of early hominids, resulting in stable traditions over long time spans. Conversely, once humans learned how to artificially stabilize the dispersal phenotype, we expect an increase in technological change. Therefore an increase in technological and cultural change ought to be correlated with the appearance of a human dispersal phenotype, ourselves.

We can also predict that the more unpredictable the environment the more individuals depend on developing solutions on the spot to a given situation and the less emotional they should be. This is reflected in cold, calculating opportunistic individuals who value ends above means. We can predict that geniuses, compared to the population average, will be far less upset and distraught by the unexpected.

Since traditions can exist only where adults transmit to juveniles their acquired knowledge, it is mandatory that the individuals be sensitive to the expected behavior of others as well as to social approval or disapproval. The behavior of someone not harmonizing with the expected, as that of a stranger, ought to be upsetting, particularly to adults. Conversely, harmonizing on mutually shared knowledge and skills must be perceived as pleasing. These are all mandatory consequences of traditions as a strong element in the evolution of an organism.

Chapter 11

The *Homo erectus* Stage

Introduction

Beyond the forest the savannah, beyond the savannah the steppe. Hominids had evolved an adaptive syndrome fitting the savannah; the open steppe was the logical next step. Other lineages of mammals had followed the same procession, as I discussed earlier, from forest to its ecotone with the steppe, to the steppe itself. If hominids went the same way, what would be evidence for this view or, conversely, how could it be conclusively disproven?

Clearly, we expect that fossil remains of the hominids next more advanced than the australopithecines will be associated with dry grasslands. This does not mean that they will be associated only with grasslands, for advanced hominids could well have replaced australopithecines or at least have competed with them in the savannah. We do expect, however, at least an increase in niche breadth beyond that of australopithecines and the occupation of grasslands.

In addition, adaptation to the steppe is most likely when steppes expand, which happens when climatic extremes increase, becoming intolerable to forests. This appears to happen when major ice ages occur. Therefore, we would expect the evolution of hominids above the australopithecines to be associated with drying climatic trends and cooling temperatures.

Also, we can play a conceptual game and ask what would happen if australopithecines were transposed from the savannah to the steppe. If we find that natural selection would indeed increase the number of characteristics we consider human, then we have support for the view that hominids did follow the evolutionary procession from forest to savannah to steppe, just as other mammals did. Here we would apply principles derived from our rather recent understanding of social and ecological adaptations of large mammals and we would keep in mind that it is behavior that is the first level of adjustment, followed by physiology, followed by morphology. Only when these levels of adjustment are exhausted does genetic adaptation set in.

Our aim here is to discover how and when those features arose that characterize human beings. Paleontology can give us an answer, at best, to when, never to how. However, even where paleontology can be helpful, it is shackled by a concept of evolution that all too often makes it doubt factual data and shape facts to fit theory. Historically, Dart's (1959) account of the reception of his discovery of the first australopithecine, the "Taungs baby," is an example of attempts to fit facts to fancy. However, the same tendency is shown in more subtle form in some modern accounts, in part because no alternative to the accepted evolutionary doctrine has been perceived. Let me illustrate this with an example from William Howells' (1973a) delightful book. On pages 65–67 Howells discusses the ages obtained for *Homo erectus* deposits in Java. If the ages obtained are valid, *Homo erectus* existed in Java from about 1.9 million years ago until some 500,000 years ago, at least. That, however, would make it almost as old as the earliest *Homo habilis* from Africa. It would also suggest, as Howells recognized, that hominid evolution was ahead in Asia, and that *Homo erectus* must have moved back into Africa from Asia, replacing *Homo habilis*. Howells is uncomfortable with these conclusions. Conventional anthropological wisdom has it that *Homo erectus* should grow out of *Homo habilis* by slow change, and therefore the two species can never be contemporaneous. Howells solves the dilemma by doubting the facts, the dates, and assuming a shorter period of existence for Javanese *Homo erectus,* and a later date for the earliest fossils, so that *Homo erectus* is now later in time than *Homo habilis*. Moreover, Howells avoids mentioning in that context the greatest embarrassment of all, the remains of Solo Man of Java, which is also *Homo erectus,* from the *late* Pleistocene. If the facts on hand are to be trusted, then *Homo erectus* existed in Java probably about 2 million years ago or very nearly that and—and this is really bothersome—changed surprisingly little in the process. Yet elsewhere, human evolution was moving right along and had reached the *Homo sapiens* grades some hundreds of thousands years earlier in Europe. How can humans evolve in one locality and not in another? The concept of gradualism and continuous evolution denies these sorts of facts; it promotes views such as Coon's (1962) that human beings reached the *sapiens* level independently in different continents.

Whether the notion that the presently accepted evolutionary dogma is at fault has been perceived in anthropology, I am not aware. I am aware, however, that the dilemma faced by Howells is no dilemma if one understands mammalian evolution using C. H. Waddington's basic ideas, and looks at the analogies in other lineages of large mammals. It is for this reason that, in this chapter, I dwell extensively on the "dispersal theory," a derivative of the Waddingtonian paradigm, although I did not know it to be such when I conceived and published this theory. It predicts that evolution is a rare event which occurs only after all mechanisms of adjustment are exhausted, so that once a form colonizes a habitat, it does not change appreciably, in essence ceasing to evolve except for minor quantitative adjustments that probably reflect more efficient exploitation of resources within the evolved adaptive syndrome. The *Homo erectus* of Java was part and parcel of a rather primitive biotic community in the tropics, one in which little if any evolutionary change is expected. I show that in large mammals the evolution of new adaptive syndromes is largely tied to dispersal into areas unoccupied by conspecifics, usually resulting in large-bodied, ecologically and socially specialized descendents far removed geographically from the parent species. Thus, if both *Homo habilis* and *Australopithecus boisei* dispersed to Asia, we would expect gigantic forms to arise, quite different in many respects from the parent forms. Do *Homo erectus* and *Meganthropus* fit the predictions of the dispersal theory? Moreover, a carnivore, which *Homo erectus* became, is expected to displace its parent species when it can disperse backward to its place of origin. That would predict a sudden replacement of *Homo habilis* by *Homo erectus* in Africa; the former, once evolved, would of course remain a fairly stable adaptive syndrome, analogous to what I just said for *Homo erectus* from Java.

The dispersal theory predicts evolution primarily at the rim of human geographic distribution, where populations are in contact with more variable environments and where mechanisms of phenotypic adjustment are most readily exhausted. Thus, the boiling pots of human evolution are the temperate and cold zones but not tropical or subtropical environments, with one exception. This occurs when a lineage conquers a new biome, in the present case exemplified by hominids adapting from savannah to steppe and giving off a new adaptive syndrome, the human one, to exist for some time contemporaneous with that from the savannah. Note that we expect little evolution in *Homo habilis* or *Homo erectus,* or *Australopithecus* for that matter, but we do expect new forms or improvements to arise at the fringes of human distribution; we expect the first *sapiens* forms from the northern areas, where they are indeed found by the middle of the Pleistocene, as recently reviewed by Howells (1973a).

We also expect, humans being hunters and acting like carnivores, that more highly evolved forms would flood backward and replace more primitive ones in subtropical and tropical areas. Primitive forms are expected to survive in the tropics and subtropics only until more highly evolved ones arrive and eradicate them. However, we are getting ahead of ourselves. In this chapter, our attention is focused on the transition of hominids from savannah to steppe and on the adaptive syndrome that this calls forth.

At first glance, it might seem a trivial matter for australopithecines to adapt to the steppe. However, a detailed examination shows that it was a step associated with a profound transformation. It changed the basically primate way of life to a human one. The crux of the matter was deciding how to live in a habitat with a short vegetation season, and among the large primates only human beings have solved that problem—and it changed them drastically. Once the problem was solved, they could adapt ultimately to temperate climates, then to the rich, productive periglacial zones, and finally to the high arctic and subalpine.

The paleontologic record of the early *Homo* forms is scanty. Interpretations by experts vary, making the task of sifting fact from fancy neither easy nor certain. All interpretations have to be examined for attempts to hammer facts to fit into presently acceptable ideas of human evolution, or the neo-Darwinian paradigm. Furthermore, anthropologists have not yet made use of the idea of growth priorities so familiar to animal scientists and so they treat variations as if they reflected genetic adaptation rather than phenotypic adjustment. It cannot be stressed strongly enough that bone shape is very much a consequence of the behavior of individuals, of the kind of muscular forces exerted on bone, as so well reviewed by Molnar (1968) for skull shape in relation to mastication. I have had to make new interpretations using the ideas described in Chapter 6, and thus my reference to population quality.

Adaptive Successions: From Savannah to Steppe

The hypothesis we pursue is that *Homo* arose from the australopithecine syndrome, the gracile one, by adapting from savannah to steppe. Therefore, we expect that remains of early *Homo* forms are associated with steppe fauna and flora, and that early *Homo* evolution is associated with a drying trend in Africa. As I indicated in the introduction, we expect a drying trend with glaciation. Early *Homo* evolution—be it of *habilis* or very early *erectus*, such as those of Java dating back some 1.9 million years (Howells 1973a)—is indeed associated with a drying trend in Africa (Clark 1970, Carr 1976, Jaeger and Wesselman 1976) and the advance of the first major continental glaciations, the Donau and Bieber glaciations, put variously at 1.2 to 1.8 million years ago (see Birdsell 1972, Cooke 1973). *Homo habilis* remains have been found much earlier, going back to some 2.9 million years ago, while primitive Oldowan type pebble tools are found apparently back to 2.6 million years ago (Leakey 1973a,b, Fitch and Miller 1970, Isaac et al 1971). These dates fall into the age of minor glaciations in Europe, the Villafranchian, a long span of unstable climate heralding the major glaciations. If a drying trend did exist in Africa paralleling the cooling trend in the north, grasslands would have spread at the expense of savannah, and this would have led to an increase in the biomass of grazers and to an evolutionary impetus for herbivores and carnivores to diversify. This idea fits in well with the paleontological data discussed by Cooke (1972) which indicate that the large-mammal fauna of Africa continued to diversify from the early to the mid-Pleistocene; this diversification included hominid diversification.

Early in the Pleistocene the face of central Africa began to change, owing to the rift valley formation which reoriented drainage patterns on that continent so that water began flowing north and south in central Africa (Pfeiffer 1972). This would open avenues for faunal dispersal into North Africa, Europe, and Asia. It may have been an element in the early dispersal of *Homo,* for we find *Homo erectus* in Java already at a very early stage, some 1.9 million years ago (Howells 1973a).

Although the driving force of human evolution, as well as of other ice-age mammals, was very likely the pulsating glaciations, the data to support or refute this idea are still scanty. Such data are mainly available for the last stages of human evolution, as we shall see, but it would be unreasonable to assume that earlier glaciations would not have affected human evolution.

When first found, *Homo habilis* is associated in the same deposits not only with the robust *Australopithecus,* but apparently also with some small hominid, maybe the gracile form (Leakey 1976b). The climate between 2.5 and 2 million years ago shows a drying trend in the *Homo* sites in Africa, as discussed by Bonnefille (1976) and Carr (1976) from investigations of pollen, fossil, and sediment data. *Homo habilis* is found with a dry savannah, but so is the robust *Australopithecus.* However, the latter lived along water courses—according to theory—and is therefore more likely to be encountered frequently in the fossil record than *Homo habilis,* which is indeed the case (Leakey 1976b).

For *Homo erectus,* we find sites of activity in dry grasslands only if we assume that the Abbevillian or early Acheulean hand-ax culture is that of this hominid. These artifacts are found associated with warm savannah and steppe (Clark 1970). However, *Homo erectus* also disperses into temperate climates. This is not surprising if we consider the fact that adaptation to dry grasslands in warm climates also demands an adaptation to cold nights. In Europe, *Homo erectus* is found in interglacial periods (Kurtén 1968a, Howells 1973a), Birdsell (1972), however, believes that *Homo erectus* even lived in cold climates, basing this suggestion on the remains associated with *H. erectus* at Choukoutien in China. To me, the same remains suggest only a temperate climate. These remains include gazelles, wild boar, and sika deer—clear indicators of short, cool, and probably snow-free winters, with little if any freezing. The many seeds of hackberry (*Celtis*), so numerous that Birdsell (1972) felt that hackberry was a staple food, argue that this tree species was not at its ecological extreme but at its ecological optimum. Here *Celtis* is associated with cool, but not with cold, winters, such as might be found in the southern United States. Coon (1962, p. 436) cites the pollen analysis data from Choukoutien of breccia collected by the Sino-Swedish expedition; it reveals a temperate forest with a predominance of deciduous tree species, such as beeches, a few spruce or willows, yew, buckthorn, and linden trees. Moreover, the skeletal remains suggest that the *Homo erectus* lived there in extreme conditions, verging on the barely tolerable. The bones themselves are apparently evidence of cannibalism, and cannibalism in human beings is associated with poor ecological conditions, as I have discussed in greater detail in Chapter 14. If ecological conditions are poor, we expect a low population quality. Indeed, the Choukoutien specimens, according to Le Gros Clark (1964), are smaller in body size than the specimens from Java. This interpretation appears to be contradicted by the larger brain size of the Chinese specimens, for brain is a tissue of low growth priority. We must note, however, that brain size is also an indicator of adversity, indicating a broader range of behaviors practiced by the Chinese *Homo erectus;* brain size enlarges phenotypically with use (see Chapter 2). Last, but not least, we find evidence of fire at Choukoutien and not in southern *Homo erectus* sites. That also indicates marginal, not optimal, conditions for this human form.

In summary, we find a greater niche breadth in *Homo habilis* and *Homo erectus* compared to the gracile australopithecines; *Homo habilis* remains and the Oldowan and early Acheulean tool cultures associated with savannah and steppe at least part of the time; and an association of very early *Homo* with a drying trend and the early glaciations. All this is consistent with the hypothesis that the adaptive syndrome of *Homo* arose in the dry steppe.

Australopithecine versus Early *Homo* Characteristics

We must now examine some differences between the gracile australopithecine, *Homo habilis,* and *H. erectus,* for it is these differences that must be explained by the adaptive strategy that shaped *Homo.*

Homo erectus was much larger in body size than *Australopithecus africanus,* as large as modern man. His postcranial skeleton—what little we know of it (Le Gros Clark 1964, Howells 1973a)—was also similar to modern man's, but it does differ in some respects. The bones have exceedingly thick walls and consequently narrow marrow cavities. On the humerus, the tubercle to which the deltoid muscle is attached is extremely well developed, indicative of powerful work with the pectoral musculature (Coon 1962). The skull is also very heavily boned. It is clearly not the light, thin-boned skeleton of *A. africanus.* The skull of the early African hominids referrable to as *Homo habilis* is still more similar to that of the gracile *Australopithecus,* according to Walker (1976). The cranial shell of these skulls is quite thin and the supraorbital tori (verfy large in *Homo erectus*) are quite small. The massive solid construction of the skull in *Homo erectus* is due not only to generally thick skull bones, and to supraorbital tori which are nearly solid bone with only tiny sinuses above the eye, but also to a thickening along the midline of the skull cap and a massive bony torus on the occiput for attachment of strong, large nucheal muscles. The nucheal area of *H. erectus* is distinctly larger, both relatively and absolutely, than that of the gracile australopithecine (Le Gros Clark 1964). The robust skeletal build is significant, as we shall see.

Although postcranial elements indicate great similarity with modern man, there are rather few postcranial elements available for study. Even these indicate that muscle attachments were somewhat different than in modern man (Howells 1973a). The significance of these differences is not known.

The teeth of *Homo erectus* are smaller than those of *Australopithecus africanus* and had begun to lose the dryopithecine molar patterns. The teeth show a further reduction in size during the course of history in *Homo erectus* (Howells 1973a) and change toward a taurodont condition (Coon 1962). However, the teeth are larger than those of Neanderthal and modern man. The canines are peculiarly enlarged and long-rooted compared with *Australopithecus* or modern man. Reconstructions show an edge-on occlusion of teeth; this finding is of no great significance if such occlusion is a simple consequence of heavy wear on the incisors, as shown by Molnar (1968). The teeth of the early *Homo habilis* have been described as being similar in proportion to those of *Australopithecus africanus* (Howells 1973a) and therefore ought to be larger than those of *Homo erectus*, particularly the molar teeth. However, the teeth are no larger than those of large-toothed *Homo sapiens*, (e.g., Australian aborigines, see Le Gros Clark 1964, p. 155).

The brain of *Homo habilis* was larger than that of *A. africanus*, and that of *H. erectus* was larger still, with a tendency toward increasing brain size from the early Javan specimens to the Choukoutien specimens and upper Pleistocene Solo Man (Howells 1973a, Clark 1970, Hofer 1972). The study of endocranial casts, as discussed by Le Gros Clark (1964) and Hofer (1972), gives a few additional insights. The brain of *H. erectus* is small in the temporal, frontal, and occipital lobes; in modern man these lobes are associated with aspects of speech, logic, comparison, recall, long-term memory, intellect, intuition, and what has been called "moral behavior," and can be more simply termed self-discipline (Guyton 1971, Lenneberg 1969, Geschwind 1970). The temporal lobes in modern man are involved with the retention and understanding of ideas, pictorial recognition, the ability to form visual gestalt patterns, accurate discrimination, and the ability to understand the meaning of speech (Milner 1956, Penfield 1959, Geschwind 1970). Here we must remember that morphological change follows physiological adjustment, which follows behavioral adjustment. This means that if *Homo erectus* increased his brain size, a morphological change, then we are dealing with a species with a vastly greater diversity of behavioral repertoires than the smaller-brained *Homo habilis*, let alone the gracile *Australopithecus*.

A very important clue to *Homo's* evolution is that the sexual dimorphism of *H. erectus* is greater than in modern or Neanderthal man (Brace 1967). It is also noticeable in *H. habilis* (Howells 1973a). The skulls of females are somewhat smaller and less massively constructed, with the tori on the occiput somewhat reduced (Howells 1973a).

There is considerable evidence for cannibalism and homicide in *H. erectus* (as there is for *H. sapiens*), exemplified by the remains of specimens from Choukoutien, Ngandong, and the Solo river in Java (Weidenreich 1939a,b, Blanc 1961, Coon 1962, Birdsell 1972). Here *H. erectus* is analogous to the large mammalian carnivores, virtually all of which have by now been shown to kill and eat conspec-ifics. This emerged from relatively recent studies of these animals in the field (Schenkel 1966b, Schaller 1967, 1972, Hornocker 1967, Rausch 1967, Mech 1970, Jonkel and Cowan 1971, Kruuk 1972, Pearson 1972, Miller 1972, Payne 1972, Russell 1972, Larsen 1972). Chimpanzees also practice conspecific killing and cannibalism on exceptional occasions (Bygott 1972).

If *Homo erectus* killed and dismembered conspecifics, we may assume that he was a hunter. The fossil record suggests the same (Clark 1970, Isaac 1968), even for the earliest of sites associated with Oldowan and early Acheulean tools. Moreover, there is evidence of camps some 2 million years ago, indicated by bone assemblages of several species of ungulates (Isaac 1976). A quantitative analysis of bones from the Olduvai Gorge of about 1.8 to 1 million years ago led Speth and Davis (1976) to suggest that early hominids had a seasonal kill pattern similar to present-day hun-ter—gatherers from arid, hot regions of Africa. Evidence in the form of a large number of smashed bones associated with Acheulean tools, or even with remains of *Homo erectus*, as at Choukoutien, has led to a widely accepted idea that these people were accomplished big-game hunters and lived a life quite similar to modern man (e.g., Clark 1967, J. D. Clark 1970, Washburn and Lancaster 1968). This is a hasty conclusion and I concur with those like Howells (1973a), who caution against it. Although the adaptive syndrome of early *Homo* does show him to be a hunter, he is by no means at the level of modern man, as we shall see when reconstructing the adaptive syndromes of Neanderthal and late Paleolithic people.

Significant is the virtual absence of evidence of fire in the African archeological sites of *Homo habilis* and *H. erectus* age (Oakley 1961, Clark 1970). There is a steady but minor improvement of hand axes from the Abbevillian to the late Acheu-lean (Oakley 1961), albeit the latter may be a product of *H. sapiens*. The tools of *H. habilis* may have been those of the Oldowan industry, an industry basic to the Abbevillian hand-ax culture.

Characteristics of the Steppe

How does the dry steppe differ from the savannah? The vegetation is primarily a blanket of grasses, green and lush during the vegetation season, yellow and dusty during the dry season. There is thus a break in the vegetation season. There may be an identical break in the savannah; however, water is more abundant here and collects in numerous pans in rolling terrain during the wet season. As the dry season sets in, the water surface recedes while a ring of green sprouting vegetation follows the water's edge until the rains come once more. Therefore, in the savannah in the vicinity of such pans, green vegetation and its associated life may continue to be available all year round. Trees frequently ring the pans, their roots penetrating deep and keeping in touch with the moisture even during the dry season. In the steppe, with its scarcity of water, trees and shrubs are comparatively rare and found primar-ily along the permanent or temporary water courses or deep water holes, or where water penetrates to the surface. Green vegetation may be all but nonexistent during the dry season. The vegetation has adapted to the periodicity of water availability and a great many species are perennial in nature and store food and moisture in bulbs, corms, and roots. For instance, Lee (1972) speaks of some 30 species of edible roots alone available in the Dobe area of Botswana and exploited by bushmen.

When the rains set in and the steppe awakes to intense throbbing life, the ancient method of foraging on the surface developed by australopithecines can be used. This is not possible in the dry season. For a large-bodied surface gatherer with a large absolute food requirement, there is little to eat unless, like the many species of ungulates on the plain, he adapts to eating very coarse-fibered grasses and has a system whereby symbiotic organisms break down the cellulose. If the surface feeder avoids feeding on what is available above the ground surface and follows by migration the green vegetation, it will return to the wooded savannah during the dry season and act very much like the gracile australopithecines could have been expected to act. Clearly, in the dry steppe during the dry season, very little can be had to eat from the vegetation layer above ground for a large-bodied primate, even when water is available.

However, there is food in the form of stored plant food in subsurface parts of plants such as taproots, tubers, bulbs, and corms. These plants burst into flowering growth during the wet season. During the dry season their vegetative parts die back and their subsurface parts become protected by a layer of hard soil. They cannot be reached unless a way is found to penetrate through the hard material. These edible plant parts would be the only reliable food source, for bushes and trees that produce crops of edible fruit or mast are expected to do so seasonally, as is herbaceous vegetation. There is likely to have been plenty of competition for fallen mast and fruit, given the diversity of suids in the African Pleistocene.

The most logical extension of food habits by hominids adapting to the dry steppe is to exploit the subsurface parts of plants. These cannot be readily extracted with human fingers. Tools to penetrate the hard soils are essential. Also essential is an ability to recognize localities likely to contain hidden food on the basis of visual clues based on the ecology of various root-forming, tuber-forming, etc., plants. This, in turn, requires the ability to differentiate rather fine clues, form conceptual gestalt patterns, be able to learn very well by imitation (trial-and-error learning here would lead to starvation!), have excellent recall, and be able to teach the experience to one's offspring. In short, if hominids adapting to dry steppe had to depend on subsurface plant parts as staples, a feeding strategy based on these would generate secondary effects, some of which are far reaching.

If, as postulated, *Homo habilis* and *H. erectus* could fall back on a secure, predictable, rich source of carbohydrate for food, then they had the time to refine opportunistic hunting into systematic hunting. This concept was plausibly applied as an explanation of the occurrence of hunting in several primate troops by Gaulin and Kurland (1976). A combination of mast and deep-rooted tubers, etc., would amply fulfill the energy demands, though not the protein demands. It would, however, free time to forage for protein, the master nutrient.

A food source rich in carbohydrate, readily accessible by females, may ensure the energy demands but not the protein demands of a gestating or lactating female. Given the long gestation and lactation periods of the large-bodied anthropoids, no expansion into a habitat with a discontinuous vegetation season is possible unless the heavily pregnant or lactating female can have ready access to protein. This requirement, as will be shown in Chapter 13, produced a social system unique to primates and is the ecological determinant of many human characteristics.

A third consequence of exploiting a quickly accessible but limited store of carbohydrate food is that such a store may be defendable. Active competition once again enters into food procurement. We shall see later that this circumstance selects for aggression and sexual dimorphism, explaining in part the robustness of skeletons and the differences between male and female skulls.

The fourth consequence I have already mentioned: mandatory tool use and tool crafting to enhance the ability of human hands to penetrate deep below the hard subtropical or tropical soil surfaces. The stone tools of the early Oldowan and Acheulean industries contain crude bifacials quite suitable for shaping branches into digging sticks, particularly for removing the fibrous endings of branches broken from trees, for these would make digging useless.

The availability of ample carbohydrate food during the dry season would not only select for systematic hunting of mammals and birds but would select also for the exploitation of all other possible sources of protein. For instance, rotting wood along watercourses contains various insects and insect larvae. A pointed stone tool, like a pointed pebble tool or hand ax, can be used to penetrate the hard outer layer of sound wood to get at the inner rotting core.

In the dry savannah and steppe, wildfires set by lightning strikes are a rather common occurrence, as we know from fire research (Komarek 1964). In its wake, a wildfire leaves singed insects, fledgling birds, lizards, and young mammals, as well as popped seeds which can easily be collected (Komarek 1965, 1967). Occasionally, quickly moving wildfires in tall grass would blind and burn large mammals. McHugh (1972) describes this happening to buffaloes, at times in great numbers, on the American prairie; such helpless creatures would not be a great challenge to kill. Wildfires could cause animals to panic and stumble into natural traps they normally avoid.

Meat could be provided on a regular basis without hunting by scavenging, for the African savannahs and steppes supported a high biomass of diverse large mammals in recent times (Bourlière 1965, Talbot and Talbot 1963, Bourlière and Hadley 1970, Foster and Coe 1968). In pre-*sapiens* time the diversity was even greater. Where many live, many die, so that many kills by predators must have been made and observable to early hominids. Whether early hominids could have displaced large predators from kills is a matter for conjecture. Schaller and Lowther (1969) demonstrated, however, that scavenging could have been effective in obtaining meat, while some societies in Africa formerly depended quite heavily on scavenging for meat (Woodburn 1968a, Cowles 1963). At the very least, scavenging would add the kills of small-bodied hunters like cheetahs or even hunting dogs to the fare of hominids, and from the kills of larger predators they would in all likelihood still be able to obtain the marrow of unbroken leg bones, and the brain and flesh adhering to a skull. A group of *Homo erectus* armed with primitive clubs would probably have been able to displace hunting dogs, dholes, a few hyenas, and even a leopard. The odd carnivore bold enough to attack would probably die of skull-bursting blows and become prey in turn. However, such postulates can only be made for the rugged, powerful *H. erectus*, not for the relatively fragile *H. habilis*. Moreover, scavanging would have selected for monomorphism in hominids, had it been a very important source of food early in *Homo*'s history, as I shall show in Chapter 13.

One can also gain some insight into probable food habits of the early hominids by noting those of present-day hunter−gatherers, provided that allowance is made for the sophisticated means of hunting they have at their disposal, their use of fire, and the impoverished megafauna of the Recent age. Such studies as those of Woodburn

(1968a,b) and Lee (1968) indicate that modern people rely mainly on gathering, and that hunting adds relatively little to their very adequate diet. This is another reason not to jump to the conclusion that early *H. habilis* and *H. erectus* were already very capable hunters, akin to ourselves and living a way of life very similar to modern people.

The dry steppe deprives the australopithecine of its protection against carnivores, for trees are absent or rare. Arboreal ability is of little use now. Yet predators are present and dangerous as ever. In the absence of selection for arboreal ability we expect the limbs, hands, and feet to reflect this loss. Bipedal running to avoid a predator is hopeless for human beings. Although fire could be used against predators, there is no evidence of fire in Acheulean or Oldowan sites from Africa (Oakley 1961, Leopold and Ardrey 1972). Such evidence would have been found had fires been used as a protection at camp sites at night. What would prevent large carnivores from picking up humans camped on the ground at night? We cannot turn for enlightenment to modern gatherers and hunters in the plains, as these use fires and are armed with weapons capable of inflicting serious damage on carnivores; similar weapons are unknown from sites of Oldowan and Acheulean people. There is no convincing answer available to the question of how humans with the limitations of early hominids evaded predation. Some observations, however, are pertinent. First, predators tend to avoid people even where they are very common and unarmed. Man-killing large cats in Africa and Asia appear to be handicapped individuals that resorted to killing humans to escape starvation. Although human flesh is consumed by large predators, the frequency of its occurrence is such that it suggests a disdain on the part of the carnivore. Do we possess a scent that protects us from carnivores?

Secondly, human beings mimic sounds and have a very loud voice. We can utter sounds similar to threats used by carnivores. Since carnivores have as a whole excellent weapons but poor defences, they ought to be very sensitive to signals denoting danger. Threat-roars would be such signals. As human beings can vary their voice to suit situations and the species of carnivores confronted, they could conceivably use the very sounds that would cause predators to withdraw. Moreover, a group of hominids could set up quite a din!

Thirdly, at night, modern humans, and very likely the earliest of hominids as well (Chapter 12), set up a barrier between themselves and their surroundings in the form of shelters. Shelter robs any carnivore confronting it of information. To a carnivore this is a potentially dangerous situation, for its prey can defend itself at times, particularly if the prey is large compared to the predator, and could injure it. Therefore, a carnivore ought to attack only if the situation is such that it minimizes the chances of injury to itself. A shelter precludes information on this point being available to the carnivore. These factors may be sufficient to deter carnivores. Whatever antipredator strategy humans used, it apparently allowed them a long, restful sleep at night, for we have not lost the propensity to snore any more than have the apes that normally rest secure in arboreal nests.

The open steppe presents a problem to hominids at night, not only one of escaping predators. Whereas during the day solar radiation imposes a heavy heat load on the body in warm climates, during the night temperatures drop very low and the open sky acts as a heat sink. Most of the heat loss is in the form of long-wave radiation. How did hominids adapt to this problem? Behavior is the first level of adjustment to an environmental stressor; we cannot, therefore, simply state that they grew a thicker hair coat! I have attempted to explain how hominids dealt with the problem posed by nightly heat loss in Chapter 12 under the topic of Evolution of Hairlessness.

However, life during the day also has its problems. Solar radiation imposes a heat load. Newman (1970) identified how this dilemma was solved and pointed to a previously unmentioned diagnostic feature of human beings, their unusual ability to sweat and thus create evaporative cooling. It is briefly discussed in Chapter 12 under Evolution of Hairlessness.

Adaptation to the steppe environment at low latitudes thus required new strategies of obtaining food, escaping predators, and dealing with temperature and radiation extremes. Let us now turn to the evolution of hunting in its simplest form.

From Opportunistic to Systematic Hunting

Within the adaptive syndrome of the australopithecines, there was no ecological requirement for hunting as a significant provider of protein. Undoubtedly, some opportunistic hunting did occur and added to the food budget, but this was probably only a little more frequent than similar hunting done by chimpanzees. Opportunistic hunting consists of killing of prey as the opportunity permits, and does not entail elaborate preparations for the kill. It consists of stumbling onto some gazelle fawn and killing it, or killing a stray youngster from a troop of savannah-dwelling primates, and it entails no specialized means of killing with tools. Chimpanzees described by Goodall (1971) practice opportunistic hunting, but they may also band together for a rare cooperative hunt.

No evolution toward systematic hunting can be envisioned for the gracile *Australopithecus* as long as it roamed in large, noisy groups across the savannah, keeping together as a means of reducing predation. Individuals could enhance their chance of meeting prey even under these circumstances. They could revisit places they had found prey previously or learn to recognize localities where prey is likely to be hidden. These are methods practiced by carnivores, and require little brilliance or insight; chimpanzees perform this type of feat very well (Menzel 1973). If opportunistic hunting becomes more important in providing protein so as to enhance reproductive fitness, selection sets in for refined perceptual abilities, such as recognizing a prey from seeing only part of it and distinguishing prey from nonprey at a distance. The opportunistic hunter can also increase his chances of success by stalking close and surprising the quarry. This, however, requires a behavior atypical for primates. It is inconceivable that stalking could be practiced if these protohumans moved about in bands of males, females, juveniles, and babies. Given the presence of the noisy troop, the males would normally find little to hunt, as most of the huntable animals would keep at a distance. Food would be obtained through precise scanning, darting forward and picking it up, or searching for it in localities which experience tells them contains food. Opportunistic killing of mammals would be rare.

In the transition zone between savannah and steppe, the situation changes considerably. During the dry season, as the grasses wither, the availability of plant food found above the soil surface declines. The troops must disperse over increasingly wider areas, and become ever smaller social units, to find enough plant food.

Dispersal of the troop increases the chance of stumbling onto prey that has not been frightened away and it allows, for the first time, successful stalking. Plant food would not be uniformly distributed; rather, it would be associated with spots retaining water the longest. For the first time a foraging area could be overlooked and would become defendable. That means that a dominant male could enhance his reproductive fitness by excluding other males from the foraging area of the females he accompanies. Note, however, that this demands that the male form a bond with females after breeding them so that the resources he defends are channeled to the benefit of his unborn offspring within the females he bred. Clearly, the subordinate males would be shunted off to the periphery and would form small bachelor groups. Concomitantly, there would be selection for sexual dimorphism, resulting in larger males with an aggressive disposition. This, in turn, would select for solid bones in skull and body, a robust body, and display organs to intimidate opponents; it would select for heavy brows and bony heads capable of absorbing blows. This process would, of course, also produce a large amount of play fighting and displays among the bachelor males, producing the physiological stimulus to the development of strong bones and strong muscles. However, these are not the only selection forces for a robustly built body in the male.

Since males are now largely peripheral to groups, they are most likely to spot prey and hunt it, largely because they are excluded from choice foraging areas. Now a refined ability to spot potential prey and a silent approach to surprise the prey, plus the ability to quickly dispatch it, are all at a premium. Here appears to lie the evolutionary origin of the structure of our legs and feet, and our superb ability to stalk. The striking feature about our legs is not that they support the body on two pillars but that we can balance our body on *one* leg for long periods of time, and that our feet are equipped with soft volar pads laden with tactile receptors.

To someone who has stalked with naked feet, the comment of Campbell (1974, p. 171) that the human foot is "unnecessarily well endowed with tactile receptors" is as surprising as the one that "during human evolution its (the foot's) functions have been reduced from three to one, that of locomotion." To an ungulate biologist who has spent years in the close presence of these animals, those statements are a revelation that anthropologists are apparently not aware that human beings are not only superb stalkers but probably also the only carnivore capable of stalking virtually within arm's reach of an ungulate. Far from having been reduced to the function of locomotion, our feet have evolved into instruments of silent approach to permit the capture of a prey unaware of the hunter. The soft volar pads, in combination with great tactile sensitivity, permit the foot to locate itself carefully on the ground, avoiding rustling leaves, dry grasses or small branches, before the weight of the body is transferred onto the foot. This also requires a superb control over muscular legs, for the body must be balanced on one leg during the process of silently placing the other leg. The sensitive toes and soles free the eyes so that they can monitor the actions of the prey, as well as search for the best passage toward the prey. The balancing ability is critical for freezing the body instantly into whatever position it is in if the prey glances up and scans an area. During the stalk, while the foot finds a spot without noisy debris, the hands are free to grasp small branches or grass blades and guide them past the body to reduce noise to a minimum. Our feet lost the old primate function of grasping organs good for arboreal life, but they gained new, entirely human, functions, namely the ability to balance the body superbly on one leg and to move the body noiselessly over the ground despite dry grass, twigs, and leaves, in a manner probably unequalled by any other carnivore. Here appear to lie the evolutionary origins of our most unusual legs and feet, not only in the economic deployment of legs during walking. The human leg and foot appear to be a product of systematic hunting.

Although other carnivores, such as the large cats (Schaller 1972, Hornocker 1970), are also stalkers, they possess a technique not practiced by humans. This is a final rush over some distance that rapidly closes the gap between predator and prey. The human form of stalking brings the stalker literally within arm's reach of its prey. It is for this reason, because we lack the lightning-fast rush at the end of the stalk, that sophisticated silent stalking is at a premium.

The third requirement, after an ability to stalk and to find and spot potential prey, is to dispatch it quickly. Since almost everything bites, stings, or pecks, it is not wise nor healthy to grasp prey with bare hands. Also, young animals may scream and attract the attention of aggressive adults who may rush in to chase and injure the predator. It is advantageous to kill instantly. This can sometimes be done by smashing small prey on the ground, as chimpanzees do (Goodall 1971). However, to the earliest of hominids this was not a possible option. They had weak skeletons and muscles, quite unlike chimpanzees, a product of passive competition compared to the requirements faced by the kin-selected chimp; this was discussed in Chapter 10. The small-bodied, weak australopithecines and early *Homo habilis* could not have swung young baboons through the air in the easy manner of the powerfully built chimpanzee male. The quickest means of incapacitating prey is to stun, not to pierce it. Therefore, a club or stone is the ideal tool for dispatching prey, provided the club can be struck downward very accurately. Note that using a club or stone after stalking close to the prey requires that the hunter pick up the club or stone *before* he commences stalking. He must use foresight. If suitable weapons are not to be found handily prior to every stalk, there is a premium on bonding with the tool so that it is carried about ready for action when the demand requires it. Therefore, systematic hunting cannot be practiced efficiently without tool use and bonding with a tool. This, in turn, predicts that if stones are being used as tools they may well be carried long distances from their place of origin—as is indeed found for the earliest of African sites (Merrick and Merrick 1976) as well as for Acheulean sites (Isaac 1968). Furthermore, small clubs and strong arms and bodies needed to dispatch small to medium-sized ungulates would also be used against conspecifics in disputes. That would select for hard skulls and very robust bones, such as we ultimately find in *Homo erectus*. Small clubs, however, are not likely to be of any use to the early forms of *Homo* in dispatching large prey such as the robust australopithecine. Therefore, it does not surprise us that *Homo habilis* and *Australopithecus boisei* remain side by side in the early Pleistocene record of Africa. The robust hominid probably disappears when *Homo erectus* appears with a new technology to kill large mammals, a view independently expressed by Howells (1973a) and Geist (1975a).

The new technology for systematic hunting is the spear. It makes its appearance in the form of fire-hardened spears of yew, such as the specimen from Clacton-on-Sea, Essex, from gravels of Cromerian age (Oakley 1949, Clark 1967). This would make this spear at least 700,000 years old, falling into the age of *Homo erectus*. A second yew specimen was found, according to Oakley (1949, p. 23), together with a skeleton of *Elaphus antiquus* at Lehringen, Saxony. In addition, one finds "spoke-shavers" in archeological deposits of that age. We must note, however, that there is a scarcity of evidence for fire in African Acheulean and Oldowan sites. Therefore, one may doubt the frequent occurrence of spears in the African Acheulean period. Nor am I aware of "spoke-shavers" being common in African deposits; here an enquiry by a specialist is needed.

The simple pointed pole used as a spear is a problematic weapon. It does not kill quickly, although it does hinder a wounded quarry. Therefore, it can be used only under rather specific conditions. It allows large-bodied prey to be killed where a simple concussion weapon would not do it. However, once a prey is pierced, the problem of escaping from an enraged still-capable quarry has to be solved. The hunter must evade the attacks of his prey by skillful maneouvers or, with the aid of companions, deflect the attention of the prey at critical points in time so that a hard-pressed hunter or companion can escape. In both cases, extreme agility and an ability to evade attacks are called for, and these call for great muscular power. Furthermore, robustness of the body is also an asset in surviving blows delivered by the prey or the joint-wrenching forces generated by the maneouvers of the hunter. If the hunter is as speedy or speedier than his prey, he is quickly out of danger after fatally stabbing a prey, for he can wait at a distance until the spear thrust has finally incapacitated the prey. Spear hunting for large prey indicates cooperative hunting by several males, for there is little profit in engaging in so dangerous an activity alone, because there would be no companions to distract the wounded prey, the amount of meat might be in excess of his own and his family's needs, and the meat not consumed would likely spoil quickly. Thus, spear hunting implies group hunting, at least rudimentary cooperation, and food sharing. However, it does not call for a great development of these traits as long as gathering, rather than hunting, is the mainstay of the species, and in warm and even temperate climates gathering is most likely to remain of prime importance. The law of least effort in maintenance activities still applies. No effort must be expended in hunting if the same gain can be had in an alternative manner at a lower cost.

The idea that *Homo* progressed from hunting with weapons causing concussion and instant disabling of prey to weapons causing internal organ damage, hemorrhage, and relatively slow death predicts an expanding range of species captured from the early *Homo habilis* to the later *Homo erectus* stages. Data indicating such a progression are only available from early to late Acheulean cultures (Clark 1970). The early Acheulean is most likely to have been produced by *H. erectus*. The late Acheulean, however, already is associated with *Homo sapiens* who, as will be shown later, is expected to be a superior hunter.

For the African scene, one does not need to invoke more than the simplest, least dangerous, forms of hunting as a means of fulfilling the protein requirement. As pointed out earlier, scavenging and the general availability of carcasses of deceased large game, plus the victims of wildfires, could produce a significant amount of meat. The diversity of large mammals during the Pleistocene deserves some attention. Guilday (1967), in speaking of large mammal faunas of the present and past, points out the high ratio of large to small mammals on the plains of eastern Africa today. Thus, according to Asdell (cited by Guilday 1967), there are 19 species of large mammals and 18 species of small mammals in Zambia. In the early Pleistocene, the ratio must have been slanted even more in favor of large mammals, given the presence of at least a dozen or more species of pigs and at least 15 additional species of alcelaphines among the antelope (Cooke 1972), to say nothing of giant baboons, sivatheres, giant bovines, chalicotheres, dinotheres, and mastodonts. If today's savannahs in tropical and subtropical Africa can produce an almost incredible standing crop of biomass in the form of large herbivores (see Bourlière 1965, Talbot and Talbot 1963, Bourlière and Hadley 1970, Foster and Coe 1968, Watt 1973), it must have been startling in the days before evolution of modern man. It is this diversity and high productivity of large mammals that had important implications for the ecology of early hominids. Not only were the opportunities for scavenging great, but also the opportunities for the systematic hunting of neonates. Hunting of neonates, however, leads naturally to the first form of cooperative hunting, confrontation hunting, in which some hunters mob the defending mother while others kill the defenceless young. Confrontation hunting will be elaborated on later (Chapter 13).

Roots of Language

The basis of language is the ability by individuals to mimic sounds accurately; it is also the basis of music. From recent experiments on chimpanzees we are aware that these animals possess cognitive abilities essential to language (Gardner and Gardner 1969, Bronowski and Bellugi 1970, Premack 1976). However, unlike modern man, chimps do not possess a vocal apparatus suitable for spoken language nor the neural capability to learn to imitate vocalizations. What we must decipher is, under what circumstances was the ability to mimic sounds so crucial to early hominids that it affected their reproductive fitness. We must conceive of situations, derived from the adaptive syndromes of different human forms, when the abilities of vocal adjustment were exhausted. Only then, when all individuals are under the dictate to adjust vocalization, are the differences between individuals largely a function of genetic potential and natural selection can proceed to alter the gene pool.

We note that chimpanzees possess a wide variety of sounds, although not the ability to imitate sounds (Goodall 1971). We note that cercopithecines from open habitats are relatively noisy (Jolly 1972) and have a vocal predator warning system that already denotes the kind of predator seen (Altmann 1967). We know from Kortland's (1967) work that chimpanzees can band together in the open and mob a predator, while screaming and using tools such as branches to create a loud din. If we place these attributes shown by subhuman primates within the adaptive syndrome of early *Homo* as a form living on grassy plains, often far from trees to which they could escape, we have one situation in which vocal mimicry would be stretched to capacity—this is when confronting predators.

Here the ability to mimic low loud sounds [which correlate with large body size (Eisenberg 1976)], together with imitation of the rhythms, tonal variations, and range of sounds specific to each predator species, would parasitize each predator species' tendency to withdraw from dangerous conspecifics. A group of early *Homo* individuals roaring in unison, choosing their sound specifically to fit predator and situation, would be overwhelming in signaling stimuli that frightened predators. There could be no evolutionary defense by predators against these sound-mimicking hominids. Any predator who ignored the danger signals emitted by the hominids would probably not be deterred by similar signals from conspecifics. This would greatly increase the predator's chances of attacking conspecifics and being wounded with the certain loss of reproductive fitness. Any species of predators that shifts its vocal threat signals would only experience that hominids mimic its new vocal threat signals. However, once we grant even a crude ability to mimic sounds, it is but a tiny step to signal the appearance of a predator by imitating its voice. This would be iconic communication, and clearly the beginning of language; chimpanzees can learn to do this, albeit with visual signals (Gardner and Gardner 1969, Bronowski and Bellugi 1970). One additional small step would be to generalize the ability to link specific sounds to specific objects in the environment. Since early *Homo* did make tools and, granted the validity of the earlier arguments about the relationship of intellect to tool use, then individuals of *Homo habilis* and *H. erectus* must have had a good ability to make conceptual representations of the world about them. Linking specific sounds to specific objects fulfills the requirements for an agrammatical primitive language, as postulated by Mattingly (1972).

At this point it is of interest that among birds that mimic sounds, some bower birds include predator calls in their repertoire and mimic predator calls, particularly if disturbed (Robinson 1975). One wonders whether the mechanism proposed above would apply in some form to bower birds as well.

The above explanation leaves us still a long way from singing or yodeling and, above all, from actual language use. What we do have is an explanation of when natural selection would select sharply for vocal mimicry, which is consistent with the adaptive syndrome of early *Homo* as elaborated so far, which cannot be fitted within that of *Australopithecus* and is obviously not present in *Pan*. The prediction arising from the concept is that in *Homo habilis* already, but not in the gracile *Australopithecus,* there ought to be some anatomical evidence of vocal apparatus similar to that in modern man.

The first step toward a specific, iconic, verbal communication system is quite conceivable in *Homo habilis.* A generalized ability to link sounds to a spectrum of objects and actions is unlikely to have developed without an environmental dictate to cooperate in complex tasks. If the evidence for spears in hunting is also evidence for a primitive type of confrontation hunting in which the safety of individuals depended on selecting only that prey, in those circumstances in which a kill could be made safely, and the safety of hunters depended on cooperation, then indeed selection would favor sophistication of an iconic vocal communication system into a crude language. In tackling a dangerous prey, ambiguity or misinformation of intent is suicidal. However, the beginnings of confrontation hunting cannot be postulated until ecological conditions made stored subsurface plant food scarce periodically, forcing hominids to tackle dangerous prey that did not run away from the hunters. For this we must go beyond the warm African steppe to the cool winters of temperate zones and forest environments.

Geographic Dispersal and the Evolution of *Homo erectus:* Applying the Dispersal Theory

During the ice ages, mammalian lineages that radiated from the warm climates to the cold ones tended to increase in body size. As a rule, mammalian evolution progresses from the warm, moist, tropical forests to habitats of increasing aridity and cold and reduced productivity, as discussed earlier. The very same trend is evident for our own genus, *Homo*.

The process by which the increase in body size takes place, with its concomitant ecological and social specializations, has been in part described by the "dispersal theory" (Geist 1971b), and is discussed in several of my publications (Geist 1966a, 1971a,b, 1974a,d). This subchapter deals with an elaboration of this theory and shows how it applies to our own genus.

The term "dispersal theory" was chosen because an increase in body size and social adaptation follows a species' dispersal into unoccupied habitat. As discussed in Chapters 1 and 6, a "dispersal phenotype" develops typical of colonizing individuals. Increases in body size and social organs may take place without noticeable ecological adaptations if the species is subject to local extinctions and dispersal due to glacial advances and retreats, as is illustrated by northern urgulates. In tropical and subtropical regions, however, the dispersal theory predicts dispersal and concomitant social specialization following ecological specialization. It predicts the geographic pattern of radiation found by Darlington (1957). The dispersal theory thus unites the divergent propositions of Matthew (1915) and Darlington (1957) on evolution and zoogeography, the former being valid for arctic and subarctic conditions, the latter for the tropics and subtropics (Geist 1971b).

Not only do species change in body size and social adaptation as a result of dispersal, they may also be subject to selection, contrasting them from closely related species, and they may "neotenize." Neotenization is apparently a consequence of selection for enhanced and plastic body growth during dispersal; it results indirectly in a reduction of secondary sexual characteristics, and morphological and behavioral ones, and in a concomitant retention of juvenile attributes so well illustrated in the evolution of caprids (Geist 1971a).

The dispersal theory predicts that evolutionary change is linked to zoogeographic distribution. Therefore, zoogeographic patterns become a tool in deciphering the evolutionary history of a taxon. A species quite different from the parental one may be found at the end points of geographic dispersal. The dispersal theory predicts that a species will experience its maximum body size after first arriving in a given region. Then the body size will decline as the maintenance phenotype develops and selection continues to refine the properties of the maintenance phenotyps as its hold strategy. It follows that it is futile to look for an ancestor of an advanced form in the same localities (except where the latter returned and exterminated the former). Note that the dispersal theory contradicts the notion of a species evolving *in situ* from primitive populations into a distinct higher form; evolution to a higher form is conceived as taking place only through a new dispersal. However, some evolution for refinement of a given adaptive strategy will occur after dispersal and colonization in the maintenance population.

Now let us consider the mechanisms operating if evolution takes place during dispersal of a species.

(1) During dispersal into unoccupied habitat, the population experiences a superabundance of energy and nutrients that can be used for growth, development, and reproduction. A new phenotype, the "dispersal phenotype," arises subsequently, a phenotype capable of dealing with new problems. The superabundance arises for several reasons:

   (1a) The dispersing animals are freed from interspecific competition for resources, so individuals can concentrate not only on ingesting food to the limits of their needs, but also on maximizing the quality of the forage ingested; they can be choosy and eat nothing but the most nutritious food.

(1b) The dispersing animals are freed from pathogens and parasites that in their place of origin had kept them in check or at least had increased their cost of living substantially (Elton 1958); this saving in maintenance cost can be reinvested in growth and reproduction; however, they may also meet new pathogens, in which case they are subject to rapid selection on the basis of their immune system; some pathogens and parasites may stop a dispersing population in its track (Elton 1958).

(1c) Because mammalian dispersal patterns have inevitably run from the tropics to the dry or cold climates, dispersing individuals were exposed both to pathogens and parasites they had not been infected by, yet to increasingly fewer parasites and pathogens, for cold, dry climates compared to tropical and subtropical ones have relatively fewer parasites; thus, dispersing populations had on average a lower cost of adapting to new pathogens and parasites than did their parent populations; the cost saved could be invested in growth and reproduction.

(1d) The cost of defence against predators might also decline, first because dispersing from the tropics to the deserts or arctic zones generally entails a dispersal into zones of lower productivity and thus lower density and diversity of predators, second because the new predators are not likely to be adept at handling the new colonizers—of course, this also works the other way: The colonizers are not adapted to escape the new predators and may readily become victims; that would stop dispersal. However, on average, dispersal from the tropics to dry or cold climates means a permanent escape from many pathogens and predators.

For the very same reasons, however, a return of species from the dry, cold zones to the tropics is very difficult. In fact, man's return to the tropics after adapting to cold zones may be considered exceptional among mammals.

(2) The superabundance of resources for growth, development, and reproduction in the colonizing population leads to an increase in the birth rate, in the viability of neonates, in production of milk, and in the growth of young, and it also probably leads to a sex ratio bias in favor of males, according to the arguments of Trivers and Willard (1973). Evidence for the preceding statement is reviewed in Geist (1971a). The superabundance of energy and nutrients, therefore, leads to enhanced exploration and physical activity of the young, permitting body tissues of low growth priority to develop maximally. Note that this leads to a phenotypically large cerebral cortex, which has a low growth priority relative to the subcortical elements of the brain (Timiras 1972). The shift in morphology and behavior of the colonizing individuals, however, is such that it maximizes the colonizing population's potential for adapting to diverse environments encountered during dispersal, and enhances the dispersal process itself by virtue of vigorous, inquisitive, roaming individuals of exceptional physical fitness and plasticity in growth, learning, and "self-discipline." Increased roaming ought to increase heterosis and rapid spread of adaptive mutations or gene combinations. Increased isoenzyme formation increases an individual's adaptability, as does enhanced heterosis, making it less sensitive to the vagaries of the

environment than is the case for homozygous inbred individuals (Mayr 1966, p. 299).

(3)   The increase in the body size and vigor of the males leads to a selection for improved combat organs, tactics and strategies, and improved dominance displays and social organs. The enhanced social play by juveniles, associated with high population quality, would develop individuals with exceptionally diverse social strategies, leading to the possibility of environmentally maximizing entirely new social strategies, which in turn would set genetic selection in motion. Moreover, the behavioral vigor of the young would probably in itself be a stimulus to faster growth and development. This can be inferred from experiments showing that mild stress enhances growth and maturation (Vernadakis et al 1967, Petropoulos et al 1968, Levitsky and Barnes 1972). Moreover, if the tendency of males to disperse is greater than that of females, then it will lead to an unbalanced sex ratio at the population's dispersing fringe. This ought to increase male competition.

An increase in the body size of the females leads to greater milk production and therefore a greater phenotypic development of the young and their earlier maturation. It would also reduce the mortality of the young, except in the case of young too large to pass through the vaginal canal; in such a case, it could lead to the death of the young and the female. Within limits, abundance of energy and nutrients during gestation leads to a quicker delivery and a quick recovery of the female from giving birth (see discussion under Population Quality in Chapter 6).

(4)   Vigorous growth by males and their greater activity and fighting will lead to hastened senility, injuries, and a shortened male life expectancy; this also applies to females, owing to the heavy reproduction and great physiological demands this places on individuals (Geist 1966c, 1971a).

(5)   Owing to the high reproductive rates, short-lived but large-bodied socially skilled individuals will be replaced quickly as breeders by equal and superior offspring.

(6)   Because there would be selection for large, vigorous short-lived males over small long-lived ones, there would be selection for physiological mechanisms that enhance body growth. One such mechanism is to keep the animal "young" or neotenous over a longer period of time; growth is a function of "juvenileness." This would, of course, drag juvenile features of the animal into adult life so that the adult of the evolved population would look and act more "juvenile" than the adult from a primitive population. This concept explains, at least in part, the neotenization in man discussed by Montague (1960); this author gives a good historic account of the concepts of neoteny, paedogenesis, gerontomorphism, and fetalization, concepts developed in their significant implications by Garstang (1922). However, as Montague (1960) rightly stressed, neoteny is too simple an explanation for the evolution of large brain size in man; it ignores the relative enlargements of the temporal, occipital, and frontal lobes of the brain in human evolution.

(7)   Neotenization is adaptive to individuals in the dispersing fringe of the population, not only because it increases body size and weapon size, which is advantageous in intraspecific agonistic interactions, but also because it retains the plasticity and adaptability of the juvenile. Thus, juveniles hold

the potential to become diverse phenotypic forms by adjusting to the environment they encounter. This plasticity is lost or sharply reduced after ontogenesis. Thus, the more neotenous an individual the better its chances to adapt to environments outside the norms encountered by the parent populations. The greater the delay of maturation or neotenization, the greater the chances of adapting to changing environments before breeding. Thus, in sexually dimorphic ungulate males, for instance, with a one-shot breeding strategy, it is adaptive to remain plastic in body growth right up to the time of becoming effective breeders.

This adaptive consequence of neotenization I failed to realize when writing my earlier works. It is most significant in that it appears to be a mechanism generating great adaptability in adulthood. Hence, the very long ontogenetic period of man would be generated through neotenization since, by virtue of neoteny, the sexually mature individual retains juvenilelike plasticity in his mental, physiological, and growth processes. That would produce the great ability of man to change morphologically in response to behaviors dictated by the environment, and this very plasticity is also the cause of consternation among those concerned with physical fitness and health because, in the absence of demand, our body quickly adjusts physically so that, instead of remaining in fine physical shape once that state has been achieved, it quickly reverts to lower physical development of muscle tissues once athletic activity terminated.

(8)   Neotenization sets back the individual in maturity from the colonizing population compared to an adult from the parent population. This applies not only to secondary sexual characteristics, but also to the behavior—and probably the physiology—of the individuals; it enhances overt aggression and sexuality at the expense of dominance displays, and creates—in conjunction with greater plasticity—new opportunities for adaptation and, thus, radiation into new ecological niches (Geist 1966a, 1971a). This is a postulate probably valid for mammals as a whole. I recognized relatively late that I had described for mammals exactly what Garstang (1922) had described as the mechanism generating adaptive radiation in invertebrates. Indeed, to use Garstang's words in a slightly rephrased version, ontogeny does generate phylogeny.

(9)   Dispersal by a few individuals leads to a number of effects on the genetic constitution of the colonizing population, quite independently of the genetic effect of r-selection which has mainly concerned us so far. It leads to an alteration of the gene pool typical of the parent population owing to the following effects.

(9a) *Founder effects,* the accident by which individuals leave the parent population and commence colonization, influence its relative frequency of genes in the gene pool, at least early in the colonizing population's career.

(9b) *Heterozygous advantage* expresses itself at several levels and alters the gene pool. First, owing to inbreeding, a large number of lethal or near-lethal genes are exposed, as well as genes leading to congenital deformities and low vigor and physical and mental development (Wilson 1975, p. 79). These genes are quickly excluded from the gene pool unless they are retained by some heterozygous advantage. We can expect a good many genes to be scrubbed from the gene pool during the early inbreeding stages of the colonizing population. Second, because vigor and large body size are adaptive, we can expect selection for *heterosis* itself, which in turn maintains a lot of neutral genes in the population's gene pool. We have reason to believe that selection for heterosis would take place, provided the findings of Craig and Baruth (1965) that dominance rank and heterosis are closely related in domestic fowl can be generalized. Third, heterozygocity in some cases preadapts individuals to environmental contingencies, as described earlier in this chapter (see also Mayr 1966). Selection for heterosis may also be enhanced if females preferentially select large, well-developed males. Thus, selection for heterosis and heterozygous advantage would conserve genes of the parent population; inbreeding would quickly eliminate deleterious genes from the population and would probably also lead to an accidental loss of others.

(9c) *Genetic drift*, or sampling effect, would lead to a loss or a fixation of genes in the early part of the colonizing population's history, as long as the population was still small.

(9d) Because the colonizing population experiences rather intensive selection pressure for increased body size, combat potential, high reproductive rates, etc., we expect *phenodeviants* to occur during intensive selection for specific traits (Wilson 1975, p. 72). This means that maladaptive genes, normally not induced into activity, are now exposed and removed. Intensive selection for specific traits, just as much as inbreeding, thus cleans deleterious genes from the population's gene pool. Since intense selection may be the product not only of r-selection but also of new environmental contingencies as the population's colonizing fringe spreads into new environments, the chance of phenodeviants appearing and being selected against increases. Should phenodeviants be the product of rare gene combinations, the affected genes will be reduced in frequency in the gene pool owing to reproductive wastage entailed by their bearers. However, such cytological mechanisms as chromosome fusion appear to be adaptations that reduce the frequency of phenodeviants; therefore, during dispersal, a species' chromosome number may shrink.

(10) Once a population has colonized a previously unoccupied habitat, by definition it has developed effective adaptations dealing with differences between its present environment and that of the parent population. Thus, just prior to the population reaching carrying capacity on the colonized habitat, its genome has been subject to change owing to r-selection, selection for heterosis, gene loss or fixation consequent upon inbreeding, and intense selection for specific traits, as well as to the founder effect and genetic drift, and to adaptations to new contingencies of the physical and biotic environment. It is this gene pool that becomes subject to K-selection once the population reaches the limit of its carrying capacity.

(11) At carrying capacity, K-selection reverses selection typical of r-selection; however, the population retains its gene pool. Under conditions of reduced resources, phenotypic development is reduced, the expression of the genes is no longer maximized, and the development of any one characteristic

reflects less the genetic potential than the environmental constraints. This process, given a heterogenous environment, randomizes selection for genes that maximized certain characteristics during r-selection. Therefore, the population, even during K-selection, holds onto its new genome. In addition, certain traits evolved during dispersal continue to be adaptive under K-selection; hence, the population retains such diagnostic features. Consequently, once carrying capacity is reached, the population freezes in its evolutionary advancement. Thus, a species dispersing across a continent may leave a string of populations that vary clinally from the origin to the end point of dispersal (Geist 1971a,b), provided its mechanisms are such that once under K-selection few individuals disperse and there is little genetic exchange between populations. We find such conditions in gregarious ungulates, for instance, in which the social system binds individuals to the exploitation of specific patches of habitat through a migratory tradition (Geist 1971a). Since K-selection only reverses r-selection, and even here it does so incompletely, the newly established colonizing population retains its distinct genetic make-up and is quite readily differentiated from the parent population.

However, under maintenance or K-selection, there is also competition among tissues and internal organs for precious nutrients. Therefore, tissues above the minimum size required for efficient maintenance are grown at the expense of resources needed for reproduction. When resources are scarce, as under maintenance conditions, excessive growth reduces the reproductive fitness of individuals. Therefore, there will be a gradual readjustment of growth priorities among tissues in the population under maintenance conditions through natural selection. This is a case of evolution under conditions of resource shortage. Such changes in species over geologic time as the slow increase or decrease of tooth sizes and tooth rows, as well as reductions in body sizes, are explained by this selection.

(12) Once an initial genetic increase in body size has taken place in the colonizing population, the increased size may be maintained by natural selection in two ways. Firstly, by virtue of large body size, the species has become preadapted to exploit somewhat coarser, less digestible forage. Bell (1971) has shown, according to the principles of bioenergetics, that large-bodied ungulates can live off forage of relatively low digestibility owing to their relatively lower metabolism. Slow digestion of coarse forage, therefore, still supplies adequate energy and nutrients for the growth and reproduction of large-bodied forms.

Secondly, if the species adapts to a less mature, less stable ecosystem in the process of dispersal, an ecosystem in which the annual pulse of free energy and nutrients is greater than in the more stable ecosystems of the parent population, then large body size can be maintained, for abundant resources will be available for growth. The dry steppe adapted to by *Homo* is less mature and complex than the more stable savannah occupied by *Australopithecus africanus*. Thus, after enlarging in body size during dispersal, *Homo* could maintain a larger body size than *A. africanus* because of the greater amount of food available for growth and lactation during the short intense vegetative season of the steppe. This argument is based on the concepts of Margalef (1963). We may note that the above hypothesis correctly predicts a small body stature for modern hunter—gatherers who do not experience a marked, predictable, seasonal pulse of superabundance in food. We find such hunter—gatherers in tropical rain forests and deserts. The clinal variation with latitude in the body size of humans, so that the smallest people are generally found in the moist tropics and subtropics while the largest generally live in arid or cold continental climates (Lasker 1969), may be a function of seasonal superabundance of forage. It should be found that the less stable the environment is in which man lives, generally, the larger people should be in body stature.

Could the increase in body size during dispersal not be due to taller individuals dispersing more readily than smaller ones? To some extent the increase in body size could be explained by this; a number of studies of modern man have shown that emigrants are indeed taller than those that stay behind (Lasker 1969). However, body size and vitality are also a function of environmental quality, as I have pointed out in Chapter 6. Thus, the tall dispersers were quite likely to have been environmentally favored individuals and not only tall by virtue of their genes.

(13) Under continuous K-selection, there is likely to be a reproductive advantage for adults that can maximize reproductive fitness at a minimum body size. We therefore expect some selection against individuals who are large, provided the maintenance cost of such individuals is high and their energy expenditures wasteful relative to their reproductive success. There is some evidence for this in natives of the highlands of New Guinea (Howells 1973b, pp. 173—4). In fluctuating environments with marginal habitats for a species, longevity may be selected for, since it permits individuals to take advantage of rare unpredictable opportunities to reproduce. This would select indirectly for slow ontogenetic development and small body size. Hence, the average body size of a species will decline after colonization, leading to the phenomenon that a species reaches its largest body size during and after it colonizes an area; thus individuals of a given species may be exceptionally large and well developed in older, rather than younger, fossil beds. This phenomenon is illustrated by a number of mammalian lineages (Geist 1971b).

(14) If the habitat continues to degenerate or shrink after K-selection has set in, and the effective population size continues to decline, reproductive wastage may again be entailed due to inbreeding. Because habitat variability is expected to decline with shrinking habitat size, so is the advantage of phenotypic plasticity. Here phenotype redundancy or canalization would be adaptive, but this is reduced severely during r-selection and dispersal into heterogenous habitats. Since phenotype redundancy permits high reproductive fitness in a stable environment, it permits a lower viable population size for species from stable environments than for animals that have been selected for plasticity. Plasticity of phenotypic development is gained only at the expense of phenotype redundancy, and vice versa. This predicts that tropical and subtropical species of mammals can maintain lower population densities and breeding populations and still survive than can species of mammals from temperate or arctic regions.

(15) When habitat conditions have deteriorated severely we may expect populations to disperse. Thus, dispersal can occur when a population is subjected to excellent, as well as to poor, habitat conditions. Caprids, for instance,

disperse when habitat conditions deteriorate, such as during drought (e.g., Nievergelt 1966, Heptner et al 1961). Insects may even produce distinct dispersal phenotypes, as is shown best by swarming locusts (Wilson 1975, p. 83). Thus, dispersal may take place not only by subordinate animals unable to find a place in the breeding population (Christian 1970), but also by the best grown individuals, as is indicated from theoretical considerations of population quality and shown to be the case in voles from increasing populations (Krebs et al 1973), or in human populations (Lasker 1969), as well as by whole populations, as illustrated by insects or caprids. The more frequently a species experiences a decline in habitat conditions, the more likely is the evolution of the distinct dispersal phenotypes that arise when resources become scarce. How environments communicate to genes that resources are about to be depleted, for such communication it is, one can only speculate about (see Chapter 6.)

From the foregoing 15 considerations a number of conclusions follow, only some of which need to be mentioned here, while others are discussed in my earlier papers (Geist 1966a, 1971a,d, 1974a,b, 1977). It is evident, for instance, that a species is not likely to disperse and colonize without changing, and that a very different animal is likely to stand at the end of a line of geographic dispersal than at the beginning. Thus, we expect *Homo erectus* to be larger than the earlier *Homo habilis*, and also to be larger earlier in its history as a species rather than later, and we expect the enlargement of biological signals implying aggression. This may be the explanation of the steadily enlarging supraorbital tori in the genus *Homo*, excepting the very last form, ourselves. The supraorbital tori reach their largest size in Rhodesia man (Clark 1970).

We also note that the dispersal theory predicts convergent evolution of forms radiating simultaneously or sequentially from the same geographic center (Geist 1971a,b). Given the many pulsations of glaciers in the northern hemisphere, it is quite conceivable that several large-bodied and large-brained hominids could have radiated from South African hominid stock. Should major cataclysms destroy animal populations but only damage their base of subsistence (like magnetic reversals of the earth?), then the dispersal theory predicts a wave of giantism will follow.

The theory also predicts that tropical forms ought to be very resistant to change due to phenotype redundancy, and thus explains the long lag between dinosaur extinction and the first adaptive radiation of mammals (p. 177). On the other hand, it predicts the inability of tropical forms to adapt rapidly to environmental change and so predicts their ready demise upon destruction of the intricate ecosystems of the tropics. It also raises the distinct possibility that many of the primitive animals we find in the tropics are not just somewhat changed descendants of ancestral forms but are the actual ancestral forms themselves. It may be that the clouded leopard, for instance, *is* the ancestor of the saber-toothed cats which repeatedly evolved in Tertiary times (Thenius and Hofer 1960), just as the Sumatra rhino *is* the ancestor of the extinct woolly rhino lineage which culminated in the large woolly rhino (*Coelodonta*) of the ice ages. It may be that the precat is still wandering around in the form of the flatheaded cat (*Tetailurus*), a felid so primitive that it lacks fully retractable claws and still has large premolar teeth.

The dispersal theory predicts a close relationship between ecological and social specialization in mammals, so that ecological specialization followed by dispersal results in social specialization; dispersal without prior ecological specialization leads to social evolution only (Geist 1971b). The more frequent the dispersal, the greater the rate of evolution and also of extinction; hence the high rate of both in ice-age mammals.

One aspect of the dispersal theory is of some theoretical importance: It does *not* explain increases in body size due to *fundamental advantages* of large body size (Stanley 1973), some of which are assumed to be improved ability to ward off predators and capture prey, greater reproductive success, increased intelligence, better stamina, expanded size-range of acceptable food, decreased annual mortality, extended individual longevity, and increased bioenergetic efficiency, as enumerated by Stanley from a review of the work of various authors. There is a logical flaw here; if there were fundamental advantages in large body size, *all* species would increase in body size, and large-bodied rather than small-bodied forms would be the rule among organisms (van Valen 1973, Stanley 1973). There is no such thing as "fundamental advantages" of large body size save as constructs to somehow explain the evolution of large-bodied forms. What should be noted about the dispersal theory is that it explains initial increases in body size as a function of intraspecific *social selection* and not as being due to direct *ecological selection*.

The dispersal theory can be used to check on the view of Robinson (1972a) and Eckhart (1972) that the robust australopithecine, variously called *Paranthropus* or *Australopithecus boisei*, is a descendant of a gigantic australopithecine from Asia. There appear to be two such giants, *Gigantopithecus* and *Meganthropus*, of which the former may be a giant ape (Kurtén 1972). Howells (1973a) points to a parallelism in Africa and Asia where during the early Pleistocene on both continents a robust and a gracile hominid existed together. In Africa this would be the robust australopithecine together with the gracile one, or with *Homo habilis*; in Asia it would be *Homo erectus* in Java together with *Meganthropus*. The affinity of *Gigantopithecus* is problematic according to Corruccini (1975), who felt on the basis of multivariate analysis of its teeth that *Gigantopithecus* was some aberrant form.

Regardless of the controversy, if robust and gracile australopithecines dispersed from their evolutionary origin, Africa, to Asia, then the dispersal theory predicts an increase in body size for both Asian over both African forms. Had *Meganthropus* been the ancestor of *Paranthropus,* then we expect *Paranthropus* to exceed, not to trail, *Meganthropus* in size. Therefore, the view of Robinson (1972a) and Eckhart (1972) is not in harmony with the dispersal theory, and I doubt its validity. Their view, for instance, also implies that Asia is the evolutionary home of robust australopithecines, a view not supported by the fossil data.

Had *Homo* been a pure herbivore, there would be many species of this genus today. The diverse populations would in time be separated; they would adapt to different habitats and plants and, upon meeting one another, would probably form a hybrid zone and disperse no further, at least not until selection against hybrids had shaped them into separate species and competition forced them to diversify their niches. However, *Homo* was an omnivore with carnivorous tendencies, as well as cannibalistic ones. He would reinvade the areas occupied by the smaller-bodied, less aggressive, and more primitive populations and would displace and eradicate them. The evidence is in line with this interpretation, for we find a sudden replacement of australopithecines by *Homo erectus* (Howells 1973a) as well as australopithecine tools attributed to *Homo erectus*, plus good evidence for cannibalism (Blanc 1961, Clark 1967, Birdsell 1972). The above interpretation also predicts that, at any one time in our history once we became competent hunters, we will find populations at much the same stage of evolution throughout the geographic range of *Homo*. That is, we will not find small-bodied, primitive *A. africanus* living as a contemporary of large-bodied *H. erectus,* or *H. sapiens* living contemporaneously with *H. erectus*—unless, for some reason, *H. erectus* or *H. sapiens* could not reach a relict population of *A. africanus* or *H. erectus,* respectively. This may have happened in Java, where Solo Man is a *H. erectus* that survived long after the evolution of *H. sapiens* (Howells 1973a). The recent evaluation of early *Homo* remains from Africa by Tobias (1973) fits the above idea, as does much of the prehistory of *Homo* in Africa, as described by J. Desmond Clark (1970). The evidence for several species of hominids in late Pliocene Africa occupying an area sympatrically (Leakey 1976a,b) does not argue for advanced carnivory in *Homo habilis*.

The large-bodied, robust hominid called variously *Paranthropus* (Robinson 1972a,b), *Australopithecus robustus,* or *A. boisei* (Tobias 1973) appears to have lived for a long time contemporaneously with *H. habilis* and *H. erectus,* if the dates of its mid-Pleistocene extinction are correct (Birdsell 1972, Tobias 1973). This supports the view I discussed earlier that the early forms of *Homo* did not possess hunting weapons of any consequence and could not dispatch prey as large and capable as *Paranthropus*. This he could have done had he possessed spears, but such weapons are not found until the first major interglacial, the Cromer interglacial, in mid-Acheulean times. It is apparently at this time that the robust form of hominid disappears.

An Overview of *Homo erectus*

The foregoing subchapters illustrated some of the consequences of australopithecine stock adapting to dry steppe. A critical assumption has to be made: In adapting to steppe the most dependable food source was subsurface plant parts such as roots, bulbs, corms, etc. This assumption is critical, for it explains why not one but a whole series of human features arose from australopithecine ones. Thus, it explains the reduction in tooth size, the need for hunting, tool use, the change to sexual dimorphism, the human leg and foot, the increase in robustness of the skeleton and skull, the very large nucheal area on the head of early *Homo*, and—as will be described later—the development of a social system and social behavior that completely breaks with the primate tradition that *Australopithecus* must have possessed. Let us look at some of the above consequences of feeding on subsurface plant parts.

Tool use in obtaining roots from beneath hard subtropical soils is essential. The nutritional state of individuals and their reproductive fitness depend on their finding of economical means of locating and digging for hidden plant parts. It becomes at some times of the year a matter of no tools no food. One could not postulate such a situation for *Australopithecus,* living in savannahs where surface feeding within green vegetation was possible somewhere all year round.

Tubers, corms, bulbs, etc., are lower in fiber and softer than all but a small fraction of the plant biomass above ground. They require less well developed molars than are typical of *Australopithecus*. When energy and nutrients for growth are scarce, then selection will favor individuals who shifted less energy and nutrients into teeth and more into other essential organs. We expect such selection in populations living at carrying capacity, where individuals are maintenance phenotypes. Moreover, the scarcity of fires for early *Homo* indicates that food would be consumed raw.

The high carbohydrate and low protein content of roots would force individuals to search for high protein foods. Intense concentration on animal foods is the expected outcome. One method of obtaining it is by hunting. If the store of plant foods is dependable, there is time for the more difficult and less predictable process of hunting.

The first consequence of changing from opportunistic to systematic hunting is the evolution of the typical human foot and leg, together with the ability to stalk, which requires great self-discipline. It also requires foresight and the permanent availability of tools to quickly dispatch prey. Hunting probably began with weapons capable of stunning relatively small prey; only in later stages did the spear come into its own. This weapon is likely to be an indicator of a primitive form of cooperative hunting, confrontation hunting. It also suggests food sharing beyond the immediate family. Early *Homo* was not likely to be an efficient hunter and coexisted with robust australopithecines. With the advent of spears, the latter could and would become an easy prey, as would other mammals *Homo* had coexisted with in previous epochs.

A consequence of meat diet would be selection for a powerful nucheal musculature, large canines and incisors, and great physical strength, as a means of tearing meat into mouth-sized bits. This has been explained in Chapter 12 in detail for Neanderthal Man. The largest nucheal area and biggest occipital tori are reached in *Homo erectus,* who—unlike *H. sapiens*—ate meat largely without the benefit of fire or frost; *H. erectus* apparently killed rather large mammals whose meat is tougher than that of small ones.

A consequence of the requirement to dispatch prey quickly would be selection for powerful musculature and robust bones. There is a premium on agility, strength, and great accuracy in grasping and striking prey lest it injure the hunter. Thus, we observe a change from the fragile skeleton of the gracile australopithecine to the robust skeleton of *H. erectus*.

Patches of ground densely covered with plants, whose roots form a staple, the carcass of a prey, a water seep, small areas of evergreen vegetation, and their diversity of food can become defendable resources. They would become contested during the dry season, for they could be controlled and defended. This would select for sexual dimorphism and for anatomical evidence of aggression. Such is seen in enlarged supraorbital tori (functioning in threat and as protection against blows on the head) and increased sturdiness of the skull. Since by now the early hominids use weapons inflicting concussion, a rugged, reinforced, thick-boned skull, such as that of *Homo erectus*, is clearly an advantage.

However, once such cultural weapons come into use, the aggressor can escape retaliation and the normal feedback that curbs overt aggression breaks down (see Chapter 4). Cultural weapons can only be contained by cultural means. We do not know what these cultural means were, but we can suspect that in the absence of biological controls over aggressive behaviors, strangers with strange conventions would be quite readily attacked and killed. It would greatly help in the displacement of one group of early hominids by another, and it would readily lead to cannibalism.

The manner in which early hominids—or, for that matter, modern hunters and gatherers such as bushmen—escaped predation in a treeless plain is problematic. One method could have been banding together and mimicking in unison the vocalizations of predators. This would have resulted in a sharp selection for vocal mimicry, a diagnostic feature of humans and a prerequisite for language and music.

Successful adaptation to the cold temperatures at night in the open steppe opens the doors to adaptation to temperate environments and, therefore, the rapid spread of *Homo* from Africa to Europe and Asia. A consequence of dispersal is selection for a larger body and a larger brain, as explained by the dispersal theory.

The steady but slow improvement in hand axes, the gradual increase in brain size, and the slight decline in tooth size in *Homo erectus* are explainable as the results of natural selection within populations at carrying capacity. As discussed in Chapter 6, maintenance phenotypes arise after colonization when resource shortages owing to intraspecific competition begin to set in. Natural selection acts on maintenance phenotypes so as to readjust the growth priorities among organs. The absence of new tool developments resulting in the long duration of the Acheulean period, the slow improvement in workmanship, and the slight increase in brain size suggest that behavioral abilities in specific tasks, consistent from generation to generation, were very closely related to reproductive fitness. Inefficiencies incurred in experimentation were not tolerable, as they reduced fitness. This speaks very much of resource shortages and a strained life at carrying capacity exercising proven means of making a living—and those means only! It is only during the initial dispersal of *Homo,* when individuals of a dispersal phenotype must have been present, that a change from one distinct tool culture to another arose, namely, from the Oldowan to the early Acheulean. Therefore, a rapid spread of the early Acheulean from Africa to Europe and Asia took place.

The development of spears and the hunting of large, dangerous mammals suggests that during some periods of the year human populations lived in habitats devoid of plant food. Without a secure food base, hunting no longer simply augmented the protein deficiency of starchy plant food but became the sole source of food. Hunting would then tide the population over from one period with abundant food to the next. Only in dire need, and in the absence of alternative food sources, would so dangerous an activity as hunting large, dangerous mammals with spears be developed.

The idea that *Homo* arose from *Australopithecus* by adapting from the savannah to the steppe is thus a prerequisite for explaining the evolution and development of tool use; vocal mimicry; visual mimicry; systematic hunting; human legs, feet, hands, arms, and dentition; and the typical human social system.

Chapter 12

The First Advance to the Glaciers

Introduction

In this chapter we shall be concerned firstly with the transition of *Homo erectus* to *Homo sapiens,* and secondly with the rise and fall of Neanderthal man. In the preceding chapter, *H. erectus* was illustrated as a product of adaptation to the steppe but also able to exist in temperate climates during interglacial periods in Europe and China. From the earliest specimens of *H. erectus* known to its last representatives in the specimens from Choukoutien, or even Solo man, from China, a span of time passed that was possibly in excess of 1.5 million years (Howells 1973a). Only minor evolutionary changes occurred in that time span (Le Gros Clark 1964). There are indications that late *H. erectus* populations from Europe were spear hunters, which suggests that they had adapted to conditions in which for at least brief periods of the year the meat of large mammals was the only food. However, it is not at all certain that all hunters with late Acheulean hand-ax cultures were *H. erectus.* They could also have been *H. sapiens,* for the first representatives of this form in the fossil record are associated with late Acheulean tools (Howells 1973a).

I shall call people within the *erectus/sapiens* hiatus "late Acheulean" people. We are dealing here with a mode of life, an adaptive syndrome; this is in large part independent of the physical type of its practitioners. We appear to be dealing here with adaptation by cultural and, to some extent, biological means to the exploitation of temperate climates with cold winters.

Homo erectus, using analogies from lineages of large mammals, had two evolutionary paths open to him, to adapt to life in deserts of low latitude or to adapt to the temperate zones. I am not aware of any evidence that *H. erectus* ever adapted to hot deserts. He did, however, advance into temperate zones in Eurasia. The first part of this chapter deals with adaptation to cool temperate zones by late Acheulean people, whatever their taxonomic grade.

In treating the transition from *erectus* to *sapiens,* as well as the history of Neanderthal man, I shall use the model already alluded to in the preceding chapter: New human adaptations arise at the geographic fringes of a species where populations are in contact with adverse environments and where populations may be trapped in pockets of temporarily favorable landscape. Once a new form evolves, it disperses away from the origin of its evolution and may move into more physically benign habitats where it displaces more primitive predecessors. The driving force of human evolution is seen to be the pulses of glaciation, and human evolution ought to be most rapid where it contacts the temperate and periglacial zones. At present this model is hypothetical. The uncertainties of dating of human fossil and archeological remains, as well as the uncertainties surrounding glacial chronology, preclude a rigorous testing of the model. It has more to offer when we look at people from the last glaciation in succeeding chapters.

The story of Neanderthal man is that of a special way of exploiting cold periglacial ecosystems. It was the first truly successful colonization of the rich ecosystems created by glaciers. Here a detailed understanding of the biology and ecology of cold-adapted large mammals is brought into play in deciphering the adaptive syndrome of Neanderthal people. I shall make hesitant attempts—about which I do not feel comfortable at all—to tie the history of Neanderthal people into the great climatic changes of the glacial and interglacial periods. Social behavior I shall not discuss here but defer it to the next chapter.

Late Acheulean Man and Temperate Climates

The transition between *Homo erectus* and *Homo sapiens* appears to lie in mid-Pleistocene times in Europe, at least so the limited data in the fossil record suggest. At the time of Peking man in China, some 300,000 years ago, there may have been rather large-brained *Homo erectus/sapiens*-type individuals in Europe, as the descriptions of Thoma (1972) of the Vérteszöllös skull suggest. This individual lived in an interstadial during the Mindel glaciations. During the following Holstein interglacial, true but primitive *sapiens* individuals are encountered in Europe, exemplified by the Steinheim and Swanscombe skulls (Le Gros Clark 1964, Clark 1970, Howells 1973a). In recent years more findings have been made of human fossils in Europe from this general age, including some from the Riss glaciation, and they reveal a mixture of *erectus, neanderthalis,* and *sapiens* features, which is not really surprising in transition populations from *erectus* to *sapiens* (Howells 1973a). The next interglacial of Europe, the Eemian, contains people with neanderthaloid features and new stone cultures, termed pre-Mousterian (Howells 1973a). The Krapina people of that age are a rather small-bodied lot with slender bones, indicative of a hard life, as is the evidence of cannibalism (Coon 1962). In the preceding Riss glaciation, few tools appear in European deposits and those that do are of the late Acheulean type (Bordes 1972). The late Eemian and early Würm glaciation sees neanderthaloid people in Europe and around the Mediterranean basin with Lavallois−Mousterian tools, exemplified by the Jebel-Ighoud skulls (Howells 1973a), while in Europe the classical Neanderthals develop as the first Würm stadial advances.

It appears that late Acheulean people inhabited Europe primarily during interstadials and interglacials. However, late Acheulean hand axes, few in number but present, nevertheless, appear in deposits from the Riss glaciations (Bordes 1972). The process of adapting to cold climate was thus far along, even with Acheulean tools, but was not very successful, to judge from their frequency compared to those of Mousterian, and especially late Paleolithic, remains.

The major problems faced by a southern form in adapting to temperate conditions are the lower biological productivity of the land, the greater seasonality and reduced vegetation season, the great extremes in temperature, and the greater expenditures of energy required to maintain homeostasis and reproduce. When we look at the temperate form compared to subtropical ones, we note these problems reflected in their biology. We note that the large mammals lean more to migration, which means exploiting distant foraging areas at different appropriate times; they may hibernate; they have more fat and heavier hair; and they show a marked preference for favorable microclimates (Geist 1971a, 1974a,d). We infer from this that a hominid could adapt along similar lines, as well as by diversifying his technology as an aid to survival. On one hand, he could adapt to superabundance and food scarcity by depositing heavy fat layers, growing a thick hair coat as an aid to heat conservation, and becoming very astute in selecting sites to reduce heat loss; on the other hand, he could improve shelters, use fire, and change his food gathering methods to become more efficient and store surpluses of food for leaner times ahead. He could, therefore, adapt either biologically or culturally, or both.

For a very long time after the appearance and dispersal of *Homo erectus*, no hominid population apparently succeeded in adapting to cold winters. The process of realizing such an adaptation is not difficult to reconstruct.

Acheulean cultures are associated with the steppe (Clark 1970). Here wooded ravines may be found or wooded river valleys which provide diverse opportunities for shelter as well as wood for fires. Tree canopies give some protection from convective and radiative heat loss at night (Moen 1968). Woody vegetation provides branches to build crude, simple but effective, shelters which can easily be built without recourse to sophisticated tools. River terraces may contain southeast-facing caves, which not only provide a radiation shield and protection from wind but also catch and concentrate the rays of the rising sun in the morning, and during the night reradiate the heat absorbed by rocks during the day. They also provide splendid cover against rain and reduce the risk of hypothermia through evaporative and convective heat loss. Moreover, a cave will support a fire during a rainstorm, and while a fire is easily maintained in the open during even a severe rainstorm it is at the expense of burning a large amount of wood. In a cave one needs only a small fire, and anyway the cave concentrates the radiation of the fire to provide rather high surface temperatures. Wood could be gathered wet and dried in the cave and would always be dry and in reasonable shape for kindling whenever the family returned.

The presence of caves or shelters is more important since the tool kit of the Acheulean people reveals no evidence that they had invented clothing or the process of tanning hides. We may assume that there was selection for increasing the effectiveness of localized body hair as a means of insulation as well as for depositing fat layers around the body. Such a fat layer could be seasonally deposited, given food surpluses during fall, and could serve to supplement the meager food intake during the cool winter. Such a fat layer would be handy to have, and no great hindrance in the absence of cultural means of food preservation and the frequent need to move to new sites, be it because dry wood ran out or the food supply did. Dry wood, obtainable by hand only without axes to chop down dry trees and cut them into handy pieces, would almost certainly limit the value of any shelter, particularly if winters were cold.

It is evident from the foregoing that during the short cool wet winter, late Acheulean man could hardly have exploited the open steppe without recourse to wooded ravines. During this season he would have had to exploit the ecotone between forest and steppe, and become increasingly competent in exploiting the resources of forests. This explains why, for instance, advanced man spreading into Africa some 60,000 years ago settled not only in the steppe but also in the forest biome, which previously had not been occupied by humans (Clark 1970).

It is more likely that Acheulean man opted for biological storage of surplus resources in the form of fat on his body, vitamins in liver and fat, protein in muscle and connective tissues, and minerals in the bones rather than for storage of food *per se*. Food storage demands security above all, so that no competition can get at it, be it cave bear, lion, wolf, lynx, fox, badger, wild boar, rat, mouse, raven, crow, or magpie. It may also demand that someone be in attendance to watch over the food stores, thus requiring extended residence—and that, as we shall see, contradicts the adaptive syndrome of late Acheulean man. It also demands efficient storage methods to reduce spoilage, as there would be no point in investing precious energy in working hard to store food only to find it spoiled or poisonous when needed. Without going into detail, it is obvious that food storage is no simple matter, and that it requires considerable skill, knowledge, technology, and the ability to withstand very large and very dangerous animals to do it. The biological storage of excess energy and nutrients is easier and safer, but of course it also has its limits.

In order to store excess food, a ravenous appetite must be developed, as well as an ability to fast. Only the former leaves indirect evidence. Since lipogenesis (fat synthesis) is exceedingly wasteful (Blaxter 1960), more than two calories must be eaten for every one stored as fat. That requires a high rate of food procurement, either by eating food when it is superabundant in fall or by making it relatively more abundant by means of equipment. The latter is detectable in the form of new technologies. Such a change is detectable in the equipment used to scrape fire-hardened wooden spears, as well as in the spears themselves, which are found fossilized in river gravel or associated with animal skeletons (Clark 1967, Movius 1950), and also wooden tools such as throwing and digging sticks (Oakley 1961, Clark 1970). In addition, there is some evidence for wooden bowls, and rather well-made shelters containing hearths (Pfeiffer 1972) as a means of energy conservation. The wooden spears and spear scrapers imply enhanced systematic and confrontation hunting which should be reflected in a greater diversity of game killed, and in increase in the frequency with which large-bodied game is killed. Some late Acheulean sites, such as those of Torralba in Spain, imply cooperative hunting or, maybe, the exploitation of natural large-mammal traps. Indeed, late Acheulean remains are associated with remarkably large and dangerous creatures, such as elephants, suids, hippopotami, rhinos, bison, horses, and giant baboons (Clark 1967, Isaac 1968, Clark 1970). Thus, a change in the expected direction has taken place if we compare the evidence for tools and hunting between early and late Acheulean man.

The anatomical evidence is also in line with the concept that late Acheulean man lived in and capably handled environments more complex than did early Acheulean man. The skulls of Steinheim and Swanscombe man are essentially those of modern man; the same can be said for the limb bones. The major difference lies in the robustness of the bones (Le Gros Clark 1964). However, here we may be looking at a phenotypic characteristic reflecting the strenuous work done by late Acheulean people; the skeletons of late Paleolithic people, our direct ancestors, are also exceedingly robust (Vallois 1961). The major differences between *Homo erectus* and these early *Homo sapiens* lie in increased brain size and somewhat reduced tooth size (Howells 1973a). Using our present knowledge of what determines brain size phenotypically (p. 31) and how morphological change occurs only after behavioral and physiological means of adjustment are exhausted, I conclude that these late Acheulean people performed a greater diversity of physical and mental tasks than did *Homo erectus*. That, in turn, fits well with the increased diversity of tools, hunting success, and occupation of temperate climatic zones.

The way of life postulated here for late Acheulean man is that of a gatherer and systematic hunter who made occasional excursions into cooperative hunting, during fall and winter. The majority of his food still would have come from gathering, but hunting played a much more important role than it did for *Homo erectus*. Much of the time he would have lived in small, widely dispersed groups, each family depending on its own ability to make a living. Acheulean sites support this notion since they indicate only small groupings of individuals (Clark 1967, Birdsell 1968, Clark 1970).

The hypothesis advocated is that adaptation to cool climates by *Homo erectus* led to the evolution of early *Homo sapiens* as sketched above. During interglacials we can expect climates to have been not unlike today's. Early *Homo sapiens* lived during the Holstein, in Europe. With the advance of glaciations, we expect these hominids to disappear from the cold parts of Europe but increase in North Africa, for by now the desert was displaced to the south and forests and grasslands cover much of the former desert region. During maximum glaciation, the Gulf Stream probably hit the west coast of northern Africa, creating a warm mesic climate there. Thus, during glacial periods, we expect late Acheulean man to be thriving in North Africa as well as penetrating along the Nile and around the west coast to sub-Saharan Africa and dispersing to the Cape. Because late Acheulean man is a superior exploiter of the countryside compared to early Acheulean man, we also expect to find more sites of late than of early Acheulean man. Both this and the geographic expectation are fulfilled (Clark 1970, p. 89). As late Acheulean man in Europe used fire habitually, we expect late Acheulean man in Africa to have fire more frequently than early Acheulean man. This apparently is so (Clark 1970, p. 101). Moreover, the use of fire implies cooking and the breaking down of fiber in food; it would select—in populations at carrying capacity only!—for smaller teeth. This could explain the reduction in tooth size from *Homo erectus* to *H. sapiens*.

During the Eem interglacial following the long Riss glaciations, human populations again prospered in warm-temperate Europe. They probably originated from populations that lived in the Mediterranean basin during the glacial periods. At that time grasslands, and even cool climates at higher elevations, dominated the area, leading to relatively high ungulate biomasses and a greater emphasis on hunting. Not surprisingly, with the Eem interglacial we find a shift from the late Acheulean to a crude, archaic, Mousterian culture (Le Gros Clark 1964, Howells 1973a) with the emphasis on flakes as tools. Flakes permit a quicker, easier dismemberment of animal carcasses. Acheulean-type industries continued to dominate in Africa (Clark 1970). This is not surprising in view of the hypothesis that cultural and biological advances of *Homo* during his later history are due to adaptations to the cool and ultimately to the cold climates. In warm climates we expect cultural and biological stability once an adaptive syndrome is formed.

In European skulls from the Eem interglacial, we find strong Neanderthal features

(Howells 1973a), an indication that meat eating had advanced in importance. I shall describe the adaptive syndrome of Neanderthal man later and show how his features related functionally to his ecology. However, the skulls from the Mediterranean region of North Africa and the Near East indicate no strong progress toward Neanderthal features, and they display archaic, late Paleolithic features, plus ideosyncracies of their own. They have large crania, robust skulls, and large teeth, but rather modern faces, as exemplified by the Ighoud and Aafza skulls and also the Skhul skulls from later deposits. Howells (1973a) considers that these people could easily be direct descendants of people similar to those represented by the Steinheim skull. These people reflecting some unknown local adaptations did not advance far toward the Neanderthal syndrome and were less dependant on large mammals for food than Neanderthal man. Given their distance from major glaciations and cold environments, this proposition appears reasonable.

Cooperative Hunting

When a species has become adapted biologically or culturally to a given environment, there is no reason to change. The adaptations may be refined, but no major alterations can be expected until the environment changes sufficiently to override the normal range of adaptability provided by epigenetic mechanisms. We have no reason to expect a major change in the manner in which a species exploits its environment unless that species finds itself in an environment that is substantially altered.

The change to cooperative hunting from primarily systematic hunting was probably brought about by an environmental constellation of factors, against which man is powerless and to which he must adapt. In temperate climates, these factors are snow, hoarfrost, and the dry, crunchy leaves of deciduous trees. When fall and winter come, crunchy leaves, hoarfrost, and snow make stalking impossible. One can no longer creep up on a prey within a few arms' lengths. Regardless of how carefully one's feet are placed on hoarfrost, crunchy leaves, or even soft, powdery snow, one makes a very noticeable amount of noise which is easily and accurately detected by alert prey. Moreover, in cold weather one must wear clothing, and clothes are noisy during a stalk. (Even soft underwear makes audible noises when one stalks.) Thus, when cold and snow descend, systematic hunting by an individual is out of the question. Under such conditions only cooperative hunting can succeed. Cooperation is forced on hunters by the inescapable facts of the winter environment, noisy leaves, frost, and snow.

Full-fledged cooperative hunting requires a precursor. It is found in the rare cooperative hunts of chimpanzee males (Bygott 1972, Goodall 1971). We have reason to believe, therefore, that the roots of male cooperation in hunting are ancient, a conclusion of Tiger (1969) with which I concur. Cooperative hunting can be disproportionately successful compared to the hunting of single individuals, as Schaller (1972) found for lions. Although snow, hoarfrost, and the autumn leaf fall must have greatly increased the frequency of cooperative hunting in the affected populations and forced its perfection, some cooperative hunting must have been practiced by hominids earlier. Dry, tall savannah grass is not very conducive to stalking, nor is dry, short-grass steppe. Some cooperative hunting may have been practiced by *Homo erectus* during the dry seasons in the steppe; it was almost certainly practiced in temperate zones.

Schaller (1972) points out, on the basis of his studies of carnivores, that early hominids could have employed various techniques in hunting, such as relay races and the encirclement of prey, all without recourse to complex communication. Such hunting would have placed the day-active hominids into a vacant niche, since all present African carnivores are night and dusk hunters; the extinct saber-toothed and scimitar cats were probably also night active (Schaller and Lowther 1969). The question to be answered is, what would be available for capture by cooperative hunters in very small groups in the dry steppe that was not dangerous, not ill with disease and so dangerous to eat, and yet quite readily foiled into dying conveniently for the benefit of the hunters.

The very beginning of cooperative hunting probably goes back to the killing of young ungulates shortly after the rainy season commenced and with it the births of innumerable gazelles, antelopes, bovines, and pigs. The steppe and savannah, granted the high productivity of the low latitudes, must have been littered with various young ungulates. It is quite likely that males, well fed at this time, would readily cooperate to kill such young animals. The enhanced success would hardly have gone unnoticed, but there would be little development of cooperative hunting since female ungulates can and do interfere energetically with carnivores attacking their young (Kruuk 1972, Schaller 1972). Such attacks would in all likelihood scatter the male group, and those who persisted in staying behind faced grave consequences from sharp horns, tusks, and hooves. There is no need to hunt cooperatively with regularity and some sophistication unless the prey significantly contributes to the reproductive fitness of males and such a contribution outweighs the reduction in reproductive fitness suffered through injuries inflicted by the prey on the hunters: Only then will cooperative hunting be selected for. Moreover, it is essential that cooperating males all share in the kill, or cooperation cannot thrive. There is no point cooperating in the hunting of small-bodied animals, since the amount of meat per hunter is small, and the very lack of temporary superabundance of meat too readily leads to the stronger benefitting at the expense of the weaker. Thus, we would find single hunters mainly going after small-bodied animals, and cooperating hunters going after large-bodied animals. This rule holds for mammalian carnivores (Schaller 1972, p. 357) with very few exceptions, such as the mountain lion (Hornocker 1970). Therefore, for cooperative hunting to develop above the level practiced by chimpanzees, it is necessary for the prey to be large, or for many small-bodied prey to be killed quickly to produce a temporary superabundance of meat, that the meat be required to support the reproductive effort of males despite injuries or death suffered during the hunt, and that conditions are such that the single hunter is considerably less successful than the cooperating one.

The easiest form of cooperation is for some hunters to distract the female defending her young while other hunters quickly move in and kill the young; wolves can do as much (Carbyn 1975). Then all could retire until the female left the dead young or she could be more easily chased away. As noted earlier, such hunting would produce safe, tender meat from animals easily dismembered.

Such cooperative hunting, under proper stimuli from the environment, readily evolves into confrontation hunting, in which not only the young ared killed but also the defending female. Here the hunters take advantage of a large animal's readiness to confront predators, making it that animal's Achilles' heel. Rather than run after the prey, let the prey attack, and be killed by appropriate means. These "appropriate means," however, are so complex and precise in their requirement that we cannot expect them to evolve until hunters were dependent entirely on the meat of large mammals for long periods of the year. More of that later.

Acheulean hunters would probably have discovered means of making a large-bodied ungulate attack, so that it fell to its death or got stuck in a natural trap. The latter would have generated a problem, however—killing the beast. It is likely that here we find the original need for a spear which could be thrust into the vital organs of the prey. A device was needed long enough to give the hunter safety from the teeth and horns of the prey and yet capable of killing the large-bodied prey with its tough hide and thick muscle and fat layers protecting the internal organs. The pointed wooden spears were effective primarily in opening holes into the thorasic cavity, leading to collapse of the lungs and suffocation of the large prey. Hemorrhage would be slight with such spears and they would be slow to kill. The foregoing hypothesis explains the occurrences of very large bovids, pigs, and even elephants in Acheulean deposits (Clark 1970), as well as the association of at least one spear with elephant remains in mid-Acheulean times (Movius 1950). Such animals are vulnerable by virtue of their ready attack on predators. Of course, it is to be expected that before Acheulean man lured large herbivores into traps, he found such animals in natural traps or encountered individuals maimed by wildfire (McHugh 1972) but still alive and so an easy prey—provided he had the means to kill them. However, cooperative hunting would remain relatively rare, and its strategies and killing weapons simple, as long as systematic hunting of small-bodied prey by single hunters was sufficient to ensure high reproductive success despite the dry nonvegetative season. At that time, it was postulated, family units dispersed, each to look for its own food. We require a mechanism to concentrate families.

In the temperate zones with their cool nonvegetative seasons called winter, it was indicated that forests and wood became indispensable for survival. Forests are found in steppe country along water courses. Thus, in winter the availability of wood would have concentrated families along rivers.

There would have been additional incentive to go to rivers, particularly to flooding rivers. The rich alluvial soils along rivers produce many species of trees and shrubs that in early fall produce crops of edible fruits and nuts. The "mast" from beeches, chestnuts, walnuts, and filberts must have been most attractive to Acheulean man. However, such mast, as well as that of oaks, must also have been most attractive to other animals, such as bears and wild boar. The mast would have been a source of food for the latter well into winter, if not spring, depending on the size of the crop. In addition to the bears and boar lured to the valleys by mast, the favorable microclimates—and especially the rich meadows, marshes, and shrub flats—would have attracted small- and large-bodied herbivores and consequently also carnivores. This circumstance, during wet days when the rustling leaf litter was silent, would have made systematic spear hunting possible in the woods. However, it would have been suicidal for a single hunter to spear the young of such animals as wild boar, bison, aurochs, rhino, moose, the large *Megaloceros,* or elephant. Cooperative hunting is called for here.

The Acheulean hunters must surely have also had to despatch a few large stags or bulls injured during the rutting seasons from late summer until late fall. Northern ungulates, with their short breeding seasons, concentrate their reproductive effort, with the consequently enhanced competition between males; wounded and dead males are not that uncommon (Geist 1971a, 1974d), especially given the high density of diverse mammals in the temperate zones during Acheulean times. Wounded males of the large species, like females defending their young, would turn on the small, two-legged tormentors only too readily; ungulates are an aggressive lot! Again, cooperative hunting, which distracts the attention of the dangerous prey long enough for some other hunter to stab the prey deeply and fatally, is a necessity here.

Opportunities for hunting dangerous ungulates by confrontation techniques would probably have been available throughout the year. For instance, during years of rich mast, wild boar would farrow repeatedly throughout the year (Oloff 1951). This means that throughout winter, and especially in spring with its poor availability of food, the hunters would in many years encounter either a nest with piglets defended by a most aggressive capable sow, or a sow leading piglets. These sows attack readily when leading young (von Raesfeld 1952); moreover, the nest stage, when sows "brood" their young (Frädrich 1967), probably to prevent hypothermia of the piglets, is a most vulnerable one for the pig and can easily be exploited by hunters. The wild boar, a very dangerous ungulate, was already a food of Acheulean man, as well as of the later Neanderthal people (Clark 1967).

Confrontation hunting of dangerous mammals is but one form of cooperative hunting. As I shall show later, we have reason to believe that it was brought to its greatest perfection using close-quarter tactics by Neanderthal man. Another form of cooperative hunting is to lure or herd animals that are readily subject to herding, such as highly social ungulates (Baskin 1974), to their death. This form of hunting would have been aimed primarily at horses, onagers, and reindeer, and occasionally bison. Wild horses of the northern type akin to *Equus przewalski* would also have been subject to confrontation hunting, granted an antipredator strategy similar to that of zebras (Kruuk 1972, Schaller 1972, Carbyn 1975), that is, by the stallion defending the mares, or the group turning on predators. The highly aggressive disposition of *Equus przewalski* is well known (Heptner et al 1961, Mohr 1970). The stallion, like the other aggressive northern ungulates, would in all probability have been only too ready to attack any hunters approaching his harem of mares and juveniles.

River valleys feature not only favorable microclimates, wood, and game, but also caves and cliffs. Some clifts create an opportunity to drive game over the edge. Good "jumps" are rare, that is, such jumps as permit cooperating hunters to spook game successfully in the proper direction and prevent it from detecting the trap. Such good jumps would be known widely and would probably attract several families during the winter. All would hope to make use of the jump and secure food. Granted the presence of several families, and the fact that a single carcass of a large herbivore becomes an indefensible resource to any individual if he is faced by several hungry ones, and granted that uncooperative activities of several families could alienate large herbivores from the jump, whereas one family could probably secure an occasional carcass by cooperating in driving, then it is quite likely that cooperation in driving game over cliffs was forged by sheer necessity.

Repeated success at a jump, however, attracts carnivores such as bears, lions, and wolves, and brings them into direct conflict with hunters at times. Granted courageous, cooperative hunters, the predators may become prey in turn. Thus, selection for cooperative hunting would have focused on four strategies of cooperation: first, the distration of the dangerous females of large herbivores leading to the killing of the young and, with increasing frequency, of the female herself; second, the distraction of wounded and partly incapacitated male ungulates during the rutting season, leading to their being killed by spear thrusts; third, the herding or luring of social ungulates over cliffs into natural traps or past waiting hunters; fourth, the simultaneous coordinated attacks on large predators that came in conflict with man over kills or who came to scavange at jumps.

The foregoing explains, in part, the following characteristics of man when he appears in the periglacial zones, first in the form of Neanderthal man and then in the form of Cro-Magnon. These people are superpredators, hunting not only the largest and most dangerous herbivores, such as mammoth, bison, wild boar, and aurochs, but also the huge carnivores such as cave bear, cave lion, and wolf (Klein 1969, 1973); they hunt highly social ungulates, such as horses, onagers, reindeer, and red deer, but rarely solitary forms such as moose and roe deer, although such animals were almost certainly present in adequate numbers (see Chapter 9); it explains why there is a tendency for Neanderthal sites to be clustered close to jumps (Clark 1970) and why they even built communal episodical dwellings (Klein 1973), which were built even more frequently by late Paleolithic people. It also explains why, in late Acheulean sites in Torralba and Ambrona, young elephants are surprisingly often the victims of man (Pfeiffer 1972).

Evolution of Hairlessness

Various hypotheses have been advanced to explain the loss of dense body hair in humans. Loomis (1967) suggested that north of the fortieth parallel solar intensity during winter is very low, as is the ultraviolet radiation responsible for the synthesis of vitamin D in human skin. The synthesis of this vitamin could be enhanced by removing the shading effect of hair. However, the very same effect could be achieved by demelanizing the skin on the face and extremities; a facial area only 40 cm^2 in size would synthesize in 3 hours more than twice the amount of vitamin D required per day. Montague (1960) explains hairlessness as a consequence of neoteny. The more neotenous man became, the less hair he grew. This could have happened provided the body hair gave early hominids no selective advantage whatsoever—a most unlikely proposition. If Montague's view is correct, brain evolution should have paralleled the frequency of the use of fire, for the more naked (neotenous) he grew the more he needed fire to keep warm. However, no such correlation exists because brain evolution runs well ahead of the use of fire. There is virtually a complete a lack of fires in Africa during the early Acheulean period, although they are used quite frequently by late Acheulean and modern man (Clark 1970, Leopold and Ardrey 1972). Desmond Morris's view (1967) that human beings lost hair in order to shed heat during running also predicts naked cheetahs, lions, hyenas, gazelles, oryx antelopes, etc. Clearly, a hypothesis is required that is compatible either with the adaptive syndromes of *Australopithecus africanus, Homo erectus,* or *H. sapiens,* and that explains the sexual dimorphism in hairiness in *H. sapiens.* I found no reason to doubt that *Australopithecus* possessed a typical primate hair coat; it was different with *H. erectus.*

We noted earlier that behavior is the first mechanism of adaptation to any environmental change. Morphological change, therefore, does not occur until the behavioral abilities to adapt are exhausted. This is a most important point. When *Homo* adapted to the dry steppe, he had to cope with greater temperature extremes than in the savannah, and also with greater heat loss and desiccation owing to greater wind speeds at night, while during the day he had to cope with the heat load of the sun (Newman 1970). The nights would be cold, particularly on clear, starlit nights when the sky acted as a heat sink. It was, therefore, important to escape severe body cooling at night. This could, of course, have been done by evolving a hair coat that was both denser and longer. However, such would not be likely to occur until all behavioral means of coping were exhausted and ineffective in preventing hypothermia. If *Homo* could adapt behaviorally, we would expect a hair coat no thicker than that of *Australopithecus*.

To cope with heat loss at night, *Homo* would require a radiation shield to cut radiative heat loss to the open sky and a place to reduce convective heat loss. These requirements are satisfied by building a shrub and grass shelter to rest under at night and placing it in some cut or below a hill to escape most of the effects of the wind. The skills required to build such a shelter are barely above those required to build shelters of the type used by chimpanzees or gorillas. In the dry steppe, material such as branches and long grasses would have been found along active or dry water courses. In lower latitudes many bushes are armed with thorns, particularly in dry areas, and the hand ax would readily double as a tool to cut and dethorn branches for a shelter. Once in a shelter, individuals could huddle during the night, thereby reducing their collective surface-to-mass ratio and thus heat loss. During the day, the warmth of the sun and the metabolic heat of common maintenance activities would reduce the need for body insulation.

Lasker (1969) pointed out the following pertinent points: Hot, dry regions can only be colonized by naked modern man under one of two conditions, either there must be the technology available to transport water or the individuals must be able to cover the body to protect it from the heat load imposed by the sun. Nakedness is a handicap under conditions of great heat, cold, and desiccation. If we grant *Homo erectus* a light hair coat or no hair, plus the ability to form shelters to reduce radiative, convective, and conductive heat loss, then *Homo erectus* could have exploited dry, warm steppes but not hot deserts. We also would expect him to be tied to water courses or water holes in order to obtain enough water daily in the absence of a technology for transporting water and to obtain materials to build nightly shelters. In fact, Acheulean artifacts are associated with steppe and with water courses, according to Clark (1970). This also fits Newman's (1970) view that during hot, dry, windy conditions in the daytime, thermoregulation would be achieved by evaporative cooling from sweating. In this, *Homo* excels far beyond anything known from mammals. Moreover, Newman (1970) points out that under these conditions both large size and erect posture are advantageous in reducing the heat load imposed by the sun. A larger body with its relatively smaller surface area stores heat more slowly; comparing an upright body to a horizontal one, both the size of a sheep, it is found the upright body receives only two-thirds as much direct solar radiation on a daily average as the horizontal one, and only one-quarter as much during the noon peak load, according to Lee (1950). One may also suspect that wind currents—and, thus, convective heat loss—is greater some distance above the ground than close to it. Thus, the propensity for humans to sweat far more than any other mammal would be part of an adaptive syndrome fitted to warm, dry steppe.

The modest level of adaptation to cool nights just described would have permitted *Homo erectus* to advance from the steppe into cool-temperate climate. When early hominids of *H. erectus* grade do appear, as in France and China, they do indeed build good shelters and have fire (e.g., at Terra Amata, see Pfeiffer 1972, pp. 140–41). Such facilities would probably have been sufficient to prevent hypothermia at night and would have counteracted selection for a heavy hair coat. However, we have reason to believe that the hair coat was being reduced by then through sexual selection.

Here I must jump ahead a little. I shall argue in the next chapter how the ecological conditions faced by *Homo* selected for a radically different social system from that found in *Australopithecus*. The system changed from a cercopithecine, in which females maximized reproductive fitness by selecting the most dominatnt males, to one in which they maximized reproductive fitness by making males support their own offspring indirectly by supporting the female during gestation and lactation. Females bonded males by freeing sexual activity from a narrow procreative function. A mandatory consequence is to tantalize the male with visual, tactile, and olfactory stimuli. I shall explain later why olfactory stimuli waned. To enhance tactile stimuli the female, in effect, mimicked the estrous swelling stimuli on her body, for she had only sex to bond the male; the naked, pudgy baby would, therefore, also become an object of tactile delight to the male. As I shall show later, the basic bond was sex and a secondary one the child, through its parodies and hence laughter-generating capability and also its tactile properties. Hairiness of females and infants would, therefore, be selected against. This explains the sexual dimorphism in hairiness in humans and the different tactile properties of the skin and bodies of males as against females and babies.

The foregoing indicates that *Homo erectus* or its ancestor, adapting to dry steppe, *lost* body insulation while experiencing a climate with greater temperature differences. Clearly, huddling at night would be selected for, made more pleasurable for all by the rapid exchange of heat between naked bodies. However, a provocative exterior would place the female's ability to reproduce in jeopardy for reasons I shall discuss in greater detail later. An exterior that greatly stimulates and bonds the female's male may also excite other males and lead to the defeat of the female's male and her rape. While active during the day, the female must not excite sexually but must do this in private with her male to strengthen the bonds and her reproductive success. In an analogous fashion to the disappearance of estrus, the female had reduced the provocative visual stimuli. Given the ability of these hominids to adapt behaviorally and transmit culture, it is not unlikely that the females covered from view sexual organs and probably buttocks. Compared to *H. sapiens,* in which the female bonds the male also with essential services, the females of *H. erectus* had to be far more provocative in external appearance. A covering for provocative visual signals would thus be the very first clothing.

In the temperate zones, with their cooler temperatures, even simple, crude clothing covering the trunk would be adaptive. Once clothing significantly contributes to reproductive fitness, selection will at once be directed against hairy individuals. Clothing, together with the dirt generated by living around a fire—used frequently by people from the late Acheulean onward—would soon lead to a matting of body hair and so to soaring ectoparasite populations. Individuals with little body hair would be favored because personal hygiene is possible by lifting the clothing and removing the ectoparasites by either individual or social grooming. As anyone who has lived for months on end by a campfire can attest, one is soon covered with ashes, charcoal, resins, bark, and dust. Individuals with a great deal of body hair would, in cold weather, be caught between the threat of freezing without cover and suffering the consequences of soaring ectoparasite populations. The latter is neither compatible with social bonding nor with efficiency in extracting a living by hunting. Severe discomforts would neither permit successful hunting nor leave the irritated male or female much time and energy for social activities that enhanced family bonds.

Prior to hairlessness, fire is not essential. It becomes essential as a source of heat when humans periodically shed clothing, even in the subtropics and tropics. This suggests an explanation for the virtual lack of fires in Africa during the Acheulean period but its more frequent use in the late Acheulean and by Neanderthaloid and modern man (Clark 1970, Leopold and Ardrey 1972). The latter, coming from cool climates, had relatively less hair than *Homo erectus* and may be the early *H. sapiens* of late Acheulean age. Once clothing and a technology to transport water are available, hominids could also invade the dry, hot desert, as was probably done by the immediate ancestor of the late Paleolithic people; we shall touch on that later. Neanderthal man did leave behind evidence that he tanned skins (Levitt 1976) and Cro-Magnon leaves not only such evidence but also remains indicative of sophisticated cloth making (Chard 1969). The sudden appearance of fires in the African archeological record suggest the spread of a naked human being, probably *H. sapiens* in an early edition. *H. erectus* was on the way toward hairlessness and the female had probably reached that state earlier than the male. If the people of Terra Amata were of *H. erectus* grade—which is not an unlikely proposition—then the lumps of pointed ocher weakly suggest naked skins for accepting the ocher.

Dictates of Cold Environments

With the adaptations of Acheulean man, life in the extreme cold and periglacial zones was not yet possible for humans. There is no evidence that Acheulean man thrived in truly cold climates, such as there is for Neanderthal and late Paleolithic man. The periglacial zones were not occupied by humans until the first of the Würm glaciations, first by the primarily biologically adapted Neanderthal man and then by the culturally adapted Cro-Magnon people. The dictates of the cold zones are harsh and precise, and the Acheulean cultures probably were not able to cope. Still, some findings of artifacts suggest that a short-lived, wide-flung occupation of cold and temperate zones may have occurred as early as the Holstein interglacial and the following Riss glaciation (Chard 1969). It may well have been only an advance during a mild climatic spell for the dictates of the cold zones are harsh and precise.

Man could not sustain himself by gathering throughout the year, nor go undressed, nor keep a heavy cover of body hair for insulation, nor exist without fire, nor live without skilled technology, extensive knowledge, and intense cooperation. Not even bears, stronger and faster than man, can continue to roam in winter and live by omnivory but must hibernate to survive. Barring hibernation, there is no means to sustain an omnivore except by making him into a carnivore. The omnivore could not be turned into a herbivore, as a very large number of herbivores already existed and had adapted themselves to the periglacial and cold ecosystems prior to man's arrival. There was no room for a slow, herbivorous ape.

There was also an abundance of carnivores roaming about, even in winter, such as cave lions, wolves, probably lesser scimitar cats, cave hyenas, and wolverines. Only the supercarnivore niche is open, since the aforementioned predators concentrate preferentially on medium-sized and small-bodied ungulates like red deer, reindeer, and onagers, because ungulates of large size, such as moose—but also horses, and presumably bison, mammoth, and woolly rhinos—are dangerous, capable opponents. Wolves, for instance, consistently select the most defenceless ungulate they can get hold of, as can readily be seen from Carbyn's (1975) work, as well as from the well-recognized phenomenon that predators shift from adult to juvenile prey when the latter become available (Carbyn 1975, p. 183). The niche of the swift carnivores preying on small and medium-sized herbivores is also closed when man arrives in the periglacial environment. The only niche open is that of the supercarnivore, who can despatch very large, slow, dangerous herbivores, and also the large carnivores themselves.

The only contender for the supercarnivore position—and a very weak one at that—would be the lesser scimitar cat *(Homotherium latidens)*, as it probably preyed on elephants and rhinos, albeit juvenile ones (Kurtén 1968a) or incapacitated ones, as did its relative, the saber-toothed cat (Stock 1956). The hypothesis that man moved into and occupied the supercarnivore niche and thus overlapped that of the scimitar and saber-toothed cats is supported by the rapid demise of these predators with the occupation by man of Eurasia and America (Kurtén 1968a, Martin 1967).

What are the characteristics of cold climates relevant to an understanding of human adaptations and evolution?

The productivity of the land is generally low and spotty; therefore food is scattered over a wide area and concentrated in spots. The species diversity of plants and animals is very low compared to the tropics; thus the choice of food items is restricted. Plant food is increasingly scarce with latitude and is highly seasonal in availability. Moreover, the availability of different types of plant food is roughly synchronous, green plant food early in the vegetative season, berries by late summer, and nuts and tree fruits in early/late fall. Migratory animals that exploit the seasonal superabundance of summer are available for exploitation for a short time only. Small mammals are available mainly in spring and summer and disappear into hibernation or below the snow blanket in fall and winter. During glacial times, the large number of migratory birds we are familiar with were almost certainly absent, because their vast breeding grounds in northern Eurasia and America were covered by ice or were polar deserts. The seasonal availability of plant food, birds, and small mammals, and also insects and fish, cannot be stressed enough. Once the winter comes, the superabundance of food is gone and the only food exploitable is the large mammals. These, however, may exist in surprising density and diversity in the productive periglacial zones (see Chapter 9).

Once snow covers the ground, it covers most of the edible plants, making it difficult or impossible to detect them. Berries and fruits above ground freeze, then thaw and disintegrate. Tubers and corms—even if discovered below the snow—are virtually impossible to remove from the frozen soil even with a modern pick, let alone with a stone ax or digging stick. Most small mammals hibernate in winter or live below the snow and are unavailable to man, as are most birds, because they have migrated south. The lakes are frozen over and fishing is impossible through much of the winter since fish are inactive and/or effective ice-chopping tools are required to get through the thick ice. At the inlets and outlets of streams on the lakes the ice remains thin and can be penetrated with primitive tools, but this also increases the risk of breaking through the ice and, even after scampering to shore, of freezing to death unless a very big fire can be made very quickly. Great areas of land are almost completely devoid of game. Moreover, the noise of clothing and the crunchy snow make silent stalking—the most elementary of hunting methods—impossible. Deep snow hinders walking and confines those not having the skill and ingenuity to produce snowshoes or skis.

In order to live in winter, one has to hunt large-bodied herbivores, almost all of which are alert, agile, and dangerous opponents at close range. Moreover, one must predict the time and place these animals will occupy rather precisely in order to hunt them, and such predictions cannot be made precisely just from watching the seasonal climatic and vegetative changes, for ungulates move on the basis of internal timing systems that are not dominated by weather factors but only affected by them (Geist 1971a).

Even if one succeeds in killing a large animal for food, one's difficulties are not over. The carcass freezes reasonably rapidly and must be dismembered almost at once. Once the carcass is frozen, a saw of steel or, under fortuitous circumstances, a steel ax can be effective in dismembering the carcass. Unless dismembered, the carcass could not be transported away from the kill site. If one wants to eat the frozen meat, one must have a fire to thaw it out, as human teeth are totally inappropriate instruments on frozen chunks of meat. One can shave thin slices off chunks of frozen meat, but that necessitates possession of a sharp *flexible* knife, and is laborious work at that. One must, therefore, have a very exact knowledge of how to dismember a large carcass. This takes the form of precise anatomical knowledge of the various joints and weak spots; these differ from species to species. What will work on a reindeer will not work on a bear and vice versa, what works well on a young animal may not work on an old one, and what works well on a female may not work well on a male.

One must have exact knowledge of how to light and maintain a fire during drizzly, wet snow showers in late fall, when the ground and trees are soaking wet and covered with soggy snow. One must protect oneself when caught by a blizzard, which tends to blow every second to fifth day in the tundra. The extremities must be kept protected from freezing, since the fingers, toes, ears, and nose cool off well before the rest of the body. One must be able to sleep comfortably when it is 50° to 60°C below zero. One must be able to read the snow lest a step into a lake overflow wets the feet.

When spring comes the creeks become torrents, some carrying large chunks of ice that may grind animals foolish enough to enter to certain death, and almost all creeks effectively block the way. Avalanches, mud slides, and rock falls are a hazard wherever steep slopes and erosion gullies occur. In summer there may be an insect plague that torments large mammals almost beyond belief. Studies on reindeer have shown that biting insects remove as much as 125 to 150 cc of blood per day from an animal and that they can cause exhaustion and even death of these animals from their running to escape the insect pest (Preobrazhenskii 1961). These insects cause enough upset to terminate the hydrochloric acid production in the true stomach, which in turn permits the rumen flora to go through the stomach alive and ulcerate the gut; they can cause reduced feeding by caribou and thus cause very poor physical condition of reindeer in summer (Zhigunov 1961, Kelsall 1968). However, in the periglacial region in the lee of the continental glaciers the dry, windy climate would make biting insects less of a problem than they are in the tundra today.

The earth is frozen in winter, limiting opportunities to dig for stone cores that can be turned into tools. Therefore, conservation of stone for tool use becomes a necessity, as well as the efficient utilization of the material; it is adaptive to produce more tools per unit weight of stone. It also pays to diversify and use a broad range of materials for tools.

When exploiting large mammals as the only food for part of the year by killing individual animals as the need arises, hunters may be forced to move long distances between successive kills. In part, this would be due to the animals moving long distances between seasonal ranges, in part to disturbance caused by hunters, and in part to the vagaries of fortune as to where kills happen to be made. Such movements without domestic animals to carry loads, in the absence of food preserved so as to be small in bulk, light in weight, and yet nutritious, would be a considerable burden, particularly to children and old people. They would have to undertake long journeys on empty stomachs through cold weather and probably build shelters at the kill site before they could eat. For a family it would be an impossible burden to do this and look after several small children. Consequently, we would expect a heavy mortality of children and old individuals if, by ill fortune, the kills were widely spaced in time and space. Consequently, only small groups can be expected to live during the period when humans began extending their activities into cold climates and learning to conquer winter; the role of a supercarnivore who does not harvest medium-sized and small ungulates will not permit a high density of individuals anyway. The death of children which adults could not look after properly would set the stage for conscious population control. Not only would the deaths be traumatic but also the adults would have had to have noticed how well developed and competent those few youngsters became that were the recipients of all the parental attention. The origin of population control, a subject we shall hit upon repeatedly, would therefore be found in an adaptive strategy of exploiting large mammals for food during long winters.

Neanderthal Man's Characteristics

It is now possible to turn to one of the most debated, almost romanticized, figures in human evolution, Neanderthal man. His build and his disappearance are to this day a source of fascination. In the following sections, I shall propose that Neanderthal man's structure can be explained as a result of a very sophisticated manner of hunting large, cold-adapted, ice-age mammals on which he depended entirely for food for many months of the year. Only a very extreme form of behavior could have produced such morphological deviations from the rather modern body form of its late Acheulean ancestors. Far from being primitive, Neanderthal man appears to be the most highly specialized human form to have evolved. His disappearance is explained as a logical consequence of his adaptive syndrome running afoul of glacial history. It is pertinent that we examine some of his attributes.

In his classic and most specialized form, Neanderthal man appears in Europe with the advent of the first pulse of the Würm glaciation some 90−70,000 years ago (Klein 1969, 1973, Bordes 1972). He inhabits Europe, parts of Asia, and Asia Minor during the following stadial. The Mediterranean basin on the African side and in the Near East have yielded fossils of people called neanderthaloid by some anthropologists but not by others (Howells 1973a). These doubtful neanderthaloids do have a Mousterian industry. According to Clark (1970), Rhodesia man is considered to be neanderthaloid. If he was, neanderthaloids did disperse into Africa. However, the classification of Rhodesia man as neanderthaloid is problematic, as pointed out by Howells (1973a). Regardless of this controversy, it is evident that neanderthaloid people did not disperse as far as their successors, the upper Paleolithic people. Neanderthal man disappears following the first two pulses of the Würm glaciation, combined in some classifications as Würm I (Bordes 1972). He was the first human to colonize the periglacial zones successfully, albeit not as successfully as the following upper Paleolithic people.

This is revealed not only by the smaller distribution of Neanderthal man, but also by evidence suggesting a low population density compared to upper Paleolithic people. Thus, in the Ukraine, where extensive archeological excavations have been carried out, one site is found with Neanderthal cultural remains (Mousterian culture) for every five sites with late Paleolithic remains (Mongait 1955, Klein 1969, 1973), although the durations of the Mousterian and post-Mousterian cultures are similar. Nor have we reason to suspect that the big-game fauna of the early Würm glaciations was less rich than that of the late Würm glaciations. If anything, the late Würm fauna of large mammals was poorer, due to the extinction of such species as mammoth, rhino, steppe wisent, cave bear, and cave lion.

Our suspicion that Neanderthal man was less numerous than Cro-Magnon in the periglacial environment and lived at a lower density than did the latter is strengthened by the findings of the "ruins" of a communal dwelling of Mousterian age (Klein 1973). It indicates clearly that Neanderthal man was capable of forming a social organization like that of the late Paleolithic people who commonly occupied "communal episodical dwellings." This term is from Schoenauer (1973) who showed that housing types are closely related to the social systems of people, and that exceedingly similar housing types are found among different racial groups, on different continents—provided they exploit similar physical environments in a similar manner. Schoenauer's concepts can therefore be used as tools to interpret the social system and environmental exploitation practices of man.

Communal episodical dwellings appear even before Neanderthal man's time, in late Acheulean deposits dated to the late Mindel glaciation in France (Birdsell 1972). Clearly, late Acheulean man could already form cooperative groups of large size, probably to exploit a seasonal food supply. There is no reason to suspect that if late Acheulean man could form such groupings Neanderthal man could not. If, however, Neanderthal man was capable of forming a social system that caused communal episodical dwellings, and the one site demonstrates that he could, then why did he do it so very rarely? The most plausible explanation, and one which fits perfectly into this form's adaptive syndrome, is that he usually lived at densities far too low to permit communal episodical dwellings to arise; conversely, he rarely reached the very comfortable level of existence implied by such dwellings and found hunting rarely productive enough to form communal episodical camps.

The tools associated with Neanderthal man are of the Mousterian tradition, and are distributed in the Mediterranean basin, in Europe, Asia Minor, and western and eastern Asia (Oakley 1949, Clark 1967, Clark 1970, Chard 1969). They are flake tools and still rather crude, but they already form several industries, their diversity is greater than during the late Acheulean period, and they show regional specialization (Clark 1970). Some crude bone tools were also made, as well as some from wood

(Mania and Toepfer 1973). There is some evidence for skin-working tools—and thus for clothing—in the form of scrapers, stone awls, and narrow strip flakes (Levitt 1976). An examination of the tool kit reveals relatively large stone points that may have been hafted to spears, given the fact that these people did work wood as well as mastic (Mania and Toepfler 1973). The many diverse scrapers with convex edges appear to be knives serviceable in rapid, but not precise and accurate cutting. Such knives are just fine for cutting sinews that bind articulating bones, making slits in carcasses, cutting off chunks of meat, etc. They would have been a great help in rapid dismembering of a carcass, provided sufficient muscle power is used in addition to the stone tools, to snap joints apart and tear off muscles. The curved edges of these knives would have been fine tools for skinning large animals and do not differ in principle from a "skinning knife" of today.

Neanderthal man probably had clothing, but it must have been crude. Note that Mousterian sites in Russia have few remains of arctic fox (*Alopex lagopus*) or hares (*Lepus*) compared with late Paleolithic sites (Klein 1973). These animals, Klein (1973) points out, were used by the later people for their fur, rather than for their meat. One finds articulated skeletons of arctic fox minus the paws, for instance, as well as single articulated skeletons of paws. This suggests that—as in our times— the fur was removed with the paws. Neanderthal did frequently kill large carnivores with fur useful for clothing, such as bears, wolves, cave hyenas, and cave lions (Vereshchagin 1967, Klein 1969, 1973). Neanderthal may have had use only for large skins, and not small ones which had to be sewn together. It has been claimed that Neanderthal man did have needles (Birdsell 1972, p. 283) and therefore could sew clothing. This is in marked contrast to his crude tools and what other authors believe is evidence for poor manual dexterity (Mongait 1955, Musgrave 1971). Yet, as Marshack (1976) points out, Neanderthal man could make some dainty implements; Howells (1973a) rates some Neanderthal tools as equal to any produced by upper Paleolithic people in workmanship.

The dwellings of Neanderthal man are poorly known. He could obviously build crude huts large enough to house several families, as the Moldova I site indicates (Klein 1973, p. 70). It implies a high degree of cooperation, and also some division of labor. Postholes in cave debris are known (Bordes 1972) and we can safely assume that some wood and skin shelters were built to enhance the microclimate of the caves he inhabited. From the scarce evidence for constructed shelters, we infer that Neanderthal man used primarily natural shelters and usually enhanced their amenity value somewhat. Fires are apparently less common than in the upper Paleolithic (Clark 1970, Leopold and Ardrey 1972) and they contain more bone ash than do hearths of late Paleolithic age (Pfeiffer 1972, p. 233). It is not clear to me from reading whether Mousterian camps are most frequent close to natural trap sites and jumps. As I shall show later, the adaptive syndrome of Neanderthal man suggests that they should be. Clark (1970, p. 142) reports such an association for Mousterian sites in Africa, and the Russian sites along rivers imply the same to me, for reasons to be enumerated later. Game was driven over cliff edges in Mousterian times, as excavations in Russia have revealed (Vereshchagin 1967).

The large mammals that Neanderthal man apparently hunted are impressive evidence for the competence of this human form. They represent not only the largest, but also some of the most dangerous animals humans ever faced to kill. They include mammoth, wild boar, woolly rhino, steppe wisent, the rather aggressive and dangerous wild horses, giant deer, bears, cave lions, cave hyenas, and wolves

(Vereshchagin 1967, Klein 1969, 1973, Pfeiffer 1972). Moreover, in total, these very dangerous large mammals, and particularly mammoth and wisent, supply by far the most meat (Mania and Toepfer 1973). Neanderthal man also hunted highly gregarious ungulates, such as onagers, red deer, and reindeer, as well as small game, although apparently less frequently than late Paleolithic people. If that is so, then it would substantiate the hypothesis I shall advance later, that Neanderthal man was far less efficient in the acquisition and utilization of his food, and could not afford the time and energy to hunt small game often.

Glaring by their absence or infrequency are remains of solitary ungulates, such as moose and roe deer (Klein 1973), yet the vegetation of glacial refugia depicted by Frenzel (1968) argues that they were not uncommon. Even late Paleolithic man rarely hunted these, and turned to them more frequently only in the Mesolithic. Clearly these ungulates are difficult to hunt. Also absent or rare is the aurochs, for reasons discussed later.

The anatomy of Neanderthal man is the key to the understanding of his adaptive syndrome. In body size he is large, though not tall; he is larger than the small-bodied Krapina specimens from the preceding interglacial of Europe (Coon 1962). His skeleton is massive, with thick, somewhat curved, long bones. He is stocky, very muscular, with short appendages, large hands and feet, and thick, muscular fingers and toes (Mongait 1955, Le Gros Clark 1964, Coon 1962). The hand is of special interest: It has a long palm and fingers with short first phalanges and long outer phalanges. The heads of the metacarpals are large. The fingers could separate in such a way as to give the hand a very wide span compared to modern hands. The insertion ridges and positions indicate that the grasp of the hand was exceedingly powerful, a view confirmed by experimental archeology, which showed that modern people could not generate the strength to duplicate wear patterns found on Neanderthal tools (Levitt 1976). The scapula is somewhat different from that of modern human beings, and at least one authority, T. D. Stewart (1963), suggests that it is more advanced than in modern man. The pelvis also differs slightly (Stewart 1960) but the functional significance of this has not yet been determined. The skull of Neanderthal man is large and diagnostic in the elongated shape, heavy supraorbital tori, strong prognathism, massive jaws, and heavy construction. The infraorbital foramina are large, as in Eskimos, indicating a rich blood supply to the face (Coon 1962, p. 534). The noses must have been large in life. The brain is large (Coon 1962, p. 529, gives a range of 1525−1640 cc for males and 1300−1425 cc for females), well within the range of presently living modern man but not as large as that of some upper Paleolithic individuals who replaced him or of some modern people (Mettler 1955). The brain differs significantly in its proportions from that of modern man (Hofer 1972). Given our present understanding of the phenotypic development of the brain, his enigmatic brain suggests that Neanderthal man performed a great variety of intellectual and, in particular, physical tasks. Studies of the anatomy of the skull indicate that Neanderthal man might have spoken a language, albeit one less diverse than ours (Lieberman and Crelin 1971). However, this conclusion has recently been challenged by Le May (1975), who showed that Neanderthal man did have the anatomical prerequisites for speech. The teeth of Neanderthal man are large, have edge-on occlusion and the molars are often, but not always, taurodont (subhypsodont) (Coon 1962, Brace 1964); the incisors and canines particularly show heavy wear. The teeth on average are larger and longer than ours, usually having five rather than four cusps on the lower molars. Taurodonty is achieved by root coalescing; this provides a wear surface late in life when the teeth are worn down. There is much secondary dentine formation as a result of tooth wear (Dahlberg 1963). However, the premolars and molars of Neanderthal man are generally smaller than those of *Homo erectus* (Howells 1973a).

One characteristic of the classic Neanderthal skulls is at once surprising and most revealing: According to Birdsell (1972) they show surprisingly little morphological variation. This means that phenotypic variability was low. In conjunction with the large size of the body and brain and the robust skeleton, it means that Neanderthal man lived in rather high-quality populations. The cultural evidence suggests much the same, for the first significant indications of religious beliefs and practices in our genus are associated with periglacial Neanderthal man. During severe population decline, humans show not only a deterioration in physical development and health but also a cultural deterioration, as will be discussed later. The low phenotypic variability of Neanderthal man suggests also that besides practicing a budding religion, they also consciously practiced population control, so that individual ontogenetic development, rather than population size, was maximized. The alternative to this hypothesis would be to postulate forever expanding Neanderthal populations in order to account for their large body and brain size. The evidence for this hypothesis is lacking; Neanderthal populations, compared to those of the late Paleolithic people, are quite low in density. Thus, conscious population control is the more likely explanation. It should be noted that the morphological uniformity argues strongly for severe natural selection for an individual of rather specific physical dimensions (Coon 1962); it also argues that nutrients and energy were ingested far in excess of immediate needs. It is difficult otherwise to conceive how such phenotypic uniformity could be achieved in the light of our knowledge of mammalian growth.

It is also noteworthy that Neanderthal man reached his most diagnostic extreme form in western Europe, the very area that generated the richest cultures of the upper Paleolithic, and that was also the evolutionary home of the massive cave bear (Kurtén 1968a). The country lay to the west of the Scandinavian and Alpine ice sheets in the more humid oceanic climate of western and southern Europe (A. L. Washburn 1973). Early in the glaciation and just following the beginning of large-scale glacial retreat, the warm Gulf Stream must have hit the coasts of Spain and France and was probably responsible for generating a rich diverse flora and fauna in western Europe. Like the cave bear, Neanderthal man is, thus, apparently associated with high productivity. We can infer much the same from the distribution of Mousterian artifacts in the Soviet Union. Neanderthal people inhabited Manchuria and Mongolia but apparently did not penetrate into Siberia; they are closely associated with mountains in Asia (Chard 1969), which implies that they took advantage of seasonal game concentrations, so predictable in mountain areas. All this implies that Neanderthal man needed a higher density of huntable game to survive than did the upper Paleolithic people. They did penetrate into Siberia and on into America, which Neanderthal people failed to do; however, upper Paleolithic people did not develop in the dry, cold regions of Asia a cultural richness comparable to that of western Europe or southern Russia and the Ukraine.

Neanderthal man had religious conceptions; he buried his dead and surrounded them with chosen bones of animals, such as the horn cores of ibex and jaws of wild pig (Mongait 1955, Clark 1967, Klein 1969). Flowers were apparently also used as gifts to the dead (Birdsell 1972). He may have practiced magic rituals, as revealed by skulls of murdered individuals and skulls of cave bears placed in ritual settings (Blanc 1961).[1] He used natural paint, such as ocher and manganese dioxide (Bordes 1972). He apparently nurtured the injured and aged. There is, therefore, evidence for an active mental and spiritual life.

Neanderthal man had continued the process of broadening the human ecological niche, a process begun by *Homo erectus*. He inhabits not only the steppe in temperate and cold climates but also forests, mountains, and the periglacial environments themselves. He also begins to occupy campgrounds all year round (Clark 1970) compared to the ever-moving Acheulean man.

Did Neanderthal-like people disperse into Africa from the Mediterranean basin as the model I proposed suggests they would? The Lavallois−Mousterian tradition, which is generally associated with Neanderthaloid people, did spread southward into the Ethiopian and Somali regions (Clark 1970). The paleontological evidence, unfortunately, is bedeviled by divergent opinions about what to call and what not to call Neanderthaloid, as well as by controversies about the dating of fossils and archeological sites (Howells 1973a). We noted that human fossils of Mousterian age from the African side of the Mediterranean basin were not considered quite Neanderthaloid nor modern but reminiscent of the very earliest of *sapiens* types such as Steinheim man. We can infer that these Mousterian people were largely meat eaters because, during the Würm advance, cold climates would settle over the Mediterranean basin.

If we assume that such people dispersed into Africa to the south during the early Würm phase and gave rise to Solo man, then a peculiar parallel to later dispersals of upper Paleolithic people is to be noted. The teeth of Solo man from Broken Hill are smaller than Neanderthal teeth and they have a high incidence of caries. The same is seen in people who colonized Africa at a later date (Gable 1965, Clark 1965). Such caries indicate a diet rich in carbohydrates, one which would deviate considerably from the predominant meat diet of the cold-adapted northern ancestors. A long period of evolution on a meat diet would presumably lead to a loss of resistance to tooth decay, a resistance that earlier human foms probably had. When modern hunters disperse southward into the warm areas of Africa and America they, too, are haunted by caries and heavy tooth wear, as are agricultural people from the Near East (Gable 1965, Clark 1965, Green 1970, Angel 1968, Dahlberg 1963, Molnar 1968).

On the Adaptive Syndrome of Neanderthal Man

The foregoing subchapter enumerated the characteristics of Neanderthal man which require an explanation, by showing how they are logical consequences of the life of these people. Given an understanding of Acheulean man, as described previously, plus an understanding of the characteristics of the cold climates and the periglacial ecosystem, the characteristics of Neanderthal man are those of a cold-adapted hunter who solved the problems of life in the cold by biological, rather than cultural, means. Neanderthal man emerges as biologically the most specialized human being to have evolved and not at all a primitive form, as Brace (1964) has suggested. The

[1]This is disputed by Kurtén (1976).

irony of the matter is that the Cro-Magnon, of our species, can be considered to be primitive compared to Neanderthal man.

Let us begin by fitting Neanderthal man's teeth into a functional context. Brace (1962) points out that Neanderthal man's incisors had an edge-on occlusion and are very large compared to those of modern man; they are heavily worn, while the premolars and molars may have only moderate or little wear (Molnar 1968). Howells (1973a) tabulates data showing that Neanderthal man's incisors exceeded those of Peking man in size; however, his canines, premolars, and molars were noticeably smaller, albeit larger than those of upper Paleolithic and modern men. We must note, however, that in some modern populations, e.g., Australian aborigines, the teeth are sometimes of a size equalling those of Peking man (Le Gros Clark 1964). Neanderthal man's incisors also show a large basal swelling, in short, they are very strong teeth.

Due to the wear on incisors, Brace (1962) suggests that Neanderthal people gripped meat with their incisors and, in the absence of refined cutting tools, worried off pieces of meat. Modern people, in contrast, grip the meat with incisors but cut off the piece of meat with some knifelike tool. They do not usually attempt to tear the meat by muscle power. To anyone who has doubts that meat is very difficult to cut by teeth alone, I suggest taking a piece of raw venison, biting into it, and trying to worry off a mouth-sized piece of meat.

However, in contradiction to Brace's (1962) proposition, Koby (in Kurtén 1972) found that Neanderthal people apparently did cut meat off with some knifelike tool in front of their teeth, just as modern man does. Apparently this process leaves tell-tale scratches on the teeth where the flint inadvertently touched the enamel. This does not, of course, rule out Brace's hypothesis, because it is addressed to the origin of Neanderthal man's teeth.

Let me emphasize again: No morphological change is expected unless behavioral and physiological mechanisms are exhausted in the solution of a problem and the morphological change is tied to reproductive fitness (see Chapter 6). Powerful incisors simply will not grow for the sake of decoration alone; they were used!

Brace (1962) did not follow through on his own suggestion. It requires more than large incisors and an edge-on occlusion to worry off pieces of raw meat. It also requires a powerful facial musculature to grind incisors against each other, shredding meat fibers and promoting the severing of the meat. Note that this implies a strong forward and backward motion of the jaws with some sideways rotation, while the incisors are embedded in the meat. Vallois (1957 in Molnar 1968) suggested a forward and backward chewing motion in Neanderthal people on the basis of tooth wear. The muscular insertions on the jaw indicate exceedingly powerful internal and external pterygoid muscles, while the temporal muscle apparently was quite weak (Loth in Coon 1962, p. 537). This also supports the idea of powerful bite forces on the incisors in conjunction with anterior−posterior motions of the jaw.

Meat tearing also requires a muscular neck and a strong pectoral girdle. In Neanderthal man, the nucheal area on the occiput is large, indicating powerful neck muscles; it is not as large as that found in *Homo erectus*. Nor are the mastoid processes large in Neanderthal man. *Homo erectus* had the more powerful bull neck of the two, and that for a good reason. Whereas Neanderthal man had to eat cooked meat for much, if not all, of the year for reasons detailed below, or to gnaw frozen meat with incisors, *Homo erectus* ate raw meat without the benefit of fire or frost. He had to sever raw collagen and elastin fibers, neither of which is amenable to breaking under the bite pressure of the incisors as is tough but cooked meat. Tearing between hands and teeth is, thus, a more profitable way of dismembering raw meat. It requires anchoring large canines in the meat and tearing it with sideways twisting motions of the head. Large canines and large mastoid processes take the strain of the meat tearing. Brace's (1962) concept thus fits *Homo erectus* better than Neanderthal man.

In comparison with Neanderthal man, *Homo erectus* had less wear on incisors, smaller incisors, but larger canines and molars, and relatively heavy wear on molars (Coon 1962, Molnar 1968). This is understandable when comparing the adaptive syndromes of these two human forms. *Homo erectus* depended less on meat and more on tough-fibered uncooked foods gathered in steppe and savannah, hence the emphasis on chewing with molars and premolars and thus the large size and heavy wear of premolars and molars.

Neanderthal man's facial musculature must have been powerful, not only to shred meat off with incisors or gnaw it off frozen joints or chew it off after biting, but also to use the incisors as tools (Molnar 1968). This hypothesis is supported by the skeletal and muscular development of the faces of Eskimos. These people, raw meat eaters and users of teeth as tools par excellence, have both the largest jaws—larger than those of Neanderthal man in the breadth of the ascending ramus (Laughlin 1963)—powerful facial skeletons and chewing muscles, and also very high bite pressure. Scott (1967 p. 145) found that Eskimos develop twice the bite pressure of Caucasians. Their facial development does suggest that a diet of raw meats and the use of their teeth as tools selects for powerful facial muscles and the concomitant skull development.

The differences between the prognathus "beak" of Neanderthal people and the flat faces of the Eskimos lie in the different ways in which Neanderthal man and the Eskimo ingest meat. While the latter eats his food mostly finely dismembered or cooked prior to its reaching the teeth, Neanderthal man emphasized the severing of cooked meat with his incisors or, equally important, gnawing rabbitlike into frozen chunks of meat, with concomitant wear on the incisors.

The reason for the meat diet lies in the lack of vegetable foods for eight or nine months of the year in cold climates. In addition, as we have seen, Neanderthal man killed some of the largest mammals, whose meat compared to small mammals' is very tough if eaten fresh. Aging of meat, that is, tenderizing it by letting it hang for several days, is not possible in winter, of course, because the meat freezes into a lump of ice. Thus, provided it can be thawed, the meat will still be tough even if eaten several weeks after the animal's death. This leads us to the next problem, how to eat frozen meat in winter.

Frozen meat need not be thawed for consumption provided one has the patience and ability to gnaw frozen meat or the tools and patience to slice thin, transparent slices off a frozen piece of meat. This requires a flexible blade such as can be made from bones. Tools of such a nature have not been found in Mousterian sites. Brittle stone tools are not suitable for slicing frozen meat, for the brittle edge breaks and leaves the meat full of stone chips, equivalent to glass chips, which would lead to bleeding gums and digestive organs if consumed. If one has fire, frozen meat can be thawed, of course.

We noted that Neanderthal's tool kit contained tools that were good for dismembering carcasses, and the evidence from bones (Oakley 1949) indicates that they did dismember carcasses. We must assume at this point that Neanderthal dismembered the carcasses crudely in winter and carried off relatively large pieces for consumption. This means that a large amount of bone and water was carried back to the camp, an inefficient procedure. Such pieces would be placed on a fire in the frozen state and thawed. They would then be consumed partly charred, partly cooked, and partly raw. As thawed meat was eaten away, the chunk was returned to the fire for further thawing and roasting until finally the bone was reached, which in its turn was roasted and cooked, with its marrow eaten before being thrown back on the fire to help in developing a large coal bed and to reduce the risk of scavanging carnivores, such as cave bears, being attracted to the camp. The foregoing explains, first, the heavy incisor wear of Neanderthal man, and second, the large amount of bone ash in the hearths of Neanderthal man.

Heavy tooth wear would be imposed by the large amount of grit, charcoal, and ashes consumed with the meat, for the large pieces of meat would not be suspended *beside* the fire for roasting but dropped *into* the fire. We shall see the reason for *large* pieces of meat a little later. The grit ingested with the meat would select for taurodont teeth and strong secondary dentine formation. We might note that Australian aborigines who roast game by placing it directly into a coal bed would also subject their teeth to wear and would select for large teeth. Such teeth they do indeed have (Le Gros Clark 1964). In addition, these aborigines eat very fibrous foods as well as chewing bones from small mammals, all of which enhances tooth wear (Molnar 1968).

He who must live off the meat of large mammals is faced by the following problem: The carcass of a large animal can be fully utilized without significant loss of meat only in winters with continuous freezing temperatures unless one has the technology, skill, and knowledge of how to preserve meat during periods of warm weather. In spring, summer, and fall, there are spells of continuous warm weather, which would rapidly lead to spoilage of the carcass due to bacterial degradation and the actions of blowflies and other insects. I postulate that Neanderthal man did not have the technology to store or preserve meat effectively, which, by contrast, the late Paleolithic people did. This will be discussed later. Some Neanderthal men did store large chunks of meat in pits within caves (Pfeiffer 1972, p. 188), and they probably did what Eskimos do to this day, store carcasses under piles of sod close to permafrost. The latter method, however, often leads to spoilage of meat and is very wasteful and inefficient (Banfield 1954). It is a very poor method if the country is teeming with large and small predators that can detect the cache and appropriate it. We have every reason to suspect that bears would have done just that. The foregoing makes it evident that, in the absence of an efficient means of preserving meat, the best strategy is to convert as much of the meat into stores of body fat as possible and live off the fat during hunting and between kills.

This strategy of conserving temporary excesses of food would select for a ravenous appetite and an apparatus to chew and digest the largest possible amount of meat in the shortest possible time. Hence the large mouth, large taurodont and five cuspid cheek teeth, and heavily muscled face. The heavy frame of Neanderthal man also fits this hypothesis, since adult individuals would be expected to carry rather heavy fat loads.

If fat were stored in the body as a primary energy store, then we expect very high population quality in Neanderthal man. This is apparently found. This must be so, since fat compared to other body tissues has a very low growth priority and develops only after the growth demands of other tissues have been fulfilled.

As a consequence of high population quality, however, one expects relatively large babies at birth. In females this would lead to selection for large birth canals and appropriate pelvic structures. Since Cro-Magnon females were also of high quality, the differences in pelvic structure between Neanderthal and Cro-Magnon man must be due to other causes. Large fetuses would lead to frequent difficulties at birth and high female mortality. This the fossil record confirms, but on the basis of a small sample only (Vallois 1961).

Since the change from metabolizable energy to fat, a process termed "lipogenesis," is very expensive in calories (Blaxter 1960), the food intake of Neanderthal man for fat storage was very high indeed, in fact, roughly twice that of the food intake needed for maintenance. For every calory of energy stored as fat, about one calory of fat is lost as heat. This is an additional factor selecting for a large eating capacity and appropriate chewing organs.

The muscular, heavy-framed body of Neanderthal man must also have been adaptive in another context. During winter, the carcasses of large mammals freeze into ice, so they must be dismembered almost at once after a kill. The chunks of meat must be carried close to the camping area in winter, where they can be protected from predators. Large chunks are more convenient to carry than many small chunks; therefore, we can expect large chunks of frozen meat to be carried back to camp. Such work requires a muscular, stoutly-framed body. This hypothesis also implies that Neanderthal man carried about a lot of water because meat is 50 to 70% water. Also, the carrying of heavy loads of low caloric density would have made roaming to distant hunting grounds unadaptive; hunting had to be confined to relatively short distances around winter camp sites. In short, Neanderthal man would have been a most inefficient converter of the meat he killed, by virtue of losing meat to decay except in winter, by storing food surpluses as fat at great cost in food, by expending much work carrying or dragging frozen meat of low caloric density to winter shelters, by excessive charring of meat surfaces during roasting, and also by incomplete cooking of edible skin, tendons, and gristle which would remain tough and hard to chew off because of cooking large chunks of frozen meat in the fire, a condition which usually keeps the center of the chunks frozen or undercooked. Therefore, he demanded more meat *per capita* than did the late Paleolithic people who followed, and he had to kill relatively more frequently. Since the hunting techniques of Neanderthal man, as described below, are dangerous to the hunter, we can expect a shorter life expectancy for Neanderthal males than for Cro-Magnon males, as is apparently found (Vallois 1961). We also expect that Neanderthal man lived at lower population densities, usually in small groups, had less permanence in his camp sites, and had a geographic range more restricted to areas of high biological productivity than Cro-Magnon man. In addition, we expect Neanderthal man to be faced with starvation more frequently than Cro-Magnon man due to these inefficiencies which, in turn, predicts that cannibalism and homicide ought to have been relatively more frequent in Neanderthal man. The preceding five deductions match with the existing archeologic record.

To understand the body build of Neanderthal man, we must assume that he depended heavily on the meat of large mammals in winter; he continued to develop along the pathway taken by Acheulean man and became increasingly more effective and competent in the killing of the largest and most dangerous large mammals. He became expert at cooperative confrontation hunting at close range. That is, rather than running after game, he let it come to him, provoked it, and in cooperation with a few other hunters despatched the prey by moving in very close and wounding it fatally. We noted the relatively high frequency of large carnivores and large ungulates in his kill, many of which were species that attacked predators, to judge from their modern equivalents.

Close-quarters confrontation hunting depends on a principle of defence, getting in so close to the opponent that his weapons become useless. It helps to grab hold of weapons, such as antlers or horns or long hair, and hang on. As long as the hunter is able to hang on to the prey, away from its weapons, he is safe. We should not forget that a man can throw a 600-lb steer without too much difficulty—provided he knows what he is doing; we see it often at rodeos. Prey with long hair, such as woolly rhinos, mammoths, or bison, can easily be held on to, so there would be an advantage in large, wide-spreading, powerful hands such as those we find in Neanderthal man, as well as a powerful thoracic girdle. He probably was able to snap the neck of medium-sized ungulates with his powerful arms and body. No wonder Stewart (1963) thought of the scapula of Neanderthal man as advanced. Once the prey is disadvantaged by one hunter holding on and distracting the prey, the others can move in quickly for the kill. Time is of the essence or a hunter may be injured. Thus, quick-killing weapons are a must.

The traditional hunting spears of the medieval European hunters give us a very good idea of what a quick-killing spear, that causes massive hemorrhages, looks like: It is about 3 inches wide, has a rounded point, which ensures quick slippage along bones into the prey's interior, and its blade is rather thick and sharp. A boar penetrated by a spear of such structure dies quickly, indeed. To this day it is recommended as a weapon superior to a rifle in thickets, when dogs hold a boar at bay (von Raesfeld 1952). Since the most difficult task of such a spear is to penetrate the hide of the prey, which is often thick and tough, it must not be too broad, nor too rounded at the point, since it would fail to penetrate in time, nor too thin, lest the blade break on hitting a bone. A spear point too narrow provides too little hemorrhage and permits the pain-stricken prey to live too long; a spear too pointed has a tendency to get stuck in a rib and thus fails to penetrate and only enrages the stabbed animal. We may safely assume that a broad-hafted point is designed to kill large-bodied prey and produce the most massive hemorrhage possible; a narrow blade is functional in deep penetration and is indicative of thrown, rather than hand-held, spears. The fire-hardened wooden spear is quite insufficient for effective close-quarters confrontation hunting, because it is not sharp enough to penetrate thick hides and does not cause sufficient hemorrhage to kill quickly, thus keeping the hunters in danger for a relatively long time. Note that the hafted points found in Mousterian remains are very close in their functional design to the heavy, hand-held European medieval hunting spear designed to kill large-bodied, dangerous game, such as wild boar and also bear. Thus, the points of Mousterian hunters uncovered by Mania and Toepfer (1973) are ideal for quick killing of large mammals. We also know that these people worked wood and used mastic (Mania and Toepfer 1973), so that we have no reason to assume they could not haft stone blades to wood. The "clumsy" points of Neanderthal man are hence "clumsy" by design, not because of incompetence.

The broad-hafted stabbing spear is not the only weapon conceivable for close-quarters confrontation hunting. A sharp, broad-edged, large piece of flint held in mastic to provide a good grip during forceful use, such as found by Mania and Toepfer (1973), could also be useful, particularly on smaller ungulates. It would essentially function like an ax which, with one or more hard blows, could sever nucheal tendons in the neck, or disembowel prey, or cleave open its thoracic basket, leading to unilateral collapse of the lung or bilateral collapse if applied to both halves chest, or slash open the carotid artery and jugular vein just behind the jaws of the prey. The advantage of such a "hand ax" is that it permits the hunter to hold on to the prey and control its movement to some extent or, conversely, allow himself to be moved about by the prey while still maintaining a favorable position relative to the prey's weapons.

Close confrontation hunting also demands very great agility and high powers of acceleration in the legs. Only then is it possible to quickly evade a charging prey by sidestepping and at once move into position to kill or hold on to it. Note, for instance, the stance taken in fencing, in which the very same requirements are made of individuals. Foil in hand, the competitor must be able to withdraw and advance quickly in order to evade or take advantage of an opponent's weakness. He must also be cool, and spot immediately any weaknesses in his opponent's defence. Note that in fencing the normal stance is abandoned in favor of a spread-legged crouch. In this position, a very uncomfortable one, a quick approach and retreat is possible; it is also possible to avoid a charge by sidestepping. When we look closely at the sacrum of Neanderthal man, we note that it differs from modern man in the greater length of the upper and lower rami of the pubic bones and a correspondingly larger size of the obturator foramen in the pelvis (Stewart 1960). This suggests at once that the adductor muscles of the thigh were at a mechanically more advantageous angle in Neanderthal man than in modern man. Together with the shorter appendages, Neanderthal man in a low crouch must have been able to move with great speed and agility—precisely what is required in close-quarters confrontation hunting. His pelvis and heavy, curved leg bones were thus highly specialized to do an important task well, namely, to dart in quickly, evade, grab, hold on, spear, and wrestle a large, panic-stricken mammal to the ground by taking advantage of that species' weakness. It is here that the thick, muscular, large hands came into use to securely grip and hold prey; the broad feet gave a secure stance and large gripping surface when accelerating; the short, heavily-muscled arms held the prey and swiftly delivered a deep, fatal spear thrust through inches of heavy hide, muscle, and fat; the short, crooked legs accelerated the heavy body away when danger or opportunity appeared, or spun the heavy body quickly into a new position. Close-quarters, high-speed manoevering exerts great demands on the bones and selects for heavy bones and massive muscle insertions to withstand the great acceleration and deceleration of the heavy limbs and body. We also have reason to believe that the massive brain of Neanderthal man may well have been the phenotypic and genotypic product of exercises and selection for many diverse, exacting, and speedily executed motor patterns of the body. I suspect the high brain volume of Neanderthal man, despite the small frontal, parietal, and temporal lobes of the cerebral cortex (Hofer 1972, p. 21), indicates a massive brain stem and cerebellum. The latter is responsible for coordinating motor patterns. A reexamination of endocranial casts of Neanderthal skulls appears to be in order to check whether the brain regions associated with motor activity are relatively large.

Studies by Brozek (1966 in Jacobson 1973) on the effects of caloric intake and work output of men revealed that the ability to perform maximally was acquired gradually. Therefore, it is essential to have a high intake of food in order to remain in peak physical condition. Brozek showed that it took some 16 weeks on a high-plane diet (3500 kcal) before men subjected previously to poor nutrition registered a measurable amount of recovery in their work performance. This implies that for a human being on which great demand is placed physically an adaptive strategy with periods of food deprivation is not viable. Brozek's study thus suggests that in order to maintain a reasonably regular large intake of food, Neanderthal people had to keep their population well below the carrying capacity of upper Paleolithic people who stored food and thus ate regularly.

The thick fingers and toes of Neanderthal man have been explained as adaptations to the great coldness. There is no doubt that a small surface-to-mass ratio would indeed reduce the risk of freezing the extremities in very cold weather. This explanation does not, however, explain the massive bones of the hand and their large insertions for muscles, leading to the conclusion that they were hands of great strength but little dexterity (Mongait 1955, Musgrave 1971). Hands of great strength presumably have a large blood flow to the muscles and would, therefore, be well vascularized. This would, of course, be most adaptive in cold weather because a little exercise of the hand would readily lead to perfusion of the fingers with warm blood. The very large nose of Neanderthal man may indeed have been primarily selected to withstand cold. At $-40°C$, one's nose freezes rapidly and very painfully; a large, well-vascularized nose may indeed have been adaptive in the absence of clothing protecting the face.

Howells (1973a) follows Steegmann in explaining Neanderthal man's protruding face as having adapted its shape to reduce frostbite. Faces prominent in the midline appear to be less subject to freezing than those with prominent cheekbones, as were large, long heads. In the absence of clothing, every adaptation to reduce frostbite on the face must have been important. Coon (1962) felt that the huge nose acted as a means of warming inhaled air (I wonder what Neanderthal man did when running), but did not address himself to the problem of why the large noses in Melanesian or australoid people from warm climates came about. However, the hypothesis does not exclude that of Brace (1962), which I elaborated upon. Moreover, facial hair may have been prominent in Neanderthal males, the sex most frequently without benefit of shelters and fires.

One benefit of lipogenesis to be mentioned is that in cold weather, the excess heat lost can be used to heat the body. Neanderthal man could also have mastered the cold by frequent exercise. He was, according to hypothesis, an inefficient converter of meat and was forced to kill rather frequently. Frequent persistent exercise could have kept him warm by day away from shelter. That Neanderthal man could survive blizzards far from shelter is doubtful, and it is likely that he never roamed far or for long to hunt in winter for this reason, as well as because of the need to carry back heavy chunks of frozen meat, which after all is mainly water. Thus, there would be little need for snowshoes or for shoes and gloves; in addition, the relatively low snowfalls were postulated for periglacial regions inhabited by man to be characterized by light fluffy snow, alternating with wind-packed hard snow. Moreover, Neanderthal man had relatively broad feet, although he was not very tall, and so probably had a superior "foot loading" to ours for moving over crusted snow.

If Neanderthal man were to hunt by ambushing game in winter, a form of hunting that requires long periods of inactivity and consequent accelerated heat loss from the extremeties, he would have had to dress in superbly warm but lightweight clothing. Bulky clothing would have interfered with the quick attack after a long wait. Since no evidence for tools permitting sophisticated clothing to be made exists, we can conclude that ambush hunting was not practiced in winter. Ambush hunting also requires a throwing spear, and the hafted points of Mousterian cultures are relatively large, suggesting a stabbing spear only.

The critical time for Neanderthal man, however, was almost certainly the winter night during a cold snap or a blizzard, provided the inference is correct that he had poor clothing. In the absence of good clothing and camping equipment, such as light blankets and light pelts for sleeping, as well as a tent to create dead air space, it is inconceivable how the poorly clothed Neanderthal man would have escaped freezing. That Neanderthal man did live in times of extreme cold is beyond reasonable doubt, as evidenced by frost-worked soil, frost-cracked rocks, and reindeer with which his remains are associated (Washburn 1973, Klein 1969). The hypothesis that best fits the available data is that Neanderthal people did little roaming in winter and selected for winter residence areas frequently visited by game, so that hunters did not need to go far and could always return to secure shelters before nightfall. This line of thought suggests that Mousterian sites should often be located close to favorable jump sites. Indeed, this appears to be found in Russia (Vereshchagin 1967) and in Africa (Clark 1970, p. 142). Cliffs would provide not only opportunity for cooperative jump hunting, but also shelters in the form of overhangs and caves, ideal "winter habitats" for Neanderthal people.

Opportunities for jump hunting would have been greater in glacial times than today. In areas of heavy loess deposition by wind, river, and creek courses would be bordered by steep, crumbly loess banks. Today these banks are eroded; in glacial times, however, frequent loess storms would keep the walls steep, just as happens today in areas of heavy loess deposition in the St. Elias Range on Sheep Mountain. Any heavy game animal grazing close to the steep loess bank risks a sudden collapse of the bank and a fall to the river bottom. This is how some mammoths were buried and preserved in the loess fields of the Taimyr Peninsula in Siberia (Garutt 1964, Vereshchagin 1967, 1970). A search along creeks and rivers in the areas of heavy loess deposition would have produced a fresh carcass periodically, and miles upon miles of steep crumbly river bank was an excellent prerequisite for jump hunting of various ungulates. Particularly in winter, driving game over the edge of loess cliffs onto river ice must have been productive, since such game inevitably fell to its death and could be retrieved from the ice.

Solitary ungulates such as moose and roe deer were not bagged by Neanderthal man, or if they were it was so rare that their remains are not recorded in Mousterian sites in the Ukraine (Klein 1973, p. 50). Roe deer are small, shy, fleet-footed creatures, difficult to approach; moose can be induced to attack and will stand their ground against predators. Why were they not taken? According to Frenzel's (1968) reconstruction of the plant communities of glacial Eurasia, moose must have been common enough in eastern Europe, particularly along river valleys. The explanation confirms the original hypothesis advocated here that Neanderthal man was close-quarters confrontation hunter. Moose are exceedingly dangerous opponents, since they use front and rear legs in a very effective defence, so that neither a frontal nor a rear approach is likely to succeed. Secondly, they have, even in winter, hair that is too short except on top of the neck to grasp and hang on to. Hence, a hunter who closes in successfully will at once be shaken if he grasps the animal. Moose can be killed in confrontation hunting only by projectiles which permit hunters to stay beyond the radius of the front and hind legs and the short bluff charges. They can be killed by throwing spears or arrows; a stabbing spear is not much good against the moose. Therefore, we expect late Paleolithic people to kill moose, since they had such technology. They did (Klein 1973, p. 50). Note that the argument made for moose can be repeated for *Bos primigenius,* as I do later when discussing modern man. Again we do not find these cattle in the Ukraine killed by Mousterian people, but by late Paleolithic people (Klein 1973, p. 50).

Acheulean man was our departure point in explaining the adaptations of Neanderthal man. The former changed gradually into the latter by penetrating into the periglacial zones along river systems. The process can be envisioned as follows: In our discussion of periglacial vegetation communities, it was noted that relatively warmth-loving plants advanced close to the glaciers in the canyons of rivers. Here, protected from glacial winds, fertilized by the rich silt of the flooding rivers, backed by the solar bowls of the river terraces in abundant sunlight, and often surrounded by moist, warm oceanic air that westerly winds moved up the river canyons, were found pockets of high plant productivity and also of low snow accumulation in winter. With the advance of glaciations, the rivers would become more fertile still, owing to the increase in rich silt generated by the glaciers. Thus rivers were bands of great productivity and excellent large mammal habitat that penetrated the less productive climax forest belts into the rich periglacial zone. Acheulean man's adaptation to the rich fauna living in river valleys in temperate zones was therefore a prerequisite for man's first penetration and ultimate adaptation to the cold, but rich, periglacial ecosystem in the form of Neanderthal man.

The Extinction of Neanderthal Man

The extinction of Neanderthal man and the subsequent appearance of the modern Cro-Magnon is a controversial subject. I shall not review it in detail but point to Brace's (1964) iconoclastic account or to Howells (1973a) for a brief review. It suffices to point out that two opposing views exist: One, championed by Brace, is that Neanderthal man did not become extinct but was transformed into modern man as technology began to substitute for biological adaptations. The second view, a more widely accepted one, is that Neanderthal man is an aberrant lineage that became extinct and left no descendants. There is some debate as to whether Neanderthal man became extinct by hybridizing with upper Paleolithic people or whether he was wiped out by them when they dispersed into areas occupied by Neanderthal man. The admixture of neanderthaloid characters in the Skhul skulls of early Würm age lends fuel to the debate (Howells 1973a). I shall present a hypothesis about Neanderthal man's extinction that demands neither interbreeding with upper Paleolithic people nor that Neanderthal man fall victim to genetic decay, as Brace (1964) would have, nor that upper Paleolithic people wipe out their specialized predecessors. Moreover, the hypothesis can be tested by applying it to upper Paleolithic people of Europe.

A theory of Neanderthal man's extinction must account for a gap between the occupation layers formed by Mousterian and upper Paleolithic people in Europe so that people of the two cultures were not contemporaneous; it must account for Neanderthal man's extinction in the interstadial between Würm phases I and II; it must explain why Neanderthal man could not cope with specific new circumstances;

and it must explain why Neanderthaloid people survived and thrived in the previous interglacial.

Let me first point out that Neanderthal man's demise could not possibly be due to a "superior adaptation" of upper Paleolithic people. There is little temporal affinity, as Brace (1964) points out, between occupation sites of Cro-Magnon and Neanderthal man; the former thus came into Europe after Neanderthal man's extinction. Moreover, the earlier picture of Neanderthal man as a hapless brute, primitive, crude, and clumsy, hopelessly inferior to our magnificent selves, has blinded us to the picture emerging from the fossil record. Neanderthal man must have been a formidable opponent, powerful, agile, quick, ruthlessly courageous, and probably far superior in hand-to-hand combat to ourselves. Moreover, he did hold his own very well, for he evolved, flourished, and became extinct while a contemporary of modern man (Protsch 1973). Moreover, in his early, nonclassic form just prior to the first Würm stadial, he may have replaced early "modern" man on the Mediterranean coast of Africa, as is evidenced by Mousterian cultures abruptly replacing a pre-Aurignacian one (Clark 1970, p. 116). As we shall soon see, the long-bladed cultural debris probably indicates rather warm climates and an absence of long winters with continuous frost. Such winters arrived with the Würm stadial, and thus the condition for Mousterian people to thrive existed.

The rather modest spread of Neanderthal people had little to do, therefore, with a superior competitive ability of upper Paleolithic people but was probably caused by limitations of Neanderthal man's adaptive syndrome itself. He depended on dense game populations; he converted food inefficiently; he avoided plains, as shown in the Soviet Russian archeologic record, and very cold climates with low productivity; he did not roam far and probably had a relatively low replacement rate, given on one hand a high adult mortality and on the other artificial fertility control to maximize individual development. The latter practice suggests infanticide. Note the predominance of males in the sample of Neanderthal people unearthed so far (Vallois 1961), an indication—a weak one—of selective female infanticide. [If selective female infanticide was practiced, a correlate ought to be weak sexual dimorphism (see Chapter 13). Whether this is the case in Neanderthal man I do not know.] Thus, the need to conserve hunters would have been a deterrent to dispersal, as would the very low surplus of individuals, if any.

The inefficiencies in Neanderthal man's adaptive syndrome are of concern here. He depended on the *continuous* presence of the largest, most gregarious, and easiest to hunt, large mammals; he had to kill relatively more frequently than upper Paleolithic people. Therefore, when the first major interstadial set in and the steppes shrank to be replaced by forests, when the loess fields turned into unproductive tundra and boreal forest, the populations of bisons, mammoths, horses, red deer, and the large predators had to decline severely. The rich periglacial ecosystem, with its diverse high-quality, large-mammal populations collapsed, leaving to Neanderthal man the option of extinction or change. It is the former course which he took.

Had Neanderthal man been able to switch to alternate food supplies, had he developed good means of conserving and storing food to tide him over in periods of food scarcity, had he developed a calendar to enable himself to intercept consistently the remnants of the now highly migratory ungulate herds, Neanderthal man might have survived. There is no evidence that Neanderthal man did any of this; on all these points there is evidence for upper Paleolithic people doing so at the end of the Würm glaciation. When a strong deglaciation arrived, Neanderthal man, the superpredator, vanished in a relatively short time, owing to a decline in the density of his prey. If this view is valid, we would expect both a decline in the cultural quality and population quality of Neanderthal man just before his extinction. Pfeiffer (1972, p. 227) states that Bordes in France did note a decline in cultural quality in terminal Mousterian occupation sites; there are too few fossil remains to support or contradict the expectation of low population quality. Kurtén (1971), however, suggests that terminal Neanderthal specimens appear to have smaller brain sizes. This is a potentially significant fact since brain is a tissue of low growth priority and sensitive to environmental quality.

If the foregoing idea is valid, surely we should find some evidence indicative of a marginal existence for people in the preceding interglacial period and a repetition, but with upper Paleolithic people following the deglaciation of the last Würm stadial. Indeed, we very nearly do. The modern people almost vanished from Europe in the terrible Mesolithic. Population size and quality declined, so did culture and craftsmanship. However, we also find a diversification of food habits, new technologies, exploitation of new ecosystems, and the terrible price of rampant homicide and cannibalism. I have discussed this in detail in Chapter 14. Compared to Neanderthal man, upper Paleolithic people adjusted themselves to the periglacial environment by cultural means, retaining a primitive body. Neanderthal man was more biologically adapted, less competent culturally. If the last of the upper Paleolithic and the Mesolithic sites tell of a terrible struggle to adjust to the postglacial conditions, a struggle that was only barely successful and which did not prevent a great population decline, if the culturally competent upper Paleolithic man almost lost the battle for survival following the collapse of periglacial ecosystems, is it really baffling that Neanderthal man did lose?

The foregoing fulfills three of the conditions a theory of Neanderthal extinction must fulfill. It does not, however, answer how Neanderthaloid people could have lived in Europe during the interglacial period preceding the Würm glaciations. A few skulls from this interglacial do show somewhat unspecialized Neanderthal features. From Coon's (1962) description, I judge their small size, slender bones, and evidence of cannibalism to be evidence for marginal conditions and thus food shortages. We may infer that these people had not then reached the highly specialized Neanderthal syndrome but lived a marginal life more like earlier people with Acheulean tools, save for a greater emphasis on hunting. During the very long interglacial, during which human beings were by no means common, they had time—thousands of years—to move from southern latitudes and to adapt. Cultural strategies of exploiting the environment are not easily formed, but they do require time for trial and error to perfect the strategy, and there must always be latitude for errors. Advanced Neanderthals enjoyed neither the luxury of time nor the plasticity of a generalized economic exploitation strategy. They were specialists, and opportunities to practice their skills ran out quickly with rapid postglacial environmental changes.

Chapter 13

On the Evolution of Modern Man

Introduction

The evolution of Neanderthal man was conceived as a consequence of archaic *sapiens* (late Acheulean people) occupying productive river valleys in temperate climates, adapting to a life in which they depended entirely on the meat of large mammals for food, for increasingly longer periods of the year, and then penetrating along the river valleys into the productive periglacial zones. While Neanderthal man evolved in the north, there is evidence to indicate that human beings of a different adaptive syndrome were in existence, humans similar to ourselves, and not to the northern Neanderthal (Protsch 1973, Howells 1973a). Unfortunately, the archeological and paleontological record of Africa is so shrouded in controversy that it is best to look for guidance on what could have happened to human evolution elsewhere. My first task will therefore be to adopt a crude model of the paleoclimate of the Mediterranean basin, North Africa, and Europe as an indicator of evolutionary opportunities. From here I shall proceed directly to the adaptive syndrome of upper Paleolithic people and contrast it with that of Neanderthal man's. This adaptive syndrome of upper Paleolithic people gives some clues as to the nature of their ancestors, a people relying heavily for their adaptation on cultural rather than biological means. Next I shall address myself to the change from the large dentition and prognathus jaw of archaic *sapiens* man to those of modern man, using essentially Waddington's paradigm of natural selection; in particular, selection in populations at carrying capacity characterized by individuals with maintenance phenotypes. This tells us under what conditions the modern human face could have arisen. Thereafter, I shall look for analogies to mammalian evolution and return to probe human evolution in Africa once more.

The second half of this chapter deals with the evolution of social adaptations in our genus. In the preceding chapters, I developed adaptive strategies from those of the australopithecines to those of upper Paleolithic people. These adaptive syndromes were compatible with the fossil record. I am now taking these very adaptive strategies and examining whether the social adaptations of our genus arise or not. I am no longer at liberty to change adaptive strategies so that they may fit human behavior. Rather, once the adaptive syndrome of the gracile australopithecine is accepted, including its social behavior, it must transform into that of the human, given the ecological strategy I described for *Homo* from its earliest to its latest stages.

Human behavior is seen as a consequence of solving the problem of how a large-bodied anthropoid, one with a long gestation and lactation period, can live in the dry steppe. During gestation and lactation, the female requires a diet rich in energy and proteins so as to maximize ontogenetic development of her offspring. For *Australopithecus*, this was not a problem, as it was shown to be tied to continuously greening vegetation in its ecology and supporting itself by surface gathering all year round. *Homo*, I proposed, exploited subsurface vegetation in which plants had stored nutrients in order to burst forth into growth and reproduction during the following dry season. If this was so, then the gestating or lactating female was faced with a starchy food source during the dry season, a food source inadequate in protein to maintain reproductive tissues. The solution to that dilemma is the primitive social system of *Homo*, a system based on the nuclear family, in which the female bonded a male and made him supply the nutrients essential to her reproductive effort (and his). It was a change that drastically broke with the rather cercopithecine social system of *Australopithecus*.

A change in behavior implies a change in a total system. One change precipitates others. The basic materials on which natural selection acts are ancestral adaptations, and one must explain how these are transformed from the old to the new, human, stage. In addition, no old adaptation is abolished without remnants, and these must form part of any explanation of the human type of behavior. The emphasis is on system—on defining logically related constructs that are testable by the paleontological, archeological, and biological record.

I have dwelt not only on the initial evolution of *Homo* but also on the consequences for social behavior as human species adapted to increasingly more inhospitable habitats. Thus, we see how the nuclear family develops into the extended family or how sexual dimorphism changes in the evolution of our genus from extreme in *Homo erectus* to less and less extreme as human beings adapt to temperate and cold and, finally, arctic environments. Part of this is dealt with in Chapter 14.

The goal in this, as in the preceding chapters, is to decipher the conditions under which traits arose that are uniquely human. Although I have devised a system in which different diagnostic human traits are tied to specific evolutionary stages of our genus, I propose these as a thesis for debate. It may well be that some characteristic of, say, Neanderthal man actually arose in *Homo erectus*, or that something I attributed to *Australopithecus* arose in an ancestral form intermediate between it and a *Pan*-like forest ancestor. In short, I am not particularly dogmatic about my proposals, although I may for brevity frame them in rather positive language. Our understanding of how ecology drives social behavior is in its infancy. Yet I believe that the infant does have something to say and if it is to grow well it cannot be ignored.

Early Environmental Settings

Although paleoclimatology is an infant it is a growing and bouncing one. We cannot ignore it in deciphering human evolution, and as a step in that direction let us briefly look at the major weather and ecological changes around the Mediterranean basin, North Africa, and Europe during glaciations and interglacials.

During glacial advances, we expect a southward movement of the northern monsoon belt and a southerly shift of the desert areas of the Sahara. Thus, Clark (1970) reports that the advent of the Würm glaciation led to a spread of forests in North Africa and a reduction in desert areas (see also Livingston 1975, Wendorf et al 1976). As the glaciation advanced, the North African climate became drier again and the Mediterranean basin became quite cold, to judge from floral and faunal remains (Bonatti 1966, Frenzel 1968). We expect a reversal of these steps with deglaciation, namely, a return to forests and grasslands followed by a spread of desert. Indeed, when the first Würm interstadial sets in there is a spread of grasslands in northern Africa some 40−45,000 years ago (Clark 1970).

For the present we need not go further than this very crude approximation, for it suffices to explain a shift of European fauna into North Africa as forests spread in the Mediterranean basin in the early Würm. At this time, red deer, brown bear, and European rhinos appeared in North Africa (Clark 1970, p. 117), and Mousterian people came as well. During the glaciation they occupied all of the cold Mediterranean basin and may well have displaced an earlier people there who practiced what Clark (1970, p. 116) called a "pre-Aurignacean" industry. To the south of the vast area occupied by Mousterian people, there were other people, at least there are artifacts there of a different nature. Clark (1970) identifies a Fauresmith tradition, and Protsch (1973) calls it a "Lavelloisian" tradition that appears to be contemporaneous with the northern Mousterian ones. The Fauresmith tradition is associated with dry steppe, which should have been widespread in North Africa during the glacial period in Europe.

When deglaciation sets in and the Mediterranean basin returns from cold to moderate climates and to grasslands as a major vegetation cover, we still expect montane glaciations in the mountains surrounding the Mediterranean Sea. Thus, we expect periodic silt-laden floods, and floodplain and swamp formation along rivers. Owing to the small size of the Mediterranean Sea, more land would have been available there for human settlement than today.

Whenever grasslands spread in the Mediterranean basin, be it due to deglaciation or reglaciation, cursorial herbivores are expected to increase in number, density, and species diversity. The concomitant climatic variability produces altitudinal vegetation zones, as well as a predictable altitudinal upward creep of sprouting vegetation in spring, the creation of highly digestible subalpine and alpine vegetation, and seasonal forage superabundance. Thus, distinct seasonal home ranges can be established for ungulates, and hence distinct migratory patterns. River valleys also contain some forest with the wood required by human populations for fires during the cool winters. These conditions, as S. R. Binford (1970)—I believe, correctly—recognized, were the precursors to the evolution of modern man.

Since, with the length of the temperate winter, gathering would decline in importance, and herbivores would increase in biomass due to the spread of grasslands, we expect a shift in favor of large herbivores in the diet of people whenever grasslands spread. This is apparently found, as illustrated by bone deposits at the sites of Mount Carmel, the Qafzeh Cave, and a site north of Beirut (S. R. Binford 1970).

Granted migrations of herbivores, so S. R. Binford (1970) argues, then such animals were presumably superabundant at certain seasons. Multiple kills could be made by cooperative hunters. Hence a superabundance of meat could be generated. Binford did not, however, fully pursue the consequences of this. Such meat could not be conserved readily in a climate that did not have long periods of subzero temperatures. Keeping meat for weeks was no problem for Neanderthal man in winter, but to the Mediterranean people, experiencing cool winters with alternating periods of freezing and thawing, conserving meat was no easy matter. Meat preservation under such conditions required a rather complex and sophisticated knowledge base and technology. The solution of the diverse problems generated by meat conservation and its secondary effects on the social organization, culture, and biology, plus the effects of cooperative hunting of migratory large or medium-sized ungulates, as S. R Binford (1970) hypothesized, led to the final shaping of modern man. However, I must point out that the cooperative hunting of modern man was quite different from that carried out by Neanderthal man. But more of that later.

Meat Conservation and its Consequences

Please note that for late Acheulean people we need not postulate the need for conserving meat; they killed relatively rarely and the band would have consumed each kill rather quickly. Neanderthal man, by virtue of being in cold climates, did not require a sophisticated meat preservation technology either. Moreover, meat drying is not easily done or necessary under conditions of extreme cold. However, meat preservation of a sophisticated nature is essential if meat becomes temporarily superabundant and is required for weeks after the kills are made, and the climate is such that long periods of frost are absent. Without either of these conditions— superabundance of meat and scarcity thereafter or relatively mild winters with only short periods of frost—a technology for meat preservation will not evolve.

The simplest manner of preserving meat is to dry it. That, however, requires the technology and manual dexterity to slit meat across the grain into thin, even strips; it requires someone's time and patience to slice mountains of meat into thin strips; it requires the use of smokey fires to keep blowflies away from the meat and prevent maggots from forming in it; and the fires are needed to dry the meat, even though it may be damp and raining during the meat drying. Once the meat is dry, it must be stored in a dry place and secured against predators, which means that somebody may have to watch over the meat store. It requires knowledge and experience to dry and handle meat properly, since improper handling can lead not only to spoilage but to the accidental culture of deadly microorganisms.

The first consequence of a shift to drying meat is a technology that permits the slicing of mounds of meat in a relatively short time into thin strips for drying. Time is of the essence! The Mousterian flake cultures do not appear to be suited for this very exacting task, but the straight edged, backed blades of the late Paleolithic cultures are ideally suited. Note, the meat must be cut across the grain to ensure, first, that water is lost rapidly, and second, that it can be chewed easily when it is dry. The meat must be cut into uniform thin slices to ensure rapid drying and the absence of moist centers that may not only spoil, but may also lead to salmonella or staphylococcus infections and, in the worst case, to botulism poisoning and hence quick and certain death. Improperly dried meat can be very dangerous; hence the need for accurate, thin, even slicing and quick drying. Hence the need for great manual dexterity in order to produce the highly coordinated movements to cut the meat properly. Here lies one probable reason for our great manual dexterity, which is apparently greater than that of Neanderthal man (Musgrave 1971). The blade cultures of the late Paleolithic people were such that large quantities of straight-bladed knives were readily manufactured from prepared flint cores. Large quantities of blades were needed for groups of people to cut the meat of many animals into thin strips for drying. Straight blades were needed to perform the controlled cut; a curved blade, such as that from a flake is not useful for that purpose. The backed blade is required so that a finger may be placed on the blade in order to maximize control over it and ensure a precision cut.

Meat drying in variable weather and dampness requires slow, smoky fires. This predicts a relatively high frequency of fires in late Paleolithic, as compared to Mousterian times, which is indeed found (Oakley 1961, Leopold and Ardrey 1972). It also predicts that late Paleolithic hearths will contain mainly wood ash. This is also found. Mousterian hearths will contain a lot of bone ash (Pfeiffer 1972). The wood is required to generate small hot or smoky fires to cure the meat; the bone ash in Neanderthal man's hearth was explained earlier. I disagree with the view that it demonstrates that bones were burned for heat.

A second reason for an increase in fires during Paleolithic times would have been the development of tanning and cloth making. Oiltanning, the quickest and easiest way to obtain a durable hide, and the only one compatible with frequent shifts of camp sites (Farnham 1950), does require the use of smoky fires to smoke the hide, and wood ashes to remove the hair from the hide. Smoking waterproofs the hide and provides a material that can be used in damp or rainly weather. Note that such tanning cannot be done in subzero temperatures, but must be confined to relatively warm days. For Neanderthal man oil tanning would not be possible during the greater part of the year, and there is little evidence that these people acquired the skill of tanning, different from late Paleolithic people (Levitt 1976). Since tanning and hide preparation requires considerable labor, and since clothing of fur and oil-tanned leather is not very durable, a lot of hard work is associated with cloth making. This suggests an explanation for the exceptionally well-developed physique of late Paleolithic women.

Dry meat must be stored where rain or damp cannot touch it, nor must it be accessible to carnivores. The first requirement predicts for late Paleolithic people with blade technologies far better shelters than for Mousterian people, as well as a relatively more frequent use of shelters, natural and man made. What evidence we do have supports both expectations (Klein 1973, Clark 1970). We also expect longer periods of residence, since food stores need to be guarded; this is also found, at least as indicated by denser occupation layers for late Paleolithic people (Pfeiffer 1972, Klein 1973).

The extended and frequent contact with smoke, ashes, and dust around fires would lead to an increase in the dirt load on individuals, particularly the hairy ones. For reasons outlined earlier, one would expect accelerated selection against hairiness. A reduction in body hair selects for improved clothing in cool and cold climates. We therefore expect a parallel development of straight-edged, backed blades and an increased variety of small hide scrapers, awls, and tools indicative of cloth making. This parallel development is apparently found, but has not been explicitly addressed in the literature. Good evidence for clothing in the late Paleolithic comes from burial sites excavated by Russian archeologists. The remain of richly decorated garments are found as early as 30,000 years ago. They represent a pullover shirt, trousers, boots, and some headwear (Chard 1969).

The hypothesis that the evolution of modern man and the development of a blade industry of the Aurignacean and Perigordian types is associated with meat drying can be tested also by the following: If bearers of such cultures radiate into areas with rich sources of plant food and few animals to hunt, then their industries, though not void of backed blades, will have very few tools indicative of meat drying. If game is available year-round, and no temporary surpluses in meat are generated—that is, the hunted game is consumed at once—there will be little need for meat drying. We also expect a decline in the manufacture of backed blades in areas with little seasonality, where the exploitation of big game is possible without long intervals between kills. On the other hand, we would expect in tropical and subtropical areas a significant retention of backed, straight-edged blades where exceedingly large-bodied mammals are regularly hunted. I might point out, though, that the discussions of late Paleolithic cultures in Africa by Clark (1970) do appear to bear out these predictions.

The consequences of shifting to a technology of drying meat are manifold and appear to match our present knowledge of late Paleolithic man surprisingly well. The availability of sharp effective knives permits individuals to cut raw meat into chunks rather than tear and worry off pieces of meat, as Brace (1964) postulated for Neanderthal man. There would therefore be less phenotypic development of a heavy facial skeleton and no genetic selection for heavy jaws, malar arches, big occiputs, and large nucheal musculature. This would retain the small "modern" face and slim neck of modern man. There are other reasons for this as well.

Given the ability to cut pieces of meat easily into small chunks which can be suspended beside a fire for slow, even cooking, several consequences follow: First, the meat is not regularly contaminated with grit, ashes, and charcoal, so there will be relatively little tooth wear compared to Neanderthal man and no selection for taurodonty and broad, long teeth; second, the collagen matrix of the meat has been subjected to "steam" inside the meat and has become soft. Therefore, the meat is tender compared to raw meat and chewed much more easily. Again, the chewing apparatus is not as heavily used, leading in ontogenetic development to relatively small jaws and malar arches. The same would be the consequence of the innovation of boiling meat and bones. Cooking softens collagen and permits one to eat the gristle around joints; it permits the boiling of skin and thus the conversion of the tough dermis into tender, easily digestible, food. Moreover, boiling would extract fats and amino acids from the bone. The discovery of boiling by early Aurignacean man (Pfeiffer 1972, p. 232) enhanced the efficiency of carcass utilization, primarily by tenderizing otherwise inedible parts. Thus, technological development maintained for modern man the relatively small face and jaw and small teeth. In advanced modern man, there is even a reduction or absence of the third molar (Campbell 1974, p. 231). Technological advances also maintained the gracile body build of pre-Mousterian man for modern man, the slender neck being a result of reduced tearing of meat or food between teeth and hands.

Dry meat, owing to its light weight and great digestibility, can be taken on hunting trips and reduces the need to rely on stored fat for sustenance. It also permits distant roaming without the need to take time out to hunt, and thus sets the stage for enhanced social contact between groups of hunters and the development of complex social relationships and more rapid cultural diffusion. Ultimately, it permitted such activities as trade, prolonged ceremonies at the tribal level, exploration of areas void of game, and crossing to game-rich areas, and it made possible epic journeys by groups of people, all of which would secondarily enrich the cultural milieu of Paleolithic people. Distant kills can be converted to dry meat and transported economically back to the camp, reducing the need to move the camp, as well as the labor of moving food to the camp, for meat is largely water. Therefore, late Paleolithic people brought relatively fewer bones into the camp than did Neanderthal people. In fact, deboning and drying meat would permit long-ranged hunting expeditions and the killing of animal whose bones would never show up in camps; it permitted a wide radius of activity. Since the meat consumed delivers energy directly, rather than by the wasteful process of lipogensis, more work can be done by hunters per unit of meat consumed. This means that modern man compared to Neanderthal man was not only more efficient in converting food to work, but also had less need to ingest huge amounts of meat for fattening. Therefore, there must have been less strain on the digestive system, metabolic machinery, and chewing apparatus. We expect a relatively small mouth and slender body build; reduced metabolic activity argues for a longer life expectancy in late Paleolithic people.

The same is predicted by the fact that stores of dried meat are a better buffer against food shortages than stored body fat. Stored dry food permits hunters to be more selective in their hunts, which at once permits them to become more productive per unit of effort and reduce the risks to themselves. The more efficient metabolic use of food permitted by dry meat also reduces the amount of meat required per individual, reducing the frequency of hunting. All these points argue for a longer life expectancy for late Paleolithic than for Neanderthal man, which is found (Vallois 1961, Cook 1972).

Efficient use of food also indicates that, per unit of game harvested, Cro-Magnon could develop a higher population density compared to Neanderthal man. This is supported by the higher density of late Paleolithic, compared to Mousterian, sites, although the Mousterian cultures lasted somewhat longer than the late Paleolithic ones. This is exemplified in the Ukraine (Klein 1973). It also follows that late Paleolithic man should occupy more marginal habitats than Neanderthal man.

Under circumstances of periodic superabundance of meat and only a few alternative or poor food sources between kills, females are freed from procuring their own food supply, but are required for the exacting task of meat preservation as time is of the essence. Periodic kills and storage of surplus food frees time for other activities, such as developing skills in making clothing and construction of shelters, and even the invention and development of temporary shelters such as tents. This also implies the development of techniques of preserving skins, other than scraping them thin and drying them; it implies the discovery of tanning techniques. The role of the female vis-à-vis the male now begins to change. For the first time, the male begins to depend on the skill and labor of women. This is especially so once modern man settles in the cold and periglacial climates, where warm clothing and tents are essential for hunting activity in winter. Previously, as will be discussed later in this chapter, the male was required to support his own reproductive effort and that of the female, but as an individual he could live very well on his own. This would still apply to Neanderthal man but would no longer hold for Cro-Magnon man. So in the latter case, the woman assumed a vital role, in that a male without female support would not be an efficient provider and probably not even a viable one. We therefore expect evidence for man's concern for the female, which we would not have expected in earlier cultures. Hence the development of female cults in the late Paleolithic people with the first naturalistic, then increasingly abstract and stylized female sculptures and images, as described by Marshack (1972b). A male could not live without predicting big game behavior and movements, nor without a female companion, and this accounts for the intense preoccupation with both shown in Paleolithic art.

Often the question is asked: Could not woman hunt big game? The answer appears to be "no," excepting some highly favorable circumstances. On average, women are some 7% smaller in linear dimensions than men, their muscular strength is only some 60% that of men, and their endurance is considerably lower; for an equivalent effort, women expend some 80% more heart beats, partly because their blood volume and hematocrit is lower (Kleine 1964). In addition, their bodies are taxed by menstruation, the very heavy demands of pregnancy, and lactation (Brody 1945). Traditionally, as anthropological research showed, women trap, hunt small game, and play a major role in communal hunts (Watanabe 1968, Tiger 1969, Gough 1971); in short, the very kind of activities that are compatible not only with the physical characteristics of women but also with the care of small children (Brown 1970). If required to work in industry, women show, apparently for equal work performed, a multiple of the rate of illness recorded for men (Kleine 1964).

Multiple kills and stores of dried meat led to the danger of hunters reducing physical fitness and being in less than excellent physical fitness when they were needed during the hunt. Later in this chapter, I describe how this led to strong selection for vocal and visual mimicry, expressing itself in music and dancing. Neanderthal man had a lesser need for these means of maintaining physical fitness since he had to hunt more frequently than did Cro-Magnon and break camp more frequently; in short, due to his inefficiency, he had to move more frequently, and was therfore kept in good physical shape.

Chronologic Time and Big-Game Migrations

It was postulated, following S. R. Binford (1970), that modern man arose by exploiting migratory ungulates, which provided him seasonally with a superabundance of meat. Social ungulates, particularly those in cold climates, exploit their habitat by moving in a precise manner at specific times of the year to seasonal home ranges. The degree of migratory specificity is probably a consequence of the scarcity of a species' habitat; movements become increasingly precise the smaller and more dispersed the patches of available habitat are. The mountain sheep I investigated made their seasonal migrations within about six days of the same date; many moved to specific seasonal home ranges on the same date in consecutive years, some 90% of the females and 75% of the males appeared in successive years on the same mountain side at the same season (Geist 1971a). Observations of mountain goats and my mule deer work—not yet published—suggest an equally precise space–time system. Reindeer or caribou, for various reasons too diverse and complex to be usefully discussed here, are more variable in their timing of movements (Kelsall 1968, Bergerud 1974). The movements of saiga antelope may be highly predictable depending on the season of the year, in particular the spring migration (Bannikov et al 1961). Bison, exploiting meadow systems, also appear to move at much the same time in consecutive years, but—as in probably all populations—weather factors can influence the time of movement between seasonal ranges (Maegher 1973). The same kind of predictability was not present for the bison herds exploiting the continuous, huge prairie regions of the American Continent (McHugh 1972). We can safely assume that, as in present ungulates, the montane and perigla-cial migrations exploited by early modern man were timed hormonally, such as by the rut or parturition, as well as by diverse weather conditions (Geist 1971a). The former kind of migrations cannot be predicted accurately by observing either the weather or the changing vegetation and snow lines. To predict these migrations, one needs a system of maintaining an accurate account of annual chronologic time; one needs a calendar. To be able to anticipate the movements of game herds may be very important since seasonal events last a few weeks at best, and one needs every day to accumulate the surplus of food from the existing herds before they vanish, almost overnight. In order to maximize hunting success, the hunter requires some means to anticipate the migratory movements, and thus be able to move their base of opera-tions so that they may be frequently in contact with huntable species. The more diverse the fauna and the greater the density of ungulates, the less the human hunter needs to depend on the system of predicting ungulate movements. Conversely, the lower the faunal diversity the more likely the hunter will have to intercept specific herds and to kill in excess of immediate need when the opportunity does arise, storing the meat for future use.

Note also that, once surplus food is stored and the hunters are forced to stay behind and guard their food stores, their mobility is curtailed and they lose contact with the species and herd they exploited. They must know when and where to intercept the next herd or find the next food source with considerable precision in order to avoid starvation. This is a contributing factor pressing for the development of a system that records chronologic annual time.

Neanderthal man, who was much more free to roam and follow herds and who primarily occupied areas of great game diversity, did not need an accurate system of keeping time. Crude indicators, such as the turning of leaves, the gathering and migration of birds, the changes in the external appearance of game animals, the break-up of ice on lakes, the greening of deciduous vegetation, the avalanche and rockslide seasons in the mountains, will all serve the purpose of keeping track of seasonal time. To spot these season timers and correlate them with the movement patterns of game animals is a demanding task, since migrations can be thoroughly obscured by movements in response to snowfalls, thaws, drought, and biting in-sects. Both Neanderthal and Cro-Magnon, under favorable conditions, could rely on these crude timers since the diversity of game animals permitted errors to be made without dire consequences; some large ungulates could still be found and hunted. For late Cro-Magnon, who saw the disappearance of mammoth, bison, rhinos, and horses and who became increasingly dependent on reindeer and migratory salmon for sustenance, leaving evidence of their dire needs in the form of hunted small game and fish, the development of light throwing spears and atlatl, and ultimately the bow and arrow, the ability to time annual migrations of food animals became crucial.

The foregoing is an ecological argument in support of Marshack's (1972a,b) view that late Paleolithic cultures had developed a lunar calendar. Since seasonal timing was of the essence, nature watching had reached such sophistication that the regularities of the movements of the sun and moon were made use of, probably much as Marshack has visualized it.

As indicated, superimposed on seasonal movements are movements by game animals caused by diverse weather factors. Sudden precipitous snowfalls have frequently been noted to trigger the movement of large mammals (Formozov 1946, Nasimovich 1955, Heptner et al 1961, Kelsall 1968, Geist 1971a, Bergerud 1974); sudden temperature increases in winter and spring can provoke much the same result (Edwards and Ritcey 1956), so can the appearance of biting insects (Kelsall 1968, Maegher 1973, Bergerud 1974) and seasonal drought (Heptner et al 1961). To anticipate the movements based on weather, keen observation of weather phenomena is required. Thus, to maximize hunting success, when operating from a few seasonal camps and possessing only primitive weapons, in an environment with rapidly changing seasonal events and complex migratory patterns of ungulates, requires a keen study of nature and the development of a system of keeping chronologic time. Here lies probably one reason for the maintenance of old people to act as stores of knowledge and experience, which the younger members of the group depended on to maximize hunting success. More of this later.

Distance Confrontation Hunting and Complex Cooperative Hunting

The hunting techniques of emerging Cro-Magnon or late Paleolithic man, although based on extreme cooperation, must have been different from those of Neanderthal man. In western Europe, the large number of large-bodied, shaggy, often solitary ungulates and carnivores were selected by emerging Neanderthal man for close-quarters confrontation hunting leading to the full-fledged Neanderthal adaptation. In North Africa and the eastern Mediterranean basin, apparently a very different fauna prevailed, one characterized by many small and medium-sized ungulates of gregarious disposition plus some larger social forms, such as the aurochs (*Bos primigenius*) and African buffalo. The hypothesized preoccupation of emerging Cro-Magnon man with these animals is reflected later in his successful hunting of and concentration on fast-moving bovids in Africa (Clark 1970, p. 138) and in the broader range of species he kills in the periglacial zones (Klein 1973). Close-quarters confrontation hunting is less likely to emerge here for several reasons. First, most small and medium-sized ungulates flee from predators and the larger-sized ones form group defences against predators (Geist 1974c). Confrontation hunting could be practiced, therefore, only on isolated individuals of the dangerous social bovids. Even here, close-quarters confrontation hunting is not likely to be very effective. For instance, consider wild cattle as targets for such hunting. They have no long hair to hold onto, but only a short-haired, smooth, loose skin. First, nobody would be able to hold on while the enraged beast spun around in its attempt to shake and gore the hunter. Second, cattle can spin around with incredible speed, as any rodeo rider or spectator can well attest. Thus, they can quite readily nail a hunter who gets in very close. It is probably more than a curious coincidence that *Bos primigenius* has not been found in Mousterian layers in Ukrainian sites, but does occur, even frequently, in layers occupied by late Paleolithic people (Klein 1973, p. 50). Neanderthal man, a close-quarters confrontation hunter, could probably not hold on to the short-haired, loose-skinned cattle, and placed himself in mortal danger hunting isolated individuals. Given long-haired animals such as wisent, he could hold on and impede the defences of the prey long enough to permit the beast to be killed. If confrontation hunting is to be practiced on short-haired cattle, it must be done at a distance and it must depend on distracting and confusing the prey. This implies that, instead of thrusting a broad-bladed spear into the prey, the hunters throw a spear into the prey and kill it quickly by multiple spears penetrating the animal. Maximum hemorrhage is caused by the deep penetration of the spear, rather than the shallow but wide penetration caused by the thrusting spear. The difficulty of penetrating thick hides, muscle, and fat layers would limit the usefulness of thrown spears to relatively thin-skinned mammals of relatively small size. We therefore expect distance confrontation hunting to arise where medium and small-size bovids and cervids are the primary source of food; cattle are almost ideal for this type of hunting. Since penetration of the skin is rather difficult, sharp points are required, and this suggests that the tipped stabbing spear preceded the throwing spear, in that a technique of hafting stone or bone points would be essential, at least for the successful hunting of large mammals. For thin-skinned, smaller mammals, the fire-hardened wooden tip would do. We expect, as a correlate of distance confrontation hunting, many small hafted points in the cultural debris. Narrow points are required to pierce the tough hide and to penetrate deeply into prey. Note that this is the very characteristic of the Aterian tool industry in North Africa of late Paleolithic times (Clark 1970, p. 126). while the Aurignacean sites of Europe frequently yield split-based bone points (Clark 1967, p. 53).

Distance confrontation hunting does not impose the same selection pressure on the human body as close-quarters confrontation hunting. It does not select for the crouched posture permitting speedy lateral movements, because time for dodging is available. Thus, it does not select for short appendages or heavy adductor femoris muscles. In short, use of a long, well balanced, deeply penetrating throwing spear which kills quickly—as eye-witnesses to spear killings of large ungulates can attest (e.g., Robins 1963, p. 115)—would preserve the "primitive" hip and leg structure of late Acheulean man.

The perfection of killing by means of thrown spears opens up two new avenues of hunting, one being ambush hunting, and the other, the killing of confused frightened game that is driven past waiting hunters by other hunters or by grass fires, as is vividly described by Robins (1963). The important thing here is to confuse and frighten otherwise dangerous animals, so that they will not attack hunters but are bent on escaping and can thus be skillfully manipulated in their movements. This kind of hunting, however, requires a great deal of precision planning, a large amount of knowledge, and much cooperation to be successful. Since emerging Cro-Magnon practiced confrontation hunting, all attributes needed for confrontation hunting were selected for, and additional new attributes were also selected for, namely, those demanded by complex, cooperative hunts that required planning, evaluation, and mental, physical, and technical preparations.

The attributes dictated by confrontation hunting are great courage; self-discipline; excellent judgement of situations; excellent learning ability; great physical dexterity, strength, and physical fitness; strong motivation; strong ability to motivate others and thus synchronize motivations; great compassion for one's companions; and thus great willingness to expose oneself to danger on the companions' behalf. It is the kind of selecting that produces the male bonding so typical of modern man, as Tiger (1969) emphasized. It need hardly be emphasized that great courage is required, as well as self-discipline and judgement, to walk up to an elephant, bison, or bear and stab him in his vitals and then proceed to "play" with the enraged, quick-moving beast so that its attention focuses onto oneself, permitting other hunters to thrust their lances into the prey, or to save an endangered companion. It takes plenty of altruism to do it; moreover, it would have to be repeated on many hunts. However, since the hunters would have been close relatives, being brothers, sons, fathers, nephews, uncles, etc., selection for altruism would have been rapid (Hamilton 1964). Altruism would not have evolved readily had humans been organized along matrilineal lines, since here the father, for instance, would have to sacrifice himself, not for his sons, but for his sons-in-law, who might be genetically unrelated. In short, confrontation hunting would have selected primarily for attributes of *tactical* importance; communal hunting selects for attributes of *strategic* importance.

Attributes of strategic significance are imagination, foresight, creativity, sound judgement of complex alternatives, stores of knowledge about the objectives of hunts and the means of attaining them, excellent memory recall, ability to communicate symbolically, excellent powers of observation and conceptualization, and the ability to socialize at the intellectual level so that plans are understood, accepted and unerringly executed. These abilities would play a role both in the early planning stages of a complex hunt, complex by virtue of a number of persons cooperating often out of sight of each other at a predetermined date of an expected situation, as well as in executing the hunt. In the latter case, one must use past experience in evaluating the existing situation; one must evaluate on the spot whether the planned strategy will succeed and, if not, introduce necessary alterations; one must assess how to maximize hunting success and yet minimize injury to hunters. One must consider the lay of the land, cover, wind direction, snow conditions, the habitual movements of the game and its response to the hunt, the animal or group of animals most likely to become prey, the experience of each hunter, the number and experience of beaters, the effect of grass fires on the game, and more. The primitive weapons of the Paleolithic hunters had to be made effective by great intellectual output and physical skill. It is in the planning and execution of complex communal seasonal hunts, rather than in confrontation hunting, that we probably find the origin of complex symbolic communication, namely language. Neanderthal man needed language in confrontation hunts as little as do wolves and lions in their cooperative hunting; much cooperative hunting can be done without language.

We think of language primarily as a vocal system of communication based largely on abstract, rather than on iconic signals. In its original form, it probably arose from iconic vocal and visual signals, in that the communicator used pantomime to bring across an idea. The ability to shape objects realistically is present just as early as the blade industries of late Paleolithic man, as is shown in the superb bone, stone, and ivory carvings of Aurignacean people (Marshack 1972b). This superbly developed skill must have had, as a precursor, simple outline sketches in the dust of the subjects of communication. Objects and some verbs can be communicated by visual and vocal mimicry of the object and its activities. We are superb imitators of animals, other people, thunder, running water, wind, and many other things. Early symbolic communication probably depended on visual and vocal mimicry; its origins, in turn, are likely to be in its adaptive significance in scaring away predators and attracting prey. However, we can rest assured that language did not evolve until the reproductive fitness associated with abstract, symbolic communication rose very high indeed.

Another necessary condition of complex, cooperative hunting is leadership, not the spur-of-the-moment leadership or one inspiring mimicry found in nonhuman primates, but institutionalized leadership. It requires not only an individual whose talents and success are undisputed, or who can generate loyalties by enhancing the egos of subordinates and protecting them from harm, but also an appeal to sanctioning powers. I have discussed this briefly elsewhere (Geist 1975, p. 189). In essence, it requires invoking the memory of actions first of great hunters of old age, then of deceased hunters, and ultimately the advice of spirits and deities. Complex, cooperative hunting sets the stage for individuals aspiring to leadership to invoke the sanctions of an organized myth. The origins of religions and shamanism are vividly described by La Barre (1970).

Coordinated hunting at the tactical level requires great discipline, for without it the success of the hunt and livelihood of the group are at stake. The systematic individualistic hunter requires some discipline, such as holding still and permitting flies to bite him when the quarry looks his way, or holding back from jumping forward and smashing down the prey until it stands just right; however, in coordinated hunting, greater strains are put on individuals. An individual may, for instance, occupy a position for many hours, even though no game even comes into sight; he may have to provoke a charge and plant his spear just right or lure the game into a trap despite powerful urges to run away. This takes discipline and willingness to subordinate one's own wishes and judgement to the judgement and wishes of others.

Ambush hunting, of course, depends on extreme self-control to be successful, and its use is limited by the hunter's ability to keep warm without heavy clothing that would interfere with the quick rise and throwing of the spear at an opportune moment. It could not be practiced in very great cold, since all but the best of clothing fails to prevent the freezing of extremities, thus precluding a long silent stay. Clothing as superb as that produced by Eskimos would, of course, allow waiting in great cold, but such clothing is probably a consequence of a long process of adaptation to the coldest of climates, and was not available to the earliest of late Paleolithic hunters. Therefore, hunting in which herds of prey were carefully coerced toward a kill site would have had to take place in relatively warm weather. Clothing would, however, permit both ambush hunting and complex, cooperative hunts to be extended into fall. This reasoning, together with the requirement of above-freezing temperatures to dry meat, suggests that major interceptions of game herds took place during seasons in which frost was rare or absent. Thus, migratory game herds probably supplied the bulk of the staple food during the winter, while hunters killed game in the winter more sporadically to vary the diet and to conserve food stocks. Thus, large quantities of dried meat permitted the occupation of communal, episodical dwellings and a rather dense winter settlement.

We began with Acheulean man in our discussion of modern man. The development of a meat-drying technology preserved the essentially primitive facial features, slender neck, and pectoral girdle of Acheulean man in modern man; the development of distance confrontation hunting preserved the gracile Acheulean hip and leg structure for modern man. Selection for greater manual dexterity, and especially for strategic planning abilities, would lead first to a phenotypic, and then, through selection, to a genotypic enlargement of those parts of the brain that function in planning, body control, language, etc. Hence, the enlargement of frontal, parietal, and temporal lobes, the very structures associated in man with human qualities, as indicated by the works of Milner (1956), Penfield (1959), Garn (1963), Lenneberg (1969), Geschwind (1970), and Hofer (1972).

Before Cro-Magnon

Compared to Neanderthal man, upper Paleolithic people were clearly creatures of culture and masters of adjusting themselves to the physical environment by cultural means. Their ancestors must have been under natural selection for this factor—behavioral plasticity or diversity in behavioral means of adjustment. Yet the biological differences between upper Paleolithic and Neanderthal man are not irrelevant in in pursuing the puzzle of upper Paleolithic or Cro-Magnon origins.

Note the difference in dentition: Compared to Neanderthal man, Cro-Magnon man's incisors are small; the emphasis in chewing shifts to the premolars and molars, and the temporal muscle is heavily used; tooth size tends to decrease. The chewing apparatus slides backward into the skull. It is as if the teeth are placed in such a posiion as to permit more efficient mastication with the cheek teeth, somehow reminiscent of the same backward slide of teeth and face in the robust australopithecine, compared to the gorilla. With the backward slide of the smaller set of teeth, the flat modern face, the arched cheekbone and the chin appear, as Howells (1973a) suggests. There must have been natural selection in Cro-Magnon man's immediate ancestors for chewing mainly with cheek teeth, for smaller teeth, and against a heavy use of incisors.

In Chapter 6, and again when discussing the dispersal theory, I pointed out that evolution occurs mainly under two extremes: when resources are superabundant, as during dispersal into unoccupied habitat; and when resources are scarce, when the population is at the carrying capacity of the habitat and individuals are of the maintenance phenotype. Under maintenance conditions, there is selection to minimize the cost of maintenance and to reduce organs to the minimum size compatible with high reproductive fitness. Therefore, there will be competition for scarce resources among the tissues of the body itself. Thus, in maintenance populations we expect a readjustment of growth priorities among organs as the environment changes. We noted that tooth size decreased in the evolution of *Homo erectus* (Howells 1973a). Yet teeth and tooth rows are rather constant in different populations of a species, as is so well illustrated in red deer (Beninde 1937, Gottschlich 1965). That is, though body size may vary, the tooth row is rather constant; therefore, teeth have a high growth priority. In C. H. Waddington's words, "teeth are highly canalized." Highly canalized organs are virtually a direct expression of the genes, and therefore variability in such organs reflects primarily genetic, not environmental, variability. Coon (1962) makes a spirited defence of the idea that teeth are under close genetic control. Therefore, these organs can be readily adjusted by natural selection in size. Thus, reduced tooth size probably reflects selection under maintenance conditions, that is, under resource scarcity where small teeth required fewer of the precious nutrients and calories, thus freeing them for purposes of reproduction.

Hence, the reduced size of cheek teeth in Neanderthal man compared to *Homo erectus* reflect an evolutionary history of parental populations under great resource scarcity, as well as the use of fires to soften foods. In the preceding chapter, the evolution of late Acheulean—that is, archaic *sapiens* man—was described. He evolved in temperate climates and, according to hypothesis, dispersed from Europe into Africa, probably during the Riss glaciations. In contrast to *Homo erectus,* late Acheulean man is frequently using fire in Africa, he is a very good hunter, and he makes utensils, which we know from bowls that have been found in late Acheulean sites in Europe (Pfeiffer 1972, p. 141). In short, late Acheulean man in Africa is likely to have been equipped with a technology to expand his activities into marginal habitats such as could not have been used by his predecessors, *Homo erectus*. In accordance with expectations, late Acheulean sites in Africa are frequently associated with dry, if not desert, grasslands and he reaches a higher density, indicated by the more frequent occurrance of late than early Acheulean occupation layers (Clark 1970).

It was pointed out by Lasker (1969) that hot, dry regions cannot be colonized by modern man except under two conditions: There must either be a technology to carry water during travel or there must be some form of body covering to protect individuals from the heat load imposed by the sun. Given the presence of bowls in Acheulean archeological sites, and therefore the idea of containers, we have reason to suspect that archaic *Homo sapiens* possessed means of occupying hot, dry lands.

Among mammals, as pointed out in Chapter 10, evolution has progressed in various lineages from warm mesic habitats with constant temperatures to those with little water and wide diurnal temperature fluctuations. *Homo erectus* evolved in the steppe but diverged to temperate climates. *Homo sapiens,* after adapting to the cold climates through enhanced technology, picked up where *H. erectus* left off. I propose that the roots of modern man prior to Cro-Magnon go back to very dry, hot, and seasonally variable steppe, if not desert. Here populations living in maintenance phenotypes, under conditions of resource scarcity, evolved those features that made the northern upper Paleolithic life style possible. This hypothesis explains the following.

THE MODERN HUMAN FACE AND DENTITION.   Under conditions of resource scarcity and poor phenotypic development, selection would favor organs just large enough to function efficiently so that the energy and nutrients saved could support reproduction. Since fire was available for cooking, it had the function of reducing the toughness of food and increasing its digestibility. Moreover, the emphasis on gathering would reduce the frequency of eating meat and with that the necessity of tearing meat in the *Homo erectus* or Neanderthal fashion. This would shift mastication to the premolars and molars and shift the tooth row backward into the skull in the interest of effective mastication.

THE DIVERSIFICATION OF TECHNOLOGY.   In changing from a largely hunting to a largely gathering life style in near desert, old technologies are not abandoned, though new ones would have to be added. Moreover, the scarce resources of the desert demand diversification of foraging techniques. This calls upon a greater ability to adjust by means of learned behaviors. If this is so, even under maintenance conditions there is a premium on brain size; some canalization toward large brain size will occur.

ASTRONOMICAL OBSERVATIONS. The desert presents each night a clear sky in which heavenly bodies can be observed and their regularities noted. In a desert or dry steppe in which food shortage is chronic, it pays to associate diverse environmental factors with the availability of food. The relationship between seasonal food availability and astronomical phenomena must have been noted. This would then give rise in Aurignacean time to lunar calendars, for which a history of detailed observations of heavenly bodies was a necessity.

AMBUSH HUNTING AND SPEAR THROWING. In deserts, ambush hunting at water holes is a very real possibility. Owing to tolerable temperatures in the shade, a hunter can wait without unbearable discomfort and freeze body movements when the situation demands. This calls for a store of accurate knowledge, very good judgement as to where and when to wait, and fine physical skills to rise and accurately throw a spear and impale an unweary prey. Such ambush hunting is the logical precursor of coordinated group ambush hunts of migratory ungulates, as postulated for upper Paleolithic people.

FOOD PRESERVATION AND STORAGE. In dry areas, preservation of food by drying can occur quite accidentally. It would give rise to the idea of drying food and preserving some from a given surplus for lean times. Scraps of meat dry readily and remain edible in the absence of rains for a long time. Here probably lies the origin of systematic meat drying.

LONG DISTANCE MOVEMENT. The scarce resources of the desert are expected to force large home ranges on people exploiting them, much as Birdsell (1972) found in Australia. Carrying water, according to Lasker (1969), would be a prerequisite for exploiting the hot desert biome. It is but a small step toward carrying food with the water, permitting long-range movements from water holes. That, in turn, would promote selection for preserved foods of light weight and high nutritive value. It would promote contact with people living at a distance and would be the precursor for trade and exchange such as apparently did take place in upper Paleolithic time (Chard 1969).

SHAMANISM. Scarcity of resources and anxiety about the necessary wherewithall promote the development of plausible scenarios about success and failure. This would only be possible with a fully developed language. Here I am following the view proposed by La Barre (1976) that religious cults arise from crisis such as would be the daily lot of people living under resource scarcity in warm deserts (see also p. 320). Full-blown shamanism and musical instruments with a musical scale are already in evidence in the earliest of upper Paleolithic cultures, but more of that later.

Moreover, gathering could be carried out by females, which would free the males increasingly for hunting, provided the females supplied some of the vegetable foods for males. This would have several consequences: first, a sharper division of labor; second, the use of males as scouts for plant food localities by females, much as happens today in bushman societies in dry thorn steppe (Silberbauer 1972); and third, legitimate conflict of interest between males and females as to where to go. It is here that a third party would be required, enhancing any previous trends toward cults and shamanism.

Note what has been done: Firstly, the adaptive syndrome of upper Paleolithic people was generated. This indicated that Cro-Magnon people arose from stock whose behavioral plasticity had become broad. Secondly, a comparison of the skull of Cro-Magnon and Neanderthal indicated that Cro-Magnon ancestors evolved their teeth, jaws, and skull shape under conditions of resource scarcity but in which small teeth were adaptive and chewing centered on premolars. Therefore, we have two rather different human stocks. Because *Homo* had evolved in increasingly colder climates but gave rise to Neanderthal man and not modern man, the analogy with mammalian evolution suggests that modern man could have arisen by adapting to desert biomes. As *Homo erectus* did not give rise to a desert-adapted form, the next best candidate is late Acheulean or archaic *sapiens* man. Archaic *sapiens*, evolved in cold temperate climates, used fire extensively, was a better hunter than early Acheulean man, and had an expanded technology, one which included containers. Therefore, late Acheulean man, after moving from Europe to Africa, was an ideal candidate for penetrating into the desert biome. However, when this takes place, it fulfills the conditions necessary for the evolution of the modern face, and the activities of people adapting to life in a warm desert become precursors for the adaptive syndrome of upper Paleolithic man. This picture, therefore, shows people with modern facial and skull features in Africa very early, some time after the appearance of the late Acheulean there. Also, in the vastness of Africa with its diverse biomes, one would expect the period from late Acheulean to the Würm stadial to be probably several hundreds of thousands of years long, with considerable cultural and biological diversification among human beings. Thus, at the beginning of the first Würm stadial, we expect archaic *sapiens* with Neanderthaloid features and a Mousterian tool kit in North Africa and the Near East, essentially modern people occupying at least dry steppe and desert to the south of the Mousterian people, and probably some archaic aberrant *sapiens* forms in other major biomes or parts of Africa. As far as I can tell, this picture does fit with the finding of Protsch (1973) and the review by Howells (1973a). An essentially modern but robust skull and skeleton found near Lake Rudolf (Omo I) and a more archaic one may date back some 130,000 years. Contemporary with Neanderthal man during Würm I are a number of African skulls below the Sahara that are thoroughly modern and date back 48−45,000 years. Protsch (1973) identifies these as *Homo sapiens capensis*. These can apparently be distinguished from a new wave of people who appeared as contemporaries of upper Paleolithic people in Europe, Asia, and Australia (Protsch 1973, Howells 1973b). According to Protsch (1973) and also to radiocarbon dates given by Howells (1973a), *capensis* is a contemporary of Rhodesia man. Upon reading Clark (1970), one wonders if *capensis* and Rhodesia man are bearers of different cultures and segregated ecologically, *capensis* with a culture found in steppe and Rhodesia man with a culture more suited to forest (Sangoan culture). Protsch (1973) claims in his review that a number of authorities link *capensis* with the Khoisan people. If this is so, this would make these people the last remnant of a radiation of pre-Cro-Magnon people. As discussed later, the nature of the sexual organs of the Khoisan people harmonizes entirely with this interpretation.

Consistent with the picture just presented is another finding pointed out by Clark (1970, p. 116). Apparently the first Mousterian sites overlie a pre-Aurignacean blade industry in North Africa. As the climate grew cooler and more rainy in the Mediterranean basin with the early phase of the first Würm stadial in Europe, grasslands must have taken the place of desert before giving way to forest. We therefore had, just at the beginning of Würm I, the same conditions as at the end of Würm I around the Mediterranean, namely, grasslands and plenty of migratory ungulates, conditions that gave rise to the upper Paleolithic culture and adaptive syndrome discussed earlier in this chapter. Modern people must have been already in Africa, so both the scenario I proposed and the Omo I and II finds of greater antiquity suggest. We see in both instances a similar blade industry develop, except that at the beginning of Würm I it disappears and Mousterian artifacts succeed it. Archaic *sapiens* of the type found in North Africa with Mousterian artifacts would not have developed a pre-Aurignacean culture because their culture—note the adaptive syndrome of Neanderthal man—did not treat meat as a substance to be dried and stored. Their prognathus jaws still tell of heavy meat eating and tearing of meat rather than cutting and drying it.

Some Attributes of Cro-Magnon

According to the hypotheses proposed, upper Paleolithic people had an economy based on the interception of migratory game herds followed by making multiple kills and storing the excess meat in a dried form for future need, together with the use of a new and probably safer mode of confrontation hunting, long-distance confrontation hunting. The latter probably was of secondary importance; it subsidized the food resources obtained by interception in the diet. This economic strategy had secondary and tertiary repercussions, and we shall examine some of these in the remainder of this book. Upper Paleolithic people originated in the Mediterranean basin at the end of the first Würm stadial when, for a period of time, climatic conditions favored grasslands and the resulting migratory herds of medium-sized ungulates. Once upper Paleolithic people began to spread, they did so very rapidly, appearing in Europe, Asia, Australia, and America, as well as in continental Africa apparently within a few thousand years. This is discussed in greater detail in Chapter 14.

How were they able to disperse into Europe when the rich periglacial ecosystem that once supported Neanderthal man had withered and collapsed? Although the periglacial loess steppes were gone and replaced by forests, the river basins leading to the Mediterranean and Black Seas must have been rich in large mammals still, even if not nearly as rich as in stadial times. Mammoths did quite well during interstadials to judge from the very large body size they attained then, compared to that of stadial times (Kurtén 1968a). However, as habitat for the highly social grazers shrank into ever more and ever smaller islands, they must have relied increasingly on complex migration patterns between small patches of habitat available during an interglacial, just as mountain sheep (Geist 1967b, 1971a) or bison (Maegher 1973) do today. This is a most important point. Neanderthal man could not cope with this situation, for it demanded possession and use of a system of keeping chronologic time in order to intercept the meager migrating populations of prey. Neanderthal man depended on the continuous presence of dense game populations but not so the evolving Cro-Magnon. With Neanderthal man's extinction, late Paleolithic people could penetrate easily into Europe and, with their calendar and ability to plan well ahead, they could successfully hunt the large grazers when Neanderthal people could not. The geography of Europe suggests that several groups could have entered Europe after Neanderthal man's extinction, and the coexistence of Aurignacean and Perigordian artifacts may be a reflection of this.

During the dispersal into Europe of modern-type people, we can postulate consequences predicted by the "dispersal theory" (p. 256). We expect, as a consequence of dispersal, selection for neoteny, large body size, and high population quality. This predicts the reduction of brow ridges, as well as enhancement of brain size above that of the parent populations. Thus, we find in Cro-Magnon a large-bodied people, although the exact dimensions are disputed (Weidenreich 1939a, Mettler 1955, Hulse 1963, Coon 1962, Howells 1973a,b). They have athletic skeletons showing exceptionally well-developed areas of muscle insertion in male and female (Vallois 1961), an indication of frequent and intense exercise in combination with good nutrition. Their bodies apparently were lean (Coon 1962). The skeletal remains reveal a remarkably disease-free body, while homicide is comparatively rare compared to Mesolithic people (Chapter 14) or even Paleolithic gatherers from Africa (Clark 1970, p. 171). The upper Paleolithic people attain the largest brain sizes on record (Mettler 1955, Hulse 1963), at times larger than those of Neanderthal man and, on average, considerably larger than those in modern populations. Coon (1962, p. 584) gives an average of 1580 cc for males and 1370 cc for females. This may, in part, be due to the large body size of upper Paleolithic people. From the point of view of population quality, there is not the slightest doubt that these people enjoyed a superb ontogenetic environment, a conclusion substantiated by the cultural remains.

Since late Paleolithic man is, according to hypothesis, a neotenized late Acheulean man, it follows that he displays some *evolutionary behavioral regression*, much as I showed existed in advanced—as against primitive—mountain sheep (Geist 1971a, pp. 332−37). Neotenized forms are expected to show many more juvenilelike behaviors which, in our case, would mean that we were more playful, curious, ready to explore and learn, and at the same time more sensitive to social feed back (see p. 333) than, say, late Acheulean man. It also means that previously fixed behaviors are deritualized and somehow made crude and inelegant and less frequently used, again shown well in mountain sheep. This evolutionary behavioral regression has nothing to do with regressive behavior in disturbed human beings, in which under great stress individuals revert to childlike behavior. Evolutionary regression by way of neotenization appears rather to break down genetic behavioral mechanisms evolved by the parent species and opens up new avenues for selection to shape new behaviors. It is probable that the development of dancing, music, humor, and even language depended to a great extent on "play" for their formation.

Although late Paleolithic people soon lived in the full-fledged periglacial environment owing to the advance of the second set of Würm glaciations, some evidence indicates that they lived in a more benign climate than do arctic Eskimos today. Had Cro-Magnon, who is exceedingly similar to Caucasoids (Le Gros Clark 1964, Coon 1962, Baker 1968), lived in extreme cold, then we would expect in Caucasoids a distribution of sweat glands similar to that of Eskimos. Eskimos show sweat glands concentrated on the face, while their body and feet have relatively few sweat glands in contrast to the condition found in Caucasoids (Schaefer et al 1974). The well-grown skeletons of the late Paleolithic people also indicate that they did not have to work as hard as Eskimos do, because the active hunting Eskimo males develop chest deformity as a result of the hard labor they experience in procuring a living under High Arctic conditions (Schaefer, personal communication). The view that the climate was rather benign in glacial Europe is supported by the reconstruction of periglacial climates during Würm II in Europe (Washburn 1973).

Like Neanderthal, so upper Paleolithic man continues to broaden his ecological niche. The number of species he kills in Africa, for instance, increases; in late Acheulean sites, one finds 8 to 12 species and in late Paleolithic, 12 to 49 species (Clark 1970). In the Ukraine, Mousterian man apparently killed some 14 species of mammals compared to 23 (plus birds) by upper Paleolithic people (Klein 1973). With the extinction of the large herbivores, Paleolithic man in Europe not only shifts readily to animals preyed upon by wolves, such as reindeer, antelopes and gazelles, but also begins to exploit salmon runs (Pfeiffer 1972). This is not too astonishing if late Paleolithic people intercepted migrating reindeer. These cervids tend to cross rivers preferentially above or below rapids and these are excellent places to watch and catch migrating salmon. Late Paleolithic man's ecological niche continues to expand (Clark 1965), particularly with dispersal, and he becomes more efficient, as is indicated by the game procured per unit effort invested in tools. Clark (1970) points out that the units of bone per unit of worked stone increases in modern man compared to earlier forms. We also find in the upper Paleolithic the very first evidence of trade (Chard 1969), a significant invention that is the economic counterpart of migration. In migration, consumers move to resources in trade, the resources move to the consumers. Trade implies a broad awareness of other people and other customs and is the prerequisite to speedy cultural evolution through the exchange of ideas.

The Early Anxieties

There were, however, anxieties in the life of upper Paleolithic people, so one can interpret the cultural evidence they left in the form of carving, sculptures, paintings, and systems of notation. They are evidence for religious cults, and these in turn, as La Barre (1970) shows, are evidence of anxieties, deep concerns, and crises. The adaptive syndrome of upper Paleolithic, as well as of Neanderthal man which I developed earlier permit new insights and extend the findings of La Barre (1970) and Marshack (1972a,b). In addition, ethological studies have given new insights into primate social signals, and these in turn permit new interpretations of some aspects of upper Paleolithic art.

Where success depends on close cooperation, the loss of a companion is especially traumatic, since close social bonds, emotional bonds, are woven in the process of being together and doing things together. The loss of a companion not only leaves an emotional void, but it is a genuine health hazard (Young et al 1963, Rahe et al 1967); it affects the success of the group, and it imposes a burden of rearranging functional relationships within a group. The escape of a loved one to "dreamland"—after all, he or she does reappear regularly in dreams to communicate—is only too readily accepted, leading to affectionate care of a body even after death. There is evidence for such care from Mousterian times, as well as some evidence that injured individuals were cared for (Pfeiffer 1972, Clark 1967).

That some anxieties were experienced during and after cannibalism by Mousterian people is probably reflected in the ritualized setting of skulls (Blanc 1961). If the dead go to some kind of a beyond where they continue to live, it is logical to conclude that one might meet there someone whom one ate. This requires either some ceremony of appeasement, or a ceremony ensuring that the victim does not go off into the world of the dead. However, such worries are likely only in times of reasonably abundant food. When cannibalism occurs due to dire need, scattered bone fragments are most likely all that remains as evidence.

Upper Paleolithic man, however, had more anxieties than Neanderthal man. His anxieties are closely tied to his adaptive syndrome. The first one was contacting migratory herds of big game at the right time and place, and the other arose out of the new relationship of woman to man, a relationship probably not in existence in Mousterian times. Scavenging carnivores were probably less of a worry, given the efficient use of meat brought to the cave—including the boiling of bones and the drying of meat in warm weather, as well as the larger groups of people available to scare off or attack and kill a predator. Cannibalism and homicide are apparently infrequent, although the preoccupation with death is great, as indicated by burial of clothed, decorated and well-equipped bodies (La Barre 1970, Prideaux 1973). But his greatest anxieties, and therefore his concerns, dealt with the animals he hunted, and the woman he lived with, an anxiety that is consistent with late Paleolithic cultures, be they Perigordian, Gravettian, Solutrean, Magdalenian, or whatever.

In the preceding subchapter, an argument was made that late Paleolithic man evolved with a culture that, by means of keeping chronologic time and organizing well-planned hunts, intercepted migratory herds of ungulates at strategic points, speared a temporary abundance of food and dried the meat in excess of immediate use to tide the group over until the next successful hunt. I fully accept Marshack's (1972a,b) conclusions that the curiously notched *bâtons de commandement,* whose notches coincide with lunar models of several months' duration, are indeed lunar calendars, carried about on a person securely tied with a thong of rawhide through a hole in the baton and notched to keep track of a chronologic time. I accept it because our present knowledge indicates that northern ungulates do indeed move annually at very nearly the same time, and that the presence of a herd at any given locality may be for a very short time indeed, lasting only a few days or a couple of weeks at most. So there is very real need and cause to worry about whether one's group will be at the right place at the right time, and if not, how to make up for the loss.

We expect such worries to be least when the population of people relative to the available game is low, and if the diversity of economically huntable species is large. This expectation may indeed be valid, since the batons are relatively rare during Aurignacean times, but quite common during Magdalenian times. In the latter period reindeer, a rather shifty migrant, was a major food source, salmon were apparently taken in great numbers, the spear-thrower had been invented, and apparently also the bow and arrow, while human populations were seemingly quite dense and sessile (see Pfeiffer 1972). This speaks of rather great anxiety, of people waiting at strategic locations to intercept the all-important reindeer migrations, and thus the keeping of a close watch of chronologic time by observing the position of the moon and the notching of batons. Hence the ritual concern with huntable animals that is expressed as the rich diverse cave art and the presence of shamanists. We can gauge such a concern by the courage it required to crawl deep into caves through narrow entrances with only a burning torch, far away from the familiar throbbing surface of the earth, often into some previously unexplored part of the cave system, and there leave evidence of one's presence in the form of painting and sculpture. No doubt mysticism and religion were well developed, as suggested by Blanc (1961), Leroi-Gourhan (1968), La Barre (1970), Marshack (1972a,b), and others. Moreover the art, seen through the eyes of an animal artist, reveals a superb power of observation and exactness of mental notation, and therefore a high degree of familiarity with the animals they painted. The same conclusion is reinforced by noting that the animals, where interacting, use the correct social signals. The keenness of observation also suggests a great concern with the object of the hunt.

And yet the animals must have been a source of mystery also, in as much as it is unlikely that any one group of hunters knew the reasons for the migrations and followed migratory herds during the yearly cycle. Reindeer quite likely moved long distances between the sea coast and the glacial moraines, wintered in boreal forests where the herds did not stay long at any one time, and suddenly appeared on their annual migrations at traditional stream and lake crossings to vanish again until another season. Moreover, knowing what we do about caribou (Kelsall 1968, Parker 1972), we are only too keenly aware that in some years the animals shift migratory patterns. Thus, in some years the Paleolithic hunters waited in vain, puzzled by the absence of the animals, frightened, and concerned with quickly finding an alternate food source. They must also have been puzzled by the sudden reappearance of migrating reindeer at river crossings apparently abandoned for several years. It is likely that with some variations the same could be said about other migratory ungulates, such as horses and mammoth. No wonder they attempted to find some control over these animals. No wonder stories appeared of the earth spewing out herds of game and swallowing them again, no wonder subterranean searches were made. Maybe cave bears knew where the game went; did they not move into the bowels of the earth each fall to reappear again in spring? Where else had they been in winter but with the big herds? Had they not found ibex and chamois in caves, often deep inside the earth?[1] He who could correctly tell when the herds came, and ensure adequate meat for all, was indeed a great, powerful man. Yet strangely enough, the appearance of the herds could only be foretold by specific positions of the moon and counting off a certain number of days thereafter. That must have been mystery enough! It was, after all, not until the early years of the twentieth century that it was explained how animals kept such surprisingly accurate chronologic time, so that year after year birds, for example, appeared at much the same time and place, and how this mechanism was set in different environments. We owe the basic discovery to Rowan (1926), who found that the light regime sets the seasonal migrations in birds, and we owe to Bissonette (1938) the discovery that dosages of light affect the activity of the pituitary gland and it, in turn, times the activity of internal organs.

The pictures of shamanists that have been brought to our attention (Marshack 1972b, p. 272–73) show us that they were considered fearless when performing their duty. All have exposed genitals, and the phallus appears to be erected in all. This does not indicate only the apparently obvious, a preoccupation with fertility, although this must also have been on the mind of the shaman (La Barre 1970). I explained earlier the significance of phallus erection as a signal of supreme confidence (Chapters 5 and 13). Clad in the dress of powerful bison bulls of large stags, they fearlessly confront something with bare hands, something that needs to be

[1]Mountain sheep may go very deep into caves, get lost, and die there (see Scotter and Simmons 1976, J. Mammal. *57*, 387–89).

intimidated and to be obedient to the shaman's command. We see again the same preoccupation with fearless bravery or supreme confidence in several sketches showing a man, without weapons, dancing with erected phallus in front of a bear; the second showing the man, phallus erect, on his back, in his most helpless and vulnerable position, lying in front of a wounded bison whose gut has spilled out (Marshack 1972b, pp. 272–77). These are powerful messages for whoever they were intended for—maybe for the same something that had to be kept in check; they signaled that here were individuals so capable of speedy evasion and so fearless as to be able to have an erection in front of terribly dangerous opponents. Such an ability is indeed remarkable! Woe to the animal that dared to confront such a hunter of such skill and self-confidence in his abilities. These phallic signals appear to me to be signals intimidating spirits, just as phallic symbols today are used for much the same purpose (Wickler 1966, 1969, 1971, Eibl-Eibesfeldt 1970, p. 42). These signals are most likely to be attempts at controlling animal spirits, intimidating them, making them behave, and inducing subservience to man's wishes, maybe so that the game herds appear at the expected time in the expected place, or so that dangerous big game loses its courage and fails to attack, ensuring a safer hunt.

The preoccupation with woman by late Paleolithic men, as revealed in the many "Venus" figurines, is not surprising, since women must have been a source of anxiety that was almost certainly not experienced by men from earlier cultures. In the late Paleolithic, men were probably for the first time dependent on women for far more than sexual gratification. The services of women provided short-term benefits, such as clothing and the care, protection, and preparation of food, as well as long-term benefits, namely the production of offspring who, by growing up to be good hunters and homemakers, would provide security in old age. As discussed later, it is with the late Paleolithic people that the three-generation family arise. We know from present-day Eskimo cultures how terribly important the skills, knowledge, work, and encouragement of women are to the very existence of a male, let alone his social success (Freuchen 1961). The sources of anxiety were several: the dangers of childbirth, particularly acute if the females gave birth to large babies as a consequence of good protein-rich diet; the higher mortality among women than men (Vallois 1961); the danger of the wife getting injured or killed by a large carnivore while the males were out for a hunt or a ceremonial occasion; the danger that a wife would become disenchanted with her man and fail to serve him adequately; the danger of a lazy or argumentative woman who neglected husband and children. Moreover, the females apparently were the sex in lesser abundance (Vallois 1961) and were probably highly desired by "surplus" males. There were probably other sources of anxiety. In essence, these anxieties arose out of the insecurity of the male, a product of the fact that he could not look after himself easily and almost certainly could not look after himself and his children without some help. Moreover, the hunters set the general rhythm of camp life, that is, when to move and where to stay, since such decisions were controlled by the need to intercept game herds. The hunters probably also directed complex hunts, and thus again imposed their will on women who had to abandon other activities to participate in the drives. Quite obviously this must have led to friction, since the interest of the hunters was not always that of their wives. If sufficient food—and particularly social success—were to accrue to the hunter, he had to have an obedient wife. This hypothesis permits an explanation of some of the more puzzling aspects of the Venus figurines.

The Venus figurines begin as quite naturalistic renderings of the female form early in the late Paleolithic, and through the various cultures become increasingly abstract, as is well illustrated and discussed by Marshack (1972b). The same, incidentally, appears to be valid for the renderings of animals and of the batons which become smaller and more graceful; this follows closely cultural ritualization so well analyzed by Koenig (1970). From Aurignacean times we even have a beautifully carved woman's head with long hair, a graceful neck, and delicate facial features (Clark 1967, Figures 38−43). Some of the realistically carved figures reveal quite beautifully proportioned bodies, perfectly compatible with present-day popular taste. However, even in the early late Paleolithic one notices an emphasis that becomes extreme in the terminal phases of the late Paleolithic: there is an emphasis on buttocks, breasts, occasionally on a belly distended by pregnancy, and the vulva. These female attributes are further abstracted into symbols, as illustrated by Marshack (1972b).

The emphasis on the buttock is shown not only in its exaggeratedly large size, or the correct rendering of the oversized fat layers which run diagonally up to the outer edges of the iliac blades, but also by the eversion of the buttock in some standing or kneeling figurines, and especially in the sketched female forms. The buttocks are presented thus, reminiscent of buttock presentation in cercopithecine primates or sexually aroused chimpanzees (Wickler 1967, Goodall 1971). I shall call it the "butt-to" posture.

The oldest meaning of the butt-to posture in primates is as a signal of sexual receptivity; it derived a secondary meaning as a signal of submission, and then of appeasement (e.g., Kummer 1968). It was hypothesized that in early hominids males probably used female mimicry to appease larger males, so the butt-to posture probably had the two primate meanings in early hominids. For early *Homo*, a case is made later for females binding males into a pair bond to support the female's reproductive efforts by means of sexual stimuli. This required a permanent sexual signal and an enlargement of the buttocks would roughly mimic the old primate estrous swelling. This may have been the origin of steatopygea. Late Paleolithic man evolved from late Acheulean man, a form that had not adapted fully to cold climates and that, therefore, is expected to carry externally all the sexual signals of earlier, warm-climate adapted hominids. Therefore one expects late Paleolithic man to continue to show strong interest in old female sexual signals. The butt-to posture, however, has probably begun to signify a submissive female, one who readily appeases and pleases. A submissive female, however, is the best requirement for a male, granted his need to direct female activity by directing the shifting and location of camps and the role females will play in cooperative hunting. The appeasing female is likely to get the work done that has to be done to keep the male and the children in good health and spirit; that it was physically demanding work is indicated by the powerful muscle insertions on the skeletons of late Paleolithic women (Vallois 1961). Willingness to work hard and obey were, from the male's point of view, highly desirable characteristics, and his anxiety about them was apparently great enough to invoke ritual. The excellent body development of late Paleolithic people in itself betrays that their life was very demanding, both physically and mentally, and we have reason to doubt that individuals accept hard, even painful, work for the sheer pleasure of it.

The butt-to posture has been and still is in use in diverse functions: as a greeting posture by women in an African tribe; as an appeasement posture against spirits; as a precopulatory posture; and it is enforced for purposes of punishment (Wickler 1969,

Eibl-Eibesfeldt 1970, p. 201−2). It therefore functions in its ancient context still.

The preoccupation with the female breast is probably related to the wish for good mothering. We must remember that these people did not have the milk of cows or goats to fall back on if the female's milk supply failed. Yet, fail it would if the woman was subjected to stress and unhappiness. To maximize milk production the wife had to be kept contented during the years of lactation lest children suffered (Newton and Newton 1950, Newton 1971a). Child mortality was quite high (Vallois 1961, Cook 1972) for reasons unknown; the skeletons of children are quite disease free. Of course, breasts have erotic functions also, but so has almost any part of the female's body, depending on the state of the male. The preoccupation with the vulva is obvious, as is that with the pregnant torso.

As La Barre (1970) points out, exquisite as the renditions of animals are in upper Paleolithic art, it is curious that in the Venus figurines the face, feet, and hands were not represented—as if purposely obscured. Only those parts of the body normally hidden from view by clothing were exposed. Were the face, feet, and hands obscured on purpose? It appears so, probably for the reason that face, feet and hands are often exposed to view and thus to scrutiny. Persons can be recognized from their faces and also their hands and, presumably, their feet. Would not a naturalistic representation of the face permit spirits to recognize a man's wife when he was engaged in ritual acts? If so, spirits could revenge themselves. Phallic cults among upper Paleolithic people indicate power and coercion. The greater the anxieties the more diverse and sophisticated the cults, the greater the attempted coercion of spirits—the less a Venus figurine ought to give clues of personal identity. This suggests why the later upper Paleolithic Venus figurines were more abstract than earlier ones.

From Marshack's (1972b) review and interpretations, we know that female figurines were kept—often hidden—in pits and hollows of dwellings, and that in East Siberia the native tribes conceive of the spirit of the hearth as a woman and keep her image in the tent. The female figurines of Paleolithic times were also carried about on the body as indicated by holes pierced through the figurines; the holes show signs of wear from a string. Female figurines were frequently manufactured.

It does not surprise us, therefore, that when a man died he was given not only weapons in his grave for service in the after-life but also *bâtons de commandement,* his calendar with whose mythical powers he could intercept the migratory herds and ensure abundant food, as well as female figurines to give him an obedient hardworking wife who was a good mother and temptress. Nor does it surprise us that when women died they were decorated greatly, for they were cherished as so very much of the welfare of the group depended on their skills and knowledge.

Megafauna and Late Paleolithic Hunters

Since late Paleolithic man increased hunting effectiveness by being able to predict migratory movements of social ungulates, and hunt these animals in coordinated hunts probably involving small communities and using effective weapons such as throwing spear, atlatl, and finally bow and arrow, it is not surprising that the large mammals that neither Acheulean man nor Neanderthal man could touch begin to vanish with the dispersal of modern man. Megafaunal extinctions sweep the continents and distant islands as modern man occupies them. The species most resistant to man's hunting are solitary forest dwellers and mammals of small size that have even prospered in modern times (Geist 1971c). I concur in part here with the concept of overkill as advanced by Martin (1967, 1973) as an explanation for late Pleistocene and Holocene megafaunal extinctions.

One may wonder, though, why such species as mammoth, steppe wisent, and woolly rhino survived for tens of millenia along with modern man and disappeared only at the end of the last Würm glaciations. The probable answer is that in the rich, fertile, productive landscapes of the periglacial regions the megafauna was able to reproduce faster than man was able to make inroads. Once the periglacial environment vanished, we would expect reproductive rates and recruitment rates of these large mammals to drop as the areas and quality of their habitat declined. Moreover, their habitat would become increasingly fractionated and their relict populations become increasingly dependent on exacting movements between these patches of habitat. Their movements in time and space would therefore become increasingly precise with the decline of their populations, and thus increasingly susceptible to interception by hunters who probably knew the habitat requirements of these animals only too well. If, because of harassment, these areas were denied to them, their populations would decline further still, as Batcheler (1968) has demonstrated experimentally on red deer in New Zealand. The extinction of large mammals and birds coincides with the appearance of modern man in Australia, America, Madagascar, and New Zealand (Martin and Wright 1967, Martin 1973, Haynes 1966, Clark 1967, Reed 1970, Hester 1970, Klein 1972). However, Grayson (1977) points out that in America the avifauna suffered a rate of extinction following the last deglaciation which is comparable to that of large mammals. No overkill scenario can account for these extinctions. This indicates that rather drastic changes occurred in the ecology of glaciated continents following deglaciation, a conclusion also developed in Chapters 9 and 14.

On the Ecology of Bonding Male and Female

It has been claimed that we shall never know how the human social system based on pair bonding, and thus family life, came about (Young 1971), and in a narrow sense this is true. We can, however, get a rather plausible picture of how such a social system arose, provided we do more than superficially examine the traditional sources of information on which speculations about the evolution of the human nuclear family are based. Gough (1971) lists these as data on primate social systems, archeological data from artifacts, and the social systems of primitive cultures. One may note that Gough (1971) fell victim here to the "phylogeny fallacy" I discussed in Chapter 4; we must look beyond primates. Moreover, it appears that up to the present no attempt has been made to use classical evolutionary theory and ecological knowledge to explain the development of pair bonding in the genus *Homo* as illustrated below.

The starting point is the social system as practiced by *Australopithecus africanus*, which probably was not dissimilar to that found in steppe-dwelling cercopithecines. It would have been a multimale system or a weak age-grade system using the classification of Eisenberg et al (1969). I described such systems in Chapter 10. It would not have been a "human" social system for a number of reasons.

In warm, productive climates and habitats, females need not have looked after males by preparing food and clothing, or by reinforcing their egos as hunters. Nor was there any great need to rely on old individuals for their knowledge. The knowledge required to keep the population in good reproductive condition was minimal compared to that required under periglacial conditions. In short, old people were dispensable. Nor did the females need males in order to feed themselves and the children. For this reason, there can be no claim made for the early gracile forms sharing food as J. Desmond Clark (1970) suggests. Such food sharing arises only when the pregnant or lactating females cannot gather enough food to maintain reproduction. This situation arose when a dry season disrupted the vegetative season and surface gathering was no longer possible. I discussed this in Chapter 11.

Granted a seasonal superabundance of food on a regular annual basis, it would appear that synchronous mating and parturition would evolve. Not so in primates with a very long, slow development, unless they are small-bodied species and their ontogeny is short relative to the season of superabundance. In ungulates or small-bodied savannah primates with a short period of growth, in which the young reach large size and independence, seasonal births are adaptive. If the major growth period of the young coincides with the period of abundant forage, then much or most of the growth can be accomplished in that period, and the following season of scarcity will find a relatively large well-adapted youngster who can fend for itself. In a large-bodied primate with a long ontogeny, however, even if the infant is born during the season of abundant food, it still depends on the maternal milk supply during the following season of scarcity. Therefore, the female must somehow find adequate forage to sustain lactation at that time.

One way in which this could be achieved I pointed out earlier: the females cluster about dominant males who exclude competitors from choice forage areas. This would select for sexual dimorphism (Geist (1974a), which was indeed most noticeable in *Homo erectus* (Brace 1973). However, if the females depended heavily on stored hidden plant resources, that is, on tubers, bulbs, and corms, they would exist on a relatively starchy diet. Such a diet is conducive neither to fetal development nor to lactation. The fetus and infant require a maternal diet rich in protein for their optimum development.

In the dry, hot steppe, much of the available protein is tied up in the biomass of grazers; insects, snails, and lizards are relatively uncommon. Meat can be obtained by scavenging, but this requires relatively large groups of hominids to chase away carnivores from their kills. In the dry season, with its dispersed, scarce resources, such groupings are unlikely. Had scavenging of the kills of large carnivores been of supreme importance, then we would expect that it would be advantageous for the females to ensure for themselves a large portion of the meat to ensure reproduction. It would have selected for large masculine females that could successfully outcompete males at the carcasses, just as it happened in spotted hyenas (Kruuk 1972). The same would have happened if *Homo erectus* had killed large, rather than small, mammals. The large carcass would have been rather immobile—just like a scavenged carcass—and could easily have been disputed over. Moreover, the killing of large game would have required several hunters; thus, conditions would have existed for intense competition for the killed animal. Had this been the case, clearly sexual dimorphism in *Homo erectus* would have been minimal. Yet the converse is found (Brace 1973).

The foregoing makes it evident that females were not able to forcefully usurp what males hunted. Rather, they depended on the male voluntarily bringing kills. Since during the dry season males were not likely to be in large groups, and could not hunt large game, clearly they must individually have taken only small-bodied prey. In exceptional circumstances only would the single male be lucky enough during the dry season to take a kill from predators, so most of the meat supplied at that season must have come from systematic hunting. This fits well with the view elaborated earlier that *Homo,* compared to *Australopithecus,* was both an opportunistic and a systematic hunter. It also explains why the robust form of hominid, *Paranthropus,* survived alongside *Homo* for a long time. For most of its long existence, early *Homo* probably killed only relatively small and medium-sized prey, and only at short distances at that. The male, by virtue of being free of the burden of a child, could hunt systematically with greater efficiency than the female. Thus, the potential existed for the male to subsidize the protein requirements of his offspring through the gestating or lactating female.

An adaptation new to primates emerged, that of systematic food sharing. It is found as such among some carnivores, such as foxes, wolves, and hunting dogs (*Lycaon*) (Tembrock 1957, Mech 1970, Kühme 1965, van Lawick and van Lawick-Goodall 1971, Schaller 1972). In rhesus monkeys, males permit estrous females greater access to food (Carpenter 1942). In chimpanzees, some food sharing occurs between female and young, exceptionally between a male and a female in estrous, and between adult males on rare occasions when one has made a kill (Yerkes 1940, Goodall 1971, Teleki 1973). The most likely way for a female to receive food, or even priority of access to food, would be by assuming an estrous condition or a condition mimicking estrous. Food sharing would increase the longer her period of true or apparent sexual receptivity. If this hypothesis is granted— repugnant as it may be—it at once explains the most unusual human sexual physiology, psychology, and behavior, of which the most striking part is the virtual emancipation of sex from its earlier limited function of reproduction to a mechanism of social bonding. For a detailed discussion, I refer to Young (1971) (see also Eibl-Eibesfeldt 1970, Wickler 1968 in Eibl-Eibesfeldt 1970).

Zoologically, probably the most striking aspect of our sexual behavior is not only the extended receptivity of the female, or the relatively high frequency of copulation, but the fact that the female remains receptive until well into the final months of pregnancy and continues to experience normal orgasms (Masters and Johnson 1966). This condition, which appears to be unique among mammals, speaks of a long evolutionary history of social bonding through sex. It is, however, only *one* bonding mechanism between sexes, as I shall show later.

If food sharing is to increase in frequency and extent in the population, the rules of evolution dictate that the male should preferentially share his meat with the mother of his children, or at least with the mate carrying his brother's children, or with his sister, but not with any female that happened to come along. A special bond must thus exist between the male and the female carrying the male's children and raising them. The roots of human cooperation appear to be found here, in the bonding of male and female, as well as in the grouping of males for defence. Secondly, the female must assume an exterior form and a behavior conducive to long-term bonding by first being sexually attractive. This would lead to an increase in the sexual dimorphism in *H. erectus* over that found in *Australopithecus.* Thirdly, since, under the condition in which a male preferentially supports the female carrying his children, the male limits the number of females he mates with, he must become quite selective in his choice of mates and pick the best possible mother for his children, and prevent any other male mating with her; while from the female's viewpoint, she must pick the best possible hunter to maximize her reproductive success. This again is dictated by the rule that adaptations are evolved on the basis of their efficiency in increasing reproductive success.

Let me describe the foregoing in slightly different form: Under the stated ecological conditions, the female evolved as the "temptress" who tied the male sexually to her, that is, by providing a *luxury* service—Maslow's (1954) conception of priority notwithstanding—namely sex. The male could live quite well without the services of the female, although he could not reproduce. The female's strategy under these conditions is to dazzle, tantalize, and please; hence the evolution of powerful sexual signals such as steatopygous buttocks, large breasts, and probably also enlarged and chromatic labia minor. She had no other powerful hold over the male since she did not provide an *essential* service. This changed drastically with the evolution of late Paleolithic man from Acheulean man, of course, since the female made the male dependent on her as described earlier, by providing essential services without which the male could not survive let alone compete in male society. The male could not simply leave the female any more or make her obey his whims. Hence this preoccupation with women, as illustrated in upper Paleolithic cultures. However, the greater the interdependence of male and female in exploiting harsh periglacial environments, the greater the amount of essential services by the female to the male. Therefore in modern man there would be less selection for secondary sexual signals in the female, hence the waning of steatopygous buttocks, chromatic labia minora, and sexual dimorphism of the face. Thus, the greater the amount of interdependence of the sexes, or the amount of essential services supplied to the male by the female, the less selection for sexual dimorphism based on sexual signals. We have jumped ahead a little here. Under the conditions postulated for early *Homo* there would be powerful selection for the female advertising her worth.

Thus, during the transition from *Australopithecus* to *Homo*, the stage is set for females evolving characteristics indicative of good "motherhood" in their external appearance and behavior. The stage is also set for prolonged courtship and competition between members of one sex for the best mates; prolonged courtship—as expected—appears to be typical of monogamous species compared to polygynous (Collias 1944). This is a radical departure from the primate condition in which males contribute no food to the upbringing of their children and form strong bonds between males and females only in forms that defend rigid pair territories, such as the gibbon (*Hylobates lar*) (Carpenter 1964). In most species mating is promiscuous, as it was almost certainly in protohumans; here, sexual dimorphism emphasizes sexuality at best, but not competent motherhood.

It must be noted, however, that it is still to the male's reproductive advantage to inseminate as many females as possible, although he must limit his food sharing to one, or at most two females lest he be unable to provide adequately for his offspring conceived through these females. It is, on the other hand, generally not in the interest of the female to be promiscuous since she runs the risk of leaving behind children of lower reproductive fitness than those sired by her mate. It is in the interest of the female, however, to maximize protein intake provided by the male and not to share it with another female. This would favor segregation of groups into pairs with their dependents during seasons of scarcity, namely the dry seasons. It would also select for intolerance among females and competition for males, resulting in the sexually dimorphic exterior discussed above.

The foregoing makes it evident that the most primitive of human families is the monogamous or nuclear family; all other family forms are deviations from it, be they polygynous or polyandrous human families, extended families or super-families, the latter based on extended kinship ties. It is not surprising, therefore, that the nuclear family predominates in human societies and is still clearly discernible even in the extended family (Gough 1971).

Effects of Monogamy on the Dominance Displays and Courtship of *Homo*

The ancient courtship strategy of *Australopithecus* was probably similar to that practiced by Old World monkeys and apes, in which sexual dimorphism is not pronounced. Granted a female nearly as large as a male, the most likely strategy adopted by males is to intimidate the female and make her submissive. If the female is exceedingly aggressive, armed with lethal weapons, and of nearly the same size as the male, such a strategy will not work, of course. It risks severe wounding of the male and the female, and thus excessive energy and nutrient expenditures to maintain homeostasis and high dominance rank. In this situation it pays for the male to use the strategy of playing "baby," and to use also the antithesis of dominance displays. We see this very well in the American mountain goat (Geist 1965, 1975a). If the female is very small compared to the male, then a show of dominance on the part of the male may scare her easily and increase the cost of rutting to the male. Hence, a courtship attracting the female is adaptive, particularly in species with long life expectancies. It should, therefore, not surprise us that in this situation we also find courtship by the male that uses the strategy of playing "baby," as we find in the mule deer (*Odocoileus hemionus*) (Geist in press). Given low sexual dimorphism, it is thus not surprising that intimidation by the male is a key component in courtship.

In the chimpanzee, the ape closest to the hominids (King and Wilson 1975), we find a series of dominance displays by the male as his courtship. This includes not only the typical criteria of dominance displays, but also an erection of the penis, and a guarding of the female, controlling her movements occasionally with threats (Goodall 1971). The male of the Japanese macaque may, in addition, attack the female outright (Hanby in Hinde 1974). Kummer (1968) writes of the hamadryas baboon that the male punishes females that move too far, and that a male's threat may bring the female running to the male with submissive signals. McGinnis (in Hinde 1974) notes that the approach in dominance display by the chimpanzee male may actually repel the female, and yet by various gestures he induces the female to cooperate. We have the paradox that "fear" and "pain" are used by primate males to induce females to come during the period of consort formation. The female's response during the consort period is sexual presentation, as well as signaling of submission. Thus the swollen and colored genitalia, screams, "grinning," squeaking or whimpering, exaggerated or emphasized cowering recognized as bowing and bobbing, submissive kissing, and touching by the female (Goodall 1968, 1971). We have every reason to suspect that much the same kind of courtship occurred in pre-*erectus* populations of *Homo*, in which sexual dimorphism was very low; indeed we have every reason to suspect that remnants of this old courtship system (sadism and masochism) are very much with us yet.[2]

At this point we must dwell on the paradox that in the courtship of primates the male uses "fear" and "pain" to attract and hold a female—a paradox indeed! We find in aesthetics a comparable paradox namely, that the "horrible" may be "beautiful." I shall contend that the above paradoxes are not only related but have a common root. In displaying to the female with his masculinity, using the same acts that intimidate other males, the male primate advertizes his dominance. His displays ought to attract the female, as argued earlier, since the displaying male is the most dominant, and it is to the female's reproductive advantage to be inseminated by him. In short, in the above context, females have been subject to selection that makes a "frightening" male "beautiful" to them. Granted this view, then the horrible should be not only attractive and arousing but also sexually arousing. Indeed, as discussed in Chapter 5, we find in animals and humans alike a close relationship between aggression and sex, so much so that visions of violence can stimulate sexual activities; in humans, public executions can lead to orgies (Taylor 1954).

The selectivity by females for mates that are most likely to succeed in hunting and food sharing imposes upon the displays of males a new requirement. As we have seen, dominance displays in mammals normally emphasize combat potential. Thus, size, mass, and weapons are emphasized in displays. Simultaneously, the displays of rivals are suppressed by the most dominant individuals. However, a protohuman male who is high in dominance and capable of winning most fights, but is a poor hunter, is a bad choice for the female in search of a bonding partner. Thus, sheer might and prowess is no longer a sufficient condition for a female to choose a mate. How can a female pick the best father for her children?

The solution, in part, is to expand dominance displays. Feats of strength and vigor are already part of the dominance displays of primates, such as the vigorous charging displays of chimps, their "dances" in which branches are shaken, their kicking and dislodging of objects such as branches and rocks, their hunching up and bristling, etc. (Goodall 1971), or the shouting, shaking of treetops, and crashing jumps of rhesus monkeys (Altmann 1962), or the chest beating and destructive charges of gorillas, described by Schaller (1963). We also perform, in principle, similar displays. What needs to be added are the components of skill required in hunting as well as the proof of successful hunting and the sharing of the kill during courtship.

The requirement to demonstrate skills rather than strength can best be satisfied in group skill displays; it is only in group displays where comparison is possible that small differences in skill can be detected. However, social primates already have a propensity for group displays, as is shown in the howler monkeys (Carpenter 1965), or the chimpanzees in their charging displays and rain dances (Goodall 1971). We would thus expect, in protohumans starting out on their new ecological profession as systematic hunters, an occasional gathering of young males and the performance of skill displays such as jumping, spear throwing, running, etc. In short, while

---

[2]Social behavior tends to be exceedingly conservative in large mammals, as I emphasized for ungulates (Geist 1974a), as can be seen from Schaller's (1972, p. 367) work on the large cats, and as is explained theoretically by Hofer (1972, pp. 32–33).

dominance displays indicative of combat potential can be shown in isolation, skill displays within dominance displays are best shown in groups on a cooperative basis; males should perform different skill displays in unison.

It should be noted that the demands of bond formation between mates demanded at least periodic social groupings and displays primarily by uncommitted males. However, the demands of food sharing during the nonvegetative season (or season of scarcity of gatherable food) work in the opposite directions. When the female relies on food brought in by the male during the season of scarcity, only small troops of individuals would be found, owing to the dictates of the evolutionary rule that such food must be applied toward the reproductive effort of the male, to the requirements of systematic hunting that game be disturbed as little as possible, and to the requirement that competition for the meat brought in by the male should be confined to individuals closely related to the hunter. The troop would consist of one or two males, preferably brothers or father and son, moving out with their mates and children. Such a dispersion of the population would also reduce the amount of traveling required by the individual, compared to the requirements in a large band. We must therefore postulate small group sizes for the late *Homo erectus* and early *Homo sapiens* stages. Larger groupings would probably occur only during seasons when plenty of gatherable food could be found.

Another victim of monogamy was the behavioral estrous of females so prominent in most primates. A short season of sexual receptivity, coupled with strong visual, vocal, and auditory displays that attract males all around, would have been counterproductive to the female's strategy of reproduction. It would have attracted strange males, excited them, and increased active competition for her mate. This would have greatly increased her chances of losing her mate, and thus losing the food supplement brought in by the male to the children. It would have reduced the female's reproductive fitness. Thus bonding, which maximizes the stimuli that keep partners together and cooperating in a common interest of enhancing their reproductive fitness, clearly is incompatible with estrus.

However, selection against estrus cuts both ways. It is also unadaptive for a male to be sensitive to another female's estrous condition, since if he responds he is likely to get into fights and increase the cost of living to himself, detracting from his efforts to support his own offspring. Clearly, selection will favor males that do not become attracted to estrous signals of females.

Primate males are sensitive to and attracted by volatile aliphatic acids in the vaginal secretion of females (Michael et al 1971, 1974, Michael and Keverne 1968). We therefore expect that human males would lose their sensitivity to these substances, and that they would not play any noticeable role in human sexual behavior. However, human females continue to secrete these pheromones, which increase in concentration during the follicular phase of the menstrual cycle (Michael et al 1974). Vaginal pheromones from humans continue to possess sex-attractive properties to a variety of primates, as has been tested experimentally (see Michael et al 1974). Thus the vaginal pheromones of women are an old, nonfunctional relict that indicates that the human mating system evolved from one in which male competition for estrous females was intense. The findings of Michael et al also confirm the hypothesis that once selection began for bonding, hominid males were selected that were insensitive to volatile aliphatic acids.

Behavioral estrus, on the other hand, is most adaptive to the female in a social system in which the male does not contribute to the development of the offspring materially, but maximizes reproduction by maximizing inseminations. If estrus attracts and excites males, they enter into competition around the female, so that the most dominant male has priority for breeding. This fits precisely into the adaptive strategy of the female which maximizes her reproductive fitness by mating her with the most dominant male. Estrus is in essence the female's mechanism for attracting and sorting out males in accordance with their dominance rank.

The foregoing theoretical considerations predict that behavioral estrus is minimal under conditions of monogamy.

Under conditions of monogamy, the female's reproductive fitness is on the whole maximized by reducing sexual competition for her mate, as outlined above. This ought to include a reduction in estrus, and also in other signals that may arouse males. On the other hand, if the female reduces such signals to her mate, the pair bond may weaken and, at the worst, the male may leave. If the female can maximize sexual signals to her mate and minimize signals to males other than her mate, she can maximize the pair bond. This argument makes it evident why clothing of sexually provocative parts of the anatomy is adaptive, since unclothed bodies produce considerably greater involuntary arousal than clothed ones, as has been demonstrated by Hess (1965).

Since the hominid offspring could hardly have gone hunting themselves upon weaning, evolutionary theory dictates that the male would have supported his children with his hunting effort until they were as capable as he was. This requirement applies only for seasons of scarcity of gatherable food. In order to maximize his reproduction, there must be means by which the male and his offspring remain bonded until the offspring can look after themselves. It is in the interest of the children to adopt the strategies of their parents, as well as to improve on these strategies by copying others. Thus, initially at least, in the nuclear family stages of *Homo erectus* populations, the parents become largely the model which offspring mimic. Moreover, the male increases his reproductive success as does the female by passing on their learned skills. The better the hunter, the longer his children can associate with him, hunt with him, and learn from him. Since huntable prey varies from locality to locality, adaptation by learning and the transmittal of cultural knowledge is at a premium. Hence the evolutionary impetus for prolonged juvenileness and neoteny in humans, and probably the concomitant evolution toward a better brain. We have a similar, but far less developed trend in lion society, according to Schaller (1972, p. 358), where a long ontogenetic period compared to other large cats can be related to diverse cooperative hunting in adulthood. This hypothesis predicts that the duration of human ontogeny, besides being a function of population quality, should be directly related to the complexity of a population's adaptive strategies. The more variable the environment, the longer the duration of ontogeny. To what extent this is valid for human populations exploiting diverse environments with primitive methods remains to be demonstrated.

The Social Feedback System

The change from life in the savannah to life in the dry steppe, with its discontinuous vegetation season, and ultimately to life in the cold climates, produced in *Homo* another profound behavioral change. It sophisticated the social feedback system of our genus far beyond anything known from other animals. The social feedback system is the totality of sounds, gestures, and tactile stimuli that, when employed by the individual, ultimately lead to an alteration of its immediate social and physical milieu, such as maintaining its physiological homeostasis; its roots lie largely in the mother−young relationship.

The social feedback system can be explained by reference to the human neonate. The baby is quite incapable of looking after its own needs so as to maintain homeostasis. It signals departure from homeostasis, the female responds; if the response restores homeostasis the young signals first the appropriateness of the response, reinforcing the female in her activities; it terminates signaling when its physiological homeostasis is restored, and thus permits the female to terminate her activity. Homeostasis may have been upset by the neonate losing heat, which causes hypothermy and ultimately death of the neonate. The baby responds by crying. The female, in picking it up, cuddling it in her embrace, restores the neonate's heat balance. It gurgles and smiles in response, encouraging the female to continue. Once it has been covered up and has fallen asleep, the female can go on with other matters. Smiling and whimpering, laughter and crying, are sets of opposing signals, the first of each signaling return to, and the second departure from, homeostasis.

An important attribute of this system obviously is that it teaches the female how to behave in order to minimize departures from homeostasis by the infant and to maximize its positive responses. It permits the individual during its ontogeny to shape the behavior of the female so that it selects those environmental attributes that enhance its development to become a successful adult. The more sensitive the female is to the infant's signals, the more inventive and industrious, the more she enhances her reproductive fitness. Conversely, the more sensitive the infant the greater its chances of developing optimally, enhancing its reproductive fitness.

The system demands that the signals of the infant indicating departure from homeostasis be unpleasant, upsetting, or punitive to the female, and that there be signals by the infant indicating return to homeostasis that are rewarding to the female. It can be postulated that smiling, for instance, activates pleasure centers in the brain; crying stimulates punishment centers. It is self-evident that these signals, as well as the recognition of these signals, must largely be the result of closed genetic programing; that is, they must be innate in the ethological sense. It is beyond reasonable doubt that the motor patterns of smiling, laughter, whimpering, and crying in humans, are innate (Eibl-Eibesfeldt 1967, 1970); it has not been demonstrated that they are recognized innately, but there are claims that human females are irresistably attracted to the smiles of infants (Jolly 1972, Eibl-Eibesfeldt 1970).

It is pertinent to note here that smiling and laughter, and also weeping and crying, are based on some very old adaptations, as is revealed by two lines of evidence. First, smiling is organized in the brain subcortically, in the brain stem. This is indicated by Vanderwolf (1969) in a review of clinical symptoms of patients with damaged cortical motor pathways. There are two systems controling facial expressions, a cortical and a subcortical one. Damage to one or the other reveals the nature of the of the control of facial musculature in emotional expression. This finding harmonizes well with the physiological complexity of laughter, its reflexlike nature, its "uncontrollability" in certain circumstances, and its innate organization. The second line of evidence comes from a comparative study of facial expressions in primates which traces smiling and laughter to some common submissive signals of Old World primates (van Hoof in Hinde 1974).

Please note that the care of the infant is based on open genetic programing, but the motor signals of departure and return to homeostasis and their recognition are based on closed genetic programing. The more care is based on open genetic programing the more the neonate benefits from the experience of the female; the more the signaling between infant and mother is based on closed genetic programing the less ambiguous is the communication, the less the chances are of the infant being ignored or treated inappropriately, and the greater its chances are of developing into a successful adult. Clearly, these deductions from the mother−young relationship and its goals predict either innate recognition of smiles and laughter and crying in humans, or quick selective imprinting on these behavior patterns, and a greater sensitivity of the female than the male to these signals.

The foregoing has been a simplified sketch, obviously, for excessive signaling by the individual during ontogeny may overburden the female and lead to negative, not positive, responses. At this point the simple sketch is necessary, lest complexity prevent one from visualizing how the above system restores homeostasis.

I emphasized that the infant's care could be enhanced by an open genetic programing of maternal behavior. Clearly, if the environment is not variable, and thus is highly predictable, there is no advantage to an open genetic program and the law of least effort favors a simple, closed genetic program; such a program would also reduce errors by the female and young in a constant environment. Conversely, if the individual is born at a late state of ontogenetic development and is thus equipped at birth with appropriate closed genetic programs, and open programs are quickly completed by way of imprinting, the demands on the female to rear the young are relatively low, and require less emphasis on open genetic programs shaping behavior. Therefore, the more varied the physical and social environment of the species, the greater its diversity of adaptive strategies; the more gregarious the species, the more "larval" the neotenate at birth; the longer its ontogenetic development, the lower its reproductive rate−the greater must be the importance of open genetic programs shaping maternal behavior. However, the more maternal behavior depends on open genetic programs, the more sensitive the individual must be during ontogeny to departures from homeostasis, the more readily he must signal it, and the more expressive—hence diverse—must be the social feedback system operating between infant and female. Thus, open genetic behavioral programing demands a sophisticated social feedback system. There is no escape from this. Conversely, the complexity of the species' social feedback system becomes a measure of the degree of dependence on open genetic programing of its behavior.

The foregoing explains at once the relatively simple maternal behavior of, say, ungulates (Lent 1974) compared to that of primates (Jolly 1972, Hinde 1974), as well as the finding that while ungulates reared by man in isolation from conspecifics still raise young in a tolerably adequate fashion (Gilbert 1974), primates deprived of social experience do not (Harlow et al 1971, Jolly 1972). Some improvement in mothering behavior in rhesus females raised in isolation does occur with repeated births, as a consequence of the infant teaching the females, but the performance remains dismal as a whole. Compared to primates, ungulates are born at an advanced stage of development, have a relatively short ontogenetic period, and a higher reproductive rate on average.

It is evident from the foregoing that under ecological conditions in which the female's work can no longer adequately supply the needs of the young, it is in the male's reproductive interest to support his offspring. This demands that the male, as well as the female, be sensitive to the signals of the young indicating departure from and approach to homeostasis, and that the male be capable of "mothering" the young.

Given the above, plus a long ontogenetic period of the young, it is necessary:

(1) for the male to select the "best" mother for his offspring.
(2) for the female to select a male most sensitive to her needs and to those of the offspring to safeguard and enhance her reproductive effort.
(3) for the two adults to bond for the duration of the young's ontogenetic period.

It is clearly in the male's interest to select a female quite sensitive to the social feedback system and thus acting alert and responsive to the male's courtship; such a female is more likely to maximize her effort to restore and maintain homeostasis for the developing young than a dull and insensitive one. It is adaptive for the female to select a male who is not only a potentially capable provider or a dominant individual but also sensitive to the social feedback system. To discover that, a prolonged courtship is mandatory, as is an initial withdrawal, coyness, by the female.

Granted a reasonably sensitive social feedback system, the adults can shape each other's behavior to maximize "pleasure," that is, homeostatic physiological functions of their bodies. Hence, there is selection pressure toward greater sensitivity and expressibility of the social feedback system the more the species' adaptive strategies diversify. Thus, the more sensitive the social feedback system, the more likely that a social bond will be maintained by the adults during the ontogeny of the children. Thus, if stimuli eliciting smiles and laughter are pleasurable, smiles and laughter are guides to the kind and intensity of behavior the social partner should perform. This system assumes that receiving smiles and laughter is a pleasurable, and hence reinforcing stimulus to the receiver. Conversely, crying should be upsetting or punishing; it is a means of strongly signaling discomfort without rupturing the social bond at once, and thus allowing for motivational variations by the partners.

In addition to the smile, the responses of the eye may be powerful subconscious signals, as is indicated by the work of Hess (1965). The expression of the face can be altered, not only by the reduction and increase in the size of the visible port of the eye by the eyelids, but also by the dilation and constriction of the pupils. The pupil of the eye increases in size in response to "arousing" stimuli and decreases in size in response to aversive ones. Moreover, human subjects are exceedingly sensitive to the dilation of the eyes of conspecifics, far more sensitive than they can verbalize. Thus, males respond positively to the dilated pupils of a female's eye; not surprisingly, it is known that women—at least since the Middle Ages, according to Hess—have used drugs to dilate the pupils. It appears that the infant is programed genetically to fix on human eyes (or their experimental equivalent) and smile in response (Pfeiffer 1972, p. 425). This implies that the eye is exceedingly important as an organ of communication, particularly at the unconscious level, as Hess's (1965) experiments suggest. Thus, fixation of eyes is the first prerequisite to gaining and giving emotional information.

The foregoing indicated under what conditions the adults parasitize the social feedback system between mother and child in the interest of optimum mate selection and pair bonding. As was indicated earlier, a second mechanism of pair bonding is sex. Clearly, the longer the ontogenetic development of children the more likely the selection for a maximum of pair-bonding mechanisms. Only strong pair bonding assures the developing young the optimum satisfaction of needs at various steps in its ontogeny. A third pair-bonding mechanism is the mutual satisfaction of egos. This can be achieved by mutual laughter, at the expense of some third individual. This is clearly satisfied in the family by the parodies of the children, whose behavior is often an imperfect rendition of that of adults and hence a parody of adults. Thus, the children in their clumsy way provide the basis for humor, and hence a third pair-bonding mechanism that obviously wanes during ontogeny and disappears at adolescence. Laughter is seen here to be a reflex based on an innate recognition of superiority, the very point I argued in Chapter 5; it is thus highly adaptive in pair bonding, and elsewhere, as we shall see.

We noted earlier that the female bonded the male initially through sexual stimuli. We also note that to do so tactile pleasures ought to be maximized. The estrous swellings of Old World primates provide not only visual but also tactile stimuli. Maximizing such tactile stimuli would, therefore, serve to bond males to females and would lead to a reduction in the hair coat—as indicated earlier. To bond the offspring to the male, it is logical that natural selection (following the line of least effort) selects for stimuli in infants that attract males to infants. Infants have certain visual and acoustic signals that help to bond males; these are not only the famous "kindchen Schema" of Lorenz (1943 in Eibl-Eibesfeldt 1967), or the parodies children perform, but also their attractive tactile properties. The hairless skin and soft resilient nature of their bodies are not dissimilar to those of females. Therefore, the totally different tactile properties of human children compared to those of other anthropoids can be explained as adaptations that attracted males and maximized bonding between female, male, and infant during the early ontogenetic stages of the latter.

Thus, in mate selection for maximizing reproductive fitness, both mates should select for alert, sensitive, humorous, kind partners, and courtship should contain many elements of the mother−young relationship. That clearly is the best test of the partner's ability to successfully raise children. It also explains the great preponderance of mother−young behavioral elements in the courtship of monogamous vertebrates in which both partners raise their young (Eibl-Eibesfeldt 1967, 1970).

At this point it should not be forgotten that, although the new ecological conditions demand the testing of the social feedback system and the use of the mother −young behavioral repertoire, the organism *Homo* at the *erectus* stage (and beyond) is still highly sensitive to the old way of courting. Hence courtship will be a compromise between the demands of the new way of life and the dictates of the inherited ancestral sensitivities and expectations. The old pre-*erectus* primate courtship incorporated dominance displays, threats, and even punishment by the male, and submissive displays by the female (Hinde 1974, p. 302). Clearly, we expect some elements of dominance displays in the courtship of men and elements of submissive behavior in the courtship of females. The old submissive behavior I postulated (p. 230) depended on sexual presentation plus the antithesis of dominance displays; hence, the male's interest in sexual and submissive signals, so clearly evident from the pictorial content of "skin magazines." Here, however, we aslo find the evolutionary origin of smiling; it is an old signal of Old World primates which originally was probably a defensive response of subordinates, became next a signal of subordination, and was then turned into a greeting or courtesy signal used by dominants and subordinates alike to appease conspecifics (Jolly 1972, Hinde 1974). In the chimpanzee it thus serves not only to reassure a dominant, but also to reassure a subordinate if used by a dominant (Jolly 1972); it is an appeasement behavior. If the hypothesis of the nature and structure of the social feedback system is valid, then signals of submission should relax the receiver; they should "relieve tensions." The closest we come to verifying this is the work of Fox and Andrew (1973). Socially positive stimuli, such as petting, induce bradycardia in canids; aggressive stimuli show tachycardia.

Given the ecological niche of early *Homo,* in which defence of resources by the male may be a necessity in some seasons (p. 267), it is in the reproductive interest of male and female to maximize the male's abilities in agonistic interactions. Hence the female should reinforce the male in his performance of dominance displays, make him "proud and self-confident." Hence, acts of support and shows of loyalty by the female, as well as those testing the male's responsiveness and ability to care, are most important in the bonding behavior of the female.

The social feedback system of our species is based on that found in primates (Jolly 1972, Hinde 1974). These mammals are characterized by relatively long life spans, low reproductive rates, gregariousness, and long ontogenetic periods, so that we are not surprised to find rather complex social feedback systems. They have a large series of behavior patterns based on tactile stimuli—the old primate body language, including grooming, hugging, caressing, tickling, kissing; there is a highly developed facial mimicry enhanced by the cephalization of display organs (Andrew 1963, Guthrie 1970a); there is a rich repertoire of sounds signaling pleasure and distress. In addition there are body postures, movements, sounds, and odors originating from dominant and submissive signals. Humans have all these behavior patterns (Eibl-Eibesfeldt 1967, 1970, Jolly 1972), but some social feedback systems have been enhanced, refined, and complexified, such as facial mimicry, but in particular the vocal responses. The smile and chuckle have been intensified into laughter, whimpering has been enhanced into crying and sobbing; the positive and negative social sounds have been enhanced in parallel, thus broadening the range of the responses, permitting more exacting matching of social stimulus and response.

Vocal signals have been transformed into the most human of adaptations, language and music. These extreme refinements can be understood as necessary consequences of adapting by cultural means to an increasingly complex, variable and unpredictable physical environment, hence a selection for open genetic programing of behavior, which in turn selects for a long ontogenetic period and larval condition at birth, to permit a maximum of environmental moulding of behavior.

Prerequisites of Cooperation: Music, Dancing, Laughter, Language, Games

Before proceeding to an examination of the biological roots of music, we must return to some fundamental functions of the social feedback system.

It serves not only in bonding of mother and young, or the nuclear family, but also of males into cooperating groups. Cooperation requires not only the transferral of information signaling intent by the leader, a synchronizing of motivations in order to synchronize activities, and the continued practice of that which the group shall cooperate in, but also a long association of cooperating individuals, which in turn requires continuous reinforcement of social bonds.

Since adult individuals, be they male or female, are sensitive to the same signals as their children, the social signals can serve to bond individuals through mutual stimulation to smile, chuckle, laugh—in short, with signals that are "infectious" and probably have relaxing qualities. We may note here that in some primates, e.g., rhesus monkey and barbary ape, real infants are used in some social transactions of males and appear to serve as the medium of bonding (Jolly 1972, p. 250). Human males tend to be tense in each other's presence, even closely bonded males. This is revealed by bomber crews in their excretion in high concentration of urinary break-down products of cortical hormones compared to that of lone males (Mason 1959). Hence, males should be predisposed to relieve tension through laughter and other socially positive signals. Hence, behavior conducive to laughter, such as joking and clowning, would bond males; such behavior would be based on visual and vocal mimicry, creating parodies. Laughter in dominance displays would enhance the group's superiority, as long, of course, as it was directed at individuals outside the group.

As in the male—female relationship, so in the relationship of cooperating males the bonding behaviors are in conflict with the old means of noncooperative male bonding based on homosexuality (p. 230). We can thus visualize a sliding of social bonding mechanisms from the homosexual behavior of the pre-*Homo* period, to bonding through social feedback systems based on the family bonds. In our present species, homosexuality does erupt to become a "normal" behavior under circumstances of stress such as in prisons and long-lasting male groupings (Kinsey et al 1948). It is here an example of "regression," that is, a retreat to a more primitive form of human, probably instinctive behavior, our pre-*Homo* legacy.

What is cooperation based on? At the most primitive level, cohesive grouping and cooperation were probably triggered by danger. Males grouped, screamed in great excitement, rushed the minor predator, brandished sticks, and hugged each other, very much as chimps do when confronted by a predator, as was shown by Kortland (1967). The danger creates strong physiologic arousal; the dominant male's be-havior serves as a model to others; loud yelling maintains arousal; a few socially positive gestures temporarily reduce tensions. Most importantly, the visible danger creates the motivation necessary to synchronized activity, that is, cooperation. This is a key point—synchrony of motivation leads to synchrony of activity (or coopera-tion).

Let us look at another situation to clarify the point. How do wolves proceed prior to a coordinated moose hunt? Mech (1970) gives a description: Prior to the hunt, after spotting the moose, the wolves come together, huddle, wag tails, grow ex-cited, and then howl. This observation, plus the one repeatedly noted by various students of wolf behavior (Mech 1970, p. 97−102) that howling by one member brings the pack together, excites wolves and makes them join in the howling, suggests a hypothesis: Howling synchronizes motivations and leads to synchronized action during the hunt. Wolves insensitive to howling would not be readily moti-vated to hunt, and would reduce the pack's efficiency at killing and increase the cost of food procurement. In order to increase the efficiency of hunting at the lowest possible cost in energy, injuries, etc., synchronization of motivation would be essential. This suggests not only an explanation of the howling of wolves, or the prehunting ceremonies in African hunting dogs (Kühme 1965, Goodall 1971), but above all it suggests a biologic function for music in man.

Music: Its Functions and Possible Origins

The hypothesis is that music functions in physiological arousal, in synchronization of motivation and thus of activity, in the creation of "common experiences," and so in bonding of individuals, as well as in a variety of ways that ultimately are reflected in enhanced health and physical competence of the participants.

Music appears to be based on heightened sensitivity of individuals to sound, so that sound has a marked immediate effect on the physiology of the body. Sounds that emphasize stimulus contrast arouse. At the primitive level we find the roots of music in threats and dominance displays, as illustrated by the noisy charging displays of chimps which include stamping, crashing through branches, hooting, etc., and which to chimps are obviously very arousing (Goodall 1971). This line of thought predicts that in wolves, howling is a transformed territorial call, that is, the aggressive signal of the solitary canid. The same thought is reflected in Eibl-Eibesfeldt's (1967) suggestion that musical instruments such as drums and horns emphasize sounds of a dominance display nature. They would therefore be an example of cultural enhancement of vocal displays. As argued in Chapter 5, such displays are "designed" to trigger and enhance arousal.

The antithesis of sounds characterized by stimulus contrast are harmonic, melodious, rhythmic sounds; we expect these kinds of sounds to reduce arousal. They can be traced to the signals of mother and young. Music can develop if the response system of individuals is very sensitive to these kinds of sounds, and I argued earlier that in humans this sensitivity ought to be great, in the interest of the developing child. Music would hence be possible, owing to the highly sophisticated social feedback system of humans, as well as to our ability to mimic sounds.

The physiological evidence is in line with the above conception. Music does indeed arouse, as indicated by elevated blood pressure, pulse rate, muscle tonus, blood glucose levels, and galvanometric skin response if it is strongly rhythmic and rapid (Destunis and Seebandt 1958, Stokvis 1958, Tränkel 1958, Traxel and Wrede 1959, Harrer 1975, Schwabe 1969). High, piercing—and so inharmonious—sounds achieve the same (Tränkel 1958, Harrer 1975). On the other hand, rhythmic, slow, and "familiar" music produces the opposite physiological responses; it relaxes and can lead to drowsiness and sleep (Destunis and Seebandt 1958) as well as to prolonged sleep (Teirich 1958, Wendt 1958), it relieves hypertension (Schwabe 1969, p. 22), and it is used as a medical therapeutic agent (Gaston 1968, Harer 1975). Children "grasp" and respond quickest physiologically to symmetrical, melodious melodies (Destunis and Seebandt 1958); little wonder that lullabies are an ancient element of human communication by mother to child, and are very similar between different cultures (Kneutgen in Eibl-Eibesfeldt 1967, p. 429).

The function of music in human biology does not end here. Together with dancing, a form of visual mimicry, it has effects that can only be elucidated if we ask why cooperation evolved to the intensity it did in humans. As indicated earlier, cooperation and altruism are prerequisites to successful hunting of large, dangerous, cold-adapted mammals by our species. Cooperation is maximized by the unambiguous signaling of intent (we assume that language evolved to fill this need), by synchronizing the motivations and behavior of cooperating males (we identified music to fill this need), as well as by frequent practice of the tasks males cooperate in (we assume that playing games filled this requirement). However, humans, unlike wolves, do not sing prior to the hunt; they do so well before and after the hunt. Obviously, music and dancing must serve some other, as yet unidentified, function. This suspicion is strengthened by noting how deeply music is woven into the fabric of human cultures.

The clue to the puzzle, however, rests again in the cooperating hunting group of males. We missed one further attribute that maximizes cooperation, an attribute so obvious that we fail to notice it, *physical fitness*. Mundane though physical fitness is, it is anything but easy to achieve and to maintain. The structure and function of the human body is maintained through exercise; without it both function and structure degenerate. Our bodies and organs are in a state of dynamic equilibrium whose size, shape, structure, and physiology is closely related to the use they receive (Cureton 1969, Larson 1973). Muscles not used shrink in size and strength, muscular coordination decreases, no new capillary beds are grown and so circulation decreases, fat accumulates in various depots, and it takes weeks to get into a peak functional state. Exercise physiology teaches that. Moreover, we know from animal experiments that frequent or continuous agitation of the body by various means, including stressors applied at low intensity, stimulates body growth and hastens maturation (Vernadakis et al 1967, Petropoulos et al 1968, Mandl and Zuckerman 1952). Not surprisingly, children exposed to athletic training grow somewhat faster than those not so exposed (Rarick 1973a, p. 220). Therefore, for maximum physical fitness and body size, a large amount of physical exercise is required.

The periglacial hunters with exceedingly primitive stabbing spears tackled ungulates that were large, exceedingly aggressive and agile, armed with sharp tusks, horns, antlers, and hooves, and protected by thick, tough hides. The early Cro-Magnon, as well as Neanderthal man before, were the top carnivores concentrating on large mammals that were almost all ready to stand and fight, judging from their modern counterparts. It is just this weakness that early man must have exploited, by making the animals attack and in the process allow the group to kill them with spears, or lure them into some natural trap. To do this required a precise knowledge of how each species of large dangerous game would respond if provoked, and how it must be "played with" to permit the hunters to thrust their spears into the prey's vitals. Granted the size, strength, and agility of the prey, it required that the hunters be powerful in build as well as in peak physical shape. Otherwise they would only too soon succumb to the attacks of a prey.

The foregoing makes it evident that stimuli promoting physical fitness must have become increasingly adaptive the more demand there was for physical fitness. I postulate three activities that simultaneously promoted physical activity, and hence physical fitness between hunts, and were highly adaptive, since the interval between kills of big game must have been so long that hunters would lose physical fitness unless it was promoted artificially. These activities are laughter, music making, and dancing. Laughter, we noted, is tension releasing (Koestler 1974), meaning that it relaxes the body through parasympathetic stimulation and hence causes blood pressure to drop; temporarily dilates blood vessels; permits maximum perfusion of tissue and blood; accelerates oxygen and nutrient transport to active tissues and the removal of metabolic wastes, thus enhancing tissue growth, permitting the immune system to act quickly and enhance renal clearance; and in females it should permit a maximum of milk flow (Newton 1961). In short, it should be conducive to good health.

Exercise enhances these functions. It also enhances the speed of motor responses by enhancing the speed of signal transmission along nerve fibers, and it increases the individual's ability to liberate free fatty acids from its fat depots to fuel muscular work (Parizkova 1973, p. 113). Moreover, from the work of Tränkel (1958) we know that exercise begins to reduce muscle tonus induced by music.

Music performs a number of functions in this context. It synchronizes the function of organ systems within the body and has a regulatory effect on breathing, circulation, metabolism, and digestion, as has been known a long time (Schwabe 1969, p. 32). As indicated earlier, it arouses and enhances muscle tonus, blood glucose levels, blood pressure, and pulse rate, and does so in proportion to the intensity of the music. In particular, rhythm enhances muscular tonus (Tränkel 1958) and probably creates a desire for activity. Our bodies can follow music physiologically to a remarkable extent (Destunis and Seebandt 1958, Harrer 1975). This should promote physical fitness and growth, as is indeed reflected by the fact that surgical wounds and broken bones heal faster in patients subjected to music and rhythmic exercises (Klose 1954). In addition, music acts to dull pain and produce drowsiness or a mild trance, which makes it possible to perform monotonous, uncomfortable, if not painful, activities. This is precisely what most physical fitness exercises are; they require a lot of repetition, exertion, and endurance. Hence the sensation-dulling properties of music are most adaptive. Music can be used to reduce dental pain (Priester 1962), the pain of surgery (Klose 1954) or of childbirth (Schwabe 1969, p. 23). Thus, granted music and singing, the unpleasant physical exertion of dancing is not noticed; on the contrary, dancing becomes a pleasurable experience.

Music has a number of secondary benefits. It promotes voluntary joining in activities, even in mentally ill persons (Brunner-Orne and Orne 1958); it depresses fears and rouses individuals from depressions (Jaedicke 1954); it enhances communication between individuals (Sutenmeister 1964); it is a therapeutic agent combating asthma, speech defects (through singing, Fengler 1950), defective heart rhythms, a number of cortical dysfunctions, as well as psychoses, depressions, and neuroses (Schwabe 1969). It increases productivity in monotonous jobs as one would expect from its property of dulling discomfort (Smith 1947). "Strange" music often brings about negative responses in individuals, such as increased blood pressure and pulse rate—as one would anticipate from the concept of pattern matching. However, it is not always so. The strange, monotonous, Chinese music played to western children produced a lowering of pulse rate and blood pressure (Destunis and Seebandt 1958). The physiological response of individuals to music is not always clearcut. Unfamiliar pieces may or may not produce responses, while some persons respond in a direction opposite from that predicted. Nevertheless, music is so powerful an emotive agent that it is used as a therapeutic agent in medicine, psychiatry, and in rehabilitation work (Schwabe 1969).

The original evolutionary reason for music and dancing is therefore to maximize physical fitness. Merriment, laughter, and dancing are thus not frivolous, but are biological adaptations of humans in the best sense of Darwinian fitness. It prepared us for the demanding, arduous work of cooperating hunters, it permitted quick recovery from oxygen debt to respond quickly with well-trained muscles and nerves, and it had many benefits for our health. Moreover, Cureton (1969) points out that physical fitness promotes such personality traits as physical courage, it reduces anxiety, and it permits rapid recovery from mental fatigue.

If the foregoing is valid, males—who did the hunting originally—should be more easily stimulated by music, song, and laughter than females; this remains to be verified. Males should also be more ready to enter into strenuous physical exercises; this they apparently are (Astrand 1952 in Timiras 1972).

If music, song, dancing, and laughter function to maximize physical fitness they should be at a minimum when caloric intake is low, and at a maximum when caloric intake is high. Under conditions of excess food, it is vital for periglacial hunters to exercise in order to reduce fat deposition, to reduce chances of fatty tissue degeneration of the cardiovascular system due to high cholesterol blood levels (see Cureton 1969), and to reduce stress that is all too easily generated by idleness and bickering. Apparently, exercise does help reduce stress (de Vries 1966), as do communal activities (Matsumoto 1971). It would be fatal for Paleolithic hunters not to remain in peak physical condition, since they have to hunt frequently, as food surpluses are short lived in a hunting economy. Yet, owing to a fortuitous circumstance, the food surpluses are just large enough to permit hunters to lose physical fitness, unless they maintain it through dancing. Here lies the ecological reason for music, song, and dancing.

Let us assume a group of five[3] hunters kills a steppe wisent in winter. How long will the meat last? How long is the interval between kills of large mammals? The group of humans consuming the bison will be about 25 people. We note that the daily meat requirement of an extended family is about 25 lb (p. 361). Hence 25 persons can be expected to consume about 80 lb of meat a day. An adult steppe wisent would provide about 950 lb of meat, assuming a live weight of 1500 lb—not unreasonable given the large size of these animals compared to American plains bison (Halloran 1960)—and a dress-out percentage of about 60 (Peters 1958), plus the weight of edible internal organs. About 500 lb of the live weight would be unavailable for consumption, in the form of gut content, bones, hide, and probably blood. Note that these are all conservative estimates. This means that the hunters would not need to go out and kill for at least 11 or 12 days, long enough for them to suffer significant decreases in heart output, increased peripheral resistance, and thus reduced physical performance (Cureton 1969). The kill of a mammoth would most likely extend the period in which no hunting would be required by some two, if not three or four weeks. Multiple kills would extend it still further, of course. Yet at the end of this time the hunter would be expected to go out in top physical condition. But without exercise, physical condition is quickly lost and very time consuming to regain (Michael and Cureton 1953, Müller 1962, Cureton 1969). Therefore, just when there is no need to hunt, when the caloric intake of fats is probably high, it is necessary to exercise to maintain physical fitness for the day in the near future when it is essential to life.

Thus the investment of energy and nutrients in vigorous diverse exercise in a social setting produces payoffs in the form of more efficient hunting and low rates of injury and mortality, thus ensuring maintenance of the hunting society. Laughter, music, and dancing would be most keenly selected for under conditions of cooperative spear hunting of the largest of ungulates and carnivores, which would mean that kills were spaced at such intervals of time that mechanisms would have to be exercised to maintain physical fitness.

The earliest musical instruments of Cro-Magnon found are pipes of hollow bone with well-cut finger holes, dating back to the Aurignacean culture (Pfeiffer 1972, p. 266). This fits well with the hypothesis elaborated earlier. The pipes, however,

---

[3]See Magic Numbers, Chapter 14.

indicate that some instrumental music must have been practiced well before the Aurignacean, since the finger holes in the pipes relate to musical scales, and these require time for their development.

A further condition for successful cooperation is the practice of skills that are important in cooperative hunting. Because general exercise does not improve specific motor tasks noticeably (Henry 1958), it is necessary to practice repeatedly individual skills. Most of these skills would have been practiced by the individuals in play since childhood. Such play we can assume would be with the encouragement and guidance of older individuals, in particular the older siblings and grandparents. Here lies the reproductive payoff, not only for *visual mimicry,* but also for practice and innovation that reduces the latency response time for each critical action in the behavior of hunting.

Visual mimicry is adaptive because it permits the individual to benefit from the experience of the successful adults, who have demonstrated their success by living the longest. The child can thus mold itself in the image of a successful model. It gains mastery of a task based on the personal experience of its instructor, provided it can mimic what it sees in relatively fine detail. It thus becomes well prepared for the dangerous task of hunting well-armed ungulates before its first hunting encounter.

Practice in play would have two additional benefits: It would permit an individual to discover simpler motor patterns to fulfill a given task, as well as, through practice, reduce the response latency for each task. Both processes would be most adaptive. In cooperative hunting, speed of response, both to dodge an attack by the prey and to place the weapon effectively in the split second of opportunity, is of the greatest importance. Since complex tasks have a longer response latency than simple tasks (Henry and Rogers 1960), simpler tasks are more adaptive in the above context.

It is tempting to suggest that games involving primarily mental exercises which can be played by groups and individuals while socializing, as well as day dreaming, are adaptive. Such games could not only increase the problem-solving ability of individuals, but also lead to the discovery of new strategies and tactics for hunting large mammals. Simultaneously they would bond individuals into cooperating groups (Gordon 1961, Singer 1974).

If physical fitness was of great importance in the adaptive syndrome of late Paleolithic people, as I have suggested it was, then the populations had to be kept well below carrying capacity so that no shortage of resources existed. For instance, in order to maximize precision of movements, let alone muscular strength, hunters had to be well fed. According to Berg (1973), a hypocaloric diet of some 1800 kcal per day led quickly to a loss of 30% of muscular strength and a 15% loss in the precision of movement. Malnourished children not only suffer severe learning disabilities but also have difficulty with the perception of forms, and lag in motor development and in eye−hand coordination (Cravioto and de Licardie 1973). Such individuals would hardly develop competence in confrontation hunting of dangerous large mammals or learn well the interception of migratory big game herds, or have the ability to form strong lasting social bonds, for malnutrition produces hearing and speech problems and impedes social and personal development. In short, the demands of confrontation hunting on physical fitness and motor competence are so great that there must always be an excess of resources, not only for ontogenetic development, but also to support such expensive activities as dancing and merriment, let alone keeping hunters well fed.

The foregoing sections revealed that the social system of man and our sophisticated social feedback behavior apparently optimize the physical and mental development of individuals. It simultaneously promotes high physical fitness and cooperation, enhances the speed of problem-solving, and transmits acquired knowledge rapidly. All this acted to promote hunting success and longevity of a people exploiting the large-mammal populations of the periglacial zones. One test of this hypothesis is the development of individuals in a social milieu quite different from the one described above, such as that of orphanages. I realize that, as a basis for comparison, people raised in modern nuclear families are a shaky example, but they still are useful, since the institutions deviate in their social milieu still more from that of the periglacial family than does the nuclear family. We find pertinent information in Patton and Gardner (1963). Also interesting in this respect are the studies on children deprived of affectionate care in some nuclear families (Powell et al 1967a,b, Glaser et al 1968). Such children fail to grow normally and may show symptoms of a malfunctioning pituitary gland. Even minor emotional neglect by strained and consequently authoritarian parents results in children with speech defects and a lowering of self-esteem and competence (Niemeyer 1974). Elsewhere I enumerate the evidence for the importance of the social milieu in the development of intellect in children (p. 370); it also supports the foregoing notion that our evolutionary environment maximizes intellectual development. The exact contribution of humour, music, dancing, games, and play to optimum growth, maturation, and health do remain partly obscure, but the evolutionary argument that is advanced, as well as the meager empirical data, indicate that they are significant. We can expect persons deprived of music, dancing, humor, and games to develop less than good health (for the converse see Matsumoto 1971).

The picture that inevitably emerges of our distant ancestors is not that of sluggish brutes, but quite the contrary. They were apparently fun loving, brave if not a little reckless, altruistic, intelligent, deeply emotional—in short, a magnificent people. The advent of agriculture in the Neolithic after the disasters of the Mesolithic was indeed a "fall from grace" in more than one way, as aptly put by Tiger and Fox (1971).

Following this discussion one must inevitably ask, "are there homologous or analogous adaptations to laughter, music, and dancing in other animals?" I pointed out some ill-developed homologs in primates, the grins of cercopithecines and the dominance displays and facial gestures of the apes (Schaller 1963, Goodall 1971). Analogous behavior to laughter and crying appears to me to be tail-wagging and at times barking, as well as whimpering and yelping, in wolves and dogs.

We dealt with music and dancing, activities normally relegated to the artistic realm if they achieve any level of sophistication. One might ask, "what about prose and poetry?" What are their origins? We experience these so much as art forms that we may find it difficult to see utility in them. Let me propose the hypothesis that poetry resulted from the discovery that poetic stances permit easier recall so that, in order to transmit a large amount of information accurately from generation to generation by verbal means, it ought to be put in poetic and good prose form. Prose and poetry are mechanisms of a verbal information transmission. We know from educational psychology that learning and recall is indeed greatly enhanced by structuring the material in either prose or poetic form (Simon 1974); in the latter, recall is of course exact. In addition, we learn that recall is greatly enhanced by recitation and by visual association; the latter suggests that acting and role-playing during recitation could serve to maximize recall—both by the actor and the audience. Recall is also enhanced by "overlearning" or repetition and the association of words with serial images (Gage and Berliner 1975). The hypothesis, of course, suggests that poetry and structured prose ought to be used in preliterate societies to transmit verbally complex subject matter. This proposition I have not investigated.

Extended Family

A further means of increasing the fitness of offspring is to prolong the imparting of tested knowledge, such as the knowledge a grandfather or grandmother can give to a grandson or granddaughter. This, however, entails the maintenance of individuals by the male during the nonvegetative season above and beyond those represented by his wife and children. It is to the male's advantage to feed, and his mate's to maintain such individuals, only if this increases the chances of survival of their children, contributes substantially to the hunting success of the male or the mothering efficiency of the female, or in some way reduces the competition between the male and other males for scarce resources. Unless old individuals incapable of hunting *significantly* contribute toward the above ends, without competing for food and care with the children of the nuclear family, the male and female are best served by disassociating themselves from the old individuals.

This is probably what happened in the warm productive habitats that suffered a relatively short period of scarcity of gatherable foods. The experience of the old would be absorbed early by their children, and would be redundant for the raising of their grandchildren; the family required only a small amount of labor, mainly because clothing was not essential and would contribute nothing to the family's reproductive success. The productivity of the land would ensure that a relatively small amount of knowledge, and learning by imitation rather than transmittal of abstract concepts, would suffice to maintain the maximum reproductive success of a family unit. In the cold north or in deserts—with their short, varied seasons, dispersed and migratory game populations, great annual variations in seasonal game abundance and snow or moisture conditions, demands on diverse technology and the large amount of knowledge required to cope with it—would the labor, skills, and knowledge of old individuals conceivably contribute sufficiently to the reproductive success of the family to make their maintenance highly advantageous? This, in turn, required the development of behavior patterns in young and old that would smooth the overlap between generations and thus increase the potential collective skills and intellect.

Paradoxically, the extended family could develop only during a period of human evolution in which resources were superabundant, so that, first, individuals could reach older age and potentially be useful to their offspring and, second, it was easy for the nuclear couple to supplement the efforts of older individuals to sustain themselves. Tolerance and generosity can only be expected when resources are reasonably abundant, not when they are scarce (e.g., Turnbull 1972); it is no different with lions (Schaller 1972, p. 358), or wolves, which I observed on kills in the wild. We thus expect a correlation in the fossil record between population quality and the frequency of healed bones, indicating that disabled individuals were able to survive. From scanning the literature this appears to be so. Note the evidence for bone healing in Neanderthal man, who, as I indicated earlier, must have lived under conditions of superabundance (see Birdsell 1972), as well as the converse evidence during the Mesolithic, when deteriorating population quality is associated with homicide and cannibalism (p. 379). Superabundance of resources would be temporarily generated if hunters killed very large mammals which would supply food for several weeks after each successful hunt. Thus, the beginning of extended family development must fall into that period when bison, horses, rhinos, mammoth, and cave bear became regular victims of human hunters. This would place the beginning of extended family formation in the beginning of the Würm glaciation.

In discussing Acheulean man's process of adapting to temperate zones, it was noted that exploiting large mammals in winter by successive kills would impose long journeys on the hunter and his family at times and that this would have mitigated against both the very young, unless they were carried by adults, and the old. Therefore, it cannot be expected that an extended family would arise unless the need for frequent rapid, long-distance movements is virtually absent. The extended family could arise only under conditions of stored food surpluses which reduce the need for old people to roam, or under conditions in which old people could be transported. There is no hint in the archeological record of domestic animals that could have supplied the labor for transport during the Paleolithic, and one can reject the transport hypothesis. Therefore, the extended family could have become a significant social unit only with the adaptive syndrome of the late Paleolithic people, but not with those of Acheulean or even Neanderthal man. This is not to deny that under favorable conditions, such as living in an area of exceptional game density and diversity, Neanderthal people could not have cared for disabled and aging individuals; the archeological record indicates that occasionally they did. However, their adaptive syndrome was not as conducive to extended family development as that of the late Paleolithic people.

We know from the study of Eskimo that after age 45, it becomes difficult for a male to hunt effectively.[4] We can safely assume that it is about that age that he would retire from hunting and be supported by the younger people. However, we can expect that, on an average, persons over 40 were grandparents. Owing to the continuing high rates of adult mortality, few individuals would have been likely to live beyond 40 years of age, maybe not more than 2 or 3 in a group of 25 people. In the conditions experienced by Neanderthal man, there would be fewer still. The knowledge and wisdom of the old probably came cheaply to the group. Still, we can expect some alteration of behavior by the old to facilitate the transfer of knowledge.

It may well be that the "mellowing" of old individuals and the extended life span of man far beyond that of mammals of comparable size, are adaptations that, firstly, held father and son together to increase the reproductive effort of both through cooperative hunting, and secondly, permitted the older individuals to influence the activities of the younger without rupturing the social bond existing between them. The same can be said for mother and daughter. We can also see here that increased life expectancy after reproductive age would be a trait selected for, provided the individuals remained mentally alert and were able to contribute to the success of the family by decreasing the rates societal malfunctions, illness, and death. In essence they would increase the reproductive fitness of their offspring, and thus their own, by living beyond reproductive age and contributing their knowledge to their sons'

[4]Dr. O. Schaefer, Charles Camsell Hospital, Edmonton, personal communication.

and daughters' well-being. This hypothesis also predicts that old individuals are better at some tasks than younger individuals; indeed, handicraft skills apparently "mature" and peak in the sixth decade of life (Rosenmayer 1975). The longer the lifespan of persons after reproductive age, provided they were healthy and alert, the greater their contribution to reproductive fitness. This makes menarch, for instance, not an "accident of nature," but compatible with enhanced reproductive fitness in humans. It appears that young women produce fit children with the greatest frequency; in older individuals point mutations, chromosomal aberrations, and reduced physiological competence would reduce the number of fit children born. It appears that *care* of offspring, rather than production of offspring, increases the reproductive fitness of old individuals.

However, such cooperation, tolerance, and strong bonding across generations can only be expected under the ecological conditions of low productivity, not in the productive landscapes. We should not forget that the time between *H. sapiens'* appearance and his occupation of the periglacial environment exceeds 150,000 years. Garn (1963) suspected a link between increases in brain size and the greater social demands of hunters. This link is supported from neurophysiological studies on the human brain (Guyton 1971), as well as by studies on the phenotypic variations in brain size which I discussed earlier (p. 31). Those regions of the brain dealing with social control are enlarged, particularly in *H. sapiens*. If the duration of human ontogeny and the concomitant increase in life expectancy, the increase brain size, and the three-generation family are a direct consequence of natural selection in highly variable environments, then hominids colonizing less variable environments ought to have a shorter ontogenetic period, shorter life expectancy, smaller brain size, and nuclear families.

From the foregoing it is evident that it is first the female that must undergo a transformation in external appearance and behavior, as well as the male in his courtship and dominance displays, followed by selection for prolonged juvenileness and strong bonding between juveniles and parents, followed by selection for bonding across generations and closely related families, and the development of demes consisting of family units. Granted the division of labor, it should not be surprising if differences are found between the sexes in humans which reflect their respective roles. Studies have shown that female children are better than males at such cognitive tasks as perceptual motor skills, identifying objects, language development; however, male children were superior in restructuring visual stimulus fields such as finding a simple pattern embedded in a complex one (Willerman et al 1970). The latter ability is not surprising in the sex historically exposed to hunting and the demands it imposes, such as spotting partially obscured game in the distance, a task that had to be performed on every hunt. Moreover, motor coordination increases more in boys than in girls after puberty (Tanner 1962).

Relict Behaviors

At the same time that new adaptations evolve, old adaptations, functional in early environments, fall into disuse, although their heritable portions are carried onward in the genomes of individuals and continue to be transmitted. These old adaptations are selectively neutral unless they demand a large amount of energy for their development, in which case they are selected against in maintenance phenotypes until they are an insignificant burden. Even if they are selectively neutral they still become increasingly nonfunctional in time, owing to random mutations affecting them and interfering with growth processes essential to their development (Brace 1963). Nevertheless, they do not disappear at once, and one can expect various "appendices" to survive from the adaptations of one environment to the next. This is so important a concept to the understanding of modern man that I must be excused for being repetitive on this point. All sorts of rudiments from the pre-*sapiens* stages of man appear to have survived the transition to *Homo sapiens,* including rudiments of what were probably old courtship behaviors and submissive and dominance displays.

Submissive displays are of particular interest. We saw that they follow three principal pathways; they are either the antithesis of aggressive displays, a recognition leading back to Darwin (1872) and rediscovered independently by numerous investigators, or they mimic the behavior of estrous females (Wickler 1967, Geist 1968a,b)—if they take place between males—or that of juveniles (Geist 1968b, 1971a). In species in which the males are social, submissive gestures tend to take the form of female mimicry, although the extent of such mimicry is clearly a function of the survivability of the lone individual. The smaller the chance of the individual surviving on his own in the absence of the troop, the stronger the selection for submissive behaviors that will permit young males in particular to remain with older, larger males. Therefore, we can expect indications of female mimicry as submissive behavior to be prevalent in many species, although only in a few will it reach the extent found in baboons or mountain sheep (Wickler 1967, Geist 1968b, 1971a). Moreover, female mimicry can be expected to evolve where males *do not* contribute to the upbringing of their children, such as in the pre-*sapiens* society postulated earlier. That is, in australopithecines from warm, productive habitats we can expect the precopulatory posture to have served as a submissive display, and mounting by dominant males to have been a display of dominance, much as is found in a number of primate societies (Jolly 1972). We cannot quite expect the genetic basis for this behavior to disappear in the *sapiens* stage of man despite the fact that now a totally new submissive display was called for.

In species in which the male contributes food to the upbringing of his children, such as in many birds or in wolves and hunting dogs, it is adaptive for the adult to mimic the behavior of a juvenile (Geist 1971a). Thus, we find in such forms juvenile behavior performed by subordinates toward dominants, or as signals of "friendliness," whereas in baboons such signals tend to take the form of sexual presentations (Wickler 1967, Kummer 1968, Jolly 1972). With the switch from females independent of males when raising offspring, to females dependent on males for food during the nonvegetative season, and the concomitant direct contributions of males to the care of their offspring, juvenile mimicry as a submissive display could be selected for. It would select for juvenile characteristics in the voice, in external appearance, in gestures, and in signals. However, full-fledged evolution of female mimicry presupposes that the dominant male reacts in accordance with sign stimuli, rather than in accordance with recognition of which individual confronts him. The above assumption is probably quite valid for many lower mammals (Ewer 1968), but is not quite valid for man. Yet traces of juvenile mimicry are found in human submissive behavior, such as crying and childish behavior of persons pleading in despair, and maybe in the high pitch of voices when pleading. It may be noted that female mimicry as a submissive behavior also presupposes the inability of dominant males to separate model (female) from mimic (subordinate) and, hence, this submissive behavior will decrease in effectiveness with increased perceptual ability and recognition of "individuals" rather than "bunches of releasers," as Ewer (1968) aptly put it. Precisely because of the superb ability of humans to discriminate, it is likely that all our submissive gestures are biological relicts of the past and a motley collection of gestures from the sexual and infantile realms, as well as elements antithetical to threats and dominance displays. It is because our biological basis for submissive displays is eroded that we find similarities between us and much of the basic social behavior of chimpanzees as described by Jane Goodall (1971). Submission in our species has moved to the cultural realm to a large extent, but not entirely, and is apparently recognized as such in the role playing of individuals by approved rules and hierarchies. We still have the biological requirement to dominate and receive signals of having succeeded, but we do it with cultural means. This is most noticeable in our use of sexual displays.

Sexual Displays

The study of displays in lower primates led Wickler (1963, 1967, 1969), and others who followed him (Morris 1967, Eibl-Eibesfeldt 1970), to propose a homology between sexual displays of lower primates and those of humans. Such displays are not performed overtly in most societies, that is, they are not universal, yet they do abound in many cultures, particularly in initiation and religious rites in the Pantheon of the gods, in charms, and in dress fashions (Wickler 1967, 1971, Eibl-Eibesfeldt 1970). We can trace these displays in the form of phallic drawings and sculpture and buttock emphasis in females down to the early Aurignacean cultures (La Barre 1970, Marshack 1972b). Shamanists, men of power, are shown with erect phalluses, as are males confronting dangerous animals, such as wounded charging bisons or bears. In the latter it is an indication of confidence, for normally the phallus shrinks during fear. A man able to have an erection when faced by danger is indeed a somebody! (Marshack 1972b, pp. 272−74). The Maya priests offered blood to the gods from a cut and mutilated penis (von Hagen 1960). The followers of Adonis emasculated themselves with swords and threw the organs to the god's statue (Smith 1952). Mediterranean cultures abound with phallic symbols and rites. According to Wickler (1967, 1969), figurines with faces and erect phallus and testes were used as guardians against enemies, evil spirits, etc., in Mediterranean and Indonesian cultures. In general we find phallic displays in cultures in which the males roam about near naked. The best examples are from Melanesian tribes in which the phallus is decorated with ornamented, colorful penis sheaths. In some Australian aborigines the penis is split down the urethra by subincision with a sharp stone, as well as being circumcised (Birdsell 1953). The adult male thus displays his status by visual alteration of his sexual organs. These are cultural alterations, and these lead one to suspect that some biological reasons underlie such alterations.

Indeed, in the Khoisan races, the bushman-Hottentot group, we note that sexual displays are at the biological level. Thus, the male's penis is suspended from a ligament and is thus held semierect. The ligament runs the length of the penis and attaches to the pubic bone. The female's sexual organs also differ significantly from those of other races. The labia minora is greatly expanded and hangs down flaccidly, but during sexual excitement it folds back and turns red (Villiers 1964). From the work of Masters and Johnson (1966), we know that the labia minora changes color in Caucasian races, albeit it is virtually hidden.

Wickler (1967) supports his hypothesis of the homology of human and primate penis displays with observations on infants who develop erections when crying in rage, as well as by the fact that human males and females tend to sit differently. Unfortunately, it was not noted by Wickler (1967) that in addition to the erection there is a cloud of scent expelled due to the exposure of the smegma when the glans protrudes beyond the foreskin; this suggests that originally a strong olfactory component entered into the phallus display which will probably be identified in cercopithecine primates. Secondly, it may be noted that the scrotum in the light-skinned Caucasians still turns dark at puberty; a vestigial purple assumes the place of pink. The erected penis, moreover, is quite red behind the glans. These color changes can be interpreted as vestiges of the full-blown ancestral phallic display.

Hofer (1972) took Wickler to task on the inadequacy of the data to support the hypothesis, and argues against the homology of these displays. Unfortunately, he treats penis displays as if they were phylogenetically determined, and quite rightly concludes that they are not, but fails to consider adequately that penis displays could evolve convergently. Thus, the fact that chimpanzees do not have conspicuous penis displays does not argue against such displays being present in humans. Chimpanzees, however, do erect during major excitation caused by a variety of events, as do rhesus monkeys (Hofer 1972). Erection is not tied specifically to dominance displays—this is the point Hofer (1972) makes. However, the chimpanzee condition is a clear prerequisite to penis displays as dominance displays. In cercopithecine monkeys, penis and vulval displays are often well developed, with the penis gaudily colored red, the scrotum blue, and the pubic hair white, as in the vervet monkey (Struhsaker 1967, Wickler 1967). Given the evidence available today, the conclusion is inescapable that sexual displays were once biologically activated and controlled in man, but that the biological means of delivering them has been lost and the displays are delivered now by cultural means.

We may ask how and why we lost the biological control (though not nearly so much the biological "desire") for such displays as are seen among the savannah-dwelling cercopithecines. We can answer the question with a self-evident little thought experiment. Assume that in the periglacial environment a male did produce a showy erection when angered. What would happen at −40°C? We need not postulate even such temperatures. What would the penis rub against when erected? Presumably against rough, filthy, hairy, rough-tanned hide. Even if the glans was not frozen, it would still rub and bang against those garments and become injured with each violent motion of the body. In fact, it is most unlikely that *H. sapiens* could colonize the periglacial environment until he had lost the biological means of delivering a penis display. When *H. sapiens* reinvaded the warm zones, following his periglacial evolution, the need for a penis display was apparently present in his psychological make-up, but the biological means of delivery was absent; hence the culturally delivered penis displays of modern man. Thus, even this line of thought leads to the conclusion of major changes in the external appearance of *Homo sapiens* as a consequence of adaptations to periglacial or cold climates.

Neoteny

In the earlier discussions of evolutionary change of hominids from the *Australopithecus* to the *Homo* stage (p. 259), it was pointed out that neoteny was a probable consequence of selection for adaptability and large body size during geographic dispersal. That is, the diversity of tasks to be performed plus the need for novel solutions to new problems in heterogenous environments, selected for individuals with retarded or extended maturational processes, and thereby a maximum of juvenilelike plasticity in shaping behavior and body structure. It is the process of retardation of maturation that creates almost inadvertently the neotenous features of humans if the benefits of maturational retardation outweigh the benefits of a morphology adaptive in parent populations. Thus, the further the species penetrated into unexploited uninhabited terrain, the more neotenous it became, that is, the more juvenile features characterized the sexually mature adult form. Therefore, where hominids colonized last we can expect them to be most neotenous; neoteny or increase in juvenile characteristics in fully grown adults becomes a measure of evolutionary change.

We can identify the periglacial and cold regions of the Mediterranean area, western Europe, and western Asia as the origin of modern man. The skeletal evidence, as well as a few carvings of human figurines, indicate that these early men did not differ appreciably from present-day Europeans (Baker 1968, Hulse 1963, Le Gros Clark 1964). From the Mediterranean region and the glacial refugia, modern man dispersed, probably in several waves, to colonize Africa, Australia, and America (see Mulvaney 1966, Martin 1967, Haynes 1966, Wendorf 1966, Howells 1973b). We expect, following the predictions of the dispersal theory, that the most primitive or least neotenous race of human beings would be found in the geographic area of origin of modern man, while advanced, more neotenous races should be found on the periphery of man's distribution area. The "periphery" constitutes the areas of latest colonization. Here we expect a retardation of secondary sexual characteristics compared to the European form (Geist 1971a).

If modern man made his first appearance in the Mediterranean basin and then in Europe, it is likely that he dispersed first to other periglacial environments along continental glaciers. This would explain the early dates for men appearing on the American side of Beringia where they are found at least 30,000 years ago (Bryan 1973). However, Paleolithic cultures noted for their richness appeared in western Europe, and to a lesser extent east of the Scandinavian ice sheet; there is no evidence for extensive Paleolithic art in Asia or Beringia. Nor do we find evidence here of dense human populations, as is found in Europe. This may not be surprising. The weather maps reconstructed for Ice Age Europe (Washburn 1973) indicate that a rather warm oceanic climate existed during maximum glaciation to the west of the Scandinavian ice sheet. We expect here a relatively richer ecosystem than should be found along the ice margins in eastern Europe, central Asia, or Beringia. It is here that Neanderthal man in the preceding stadial reached his classic form. We thus have reason to suspect that the European Paleolithic populations lived in more productive areas and more benign climates than did the Asian and Beringian ones. We would therefore expect in the latter areas the evolution of people better adapted to cold and with the largest brain sizes. The latter we expect to be a product of neoteny; the more neotenous the population the better the chances for maximal growth of body and organs. This interpretation explains the gradient across Eurasia into America of people with increasingly neotenous features.

Maybe the last major group of people to evolve are the classical Mongolians (Chang 1962), who probably are products of the cold interiors of ice age Asia and are therefore some 20,000 years old at best. This time depth is supported by Laughlin's (1963) suggestion that it only some 10−15,000 years. They show the typical neotenous features such as a very large brain and a reduction in secondary sexual characteristics, as well as a reduction in sexual dimorphism. We expect enlarged brain size to be a product of extended ontogenetic growth, and Jorgensen and Laughlin (in Laughlin 1963) do indeed suggest that Aleuts and Eskimos have the longest ontogenetic growth among human populations.

Recent work by Schaefer et al (1974) indicated that the distribution of sweat glands in Eskimos is confined more to the face and reduced on the body compared to Caucasians. It suggests that selection has taken place, so as to reduce sweat glands where they could moisten the parka and destroy its insulation quality. It may well be that distinct differences of the kind indicated by Schaefer et al (1974) speak for Caucasians evolving in relatively benign climates. This suggestion can only be validated by the discovery of a sweat gland distribution in Mongoloids similar to that of the Eskimo. This is not to deny that the total number of sweat glands may be a phenotypic characteristic, and therefore subject to environmental influences.

Aboriginal European man is characterized by a heavy growth of facial and body hair after sexual maturation, the development of frontal baldness in a high percentage of cases, and the appearance of gray hair in middle age. Caucasian females develop relatively wide hips and large breasts after sexual maturation. By comparison, aboriginal man from America is underdeveloped in these very features. There is little body and facial hair, frontal baldness is exceptional, gray hair is present but appears at a later age, the females have less pronounced breasts and buttocks and are closer in body size to adult males than is the case for European man. The rules of zoogeography, as well as those directing the evolution of ice age mammals apparently did not bypass us.

Chapter 14

From Periglacial to Artificial Environments

Introduction

A review of human adaptations to the cold and periglacial zone makes one conclusion inescapable: Our species has for a long time, probably for 100,000 years or more, structured its adaptive strategy by cultural means in such a fashion that it maximizes the individual's physical, intellectual, and social development at the expense of population size. This can only be done by artificially curtailing fertility—unless the resources of the habitat are infinite, and no such habitats exist. To put it into the jargon of population biology: Under cold and periglacial conditions *Homo sapiens* experienced continual r-selection that he brought about artificially under K-conditions. This was an incredible step, since it freed the species *Homo sapiens* for the first time from biological selection under K-conditions, or carrying capacity. Selection under r-conditions had been experienced many times before by populations of the genus *Homo*, but it had never been fixed culturally before and thus experienced on a continuous basis. Without artificial population control, a return to K-selection was inevitable. To maximize physical, intellectual, and social development phenotypically was essential, since only individuals of that phenotype could successfully tackle the harsh and variable northern environments, and even these phenotypes experienced a rather high mortality, a testimony to the demanding conditions of periglacial existence.

We can restate the above conclusions in yet a third manner, as it does throw additional light on the subject: During colonization of the cold zones *Homo sapiens* artificially fixed the *dispersal phenotype*. We discussed the dispersal phenotype under Population Quality (Chapter 6), equating it with high population quality, and showing that the dispersal phenotype is a mechanism by which genes maximize their survival under conditions of uncertainty—which do, after all, prevail when individuals disperse into habitats unoccupied by their species. The dispersal phenotype, probably universal for all living things, although we spoke of it only in relation to mammals and birds, has been designed by natural selection to be large-bodied, vigorous, roaming, exploring, able to control appetites and aversions, tenacious—in short, a phenotype that is most likely to deal successfully with the unexpected, the diverse, and the variable. The more uncertain, fluctuating, variable, and complex the adaptive strategy of the species must be, the more adaptive the dispersal phenotype will be. Thus, at a species' peripheral geographic range, we expect to encounter dispersal phenotypes more frequently than at its center, and we expect to find attributes of dispersal phenotypes fixed by genetic selection in species exploiting highly variable habitats. Since the periglacial zones are highly variable and uncertain, we expect mammals inhabiting such habitat to have many attributes of the dispersal phenotype. In particular, we expect strong development and genetic fixation of tissues that have low growth priorities, such as cerebral cortex, horns, antlers, fat, and display organs. For large mammals this is indeed so, since periglacial forms tend to be relatively large, large-brained, and with elaborate social organs, as was discussed earlier. We have another check on this hypothesis. If natural selection did indeed favor the dispersal phenotype in the genus *Homo* as it colonized the cold and periglacial zones, then we expect very rapid dispersal by *Homo sapiens* and colonization of the globe. This did indeed happen, beginning with the adaptation of *Homo* to cold climate, and culminating after the decline of the last glaciation. We expect *Homo sapiens* to be an exceptional disperser, and this he certainly is.

Paleolithic people thus began taking advantage of the hidden epigenetic mechanisms through which our genes can express themselves in order to maximize individual adaptability to the highly diverse, demanding periglacial and cold zones they colonized. Not only would this produce inquisitive, vigorous, and intellectually and physically competent individuals, but also such people as could control themselves in an exceptional manner. We noted in Chapters 1 and 6 that theory dictates that dispersal phenotypes ought to be able to suppress and control appetites, that is, quite readily to forego temptation as well as aversion—in other words, to do something despite the protests of their emotions and internal monitoring systems. This ability is needed to adapt individuals to new conditions. This means, in the final analysis, that individuals responded to the dictates of reason far more readily than under maintenance phenotype conditions. They could control panic when danger threatened, stand while a prey charged, side-step it, pause when in a tight situation, and change tactics on demand. It also means that humans are born to perpetual conflicts between internal drives they are born with and the dictates of reason. The closer individuals develop ontogenetically to a dispersal phenotype, the less difficult self-control and self-discipline ought to be for them because the threshold for pain and pleasure ought to be higher (see Chapter 6).

The conclusion that periglacial populations of humans maximized physical and intellectual development rests on a number of facts. The adults tend to display an exceedingly well-developed physique. They are usually large in body size, their bones are robust and have strong muscle insertion ridges, and the bones are remarkably free of pathologies. Such bodies cannot develop without excellent nutrition to provide the building blocks and energy for this exceptional development; such bodies cannot develop without much intense physical exercise regardless of the nutrition. It was continuous hard muscular exertion that developed the fine athletic bodies, robust bones, and strong insertions for muscles. Last, but not least, such growth cannot take place except in a most favorable, supportive, social milieu, since ontogenetic growth is very much affected by the kind of home life children experience. The large brain case of individuals from the upper Paleolithic supports the past conclusions and permits deeper insight. Since the cerebral cortex is a tissue of low growth priority, a very large cerebral cortex implies excellent nutrition. Since brain size is maximized phenotypically by the diversity of intellectual and motor tasks performed, we can safely conclude that these individuals had amassed a great deal of knowledge and many skills during their lifetimes, so as to survive the rigors of their environment. These conclusions can be tested. They predict large body and brain size in persons heavily or exclusively dependant on hunting in cold environments. In the Eskimos we do indeed find athletic development, exceptional physical fitness, and the largest brain size among living human populations.

Other facts support the above contentions. We noted that if the dispersal phenotype was selected for during occupation of the cold climates, then this occupation marks the beginning of extensive colonization by the new species. This we do find, beginning at least as early as the dispersal of Neanderthaloid people, and maybe as early as dispersal of late Acheulean people into Africa, some time after the Holstein interglacial. It speaks of enhanced curiosity, exploration, daring, and competence in dealing with the unforeseen. We may note that the evolution of complex vocal communication can take place only because of the necessity to cooperate in very complex ventures, and that music and dancing evolve under the dictates of physical fitness maintenance. This supports the idea that the periglacial environment was variable and complex and that hard physical work, exertion, and agility were mandatory abilities for individuals of late Paleolithic technology. We must note the relatively high mortality of Paleolithic people, an indication that selection was ruthless even among individuals of excellent development, hence enhancing all mechanisms favorable for the production of the dispersal phenotype or excellent physical, intellectual, and social development.

The key to maximum physical and intellectual development is the constant availability of resources that can be invested in ontogenetic development, as well as in the very expensive intellectual and social development. Excellent nutrition, readily available, must be there from conception onward. The foregoing does not downgrade the importance of a supportive social milieu, but such a milieu can only flourish under conditions of relative abundance of excellent food. Body growth requires much protein, minerals, and energy, such as is supplied by a meat diet. As we noted, such a diet is eaten not by choice but from necessity; there is little else to eat during the long, cold winter. Periglacial people had to depend on the meat of large mammals. Such a diet, however, requires very hard, disciplined, and exacting work, for a large amount of meat must be procured in order to meet daily food requirements. The work regime of people from the tropics is no guide to the amount of work individuals performed in our evolutionary environment. To perpetuate a dispersal phenotype requires resources in excess of need, and this situation can be maintained in a finite environment only artificially by keeping the number of mouths relatively low. This implies that even the earliest cold-adapted populations practiced population or fertility control. It is quite impossible to rely on physiological means of fertility controls found in other mammals because they affect individual development negatively. In species with normal mammalian mechanisms of population control as food declines and birth rate follows, the individuals born are not of the dispersal type but are relatively poorly developed individuals. Low birth weight in humans has disastrous consequences on individual development, as seen from the reviews of Jacobson (1973) and Wallace (1973). Humans appear to be rather free of physiological mechanisms significantly affecting reproductive output with a decline in food resources. This is another observation supporting the idea that we have depended for a long time on cultural—not physiological—population control. To maintain the degree of physical development found in Neanderthal and late Paleolithic people simply is unthinkable without artificial population control which ensures abundant resources for developing individuals. Thus, I concur with Dumond (1975) that birth control is probably an old cultural process, but I disagree that there is reason to believe it existed prior to the *Homo sapiens* stage.

The "evolution" of artificial fertility control can be envisioned as follows. During the early evolution of the family, sex was identified as the major bonding mechanism that tied male and female together in order that they would mutually invest in their offspring. Selection must have been against females that conceived during lactation because reproductive investment in the second offspring was at the expense of the first, even if the second child were killed at birth. Gestation would have depressed lactation, leading to underdevelopment of the suckling infant. Moreover, the female would have increased the risk of death due to birth complications. We cannot assume that resources were sufficiently abundant and easily procured to raise two children simultaneously; such conditions would have mitigated against the evolution of the family in the first place. There was therefore a requirement for a reduction of conception during lactation, and indeed we still find in humans that lactation depresses conception. The length of lactation and thus of early ontogenetic development was probably dictated by the overall adult mortality rate, so that lactation was short enough to ensure sufficiently frequent conception during a female's lifetime to maintain the population. After weaning, both parents invested in a new offspring while the weaned one probably relied on its own wits to gather its daily food. This hypothesis would predict low phenotypic development of body build, intellect, and social competence in individuals. That is, tissues of low growth priority would be poorly developed and tools would continue from generation to generation with no noticeable change, while the geographic range of the form would expand at best very slowly. What we know of *Homo erectus* and the Acheulean tool tradition fits these predictions. We would also expect *Homo erectus* to be fairly aggressive, and the strong sexual dimorphism, strong brow ridges, and cannibalism indicate that this was so; the prediction of aggression is based on the low quality phenotypic syndrome of individuals under K-selection (see Chapter 6). With lactation acting as a contraceptive, we need not postulate infanticide or any other form of artificial population control for populations in which the nuclear family had evolved and was the effective reproductive unit.

During dispersal and adaptation to cold climates, however, conditions changed. Initially, of course, r-selection began in the presence of abundant resources. It led to better phenotypic development physically and intellectually. Abundant food reduced aggression and led to tolerance, promoting cooperation among adults and enhanced body growth of the young. It also must have increased life expectancy. This would have increased the population's reproductive rate. This in turn would soon have changed r-conditions to K-conditions were it not essential that adults cooperated and practiced skills possible only with high physical development and fitness. Moreover, under conditions of abundant food, the old mechanism reducing conception—lactation—may not have been as effective as under conditions of relatively low food availability. This would have resulted in babies being born at inopportune times.

We noted earlier that one of the conditions of relying on large mammals for food by confrontation hunting is to be able to move rapidly over long distances to consume carcasses of kills made many miles apart. This imposes a heavy work load on the hunters and, especially, their families. There is no evidence that they stored food and had provisions for these trips. If so, the family would at times have had to cover long distances between kills while all—including children and older individuals—went hungry. While adults might span such periods, children and old people could suffer and even die. Under these conditions, looking after more than one small child at a time would probably be an unbearable burden to the female and to the male. Small children and at least some clothing would have to be carried long distances on hungry stomachs through cold, wintry country. Inevitably, this would lead to mortality of excess small children and old individuals and would set the stage for a conscious reduction in births and in children. However, the result—enhanced physical and intellectual development of the surviving offspring—could hardly have gone unnoticed. This "child and old people control" began as an accident of moving without food stores across long distances in cold weather between kills of large mammals. Population control, therefore, reaches back to the very roots of time when humans began colonizing cold climates while living off large mammals exclusively during a significant portion of the year.

Contrary to the situation in warm climates, the female in cold climates begins to extend essential services to the male. She assumes a work load beyond that of child rearing, namely looking after the camp and food preparation. This would entail watching the fire, without which life is not possible for humans in cold climates, collecting fuel, preparing shelters, and in winter, thawing meat for consumption. These are essential services to the male during cold weather because males would not be able to hunt cooperatively *and* look after their food and comfort simultaneously. Males can do this in warm climates, but in cold climates they would be faced by great cold at night, no fire, no shelter, and large pieces of frozen meat, were it not for the service of the female. These services are not a luxury, but are essential to life for the male, and indirectly to the female and her progeny as well. Under these conditions of work, a second suckling child would indeed be an intolerable burden and infanticide would likely result.

Since the female begins to extend essential services to the male, she bonds the male not only with sex, as in earlier stages of human evolution, but with the satisfaction of essential creature comforts as well as ensuring warmth, comfort, and satisfaction of hunger and thirst. Sex now loses its primacy as a bonding mechanism. If this happens, however, not only infanticide but also cultural control over the time, place, and frequency of coition can be practiced without weakening the bond between male and female. This in turn predicts that lactation as a contraceptive becomes subject to genetic decay and becomes less effective. Thus, an old biological mechanism of fertility control loses its effectiveness as cultural mechanisms of population control take hold. We still find its remnants, however, in humans today, just as we find remnants of other old adaptations.

Since sex begins to lose its primacy as a bonding mechanism, we expect sexual signals to begin to regress. Secondly, the female takes on physical labor in caring not only for her children but also for the male. We expect the female to grow larger in stature relative to the male. In total, this predicts a reduction in sexual dimorphism in populations living in the cold and periglacial zones compared to earlier forms such as *Homo erectus*.

Moreover, as the harshness of the environment increases, the female has to assume more and more work to service her family. Ultimately, this includes procuring much the same food as that provided by the male, since a point can be reached when the male can no longer provide sufficient food to support the family. Not only will the female become increasingly similar to the male in physical type—that is, sexual dimorphism becomes reduced progressively as humans adapt to increasingly harsh climates—but it is also to her reproductive advantage to capture and hold a second provider. Polyandry is the consequence. This can, of course, be maintained only with selective infanticide of female babies. The female now becomes the sex in lesser abundance and socially the more important one. Fatherhood becomes uncertain, but not motherhood. Hence, social accounting by maternal descent must follow, while females occupy increasingly more important social positions. It is most likely that the foregoing sketches give some essential features of the Mongolian race, which apparently evolved under the harshest of ice-age conditions in central Asia.

The genus *Homo* thus underwent a reversal in sexual dimorphism in its evolution, starting with low sexual dimorphism in the gracile *Australopithecus,* when the female alone cared for offspring, to fairly strong sexual dimorphism in *H. erectus*, when the nuclear family became the effective reproductive unit and sex was the main bonding mechanism, to increasingly weaker sexual dimorphism in *H. sapiens* as the female was responsible for more and more work in servicing the family, and the extended family became the effective reproductive unit. Female "seductiveness" must have been at a peak at the *Homo erectus* stage, declining progressively in the evolution of *Homo sapiens* with the associated reduction in erotic secondary sexual characteristics. Concomitantly, neotenization reduced the secondary sexual characteristics in males, reducing secondary sexual dimorphism still further. The beginnings of polyandry brought the basic social organization of humans closer to that of social large-mammal-hunting canids, in which, in effect, a female captures a band of adults and makes them work on behalf of her reproductive effort.

I have dwelt so far on artificial population control as a means of maximizing phenotypic development in physique, intellect, and social competence. Even under these conditions very hard physical work and a high mortality were extracted from the adults to pay for the very high cost of ontogenetic development. Every factor promoting favorable ontogenetic development had to be made use of in order to make efficient use of the resources procured, and this included the shaping of the social milieu so as to promote growth and intellectual and social development. Yet all this work and population control would have been in vain had random mating or incestuous mating prevailed. Inbreeding is not compatible with maximum physical and intellectual development, since it does depress body growth, neuromuscular ability, academic performance, vigor, and life expectancy (Schull and Neel 1965, Wilson 1975). Moreover, the damage of close incestuous matings is present in almost half the children, and is so noticeable that it is most incredible that humans would not have noted its relationship to marriages between close relatives. Because under cold climate conditions the demands on individuals were so very great, and because humans are so very observant, they would have quickly discovered the positive effects of outbreeding and institutionalized them. Hence the universality of incest taboos humans, replacing the biological mechanisms that naturally reduce incest (Wilson 1975, p. 79). Thus, it is likely that the earliest populations of *Homo sapiens* already maximized heterosis culturally in order to maximize body size, physical abilities, intellect, etc., during ontogeny. If this is so, by combining population control with mating control in the form of incest taboos, and thereby maximizing genetic selection for the dispersal type. *Homo sapiens* is a culturally made species. Our very species must therefore be a product of conscious cultural, rather than biological selection.

Let me now place the above discussion into the context of the evolution of upper Paleolithic people. In the foregoing chapter, I argued that they originated from desert-adapted people from Africa, from a group whose skull and tooth characteristics were shaped by selection under resource scarcity. It was proposed that archaic *sapiens* forms, originating in cold-temperate climates in Europe during the Holstein interglacial and practicing a late Acheulean culture, moved to Africa as the latest model of human being and established themselves there. From this archaic stock, people adapted to deserts arose and in the process assumed the form of modern man.

Note that such people would therefore arise from a people that had at least the rudiments of population control acquired in their northern place of origin. It would indeed make life in the desert easier, in which resources are scarce and where a great deal of mobility is necessary. Also, in the archaic *sapiens* there would have been a start to females supplying essential services to males. This, too, would have been adaptive under desert conditions. It would have made the female support the male in much the same fashion as females support males in modern hunter−gatherer societies in the warm deserts. The males could then concentrate on getting the protein fraction of the food and invest time and energy in sophisticating his hunting strategies and tactics. It would have set the stage for the "early anxieties" of the upper Paleolithic cultures because males now depended on females for food, and the two sexes had legitimate differences of opinion as to where to go when. This would have encouraged shamanism just as much as anxiety about resources.

Once cultural population control was developed, any desirable development could be attained during ontogeny regardless of the resource base. If large body size was a prerequisite for survival, it could be had, albeit at the cost of population size. If Africa had indeed a motley collection of populations arising from archaic *sapiens* stock and coexisting by exploiting different biomes, then I have little doubt that large body size may have been essential for survival, in order to discourage predacious neighbors. Certainly the coexistence of large-bodied *Homo sapiens capensis* (modern) and large-bodied Rhodesia man suggests this. Moreover, the huge brows of the latter suggest aggression and a premium on intimidation. Modern man was essentially shaped by the late Paleolithic into what he is today. Yet, not quite. When the last glacial advance began to wane and the once extensive periglacial environments shrank, fragmented, and disappeared, new selective forces began to act on human populations. Agriculture appeared and began to alter the genome of populations practicing it. Culture and technology increasingly supported physical and intellectual effort. Dispersal brought the last land masses under human habitation. The environments humans lived in departed increasingly from those which had shaped their major features. The argument is pursued that economic strategies shape behavior, social values, biological attributes, and phenotypic development. This is an important idea for it indicates that societal values that are cherished or healthful phenotypes can be preserved best by structuring an economic strategy that harmonizes with them.

Hunting Economics

The foregoing discussions were predicated on the assumption that under cold-climate conditions in which the last salient features of our genome were shaped, individuals worked exceedingly hard. The practical implication of this is obvious: Our condition of maximizing attributes such as good physical development, health, and even intellectual development, is frequent hard physical exercise. This is not a new conclusion because the benefits of high physical fitness are well recognized (Cureton 1969), but it stands in contradiction to the practical implications of Boyden's (1973) claim that under "natural" conditions human beings work relatively little. This is certainly true of hunter–gatherers from warm climates (Sahlins 1968, Lee 1968), but these people do not have the physical development of humans from the evolutionary, periglacial environment. Rather, people from tropical and warm climates are of generally small stature and slender build. They are secondarily adapted to the warm climates, having originated almost certainly from Paleolithic cold-climate hunters.

On the other hand, the very hard work done by Caribou Eskimos may also be misleading as to the amount of work done on average by individuals in Paleolithic populations (Balakci 1968). There is no evidence in Paleolithic skeletons of the chest deformities of old Eskimo hunters brought on by the extensive breathing during labor,[1] nor of the extreme tooth wear in women, a sign of how much work had to go into making clothing. Still, large mammal hunting would be very demanding, since so very much meat is required if it is the main or sole source of food. There is the work entailed not only in searching for prey and the hunt itself, but also in butchering and hauling meat, preparing hides, preparing and caching meat, making clothing, implements, and weapons, hauling camp equipment about, searching, breaking and carrying fuel for the fires to dry and smoke meat and hides, etc. Given these requirements, which were not faced to the same extent by hunter–gatherers from the warm zones, three to four hours of work a day does not appear adequate. That the work was hard is evidenced by the strong muscle insertions on the skeletons of Paleolithic people (Vallois 1961). How much food is required in a hunting econom?

Let us assume we have an extended family of 7 members, including the young man and woman, an old man and woman, and 3 children aged 14, 10, and 6 years. What is their annual requirement for food and clothing? The food will be assumed to be meat from ungulates with a caloric density of about 180 cal per 100 g, and a fat content of about 8% (Field et al 1973a,b, 1972). Following the recommended daily allowances of the US National Research Council (Brody 1945, p. 786), an active 70 kg male should take in 4500 cal daily; a pregnant or lactating wife will require some 3000 cal per day, the old male some 3000 cal per day, and old women 2500 cal per day, and the children 3200, 2500, and 1600 cal per day, respectively. The total daily caloric requirement for the family is 20,600 cal, the equivalent of about 25 lb

[1]0. Schaefer, Charles Camsell Hospital, Edmonton, personal communication.

of meat—note, this may be a *minimum* requirement! The annual meat requirement of the family is 9125 lb. The carcass of a tundra reindeer—a common prey of Paleolithic man—yields about 75 lb of meat, according to Parker (1972). The total annual meat requirement of the family would thus be covered by about 120 caribou carcasses, if we make no allowance for waste. They will need at least one caribou for food every 3 to 4 days.

The above is a conservative, in fact unrealistically low, estimate, as is evident from Parker's (1972) review of the subject. Parker and others concerned themselves with the ecology of Eskimos, the majority of whom had also to feed dogs. Yet, making some allowance for the fact that the estimates based on field work are low, all suggest a kill well in excess of the above calculation. Lawrie, Harrington, and Wright all give figures that exceed 200 caribou per family per year; Parker (1972) concluded conservatively that under aboriginal conditions an Eskimo family killed some 150 caribou per year.

Banfield (1954) also gives figures on the number of caribou required for clothing. It takes 12 caribou hides to produce one complete outfit of inner and outer parka, mitts, boots, stockings, etc. Such an outfit would last, at the very best, 2 years, and is usually replaced after 1 year due to wear and tear. Therefore an extended family requires some 84 caribou per year for clothing alone, and at the very least 42 caribou. This does not take into account requirements for sleeping bags, or skins for leather tool-shafting, constructing sleighs, etc. If clothing were made from small creatures such as Arctic foxes, the number of skins required becomes astronomical, as does the work associated with trapping, skinning, tanning, sewing, etc. Nevertheless, some small furs were apparently used by Paleolithic hunters (Klein 1973).

From the foregoing figures, it is evident that only very large and dense ungulate populations can support a pure hunting culture. Granted a sustained kill of about 10% of the full caribou population, a family of 7 requires a minimum population of about 1000 caribou for its survival, if we take the calculated figures, and 1500 caribou if we use Parker's (1972) figures. Thus, a band of about 25 people requires a standing population of some 3500 to 5500 caribou for their survival, and a tribe of 500 people requires some 70,000 to 110,000 caribou. The above kill figure of 10% of standing crop assumes intense exploitation; it is questionable whether Paleolithic people were capable of such exploitation. A more likely figure for a standing population of caribou to support a tribe of 500 hunters is 150,000–200,000 caribou or more.

Even these figures are conservative. If the hunters concentrate on eating mainly muscle meat and little fat so as to be on a low-fat diet or if they had to subsist on fish, birds, and small mammals, the amount of food they would have to obtain increases over that used in the above calculations. Lean ungulate meat has only about half the caloric density I used in the calculation, while that of most fish, birds, and small mammals is somewhere between 126–178 gross cal per 100 g of edible portion (Farmer and Neilson 1967, Farmer et al 1971). The hunting economics shown here are thus on the conservative side.[2]

---

[2]For data on meat consumption at fur posts and on journeys in early 19th century Canada by families of traders, voyageurs, hunters, etc. see F. G. Roe, The North American Buffalo 2/e, 1970, U. Toronto Press, Appendix M, pp. 854–55.

The Vitamin C Problem

A pure meat diet, especially of muscle meat and attached fat, is inadequate to supply some of the essential vitamins such as vitamins A or C. An active man, eating about 4.5 lb of muscle meat a day, would receive only about 400 IU of vitamin A per day, or about 8% of the required intake. If he were to eat about 50 g of liver in addition, he would take in close to 5000 IU of vitamin A, covering the daily requirement. In order to eat about 50 g of liver per day, and give the same ration to each member of his family, the hunter would have to kill about 100 caribou a year, assuming caribou livers weigh about 4 lb each. Again, we reach a very high kill figure of caribou per family. However, the same amount of liver will supply only one-quarter to one-third of the amount of vitamin C recommended (Brody 1945, Table 20.1, p. 786). This brings us to an interesting controversy, and a splendid example of how *not* to use data pertaining to "natural" man.

L. C. Pauling, in a recent book and in several articles in reply to critics (1970, 1971, 1972), has argued for an intake of vitamin C (ascorbic acid) far in excess of that recommended by both the Food and Nutrition Board in the United States and its equivalent in Great Britain. The recommended intake per day for a 70 kg adult varies between 30 and 75 mg per day (Pauling 1972, Brody 1945), while for United States soldiers in 1940−1942 it was about 86 mg (Brody 1945, Table 20.1a, p. 787). Stare (1971), in responding to Pauling, cites studies that showed that only 40 mg of ascorbic acid per day can be metabolized by men, while in a Japanese study the human subjects reached a maximum sustained level of vitamin C in their plasma on an intake of 90 to 100 mg per day; also, if ascorbic acid is taken in excess of 25 to 30 mg per day, it is excreted in the urine. This finding stands in apparent conflict with the studies of Cowan, Diehl, Baker, and Ritzel (Pauling 1971) that a daily intake of 200 to 1000 mg per day significantly reduces the incidence and duration of colds. It stands in contrast, because Pauling argues that the recommended intake of 70−85 mg per day is unnaturally low and that we should take in a much greater amount.

Pauling's argument follows much the same thought processes as I do, and yet it is a false argument. Humans, along with other primates and guinea pigs, are an oddity in that man does not produce vitamin C in his liver (Chatterjee 1973). Other animals do have this capability, and this is one reason why liver is a good dietary source of this vitamin; in fact, 200 g of liver almost cover the daily requirement. If we calculate the amount of ascorbic acid produced by animals, then one gets a daily production of 2 to 15 g per day for a 70 kg animal (Pauling 1972). From this, Pauling concludes that humans also require the same amount for optimum health. He now notes that gorillas, feeding on sprouting shoots which are already rich in vitamin C, consume about 4.6 g of ascorbic acid per day. If men were to eat the same food, they would ingest 2 g of vitamin C per day; if men concentrated on vegetation especially rich in ascorbic acid, they would consume 9 g of the substance in a food intake of 2500 cal. Pauling invokes the "natural" nutrition of the gorilla as a justification for raising the natural intake of vitamin C in man to a level he considers desirable. The correct procedure would be to examine the "natural" nutrition of man instead. Such an examination would show that periglacial hunters could not exist had they the same ascorbic acid requirements as gorillas. Eating only muscle meat and 50 g of liver a day, an individual would ingest less than 20 mg of ascorbic acid a day. Clearly, they must supplement this somehow. In summer, vegetation would be an obvious source; indeed, roots from *Pedicularis* and *Oxytropis*, willow shoots and leaves from *Rumex, Polygonum,* and *Oxyria,* boiled or raw are consumed by Eskimos (Balakci 1968, Nickerson et al 1973). Also, the content of fermented reindeer rumen is eaten and its vitamin C content, presumably, is high. In view of the kill statistics rumen content may be a common food. Moreover, Inuits of Alaska did store berries and greens in seal oil and kept it in permafrost so that the availability of these summer foods could be extended. In northwest Canada, Indians used to brew spruce-needle tea in winter as an antidote to scurvy. This is skimpy evidence for the notion that vitamin C is a nutrient not in abundant supply in the meat diet of cold-adapted hunters—a premise also stated by Schaeffer (1964, 1971b), who has been a long-time student of Eskimo nutrition. It is likely that gnawing gristle and cartilage and eating kidney, eyes, and tripe would increase the vitamin C content, but it is unlikely that it would exceed 80 mg a day, and it is more likely that it would fall between 30 and 60 mg per day. These values, however, fall within the range discussed by Stare (1971) as being adequate for human nutrition. Since the vitamin content of various tissues of caribou remains unknown, and the actual vitamin C intake of Cariboo Eskimos remains unmeasured, the foregoing does not contradict Pauling, but makes it unlikely that humans, under natural conditions, require the ascorbic acid intake Pauling proposed. Note, I said *natural* conditions. People living in densities of hundreds and thousands per square mile are living in *unnatural* conditions! Viral infections are easily contracted and spread; repeated infections by diverse viruses at rates unknown to natural man are likely. Hence, an increase in ascorbic acid consumption could, quite likely, protect one from the common cold. Under periglacial and arctic conditions, however, with their low population density of people and low rates of bacterial and viral diseases, a low consumption of vitamin C was probably all that was required to stay healthy. Conversely, it may well have been impossible for man to go on an all-meat diet and live in the cold and periglacial zones until his requirement for vitamin C was much, much lower than that of his primate relatives. A low vitamin C requirement probably evolved in *Homo* as an adaptation to dry steppe.

The Vitamin D Problem

The same argument, but in reverse, can be made for vitamin D. This vitamin, essential for proper calcification of bones, is synthesized in the human skin by solar rays from the ultraviolet region acting on 7-dehydrocholesterol; sunlight is not effective in supplying the effective dose if the sun is lower than 35° above the horizon (Daniels et al 1968). The daily required dosage is about 400 IU (10 mg), almost all of which is synthesized in the skin, while a trivial amount comes from the food supply. Loomis (1967), in his account of vitamin D and its relation to human evolution, indicates that only a few foods, such as the oils from some fishes, have significant amounts of vitamin D, while our normal foods have little, especially those foods available in winter. Individuals with white skin can synthesize large amounts of the vitamin since, according to Beckemeir (in Loomis 1967), 1 cm² of white skin synthesizes 18.1 IU of vitamin D in 3 hours, and 20 cm² of white skin will supply the daily requirement if exposed to the sun for less than normal daylight hours. Negroes, on the other hand, synthesize very little vitamin D per unit area of skin, since melanin is an effective filter of ultraviolet radiation. Thus, dark skin protects them from excessive vitamin D synthesis and the consequent maladies and death from pathological calcification of tissues. This permits a Negro to expose his body to the equatorial sun. This should be suicidal for white-skinned persons, as it would lead to the synthesis of about 800,000 IU of vitamin D. However, such overproduction has not been reported; in fact, vitamin D synthesis may be self-limiting, as Daniels et al (1968) suggest. Negroes going about clothed, and exposing as little skin to the sun as do clothed whites, synthesize only 5 to 10% as much vitamin D as do white-skinned persons. Hence, Negroes would be susceptible to rickets. Clearly, the ability of humans to wear heavy clothing in the cool climates depended on a reduction of melanin in the skin in order to synthesize sufficient vitamin D (Loomis 1967). Therefore, dark-pigmented populations of hominids from warm climates would have had difficulties living fully clothed in the long winters of the north and subsisting on a meat diet virtually free of vitamin D. Granted the above argument, periglacial life is possible only to hominids with very small vitamin C requirements and a lightly pigmented skin.

Loomis (1967) argues that white skin would evolve in populations once these groups crossed north of the Mediterranean Sea, since the low angle of the winter sun above the horizon, coupled with the filtering action synthesis. The effect would be greater still if clouds obscured the sky. He advances the intriguing argument that "nakedness" in hominids could be an evolutionary response to increased needs of vitamin D synthesis in northern populations, citing Cruikshank and Kodicek's experiments, in which shaving increased the vitamin D synthesis of rats. He also points out that in present-day races, skin color varies directly with the annual intensity of sunshine and the shading effect of the habitat. Thus, northerners are, by and large, light-skinned compared with equatorial populations, as are forest dwellers compared with dwellers of the open steppes. Only coastal people with a high diet of fish oils would obtain sufficient vitamin D to be immune to the selection for white skin which would befall populations in northern terrestrial ecosystems. Loomis's (1967) view does not, of course, explain the evolution of dark skin, since an overproduction of vitamin D in light-skinned persons exposed to the tropical sun has not been reported. The paper by Daniels et al (1968), however, does offer such an explanation. In essence, melanin depositions protect the skin against damage by ultraviolet radiation and reduce the incidence of skin cancer. It appears that the views of these authors adequately account for dark skin in tropical areas and increasingly lighter skin in populations of increasingly higher latitudes. It may be noted here that human fossils of Würm age show a complete absence of evidence for rickets (Vallois 1961), an indication that these people had no problem synthesizing or ingesting sufficient amounts of vitamin D.

Parasites, Diseases, and Periglacial Hygiene

The incidence of viral, bacterial, and parasitic disease under cold and periglacial conditions was likely to be very low; however, no adequate data are available to this writer from arctic and subarctic hunters. Dunn (1968) presents some interesting thoughts in his review on mortality, diseases, and epidemics in hunter−gatherer societies. He proposes that parasitic and infectious diseases in hunter−gatherers are related to ecosystem diversity, and he shows that, as predicted by this hypothesis, the highest helminth and protozoan infections are found in people from tropical forests and the lowest from deserts and shrub deserts. This is also confirmed by limited coprolite studies of ancient populations of North American desert tribes (Heizer and Napton 1969). This view predicts chronic infections in complex ecosystems and rare, but disastrous, epidemics in simpler ecosystems, implying that periglacial and cold-adapted hunters would be relatively susceptible to a variety of parasites and diseases if they were transported to warm climates. Where people were few and roamed far, infections would be rare, but they would be common where people crowded, as in agricultural communities (Garn 1963). Neel (1970) expresses some thoughts and observations that are most relevant to this matter. Granted adequate nutrition and long lactation periods, as observed in hunting−gathering cultures, then the child is protected to a considerable extent by the female's antibodies ingested with the milk. Malnutrition, short lactation period, and reliance on milk from nonhuman sources would decrease the infant's protection in the agricultural society and thus lead to high infant mortality. Secondly, children playing in the proximity of dwellings readily infect themselves and develop their own antibody responses early.

Studies by Schultz (1961) on pathologies in wild primates and man are quite revealing. He showed that there is a very high incidence of arthritic changes, bone fractures, dental pathologies, sinus infections with bone resorption, and congenital deformities in wild apes. His review also indicates a high reproductive wastage in primates, be they captive or wild. Humans, by contrast, appear to be more healthy, at least judged by the lower incidence of various diseases. Unfortunately, Schultz's work does not permit a ready correlation between pathologies and environment, but one cannot help but wonder whether most of the pathologies reported are concentrated into "poor quality" populations, as elaborated in Chapter 6.

A fair case can be made for pathogens and parasites being largely contained in the periglacial environments, as well as the tropical ecosystem, though for different reasons. In the cold environment, the actions of frost and snow, the flooding of the floodplains each spring during melt-off, and the long seepages of water from snowbanks would flush many pathogens downriver, or at least decrease their effective concentrations. Some pathogens would be frozen into permafrost or spilled into underground aquifers; pathogens contained in carcasses would be subjected to digestion in various stomachs of birds, mammals, and insects, and dispersed. The relatively low population density of humans would reduce the incidence of infection.

In addition, in simple or complex ecosystems that still possessed their complement of large mammals, it is most likely that the outbreak of serious diseases, such as rabies, would be quickly controlled. Assuming rabies broke out in small carnivores, infected individuals would soon be eliminated by becoming prey to larger carnivores. They would normally not pass on the viral disease to the larger carnivore since such animals are well adapted to staying out of harm's way from a capable, healthy prey, let alone a sick one suffering the disabilities of rabies. Normally, the rabies infection would be quickly contained by the rapid elimination of sick animals and the reduced chances of contact between them and members of their species. However, should a small carnivore pass on his infection to the larger carnivore, the infection would most likely be quickly contained for the following reasons: The infected large carnivore, acting ''abnormally,'' would be killed, either by members of his own group or, if he leaves the territory of the group, by the neighboring pack. The data for this hypothesis are scanty as yet—we have only begun studying the life histories of large carnivores—but the assumption should receive some attention owing to its implications for public health. The studies on wolves (Jordan et al 1967, Rausch 1967, Mech 1970), hyenas (Kruuk 1972, Goodall 1971), and lions (Schenkel 1966b, Schaller 1972) make the foregoing hypothesis probable. One also notes that rabies occurs, or is most frequently recorded, in areas with severely disturbed ecosystems—that is, the densely settled areas such as southern Ontario, Canada, or some areas in Europe (Johnston and Beauregard 1969). This may be due to the fact that rabies would be most closely watched and more likely to be recorded in densely settled aeeas, and/or due to its greater rate of spread among small carnivores in the absence of large carnivores. The subject warrants investigation. Also, if large predators reduced the density of small predators, this would also reduce the incidence of rabies and extend the rabies cycle, as is the case in areas with poor habitat for small predators (Preston 1973).

Periglacial hunters would most likely be subject to botulism poisoning, to rare outbreaks of trichinosis, and infection by *Echinococcus* if they handled canids with any high frequency. There is evidence indicating that natives of the Arctic have some immunity to trichinosis (Ozeretskovskaya et al 1970). On the whole, they would be free of debilitating viral, bacterial, and parasitic diseases, and such low immunity would make initial colonization of warmer climates by periglacial hunters a slow process.

Wastes and the potential for pollution, the transmission of parasites, and the development of toxic substances in meat residues that could be picked up by children, would be quite scarce for periglacial hunters, or for other hunters for that matter. Wastes would be quickly removed by various forms of wildlife, in particular after a camp is abandoned. Foxes, coyotes, jackals, wolves, bears, ravens, crows, magpies, jays, chickadees, shrews, squirrels, and mice would quickly remove dry or rotting meat scraps, fats, and bones, as well as human fecal matter. Unburied human corpses would soon be consumed by the scavengers, and any pathogens or parasites they contained would be digested or dispersed. We find the disposition of wastes by scavengers, or by tamed hunter–scavengers such as the dog, a constant adjunct to camp life in cultures as diverse as the Eskimos and the Australian aborigines. The latter offer a most interesting example of how early man and canids could have coexisted (Meggitt 1965). The biology of diseases and the waste disposition practices in intact ecosystems, and the fate of waste in primitive cultures, deserve some study.

How to Maximize Intellectual Development, Speed of Pattern Matching, and Creativity

I argued that, from the earliest populations of *Homo sapiens* onward, our species practiced population control in order to maximize the physical development of individuals, their intellect, and their social competence. Neither birth control nor cultural incest taboos are therefore likely to be new to our species; in fact, they may have been the key elements that permitted us to evolve our *sapiens* features. Without birth control and incest taboos, phenotypic development could not have been maximized and no vigorous, bright, inquisitive, resourceful dispersal phenotype of humans could have arisen. Whereas the skeleton does tell us that physical development was maximized, it tells us much less of the individual's intellectual competence and even less of its social one. There is no doubt, however, that beginning with early *Homo sapiens* the tools began to diversify, and there were other signs of intellectual competence—one sign of intellectual awakening was the manifestation of religion and art—while humans began to disperse more frequently and widely. This, too, I interpret as evidence for enhanced curiosity and intellect. Still, these signs by themselves are not satisfactory.

We can ask, for instance, what environmental factors are known to maximize intellectual competence. A simple question without—alas!—a simple answer! On one hand we can equate intellectual competence with *diversity* of intellectual tasks mastered. In Chapter 2, I cited papers that showed that the brain enlarges in response to specific exercises well learned; for each intellectual or motor skill learned, there was a corresponding size increase in a particular region of the brain. Therefore, the more physical and intellectual tasks mastered by an individual the greater the likelihood of a phenotypically large brain. Brain size would thus be proportional to the demands placed on an individual during ontogeny. This hypothesis explains at once the finding that domestic animals have brains $25-50\%$ lighter than those of wild counterparts (Kruska 1970a,b). Also, granted that a large amount of diverse knowledge and skills was required of individuals living in the cold climates with poor technology in order to master the hostile environment, then we expect exceptionally large brains in Paleolithic populations. This is found, of course (Mettler 1955, Coon 1962); it is also found in present-day cold-climate hunters and in the Mongolian race, the latest to emerge, probably from the harsh periglacial zones of central Asia. The hypothesis also exposes the folly of expecting a close positive correlation between genial performance and brain size (Mettler 1955). Genial performance, as well as the Intelligence Quotient, are based on excellence in a *very few* tasks; there cannot be a close correlation with brain size, which is based on mastery of very many tasks. There has been no close relation found between genius and brain size, or between IQ and brain size—and we need not expect to find one for the reasons just elaborated. A weak relationship, however, apparently does exist (Tyler 1956).

Intellectual competence can also be equated with virtuosity in a few intellectual skills. Such competence is, in essence, measured by the Intelligence Quotient, which is a speed test of linguistic, mathematical, and perceptive skills. Although the IQ tests have been used with questionable judgement and have fallen into academic disrepute, particularly in relation to the ever-present nature/nurture argument (Scarr-Salapatek 1971a, Layzer 1974), studies in IQ still give penetrating insights into environmental factors affecting IQ. What does IQ measure? Let me explain it as follows: When a muscle is trained and hypertrophies, the speed of signal transmission along the motor nerve to the muscle increases. This occurs because the myelinated nerve fibers to the muscle increase in diameter while the myelin sheath thickens, causing more rapid conductance of the nerve impulse (Edds 1950, Timiras 1972). Frequent use of a nerve fiber leads to its enlargement and enhanced performance. We are looking at a universal property of nerve fibers, be they located in the arm or in the brain. An increase in fiber diameter and myelination in the brain is associated with enhancement of brain function during ontogenetic development; conditions of hypoxia, that is, reduced oxygen transport to the brain, are associated with a decrease in lipid and protein content of the whole brain due to reduced lipid metabolism and myelination (Timiras 1972, Chapter 14). This is a finding complementary to those cited earlier (p. 31) that brain growth can be a function of intellectual and motor exercise because stimulation of brain cells is associated with vasodilation and the influx of oxygenated blood to the stimulated brain area. Clearly, the more exercise and resultant inflow of oxygen to the brain the better the myelination, the larger the nerve fibers, and the speedier the transmission of signals along the fiber tracts—provided the nutrition of the individual supplies the necessary building blocks, such as fatty acids, proteins, and energy.

It has been shown that IQ is closely related to the speed with which the brain processes perceive stimuli (Ertl and Schafer 1969) and to the inverse of response amplitudes of electric brain potential to expected stimuli (Schafer and Marcus 1973, Begleiter et al 1973). The latter measure indicates that the brains of individuals with a high IQ—or of sober individuals—process information more *economically* than those of individuals with a low IQ or under the influence of alcohol. In all individuals, sudden unexpected stimuli produce big amplitudes in the electric brain potential, but expected stimuli produce small ones (except in individuals with low IQ and in drunk individuals). These finds are supported by the fact that reaction times in mentally retarded children are longer than in normal ones (Rarick 1973b). In undernourished, poorly developed children with a low IQ, the alpha waves in the EEG were of low voltage, poorly developed, and of longer latency compared to those of normal children (Stock and Smythe 1967). In experiments with albino rats, it was shown that the lengthening of the latency between light flash and the resulting electric brain potential was accompanied by a decreased ability in solving brightness discrimination problems (Edwards et al 1969). In this study, rats raised in ''impoverished'' environments had longer latencies between stimulus and brain response than did rats raised in enriched environments. Edwards et al (1969) also cite studies showing that rats raised in an impoverished environment differed in brain structure and physiology from rats raised in an enriched environment. A ''superenriched'' postweaning environment in rats compensated for the experimentally inflicted damage resulting in cretinism (Davenport et al 1976); in short, a proper environment can overcome damage to intellectual competence. From the foregoing studies, we can conclude that IQ is a measure of the speed and economy with which the brain processes stimuli and that these brain functions are enhanced by good nutrition and brain exercise. There is ample support for this contention.

The foundation for optimum growth of the central nervous system is a high-quality nutrition, rich in protein, beginning with the developing individual's intrauterine life. The brain is very sensitive to prenatal and postnatal nutrition, as has been amply documented (Brown 1965, Cravioto and De Licardie 1973, Stock and Smythe 1967, Davison and Dobbing 1968, Dubos 1968, Eichwald and Fry 1969, Williams 1971, Brace and Livingstone 1971). There is evidence for the view that breast feeding, but not substitutes for mother's milk, maximizes the intellectual development of children (see Newton 1971a, p. 1000). A high-quality nutrition rich in protein would also favor large body size. Therefore, we expect a correlation between body size and IQ. This is indeed found (Tanner 1962, Young 1971, p. 242). This finding is supported by research showing that, in boys of low IQ, muscular strength was lower than in boys of normal and high IQ (90+), when height, social environment, and geographic location are kept constant (Heboll-Nielsen 1956 in Asmussen 1973). There is also a positive correlation between IQ and most parameters testing motor skills, in particular, the performance of complex tasks requiring enhanced coordination (Rarick 1973b). Optimum physical and mental development appear to go hand in hand. As expected, mentally retarded children are shorter and lighter in weight than normal children of comparable ages (Rarick 1973b). Inbreeding affects not only physical development but also neuromuscular and, even more so, psychological development (Schull and Neel 1965). A study of Terman and Oden (1959) showed that exceptionally high IQ is associated with exceptional physical and mental health and consistent success in career life. Conversely, the studies of Brown (1965) and Scrimshaw (1968) showed that suboptimal brain development, accompanied by a low IQ due to malnutrition, is associated with poor learning ability in adult life. These studies suggest that the large bodies and athletic build of Paleolithic people is indicative of superb intellectual development, not only superb physique. These people lived primarily on a meat diet, an excellent nutrition for brain any body growth.

There are other conditions enhancing IQ. A study of Belmont and Marolla (1973) suggested that the IQ of children declines with birth order, but develops most highly in children from families with two or three siblings. If IQ declines with birth order, then body growth should also decline with birth order. It does (Tanner 1962, p. 138). There is, however, no indication that body growth peaks in children from families with two or three siblings. It has been noted that the IQ of children from unmarried mothers is lower than from married mothers (Willerman et al 1970), which suggests that the family milieu is conducive to enhanced intellectual development. Note that the very same is implied by Belmont and Morolla's (1972) study, which shows that IQ peaks in families of two to three siblings. Zajonc's (1976) review showed that intellectual development of children improves significantly with increased spacing of siblings, even cancelling out the negative effects on intellectual development of large family size. Short birth intervals, such as in multiple births, contribute to reduced intellectual development, as does the absence of parents; conversely, the death of a sibling apparently contributes to the intellectual development of the remaining sibling from multiple births, while enhanced tutoring increases intellectual development (see below). Impoverishment of experience, even in the early months of ontogeny, can slow down the development of intelligence (Hunt 1961). That a permissive, friendly, supportive social milieu enhances intellectual development is suggested by the findings of Ney et al (1970) that affective exchange decreases among members of a nuclear family when the number of children increases beyond two. There are also data showing that intensive tutoring—possible only when just a few children vie for the attention of one adult—considerably improves the intellectual capacity of children (Scarr-Salapatek 1971b, Caldwell et al 1973), and so can environment enrichment programs (Horowitz and Paden 1973). Studies on creative individuals indicate that critical parental concern with the child's performance, behavior, friends, and daily activity enhances IQ (see Crosby 1968, p. 88). Again, we expect that intense, friendly, supportive individual attention to the child should not only improve his intellect but also enhance his physical development. This is indeed found (Rarick 1973b, p. 248). Conversely, a harsh, unfriendly, frustrating, fearful social milieu should stunt physical development. It does (see Chapter 6). Is it surprising, then, that the upper socioeconomic classes, which can afford nannies, tutors, expensive travel, sports, and schools that provide individual attention, and place emphasis on performance, are characterized by larger body size, earlier sexual maturation, and higher IQ compared to the lower socioeconomic classes (Tanner 1962, Young 1971, Deutsch 1973)?

Artificial population control, practiced by all hunter−gatherer societies be it through infanticide, geronticide, or various intercourse taboos (Birdsell 1968, Neel 1970, Dumond 1975), is the only mechanism to ensure an abundance of resources for ontogenetic development, as argued earlier, and Paleolithic people must have practiced it. Meat is an excellent diet for growth and development, and meat was the main or only food source for Paleolithic cold-climate populations. The best possible diet for early ontogenetic development is mother's milk, not only because it has the proper balance of amino acids, fats, carbohydrates, minerals, and vitamins, and children utilize mother's milk better and grow better on it than on any substitute (Jackson et al 1964), but also because it supplies antibodies activated by the suckling child and acts as an external immune mechanism (Stevenson 1949, Grodums and Dempster 1964, Warren et al 1961, Hodes et al 1961, Campbell et al 1957, Mata and Wyatt 1971, Schaefer 1971a). This immune mechanism is of particular importance in cold climates where the incidence of otitis media and infections of respiratory passages is particularly high, owing to climatological conditions (Schaefer 1971a). Extensive breast feeding is, thus, not only conducive to maximum growth and development for nutritional reasons but also because it reduces illness and its damaging effects on children. How inadequate bottle feeding is compared to natural breast feeding has been discussed in a number of studies (Cook 1966, Schaefer 1971a, Jelliff and Jelliff 1971). Moreover, breast feeding does reduce ovulation in human females and thus reduces conception (Baxi 1957, Cronin 1968, Potter et al 1965, Hildes and Schaefer 1973). This, in turn, maximizes resources available for lactation and development of the suckling child. Furthermore, breast feeding can only proceed in a psychologically supportive family environment, for otherwise the maternal milk letdown reflex does not operate well and milk is withheld from the baby (Newton and Newton 1950). Conversely, large, well-developed breastfed babies tell of the excellent social milieu of its parents (see Newton 1971).

Another parameter of intellectual competence is creativity, the ability to rapidly see and solve problems, the individual's ability to deal successfully with change. I am basing the following discussion on reviews provided by Shapiro (1968), Crosby (1968), and Gilchrist (1972), particularly on the first. Creativity can be understood using the images of pattern matching discussed in Chapter 2 of this book. I showed that IQ essentially measures the speed and economy with which patterns external to the individual can be recognized, that is, matched. In creativity, however, we are dealing with an entirely different process, namely, the speed with which patterns can be dissolved and reorganized into new forms. Thus, creativity is based on a cognitive process quite different from perceiving and filing a pattern against an internal cognitive structure. Creativity is based on a process that in essence interferes with pattern matching, as it opposes the stability of internal cognitive patterns. This hypothetical view leads to three predictions: Firstly, creativity interferes with strong memory formation and creative persons should not be noted for their feats of memory but for the opposite. Secondly, creativity tends to interfere with the process of pattern matching by generating somewhat unstable cognitive patterns resulting in an enhancement of pattern mismatch. Therefore, the IQ of highly creative individuals, though high, does not reach the peak. Moreover, there should be little relation between creativity and IQ. Thirdly, creative individuals, because they suffer frequent cognitive pattern mismatching or nonmatching, experience anxiety more frequently, as is dictated by the theory of emotionality I discussed earlier (p. 33). It follows that, to be functional individuals, they ought to have acquired superior means to cope with anxiety, frustration, and uncertainty. Thus, creative individuals, by virtue of frequently discovering asymmetrical incomplete cognitive patterns (which we normally call "problems"), are often confronted by their internal emotional upsets and arousal, but are also frequently rewarded by the ecstatic relief of pattern matching. The above-named reviews marshall evidence that supports the above predictions, but one must recognize that the study of creativity must not yet lead to dogmatic assertions because of conceptual and methodological difficulties in its study.

A process so fundamental as dissolving and reformulating cognitive patterns must have some biological significance. An examination of environmental circumstances that lead to creativity in individuals suggests an answer: The neural process responsible for creativity is an adaptive mechanism that enhances the individual's ability to deal with changing, usually deteriorating or stressful, environments and it is developed during ontogeny. Creativity is the consequence of a mild handicap suffered by an individual during ontogeny. It cannot flourish, however, where resources for ontogenetic development are greatly throttled; under these circumstances we expect the stodgy, lethargic, small-bodied K-adapted individual to develop who is an imitator rather than an innovator. Creativity can only flourish where resources are adequate for ontogenetic development, but where life is nevertheless tense and uncertain and very demanding. We noted in the discussion of population quality (Chapter 6) that violence is enhanced in individuals who developed under K-conditions. It is therefore not surprising that conditions favoring creativity—a process of adaptivity called out to serve under deteriorating environmental conditions—also foster criminality and mental breakdown. The home life of individuals noted for their creativity is characterized by high ideals on the part of the parents, a permissive attitude, and strained parent–child relationships owing to the parents practicing "benign neglect." This, of course, sets goals for children and lets them discover how to cope with parents and siblings. Such individuals are exceedingly sensitive to environmental cues and readily make use of them in the process of problem solving. They cannot practice set patterns of conformity due to the inconsistency of their parents' advice and behavior. The process of daily problem solving may make them temporarily unobservant and lead to remarks and observations in company which are incongruous with those of others in the group. So social relations suffer, for they are built on "harmonizing" (p. 35). The handicap hypothesis readily explains why first-borns tend to fill the ranks of achievers and geniuses (Bowerman 1947): It also explains why a disproportionately high fraction of short persons are counted among the geniuses (Bowerman 1947), while superior ontogenetic environments account for the even greater proportion of tall persons in the rank of geniuses. Thus, we can visualize that social or environmental stress translates itself into somewhat reduced parental affection and concern, stimulating mild stress in the children and they in turn learn to cope with it, becoming in the process resourceful, assertive, and often creative individuals. However, social relations suffer overall, including filial and parental affection. Under aboriginal conditions, this should lead to reduced group sizes and dispersal of individuals. At this point we pick up the idea developed by Chance (1969) that the *absence* of group pressure during ontogeny enhances intellectual development, including problem solving ability. Chance based this theory on ethological observations of African primates which suggested that species not known for their problem solving ability did as well as chimpanzees, provided they were *not* raised with peers but in a social milieu devoid of antagonism. This suggests that privacy is important as a means of aiding intellectual development, while awkward social relations by creative individuals tend to enhance their retreat into privacy, probably enhancing their creativity even further.

The foregoing hypothesis suggests that periods of great creativity ought to be associated with increased violence and reduced population quality. We shall see that this is so during the Mesolithic period. As noted later, it was a period of severely deteriorating environments in which great technological changes occurred, terminating with the rise of agriculture, but in which violence was rampant and population quality was low. When we examine the last of the Paleolithic cultures, the Magdalenian of Europe, and note the magnificent creativity in art and tool development, we could suspect in view of the above hypothesis that environmental conditions were beginning to deteriorate. There is substance to such a suspicion because the megafauna at Magdalenian times was already impoverished and salmon had become an important item of diet of these last periglacial hunters. Creativity is thus not associated with peaceful complacency and resource abundance, but with its opposite. Not surprisingly great historic wars and crises give rise to a disproportionate number of geniuses (Bowerman 1947).

From the foregoing it is evident that creativity can be maximized only at the expense of cooperation and group harmony. Where the latter were selected for, creativity had to be dampened, and indeed we see little evidence of it throughout much of the Paleolithic. Moreover, the desirability of high creativity can be questioned if the same environmental conditions foster criminality and mental breakdown. Such conditions also condemn an individual to a relatively lonely life and enhance the probability of a lonely old age. Creative children are not particularly liked by teachers nor by their peers and their peers continue to dislike them throughout their lives. Frequent social tensions with their colleagues are their lot. All this is because of an adaptive mechanism that promotes population dispersal and enhanced competence in dealing with a variable, stressful, yet rich environment, a mechanism probably triggered in its well-developed form from genes by messages from the environment. We may be dealing here again with an epigenetic mechanism.

A third element of intellectual competence is memory. In the framework of the pattern-matching hypothesis it can be defined as the physiological mechanism of cognitive pattern storage and recall. There are several forms of memory formation (Hinde 1970) and these need not detain us here. We are concerned with what environmental factors enhance learning and memory formation. Because memory formation appears to be a function of protein synthesis (Agranoff 1967), clearly a nutrition rich in protein, vitamins, and minerals that is conducive to growth of tissues is also conducive to the development of a good memory. Memory formation is enhanced by arousal and by motor responses following the external stimulus (Hinde 1970, p. 570). Therefore, memory formation ought to be best if intellectual tasks contain motor tasks as essential components; this has been verified to be so (Hecht 1975). The Paleolithic environments, in which individuals depended heavily on meat for food which was obtained only with much physical, emotionally rousing effort, that in turn depended on detailed knowledge, would therefore foster memory formation to the limit of individual capacity.

What, then, is the best strategy to environmentally maximize the intellectual development of a child in a natural population of human beings? The empirical data suggest the following.

(1) The female ought to be in a physiological state in which maximum resources can be diverted from maintenance toward the demands of gestation, lactation, and maternal care. Therefore, the female ought to have completed body growth prior to conception, be continually on a high nutritional plane characterized by a high protein intake with abundant critical nutrients, and, if there was a previous child, conceive only after she has recovered from the previous demands of lactation and child care. In practice, this means avoiding adolescent pregnancies, as well as pregnancies after the age at which vital capacities begin to decline (about 35 years of age), and spacing children at least 4 years apart (Wallace 1973, Osofsky and Rajan 1973).

(2) The female ought to live in a supportive social milieu where each child is wanted by all adults, is eagerly expected, and in which the pregnancy is recognized as a special social event when there is little anxiety or insecurity introduced due to social or economic problems. The prerequisite here is material security and recognized, respected, social roles by the adults (Nuckolls et al 1972, Lemkov 1973).

(3) The infant ought to obtain all his food from breast feeding for as long as possible and as much as he can ingest. This maximizes not only ontogenetic development but also reduces the risk of infections. Breast feeding maximizes contact between mother and child and holds the greatest potential for early, varied, and meaningful intellectual stimulation of the child. It maximizes the security and personal attention so important in ontogenetic development.

(4) The child ought to be able to explore, mimic, learn from face-to-face dialogue, be encouraged to play in social games, role play for entertainment, exert itself physically, be subject to changes of residence, but also to know that change follows a regular pattern, and it should be on the same high plane of nutrition as its parents. Parents must set boundaries but let the individual learn on his own within these; this implies a policy of "benign neglect" with the intention that that the child will master the social, physical, and intellectual expectations set for him.

Last, but not least, if intellectual competence is to be maximized, marriage between siblings, cousins, and other close relatives should be discouraged. Inbreeding through consanguinous marriages does depress physical, neuromuscular, and intellectual development (Schull and Neel 1965).

Clearly, these conditions call for a high order of social control over reproductive functions and are totally incompatible with the known biological mechanisms of population control. If nutritional stress were permitted to act as a controlling factor, it would result in infants of very poor development, in irritated, frustrated, poorly lactating mothers, in increased congenital deformities, and mental and physical retardation—in short, in disasterous ontogenetic development. Though malnutrition may reduce sexual libido and the motility and viability of sperm, it has not historically been a serious check on population growth. For details I refer to the excellent review of Sadleir (1969) on the whole subject of mammalian reproduction. No evolution toward human characteristics could have been conceivable had we been subject to biological mechanisms of population control. Neither Neanderthal man nor the late Paleolithic people could have exhibited the kind of physical development they did without cultural control over reproduction and a very directed effort at maximizing individual development during ontogeny.

Magic Numbers

In the study of human interactions, be it of adults or children, it has been found repeatedly, and is now common knowledge, that the best group size for discussion and cooperation is 5 ± 2 individuals (Gordon 1961, Cohen 1971, Wohlin 1972). Since cooperation was of paramount importance for Paleolithic man, we can assume that the average hunting band consisted of some 5 men. For maximum efficiency these individuals would have to stay together and learn to capitalize on each other's strengths and cover for each other's weaknesses. They would live together. There would, therefore, be some 10 to 12 adults in the band, and some 20 to 25 individuals in all if we count the children. The numbers—12 adults, 20 to 25 band members— are found with surprising consistency, not only in modern hunter–gatherer bands (Birdsell 1968), or even in modern American society (Calhoun 1963a), but they also have been deduced for Paleolithic hunters on the basis of archeological studies of occupation sites (Lee and de Vore 1968, pp. 245–48).

We appear to be dealing here with basic social numbers, and Calhoun (1963a) argues from theoretical considerations that 12 is the optimum group size for life in higher mammals and man. His review of the literature supports this view, as also do the writings of Collias (1944) for animals. Calhoun uses the work of Zimmerman and Broderick (1954) and Zimmerman and Cervantes (1960) to argue this point in relation to the American family; "successful families" (those with the lowest incidence of societal breakdown) associated with 5 other families with whom they shared a majority of social values. "Unsuccessful" families had a smaller number of peer families they associated with. Calhoun concluded that the ideal state was 6 friend-families comprising 12 adults. The limit of social contact was set by about 26 families (some 130 individuals).

There is no satisfactory explanation of these magic numbers, despite Calhoun's brave attempt, since he bases his theory on empirical data of how small mammals use space. I cannot translate it into any terms other than small mammals, except maybe that Calhoun discovered some basic neurological limitations common to all mammals. Mammalian ancestors were, after all, very small animals, of much the same adaptive syndrome as the mice, rats, and shrews studied by Calhoun and his students. Some support for this notion comes from Otto Koehler's (1952) study of "language without words." He found that individuals of different vertebrate species could identify simultaneously or sequentially numbers from 5 to 9. Human beings in these tests did no better than squirrels, and less well than ravens. Whatever the explanation, optimum group numbers for discussion and cooperation are 5 ± 2, optimum band size is 6 families, while fission into smaller groups sets in when the group has reached 19 to 20 peer individuals. Moreover, because of their universality, we can safely assume that these numbers dominated our social environment during evolution and have very real utility in making social processes less stressful.

There has been discussion of the number 500 as a valid "magic" number of human social grouping, one representative of the tribal level (Birdsell 1968, Lee and de Vore 1968, Pfeiffer 1972). This number may be required for successful outbreeding, granted that marriage takes place between relatively young partners (Pfeiffer 1972, p. 379); this appears to be the most likely explanation.

The Social Environment

We explored the life style upper Paleolithic hunters probably followed in this and the preceding chapters. We saw in this chapter that empirical evidence from hunting, evidence relating to intellectual development, as well as physiological and psychological attributes, harmonized well with earlier reconstructions. Together this permits us to sketch the probable social environment experienced by upper Paleolithic people, and this we want to know in order ultimately to understand how to maximize health. It is very likely that cultural mechanisms existed for spacing children just as they exist in hunting cultures of today (Birdsell 1968, Deevey 1968, Dickeman 1975); the sex ratio was nearly equal; adult mortality was high, but not as extreme as in the Mesolithic, among Neanderthal people, or even among some present-day populations. On the basis of old data by H. V. Vallois (G. Clark 1957, p. 245) for Paleolithic man, about 15.6% of those who survived to the age of 20 reached 40; comparable figures for Neanderthal and Mesolithic populations were 8.3 and 6.3%, respectively. Data by Wells (Cook 1972, p. 595) bear out the same relationship. Present evidence suggests that the life expectancy of hunter—gatherers is a function of the availability and diversity of food, as is partly indicated by tooth wear and osteologic evidence of population quality (Laughlin 1963, 1967, Balakci 1968, Lee 1968, Clark 1965, Gable 1965). Thus, we can assume that the basic reproductive unit of Paleolithic people was a 3-generation extended family, focused on the nuclear pair. There is further evidence as we saw that extended families clustered in groups totaling some 20 to 25 individuals.

The social milieu, however, is a function not only of the size of the extended family—probably 5 to 7 people—or the proportions of young, middle-aged, and old persons, but also of the "stressfulness" of life in general. Thus, where procurement of food is difficult and barely adequate to meet the demand and allow the population to hold steady, we can expect increased infanticide, killing of old people and invalids, cannibalism, wounding, accidents, and prolonged healing and sickness; we can expect poor clothing, poor tools, little free time or respite from the requirement to produce food, reduced social contacts with other bands, increased hostility to strangers, reduced cultural activity and reduced opportunity for learning, as well as possible low intelligence due to poor brain development as a function of prenatal stress and poor nutrition (see p. 142−144). This would result in a society reminiscent of the Ik studied by Turnbull (1972). We have little reason to assume such a social milieu in Paleolithic man from Aurignacian to Magdalenian times, since it would preclude excellent physical development and healthy bodies, the development of sophisticated art, a calendar and a system of notation (Marshack 1972a,b), religion (Leroi-Gourhan 1968, La Barre 1970), a sophisticated technology which requires time for experimentation, ornaments, and a population structure superior to

Mesolithic man with little evidence of the violence, murder, and deformities so well represented in Mesolithic sites (Clark 1965, 1957, Vallois 1961). Therefore, we have reason to assume that the rate of social turnover was relatively low for periglacial hunters, as was the frequency of traumatic events brought on by death, illness, or disturbances such as warfare. Conversely, the predictability of social events was high owing to the length of time in which individuals knew each other, and the rich development of culture in which individuals participated. We can assume, therefore, for much of the Paleolithic, the existence of a moderately complex, but stable and highly predictable, social milieu. Such a social milieu would be low in the frequency of life changes experienced by individuals, and thus fairly low in psychosomatic disturbances.

We also expect that aggression within and between groups of periglacial hunters would be greatly reduced and transformed into relatively harmless form. Hunters are providers and are too valuable to the group to be lost through warfare or alienated by intragroup quarrels. We see much the same values operating in Eskimo societies where violence is reduced, redirected, or ritualized into games, sporting contests, or song duels which settle disputes between individuals (Eibl-Eibesfeldt 1970, Hoebel 1966). Warfare can flourish where males are expendable and females with children can adequately provide for themselves, and where the resources are either defendable or game can be alienated from a hunting group's domain.

It was noted earlier that Paleolithic hunters were on the move at regular intervals, dictated by the movement of game. Granted the great space−time constancy of large social herbivores, we can expect the hunting groups to have reoccupied much the same campsites in successive years, leading to the accumulation of cultural debris described by Birdsell (1972). It does not permit the inference that very large groups of people lived there at one time; that would have seriously impeded success in hunting the wary, fast-learning ungulates Paleolithic man subsisted on.

I have noted that Paleolithic people maximized individual development. They largely succeeded, as the robust large skeletons of most individuals unearthed so far tell us. This is true for Europe (Vallois 1961) as well as for Melanesia and Australia (Howells 1973b). Still, adult mortality was relatively high—an indication that, even with their well-developed bodies, these people were barely adapted. It is puzzling why not only adults but also children continued dying at a relatively high rate. The answer may be as follows. Children mimic parental roles during ontogeny. Granted that to fulfill these required extraordinary physical and intellectual skills, as well as a large measure of courage, then it follows that children began early in their lives to enter dangerous adventures with the thorough approval of their parents. That could have resulted in many accidents, particularly since children are less skilled than adults and can be in great danger in situations judged as innocent by their parents.

The social structure and social milieu are significant from the evolutionary viewpoint: We noted that under periglacial conditions the individual's physical and intellectual development as well as his social competence were maximized. This process is so costly in food, and food can be obtained only at the cost of great labor and mortality of adults, that we can expect people to take advantage of every mechanism that maximizes the above attributes at the lowest cost in food. We can therefore expect the social organization and social relations to be such as to maximize ontogenetic development and the health of adults. If this is so, then we expect that even nowadays a similar social organization and intense social milieu are supportive of individuals, and social, mental, and physical breakdown is characteristic of the absence of this social organization in adults and children. Is strong supportive bonding between 5 to 7 individuals, and somewhat weaker long-term bonding within a group of 20 to 25 persons, of benefit to man in modern times? Empirical evidence suggests that it is.

It has been found, for instance, in many studies reviewed by Hughes and Hunter (1970), that throwing people into strange environments, such as persons in Africa adapting to urban life, led to various breakdowns among adults. The tribal structure had a most positive influence on individuals in reducing psychobiological problems (Dawson 1964). However, this evidence is counterbalanced by findings that in some instances the move to an urban center was an escape from powerful social sanctions for the individual, and was rated as a relief and freedom from fear by the individual concerned. Agricultural tribes are of course very much subject to oppression and fear, as we shall discuss later. Holmes (1956) showed that even such a disease as tuberculosis is greatly affected by the social background of a person. The TB bacillus has its greatest effect on people who suffer broken marriages, change jobs and residence frequently, have few friends, know few neighbors and maintain few kinship relationships, have difficulty coping socially, and lead a lonely life (Cassel 1971). The study by Holmes is one of many that imply great benefit from a strong social organization (Katz 1959).

Another striking study is that of Matsumoto (1970) on the social organization of the Japanese worker and the prevalence of heart disease. Strong social bonding appears to reduce heart disease drastically; the data are circumstantial—as in most studies of humans—but they are entirely in line with our theoretical prediction that intense, intimate, strong social bonding of long duration is most conducive to good health. We shall encounter further examples of benefit to ontogenetic development and adult health from a social milieu as found in the extended family. However, one must also note that such a milieu tempers individual expression and can be a tyrant, since the group demands conformity and can generate reprisals against nonconformists. As such, the group can bring suffering to some individuals, and in hunting societies, such as that of Eskimos, it can bring death to the nonconforming individual (Freuchen 1961).

The Mesolithic

The first age of culture flowered in the upper Paleolithic of western Europe among populations of periglacial hunters. They developed art, religion, and a system of notation; they diversified their tools and pioneered trade; and they reached a population density that was higher than that encountered there previously. They were a healthy, large-bodied, athletic people, a consequence of successful adaptation to a harsh but rich ecosystem. All this came to an end with the retreat of the last Würm glaciation.

This first blossoming of culture was probably made possible by a climatic accident. In western Europe the glacial masses advancing from the north and east were met by warm Mediterranean air masses that probably ascended the river canyons from the west. During early glaciation and deglaciation the Gulf Stream, now north of Britain, hit the coast of France and Spain, while during the glacial maxima it must have hit North Africa on the west coast. For this reason, as well as increased sunshine, loess, and water from glaciers, a rich, diverse biota could flourish in western Europe as nowhere else in Eurasia. It set the stage for the evolution of man, as I described earlier, and also for the first cultural achievements.

The Paleolithic cultures of western Europe appear to follow glacial advances and retreats. The Aurignacean and Perigordian cultures spread into western Europe apparently during the interstadial that followed the retreat of the first Würm advance. Unfortunately, dating based on O^{18}/O^{16}, Th^{230}/Pa^{231}, and C^{14} analysis (Dansgaard and Tauber 1969, Emiliani 1970) is too imprecise to fix the glacial advances accurately. Therefore, it is best if I relate cultural changes to glacial advances and retreats, whatever their dates.

The Gravettian culture appears to coincide with the last major advance of the Würm glaciation; the Solutrean culture caps the glacial maximum; the rich Magdalenian culture thrives during the early part of the deglaciation. It is based on an already impoverished fauna of large mammals and takes much of its sustenance from salmon that run up the rivers to spawn (Pfeiffer 1972, Clark 1969). The increase in the frequency of *bâtons de commandement* may indicate that the items people kept track of proliferated. This speaks of insecurity. Resources were, therefoe, probably exploited more critically; this harmonizes with the broadening of food habits and technology. The Magdalenians depended increasingly on reindeer, a species whose migratory patterns tend to be somewhat erratic, as evidenced by studies by Kelsall (1968), Parker (1972), and Bergerud (1974). It may well have been that intercepting migratory reindeer became both increasingly more necessary as the populations of other large herbivores declines and at the same time less predictable. This must have caused more worry and led to increased attempts to control the animals with magic, as evidenced by the richness of Magdalenian art. It was also necssary to maximize the number of kills per opportunity afforded; hence the increasing use of light throwing spears hurled by spear throwers, and the use of bows and arrows. The proliferation of art, and particularly the extensive decoration of weapons, suggests that social tensions were higher than in previous cultures, and that a considerable amount of social competition was going on. Decorated weapons speak of weapon cults and of male competition, of anxieties, and of a rule-governed society less dependent on cooperation and goodwill than previous cultures. Finally, owing to the impoverishment of the megafauna, the reindeer hunters began to follow the migratory herds, probably giving rise to the reindeer hunters of northeastern Europe who continued to struggle as a hunting culture (J. G. D. Clark 1967, 1975). They may have developed to some extent into reindeer herders using the tactics described by Baskin (1970, 1974), a modern student of reindeer and their behavior.

In the south, the difficulties of the Mesolithic were foreshadowed by the Azilian culture, which succeeded the Magdalenian there during the terminal phases of the deglaciation and the return of warm climates. Like the reindeer hunters, the Azilians were not numerous; they were culturally impoverished; their diet diversified to that of gatherers and now included snails, while the large mammals they killed included those from temperate forests, such as wild boar and red deer (Clark 1965, 1969). Then the temperate forests of the interglacial closed in, and in Europe the Mesolithic, the first dreadful dark age of modern man, set in.

During the Mesolithic, cultures and populations deteriorated, while new technologies and exploitation of previously untouched ecosystems speak of a desperate struggle to adapt in Europe, Africa, and America. In Siberia, however, we could expect a continued exploitation of reindeer and salmon, since here the change from periglacial to tundra vegetation took place on a vast scale and was not followed by the total encroachment of forest. Big-game hunting could, and apparently did, survive in the high mountains above timberline into the Holocene (Chard 1969). However, reindeer hunting changed to herding, probably as outlined by Baskin (1974). The Siberian tribes could culturally be considered an extension of late Paleolithic northern cultures and their myths from this viewpoint are very valuable in interpreting Paleolithic beliefs. A detailed account of the mythologies of Asiatic and American tribes is given by La Barre (1970).

As pointed out in the introduction, the populations of central Asia evolved under harsher conditions than did those in the periglacial environments of southwestern and southeastern Europe, and changed toward greater neoteny, reduced sexual dimorphism, and cultural institutions that enhanced the social role of the female. This was adaptation to a harsher environment that extracted heavy work from both male and female. This has some significant implications to the study of humans. For instance, in Caucasian populations one finds a sex-dependent difference in field dependence as measured by psychological tests, but one finds no such difference among Eskimos. This does not permit the conclusion that the differences found in Caucasian populations are cultural (MacArthur 1967)—not if the reduced sexual dimorphism in Mongoloid populations is biological in origin.

In Europe during deglaciation the rich fauna of the periglacial ecosystem had vanished and with it the primacy of big game hunting. The long bow came into its own to reach further and increase the chance of killing (J. G. D. Clark 1965, 1975). The dog was domesticated (Oakley 1949, J. G. D. Clark 1965, 1975). Eating habits diversified; fish and waterfowl were commonly relied on (Binford 1968a). The sea coast, with its resources of invertebrates, fish, waterfowl, and sea mammals, became subject to intensive exploitation (Waterbolk 1968, J. G. D. Clark 1965, 1975). To some extent this is due to the increase in the size of estuaries, as well as to bird migrations during the interglacial. Most settlements of the Mesolithic people were along oases of biological productivity such as deltas, estuaries, and shallow lakes. It is in such water-disturbed environments that productivity is high and greatly augments the food resources available on land. This is well illustrated by Birdsell's (1953) classic study on Australian aborigines, of their population density and land ownership in relation to resources. Rich estuaries also become defendable resources, which migratory game herds are not; we expect aggression to flourish. In Europe, much of the forest area became uninhabited during the Mesolithic, and population levels did not rise again until the advent of agriculture (Waterbolk 1968). We see on other continents similar developments to those in Europe (Lubell et al 1976, Fairbridge 1976).

In America specialized big game hunters largely destroyed the megafauna after deglaciation (Martin 1973, Bryan 1973). Populations reverted to gathering. The high mountains of South America were colonized, producing specific anatomical and physiological changes in the populations living there (Baker 1969, Lasker 1969, Bryan 1973). In South America and Africa the tropical rain forest became occupied by man (Clark 1970, Bryan 1973). In central Asia the high mountain ranges were colonized by hunters almost 10,000 years BP (Chard 1969).

During the European Mesolithic, life expectancy of individuals declined. The remnants of populations, excepting the Maglemosians on the northern fringe of settled Europe, were of appallingly bad quality, and the exception, such as the

Mesolithic people exploiting the rich fish resources at Lepinski Vir (Wernick 1975), only emphasizes the rule (Clark 1965, 1969). Individuals tended to be small in stature, and their bones were riddled with osteolysis, arthritis, rheumatism, and exostosis, as well as tooth caries and abscesses (Clark 1965). Similar tooth wear and injury is found in African and American populations and is associated with high adult mortality (Gable 1965, Dahlberg 1963, Angel 1968, Molnar 1968). Coprolite studies reveal an incredibly coarse diet for American Indian populations from the North American desert region (Heizer and Napton 1969). There may have been some selection for an increase in the size of teeth and jaws and a concomitant increase in the strength of cranial bones, judging by the work of Heinzelin (1962). The enamel hypoplasia speaks of episodes of starvation (Angel 1968). The incidence of violent death in Eurasia and Africa is high and dramatic (Clark 1957, 1965, Heinzelin 1962), and cannibalism appears to have been not uncommon (Armelagos 1969).

The archeological evidence from the Mesolithic thus indicates a concurrent decline of population size, population quality, and culture, while violence and cannibalism rose as environmental quality deteriorated, or at least changed so fast that populations would have had difficulty in adapting. One finds a similar positive relationship between environmental quality and cultural complexity, longevity, and child survival in Aleuts and Eskimos studied by Laughlin (1963). It is interesting to note that during the earlier Neanderthaloid dispersal Rhodesia man's remains also suggest a change toward poor quality. Note the bad teeth, the evidence for homicide, and the powerful supraorbital torus (Clark 1970). I interpret the latter as an enhancement of the brows, a display that appears to me to be instrumental in threat. Be that as it may, archeologic evidence as a whole suggests that homicide and cannibalism are associated with marginal, if not actually poor ecological conditions. This may be valid not only for the Mesolithic remains, but also earlier ones. Late Paleolithic remains show little evidence of this, excepting those found at Choukoutien. These murdered individuals show extensive tooth wear, arthritic vertebrae, and healed fractures, while their artifacts indicate no high cultural developments (Weidenreich 1939a). The evidence again suggests that homicide is associated with a marginal existence. The same can be said for the "Sinanthropus" remains also from Choukoutien, as I argued earlier (p. 245). Although many remains of pre-Mousterian and Neanderthal man show evidence for homicide and cannibalism (Weidenreich 1939b, Blanc 1961), the ecological and anatomical evidence is not always at hand to indicate whether such individuals lived under marginal conditions or not. Such evidence, however, appears to be present wherever homicide is extensive in late Paleolithic and Mesolithic people, such as those along the Nile valley in late glacial and early Holocene times (Clark 1970, p. 169). We find here increasing emphasis on grain food, as well as indirect evidence for strong cultural cohesian, implying defensible resources and violence. The latter is confirmed by the high percentage of burials due to violent death.

Geographic Dispersal of Modern Man

Geographic dispersal by modern man had probably begun some time after the peak of the first Würm glaciation, with the encroachment of forests in the Mediterranean basin and the vanishing of the cold-adapted Neanderthal populations resident therein. This is most speculative. It has to be, granted the scarcity of reliable dates and, in particular, the poor knowledge of Pleistocene climates and vegetation changes. As long as Neanderthal people flourished in the Mediterranean basin, it is most unlikely that any human group would slip past them to disperse eastward. However, during a warm period in which the large mammal populations on which the Neanderthaloid people depended began to wane and forests encroached on the land, we would expect Neanderthal people to decline in frequency. Hunter—gatherers could now advance northward and develop into distance-confrontation hunters, spilling eastward into Asia and ultimately into North America.

The dates of early men known from southern Asia, Australia, and the Americans tend to support this picture. Late Paleolithic tools of Aurignacean type appear in Afghanistan some 38,000 years ago (Clark 1967). In Melanesia and Australia, the earliest dated evidence of man goes back to about the same age, i.e., the skull from the Niah Cave in Borneo, approximately 37,600 to 39,500 BP (Harrison 1967 in Howells 1973b, p. 177). Although Howells (1973b) argues that early man in southern Asia and Australia may have been present as early as 50,000 BP, the actual dates are somewhat later. In Australia, early man is present at least 30,000 years BP (Mulvaney 1966, Birdsell 1972, Howells 1973b). In North America, the first evidence of man is also almost 40,000 years ago (Bryan 1973, MacNeish 1976), prior to the last advance of the Würm and Wisconsin glaciers; in South America, man appears at least 22,000 years ago (MacNeish 1971, 1976). Four stages of early cultural development can be dimly discerned in America, of which the fourth is associated with the disappearance of the megafauna, the earlier stages reflecting a hunter—gatherer, rather than hunter, existence (MacNeish 1976). Significantly, central Asia is virtually devoid of typical late Paleolithic sites, and not until the last Würm glaciation is well past its peak do we find the occupation of central Asia by modern man, some 16,000 years BP (Chard 1969). Thus, the first thrust by modern man appears to go southeastward into southern Asia and Australia, well before the last major glaciation, while occupation of central Asia and Siberia did not occur until after the last glacial maxima. Apparently these late cultures, though distinct, are similar to those of the aurignacoid big-game hunters of Russia (Chard 1969).

Who were the early migrants? There is no certain answer. If we assume that the Negrito characteristics of the Andaman Islanders and those of some hill tribes of India (Lal 1974) are homologous with those of the Khoisan people, then the geographic evidence sueggests that the first wave of migrants were pre-Cro-Magnon people. We expect pre-Cro-Magnon to come from hot, dry regions of North Africa where they lived essentially a late Acheulean way of life, except for more frequent and diverse use of fire. Steatopygea, it was argued, was adaptive in the context of Acheulean man's adaptive syndrome; it became less significant among people from cold climates, especially under very harsh conditions which welded male and female into a unit of intense cooperation. The fine, kinky hair is evidence for desert or warm climate dwelling, being analogous to the fine, kinky wool found in camels, where it also serves as insulation against solar radiation. Pre-Cro-Magnon entering the Mediterranean basin would have been kinky haired and steatopygous. Long hair covering head, ears, and neck I interpret as products of selection in cold climates. Thus, Aurignacean and Perigordian people, the earliest of the European late Paleolithic people, not only have Caucasoid skeletal features (Baker 1968, Coon 1962), but also have long hair, as illustrated in the sculptures of women, and they also lack the large labia minora (Fig. 202b in Marshack 1972b). If we adopt the hypothesis that modern men spread repeatedly from centers of adaptation to increasingly hostile environments, then successive waves of migrants ought to be progressively less steatopygous to be sexually dimorphic, and to have a penis pendulance, reduced labia minora, and increasingly heavier, long straight hair. On the whole, the evidence conforms to this hypothesis, as witnessed by the zoogeographic distribution of races, in particular, remnants of people that survived successive waves of migrants in areas difficult to reach and settle.

However, the Negrito characteristics of people from southern Asia and Australia are subject to considerable dispute among anthropologists (Howells 1973b). One can make an excellent case for the hypothesis that such Negrito characteristics as are found may be convergently evolved to those shown by African Negritos. Moreover, upon closer examination, the Negrito traits of Africans and Melanesians or Tasmanians are only superficially similar, a point stressed by Howells (1973b) in his very careful analysis of the races and prehistory of Melanesia and Australia. The African Negrito—in contrast to the Khoisan—are confirmed first about 20,000 years BP (Clark 1970). Thus, clearly, those populations that may have strong affinities for people from Africa owe these affinities to the Khoisan people if they qualify as the earliest migrants into Asia. The present-day Melanesian and Australian stock are probably a somewhat later and more successful stock of immigrants, which evolved considerable diversity in response to local adaptations. I am tempted to see them as Ur-Caucasoids that spread from the Mediterranean basin eastward, wiped out the first wave of migrants in all but a few localities, and then adapted to local conditions to produce the diversity of physical types we now find. The rather Ainulike "Murrayans" of Australia (Birdsell 1972) could well be the least altered group; during much of the Pleistocene they may have lived in cool grasslands in Australia where now we find desert, and are therefore least changed toward the Negrito characteristics found in tropical people. It is interesting to note that the skulls and bones of the prehistoric people of Australia and Melanesia are large brained, robust, and large compared with most individuals from present-day populations (Howells 1973b). They were largely big-game hunters who probably played a role in the demise of the megafauna, developed later agriculture based on plants, and suffered reduction in body size in parallel with populations in Europe in post-Pleistocene times. Here we also find the diversification in food habits, and the increase in tooth wear as found in European, African, and American post-Pleistocene populations.

It must be noted that late Paleolithic people colonizing the rich periglacial environments had no option but to exploit large mammals for food. This was a dangerous and exceedingly demanding task. During recolonization of warmer climates, the colonizers had not only the option to continue hunting large mammals but also to exploit food resources less demanding in skill, knowledge, effort, and injury. Therefore, following the principle of least effort, recolonizers would tend to shift from big-game hunting to gathering. Thus, gathering would be secondarily derived from hunting among modern hunter−gatherers.

As human beings occupied habitats with continuous marginal productivity or low pulses of biological productivity and filled them to capacity, selection would at times favor small body stature. This is so if low priority tissues (e.g., the brain) are required in full expression but calories and nutrients are in short supply. Under such conditions, genetically large people may develop large tissues and organs of high growth priorities but starve tissues of low growth priority. The result would be selection against large body size and for a body size which, under prevailing cultural practices, still permits adequate growth of the cerebral cortex to muster the cultural adaptive strategies. Well proportioned but small humans, pygmies, are the result under this condition of maintenance selection after colonization.

I hypothesized that successive waves of human beings emerged from areas where humans adapted to progressively harsher environments. The last major racial wave to emerge were the Mongolians (Chang 1962); they may have come in several waves from central Asia and replaced earlier Caucasoid waves, but not uniformly. The zoogeographic distribution pattern of Mongolians points to central Asia as their homeland. So does the evidence from archeology, since we find in the Paleolithic people of Asia mongoloids with many Caucasoid features, much as is found in American Indians (Laughlin 1963). The third line of evidence is the occurrence of polyandry and the counting of descent by the maternal line, which I explained earlier as a consequence of adaptation to very harsh environments, along with the neotenization and the reduction in sexual dimorphism (p. 359).

The closer we come to modern times, the more evidence there is for repeated movements of people of diverse origins. Howells (1973b) discusses such movements for Melanesia, Indonesia, Micronesia, and Polynesia. Lal (1974) speaks of mongoloid tribes occupying the high mountain areas at the north of India, while continental India was invaded by agricultural Caucasoid tribes from the northwest and displaced the earlier Veddoid hunter−gatherers. In Africa, there was also tumultuous movement. The most important displacement was that of the Khoisan people by modern Negritos, apparently a forest-adapted people (Clark 1970).

America was probably repeatedly colonized. Little is known of the first people that came to Beringia and South America prior to the last Würm advance. The zoogeographic pattern of dispersal at that time suggests that they were ancient Caucasoids. Culturally they were very primitive compared to their Mediterranean counterparts (MacNeish 1976). Following the last glacial maximum, specialized big-game hunters became active in North America by about 13,000 BP south of the Laurentide ice sheet (Martin 1973, Bryan 1973). MacNeish (1976) argues that theirs was probably a native American culture derived from an earlier Aurignacean-like stage. Missing is any indication that these people kept chronologic time and used interception of migratory herds as a strategy to hunt large herbivores. Rather, they were specialized distance-confrontation hunters, to judge from their tools. According to Birdsell (1972), the early colonizers were somewhat more Caucasoid in skull characteristics than the present Indians; MacNeish (1976) describes them as somewhat primitive Indians. In view of the physical variability of late Paleolithic people (Weidenreich 1939a), this generalization may or may not be valid. Moreover, American Indians, Polynesians, and Europeans are rather close in skull structure (Howells 1973b). The Aleuts, the parent group of the Eskimos, make their appearance in North America some 8700 years ago and settle coastal Alaska (Laughlin 1975). The high arctic coastline was settled some four times by Eskimos beginning some 4000 years ago. The continental tundra was settled first by Indians about 5000 years ago, and subsequently by Eskimos some 500 years ago, probably in response to the "Little Ice Age" of the Middle Ages (McGhee 1974). In the meantime the last large oceanic islands were settled by man, New Zealand and Madagascar, about AD 1000 and 800, respectively (Battistini and Vérin 1967, Howells 1973b).

The dispersal of people who conquered the preceding wave of immigrants has, of course, never stopped, and may well have increased in historic times. Remnants of earlier migrants survive in inhospitable or inaccessible places; hence the Ainu on Hokkaido or the Negroid people on the Andaman Islands. The Khoisan people of Africa fell victim to the Negroid and later to white agriculturists, and remain today only in the form of the diminutive Bushman of the desert regions of southern Africa (Clark 1970). One could list more examples. In general, modern man, with increasing frequency, dispersed into regions inhabited by more primitive cultures and, in the process, the conquered vanished or occasionally were absorbed by the conquerors.

On the Origin of Medical and Pharmaceutical Knowledge

The periglacial and cold ecosystems last conquered by man were in may ways a favorable environment. The abundance of resources and demand for intellectual and physical competence permitted excellent body development. It promoted intense cooperation and must have fostered the care of individuals physically incapacitated. Even among Neanderthal men there is evidence for individual care in the form of healed injuries, as well as burial with flowers and artifacts (Birdsell 1972, Pfeiffer 1972). There is little evidence in late Paleolithic cultures for homicide, and little theoretical reason to believe that overt aggression between individuals was common. There would therefore be few wounds, except those suffered during hunting. Moreover, in cold and periglacial conditions there are relatively few parasites, as discussed earlier.

With the spread into new biomes during the Mesolithic period and thereafter, and the exploitation of new food resources, not only new technologies were necessary. It was also necessary to alleviate the discomforts suffered by living in habitats only partially suited to occupation by man. We need only consider the disease and parasite load of people from tropical lowland forests (Dunn 1968, Lowenstein 1973, Bourlière 1973a), or note the terrible condition of teeth that occurred in Mesolithic populations in Europe, Africa, and America, their disease-riddled bones, and the evidence for wounds inflicted in combat. We expect some forms of medicine and drug use, therefore, to alleviate suffering in tribes from marginal habitats. We indeed do find evidence for this.

Indians from the higher elevations of the Andes take alcohol and coca as an antidote to the discomforts, distress (Chapman and Mitchell 1965), and pain from the low O_2 intake and the requirement for hard work. At these elevations ontogenetic growth is retarded, except for the development of the chest, a sign of strenuous living conditions under low O_2 tensions (Lasker 1969). Australian aborigines from the cold interior may be able to let peripheral and core temperature of their body drop at night, partly owing to the chewing of *Duboisia* leaves which contain an alkaloid poison (Lasker 1969). The treatment of wounds, infections, and various injuries was well developed among American Indians, and, as is being belatedly recognized, was surprisingly effective (Vogel 1970, de Montellano 1975). The great longevity of the Aleuts has been ascribed to their great medical skills and herbal knowledge (Laughlin 1961, 1963). Laughlin's (1961) paper, in particular, is revealing. He discusses the rather astonishing knowledge Aleuts particularly—but also other tribes—had of anatomy and the manner in which such knowledge was obtained and cultivated. These people had detailed accurate knowledge of human structure and function as well as that of other animals; they practiced autopsy of humans and animals, applying such knowledge to diagnose illnesses; they practiced acupuncture and occasionally surgery, in which they used sinew for suturing, and also blood-letting and massaging, particularly during and after pregnancy to put organs in place; they prescribed fasting, rest, and a great diversity of herbal extracts toward effective alleviation of suffering. They understood the reproductive functions of coitus and castration, and had a basic grasp of heredity. They had deduced the existence of microorganisms from careful observation. They also kept animals in order to observe and study their natural history, following this by autopsy and thorough examination. The Tungus, a tribe of East Siberian natives, clearly recognized the concept of homology and labeled homologous bones and parts in diverse mammals with the same terms. This sounds astonishingly modern, and yet, as Laughlin (1961) emphasizes, it is a logical extension of the hunting way of life of people that were prosperous enough to have time for curiosity. Shamanists and magic were important in Eskimo society, primarily where economic conditions were poor; where the economic conditions were good, greater reliance was placed on rational treatment. Eskimos, Aleuts, and American Indians in the northern latitudes of North America all had an extensive knowledge of herbal medicine (Smith 1973).

We have reason to suspect that the postglacial occupation of marginal habitats was, in part, made possible owing to developing new technology, and in part also by the development of pharmaceutical and medical knowledge. We can safely assume that music in the form of rhythmic chants and instrumental music played an important part, with empirically derived methods of psychoanalysis to make life bearable (see p. 342). The witch doctor and his magic are probably important elements in human adaptations to suboptimal natural environments. We have even more reason to suspect this when examining agricultural societies.

The Beginning of Artificial Environments and Some of Their Biological Effects

The Mesolithic period was terminated by the invention of agriculture which changed the face of the globe, beginning with the spread of man-made plants across the earth (Anderson 1952). Much of the plant world of today is different from that of the preagricultural ages by virtue of new plant species that arose, often accidentally, as a product of human activities. Agriculture also changed social systems and even man himself. Economic strategies dictate cultural and biological adaptations.

Agriculture forced some biological adaptations on populations practicing it. Dahlberg (1963) and Green (1970) argue that in the Caucasian race, the reduction in size and complexity of teeth is an adaptation to the high carbohydrate diet of grain farmers. A reduction in size and complexity reduces the surface area of teeth exposed to contact with decay-promoting sugars and starches. The smooth surface of Caucasian teeth reduces bacterial adhesion to tooth surfaces (Sherp 1971). There also appears to be a reduction in size and complexity of teeth among populations of American Indians exposed to agriculture (Brace and Mahler 1971). Primary hunter—gatherers such as the Australian aborigines, Melanesians, or Eskimos retain large, complex teeth. Such teeth rot quickly if exposed to a high carbohydrate diet, as has been amply demonstrated for Eskimos (Schaefer 1971b, 1973b) and Australian aborigines (Molnar 1968, pp. 41−43). Brace (1963) advances the view that teeth fell victim to the mutation effect akin to eyes in cave fishes (Wilkens 1971), and were thus reduced in size. Brace's hypothesis predicts, however, not only a reduction in tooth size but also an increase in the frequency of tooth anomalies. Such an increase has not been reported, however.

The ability of Caucasoids, although not of Eskimos, to readily metabolize large dosages of blood sugar (Schaefer 1969, 1970), and to metabolize alcohol more rapidly than do Eskimos and Indians (Fenna et al 1971), may also be an adaptation to agriculture and its products. Also, Caucasians and many Negro populations can use whole milk as food and have enzymes to metabolize lactose, which is also an adaptation to the availability of milk through agriculture (McKusick 1969, Bayless and Christopher 1969). Chinese populations have a better ability than ''Americans'' to metabolize monosodium glutamate, a common seasoning in Chinese dishes, but not in European−American cuisines (Lasker 1969). Some populations, probably due to genetic adaptation, require less of various essential nutrients than do others (Lasker 1969).

Agriculture has also promoted an increase in genetic load, a consequence of relaxed genetic selection. Caucasians have a high incidence of hypolactosis, breast cancer, myopia, and colorblindness; the former two are clearly related to the practice of feeding infants on the milk of domestic animals if the mother's breast milk production failed (Newton 1971b), and myopia is probably due to the survival and utility of persons with this defect as craftsmen (Brothwell 1971, Post 1971). Other organs which appear to have fallen victim to relaxed natural selection are tear ducts, nasal septa, ears, and eyes. This subject has been pursued by R. H. Post (Post 1971).

Early agriculture was no boon to physical development. This was told only too plainly by the bones of most deceased agriculturists which reveal a short stature, diseased and malformed bones (Angel 1968, Armelagos 1969, El-Najjar and Robertson 1976), thin skulls, and reduced brain size (Hulse 1963). However, some of the Neolithic, Bronze Age, and early medieval populations did revert to a most respectable physical development, as evidenced by the stature of men that on average exceeded that of modern Americans, presently one of the tallest people (Huber 1968). It is interesting that, as in modern times (Tanner 1962), so in an early medieval population Huber (1968) noted a positive correlation between body size and social status; the latter can be inferred from the amount of ornaments and grave goods buried with individuals.

Since the Paleolithic, there has been a shift from an elongated head shape to a spherical one in most human populations. This shift is termed brachycephaly, a change from the dolichocephalic head form. We see it in people as different as Europeans, Greenland Eskimos, Aleuts, Japanese, and American Indians. Almost everywhere one finds in the early populations long-headed individuals, and more or less brachycephalic ones in their modern descendants. Exceptions to this trend are Melanesians, Australians, and many African tribes which have remained dolichocephalic. Brachycephaly has been upheld by physical anthropologists as an example of post-Pleistocene human evolution (Hulse 1963, Laughlin 1963). However, one can make a case for the possibility that we are dealing here not with a genotypic, but with a phenotypic, change.

Brachycephaly, I shall argue, is caused largely by a shift in life style from one in which individuals had to master a great diversity of motor skills and roles to one in which relatively little is required of individuals owing to a great division of labor and the consequent reduction in the skills and roles performed by an individual. Dolichocephaly is the product of extensive training in bodily skills demanding an exacting performance of the body in time and space, that is, exact coordination between vision and body movements, while deploying—as the occasion demands—great physical strength or speed. Brachycephaly is the product of a low emphasis on body coordination, strength, and speed, but a strong emphasis on verbal skills and knowledge. I hypothesize that the head form reflects the differential development of the brain, as produced by the adaptive strategies of life styles.

As discussed in Chapter 2, brain size is altered phenotypically by the diversity of intellectual and motor tasks performed. In general, this is a direct relationship. The portion of the brain involved in the control of motor activity, particularly in coordinating, correcting, and mediating, is the cerebellum. If individuals acquire and maintain a large diversity of motor skills in which speed and precision of execution is vital, then we expect not only a phenotypic enlargement of the motor cortex and the occipital lobes of the cortex, but also of the cerebellum. This enlargement ought to lead to an enlargement of the rear portion of the brain, resulting in dolichocephalic skulls. Reduced emphasis on diversity in motor skills or speedy execution of motor skills mastered ought to reduce to size of the cerebellum, as well as the occipital lobes and motor cortex. Whether this is so remains to be tested in humans; the deductions made were done from first principles.

Granted the validity of the "phenotype hypothesis," it explains at once a number of puzzling attributes about skull form. Note that it predicts an extreme dolichocephalic skull in Neanderthal man, in conjunction with the adaptive syndrome, as discussed in Chapter 12. This is, of course, found; in fact, it is one diagnostic feature of Neanderthal man. In pioneering populations in which plenty of resources for growth and development are available, but at a cost of strenuous, diverse, and exacting work, it predicts dolichocephaly; once the population has settled into a traditional way of exploiting the environment on the basis of division of labor and specialized roles, which can develop only at relatively high population density, there ought to be a shift toward brachycephaly. Hence, the larger the population, the greater the shift to brachycephaly. I believe this explains Laughlin's (1963) findings of brachycephaly in Aleuts and Eskimos. In a shift from a subsistence to a surplus economy with concomitant reduction in diversity of skills practiced by individuals, we expect a shift toward greater body size concurrent with an increase in brachycephaly. This may explain the case Hulse (1963, p. 403) refers to of concomitant increase in body size and brachycephaly in American Indians. Note that this concept predicts cultural diversification to accompany brachycephaly as well as a reduction in participation in war by individuals, for reasons explained below. The explanation advanced for the case of American Indians appears to me also applicable in explaining the concomitant increase in body size and brachycephaly in Hawaiian-born Japanese, and American-born Chinese (Hulse 1963, p. 402). Good nutrition promoted body growth; an emphasis on clerical and intellectual tasks rather than motor skills promoted brachycephaly. Within a population, brachycephaly ought to be inversely correlated with body size, provided there is good nutrition. From exercise physiology we know that large body size is promoted by intense physical exercise. Therefore, classes of people who lay great stress on sports and a diversity of physical activities as part of their social behavior ought to be tall and relatively dolichocephalic. One can predict that, for instance, the upper class of British society is not only taller than the population mean (Tanner 1962) but probably also more dolichocephalic. In general, body size correlates with dolichocephaly (Hulse 1963).

The "phenotype hypothesis" also appears to explain dolichocephaly in at least the Melanesians and Australian aborigines. In the latter, we expect dolichocephaly due to their life style as hunter–gatherers, depending on a diversity of motor skills to compensate for a low level of technology. In the Melanesians, we find agriculturists living at a rather high density. On the face of it, we would expect brachycephaly, since agriculture that centers on plants is less demanding of diversity in physical skills than a hunting–gathering life style. However, the Melanesians are a warring people (e.g., Matthissen 1962) that fight with spears, bows and arrows, and a variety of striking instruments. In order to use them, as well as to defend onself against them, one must practice from early childhood onward. A common form of defence against arrows, spears, and blows is evasion, which requires split-second coordination. Thus, a high intensity of exercise in performing evasive maneuvers against spears and clubs, etc., ought to be reflected in enhanced dolichocephaly—even in agriculturists short of protein to grow a large body. Note that the same hypothesis predicts dolichocephalic skulls in individuals in the European Middle Ages prior to the development of armies as we know them. In the early Middle Ages, a farmer had to double as a soldier in case of national emergency, and was thus a generalist. Once massed armies were deployed, soldiering became a profession of its own, while the development of firearms reduced the emphasis on close-range weapons and thus the need to master sword, lance, battle-ax, mace, club, shield, crossbow, bow and arrow, and dagger—as well as being skilled in avoiding or blocking these weapons. Granted good nutrition, it is not surprising to find relatively dolichocephalic individuals (Hulse 1963) of surprisingly tall stature in the early Middle Ages (Huber 1968). Conversely, with the advance of European civilization, the advent of organized armies, the drop of status for the common farmer to that of peasant, the increase in trading and thus the demand for intellectual skills, and ultimately the industrialization of the landscape, should all lead to a reduction in the diversity of motor skills performed by individuals, and hence to an increase in brachycephaly. Finally, the "phenotype hypothesis" is supported by data from Ceylon as reported by Hulse (1963, p. 409), which showed that the urban population was more brachycephalic than the rural one. This follows from the explanation proposed that rural people, presumably, have to master more motor skills to handle domestic animals and agricultural tools than do urban people, and therefore they ought to have a more dolichocephalic head form owing to an enlargement of occipital lobes and cerebellum.

The effect of agriculture and subsequent formations of states and urban centers also extended to the social behavior of people under its influence. Agriculture brought about selection pressures quite different from those of periglacial society, some antagonistic to humane values. Selection emphasized overt aggression in males. Agriculture produces at least a temporary surplus, which becomes a defensible resource, attractive to others. Even where agriculture generates no surplus, the agriculturist is enormously vulnerable compared to the hunter and gatherer. A crop planted on a plot can be readily destroyed by humans or animals. Anyone destroying the crop can bring starvation to the family that planted it. The agriculturist is thus vulnerable to intratribal, as well as intertribal, strife. He is a ready victim not only of marauders but also of blackmail. He therefore requires strong security measures within the tribe and against other tribes to safeguard his livelihood. It is this extreme vulnerability that generates suspicion, hostility, intolerance, and aggression toward others. Not surprisingly, defensive structures can be identified in the earliest of agricultural settlements (Rowlands 1972, Tringham 1972).

The agriculturist is also a victim of the weather, be it too much sun, wind, rain, or hail, as well as the depredation of wildlife. His view of nature changes. If he has milk, and therefore an assured protein supply, animals are no longer brothers, but enemies to be hunted and killed as an exercise for war. The natural forces become personified as deities that must be appeased and controlled. All must behave correctly toward these natural forces, lest they alienate them and endanger the crops and thus everyone's survival. Conformity to religious views is harshly enforced and tolerance vanishes. Only if wild animals continue to supply a significant amount of protein to the agriculturist do they not fall in contempt, as exemplified by the Zapotec, for instance (Flannery and Marcus 1976).

The rigorous requirements for knowledge and diverse skills typical of life in the periglacial ecosystem are relaxed with the advent of agriculture. Fewer individuals with a less diverse and demanding effort can sustain the population. This is true not only under agriculture, but also for life as a hunter–gatherer in warm climates. Women can easily sustain the whole tribe with their labor, as is well known from tropical and subtropical hunter–gatherers, and that with relatively few hours of work (Pfeiffer 1972). Males become less essential for the support of the reproductive effort of females; they are no longer required to be the hard-working providers, and have free time on hand. They become expendable. This is true even in agricultural societies in which males do heavy labor with animals in tilling fields.

Agricultural systems based on plants or a mixture of plants and animals are fixed geographically. The requirement for frequent movement and physical activity vanishes. A hunter who must be mobile cannot amass possessions, since they are a burden to him, but with a sedentary existence possessions can be hoarded and become symbols of prestige, prowess, and power. Material goods, agricultural surpluses, and also possessions in the form of livestock, can be hoarded to become incentives for raids and warfare—particularly in times of food shortage—as one commonly observes among nomadic herdsmen (Leeds 1965, Sweet 1965, Turnbull 1972, Gauthier-Pilters 1974).

The discovery of animal milk as food, and its consequent husbanding as a reliable supply of high-quality protein, must have led to fundamental restructuring of a society in which agriculturists previously depended on domestic plants plus the meat and eggs of wild animals for food. First, animal milk can be used directly to feed the human infant, relieving the male of the necessity to structure a supportive psychological milieu for his wife or wives. We noted earlier that human females may respond with a cessation of lactation to a stressful milieu. The first consequence of the availability of animal milk is to separate male and female, or at least make a supportive social milieu for the female nonessential. The role of the mother *per se* is degraded. In her absence or inability to produce milk the infant can still be raised on animal milk. The availability of animal milk permits an increase in reproduction, an increase in phenotypic development of individuals, and thus production of more and better soldiers to protect the society. It also increases the expendability of males and permits warfare to flourish more frequently and with greater loss of life; warfare can be less ceremonious. The availability of animal labor, of course, also aids warfare as well as permitting greater exploitation of the soil for crop production. The foregoing concept explains, for instance, why in Mesoamerican Indian agriculturists one finds a rather high social role for women, reverence for wild animals, the presence of ceremonial warfare, instigation of war to obtain slaves, be it through conquest or tribute because, in the absence of milk-producing domestic animals, each tribe or state could not produce enough males for work and sacrifice (von Hagen 1960, 1967, Flannery and Marcus 1976).

The natives of the west coast of North America developed a complex cultural system on the basis of temporary surpluses harvested from the seacoast, salmon streams, or ocean. The relationship of these cultures to the somewhat precarious nature of the resources is discussed by Ruddelle (1973). Not resource surpluses, but localities from which resources could be readily harvested, were defended. The somewhat unpredictable nature of salmon runs, herring spawns, and whale hunting imposed a mobility on the lower social strata, which shifted to areas of greatest resource abundance, each held in the possession of a hereditary chief. These resources could be harvested with the chief's permission and by payment of a tribute, which enlarged the surpluses acquired by the chief and which served as a measure of his rank and importance.

The Ainu of Hokkaido exploited a resource base quite similar to that of the American Indians on the northern Pacific coast. There are similarities to that of the Indians in their social system and ownership and defence of localities from which resources could be harvested. The Ainu along river systems exploited rather predictable resources, and one does not find the mobility of lower social strata here. However, one finds a society with a complex religious system that does express anxiety over availability of animal foods (Watanabe 1972).

Where surpluses are generated readily, particularly by female labor, males have free time on hand and can devote much of it to rank competition and ego displays. Material possessions become symbols in that competition and, along with the surplus, incentives for warfare. This, in turn, endangers the lives of women and children. Thus, the males become protectors of the tribe, much as one finds in lion society (Schaller 1972). Values supporting aggression must therefore become dominant and subject society to their ends. In contrast to hunter–gatherers (Steward 1968), great premium is laid on warfare. Due to warfare, males suffer relatively high mortality, becoming the sex of lower abundance and thus of greater social importance, since they become the subject of female competition.

In contrast to the continuous vigorous physical exercise that is demanded by the hunting and gathering way of life, the sedentary existence of plant agriculturists is less demanding of physical work, as well as being less regular in its work demands. The intermittent form of activity is more likely to lead to a deterioration in health, as is known from the findings of exercise physiology as it relates to health (Cureton 1969). In addition, the sedentary way of life at high density creates sanitation problems, which must lead to a higher incidence of parasitism and disease in agricultural populations compared to hunter–gatherers.

The monogamous family is in jeopardy wherever the sex ratio shifts strongly in favor of one sex. Among hunters, for instance, only rarely could one male support more than one family. Where food surpluses are generated by female labor on land held by males in possession, where females are the important labor force, and where females are mainly of low social status and become symbols of the male's wealth, polygamy is almost inevitable.

Whereas in the Paleolithic society the female had to select not only the socially most successful male possible (dominant by virtue of strength and size) but also the best and most loyal provider, in the polygynous surplus society she needs to concentrate only on the most dominant male. Wealth follows dominance. She must select the male with proven aggressive abilities. Note that the female need not maximize the duration of marriage, as is essential for cold climate hunters, since even if her warrior-husband dies her children can readily be supported by her husband's wealth. Moreover, her offspring are likely to be cared for by her husband's wives if she dies. The female's reproductive success no longer depends on her physical ability to work on behalf of her children and husband as in cold climate hunting societies, but depends largely on the rank of the male she marries. She must compete with other females for the very limited number of dominant males, and the greater the population and choice of females to the dominant males, the more severe such competition and the greater the emphasis on female advertisements. Muscular development and physical size are no longer of prime importance to the female's reproductive success. Nor does her social and reproductive success depend on her physical skill, but to a large extent on her sexual attractiveness.

This, however, fits into the male's interest. He maximizes reproductive success now by maximizing inseminations, while the labor of his wives and dependants ensures the necessary resources for the ontogenetic development of his offspring. He need not select critically for competent mothering; the polygynous family can compensate for deficiencies of individual mothers. Nor need the male worry if his actions are so stressful to his wife that she prematurely terminates lactation. The deficit in mother's milk can be compensated for by milk from domestic animals; cow's milk seen in that light is no boon to humanity or the individual, but the obverse. The female ensures the male's attention by maximizing attractiveness in form and behavior, and she must maximize the duration of that attractiveness, since her reproductive fitness now depends on maximizing the number of offspring, rather than maximizing the phenotypic development of her offspring. The mortality of adults and children is high due to unsanitary conditions; milk from domestic animals reduces the need for the female's lactation, and leads to rapid conception provided she succeeds in attracting the male. In order to maximize her attractiveness the female ought to become more paedomorphic, that is, extend her youthful appearance by enhancing those physiological mechanisms that keep the body youthful. Thus, in agricultural polygynous society there is selection for an extended period in which the woman retains her womanly "beauty." It is also in the interest of her reproductive fitness to attain puberty early in life.

The male maximizes his reproductive fitness by maximizing those attributes that lead to dominance over other males—strong body, combat skills, courage, self-assertion, and an enhancement of secondary sexual characteristics that differentiate juvenile from competent adult males. Hence, there is powerful selection for sexual dimorphism under agricultural conditions. We anticipate from the foregoing, and from earlier discussions, that sexual dimorphism is minimal for *Homo sapiens* populations living under marginal ecological conditions, while humans who have been subject to selection in productive landscapes with surplus agriculture should have the greater sexual dimorphism, the more aggressive, self-assertive males, the greater individual diversity, smaller body size, lesser physical development, the greater genetic load, and by and large, the lesser population quality.

Diversity under agricultural surplus conditions is produced by reproduction of individuals that are capable of being skilled in only a few abilities, but that suffice to make them desirable and even prestigious members of society. Such individuals have as much a chance to reproduce as individuals who have a wide range of faculties well developed, as was demanded by Paleolithic conditions. This promotes genetic diversity, as well as an increase in the genetic load. The process can be termed self-domestication. It explains, in part, the reduction of brain size in humans since the Paleolithic (Mettler 1955, Hulse 1963). We find the very same in domestic animals (Kruska 1970a,b).

It appears that one can also predict accurately the type of personalities favored by natural and cultural selection in periglacial and postglacial societies. Under personality, I shall consider the neuroticism and extroversion axes (Gray 1971). Under circumstances where meticulous planning of one's actions is essential to survival, such as that practiced by hunters in cold, hostile climates, introverted personalities should be selected for. Such individuals would be calm, studious, mistrustful of impulses, and fond of well ordered, predictable execution of activities; they would avoid excessive excitement and be serious and dedicated to their task; they would be a little pessimistic, and proponents of ethical and moral standards. By contrast, persons who are impulsive like taking chances and dislike contemplation, and they tend to be aggressive and disrupt activities; they are egocentric and undisciplined, as well as losing their temper quickly, and are unlikely to be a success in a primitive hunting economy. Their condition would be worse still if they rated high on the neuroticism axis and were moody, touchy, restless, and easily aroused (Eysenck and Rachman in Gray 1971). We would expect that on the neuroticism axis, stable individuals, and on the extroversion axis, introverted persons would do best under periglacial conditions. One may take note that Eskimos killed individuals who would not cooperate but disrupted the harmony of their society (Freuchen 1971).

However, the extroverted individual is likely to do best where bravado, showmanship, reckless aggression, rage and spur-of-the-moment decisions are required to ensure the population's survival. These conditions would apply to sessile societies with surplus economies, as described above. Here men would not play a crucial role as providers, but as defenders of their families and possessions. Warriors who prove themselves in such societies are likely to be strongly extroverted as well as relatively calm. However, we can also expect that excitable, touchy individuals would do well. Hence, there would be some selection toward neuroticism in the males of such societies.

Such selection would have far-reaching biological and cultural consequences, since it would enhance sexual dimorphism and psychological and biological attributes enhancing aggression, favor showy, loud, boisterous dress, decor, and behavior, and complexify cultural attributes of the society. To understand why this follows, we must first examine Gray's (1971) hypothesis about the neuronal correlates of extroverted personalities. These people, Gray argues, are less sensitive to punishment (in the widest sense) than are introverts. He postulates a higher threshold of sensitivity in the "punishment centers" located mainly in the hippocampus. Thus, actions perceived by the sensitive introvert as punishing would not be perceived by the extrovert as such.

This view harmonizes well with the finding that introverts apparently salivate more than extroverts in response to acid stimulation on the tongue (Howarth and Skinner 1969), and that introverts tend to sit further away from strangers than do extroverts (Fesbach and Fesbach 1963). It appears to me that one must go one step further than Gray did, however, and postulate a higher threshold of sensitivity for the "pleasure centers" in the CNS as well. Thus, the extrovert requires strong stimuli to trigger the pleasure centers. In order to generate these, he acts extroverted, or rather vigorous. Clearly, an individual with highly sensitive pleasure centers does not need strong stimulation to trigger these, and can be satisfied with far less stimulation. Thus, the extroversion axis appears to represent a gradient of sensitivity in the pleasure and punishment centers; the neuroticism axis probably represents sensitivity of the arousal centers.

Granted this, there will be, of course, selection for sexual dimorphism, not only for the reasons elaborated earlier, but also because it will take relatively strong sexual stimuli to arouse the interest of male and female. Individuals with poorly developed secondary sexual characteristics are not very attractive in the society where most individuals are strongly extroverted. Thus, it takes a relatively strong "female form" to arouse the interest of the most successful male, not only because he has so broad a selection of females for his family, and is thus habituated to unexceptional female figures, but also because he requires a high stimulation to become interested in the first place. It appears that this explains the existence of "sensuous female dances" in the tropical agriculturists, as discussed by Lomax and Berkowitz (1972), not only due to selection for extroversion, but also because of habituation to the naked female form.

Another aspect of the human male needs an explanation, namely, that which almost universally leads males to dominate females in most human societies. Domination may not be great, as in societies of most hunters and gatherers (Freuchen 1961, Pfeiffer 1972), or it may be extreme, as among some agriculturists. Why, one may ask, are women not accepted as equals?

The answer appears to be that domination is a consequence of relatively higher self-assertion by the males due to a greater requirement for social reassurance, reinforcement, or approval. Its evolution can be explained as follows: In cold climates, close cooperation among hunters was indispensable in order to provide carcasses of large mammals continuously, for food and clothing. Individual hunters had to respond quickly to opportunities and to each other's actions as they closed distance with a dangerous prey, and hence they had to observe each other closely, and they had to practice precise, skillful maneuvers in order to kill the prey. To be able to attain both the skills and close cooperation, natural selection favored individuals that were sensitive to criticism, willing to adjust, and willing to please others. They were individuals craving approval, or even admiration. The converse of this is that without approval and frequent positive social reinforcement, such individuals are relatively insecure. This must lead to frequent self-assertion, which in each instance is a test of whether one's actions will win approval or an outright fishing for a complement or reassurance. The individual who frequently asserts himself in comparison with another individual who does so rarely will soon dominate, if he is also physically stronger. Clearly this leads to male dominance, unless environmental conditions—normally imposed by harsh environments—force the male to closely cooperate with his spouse or other women. If the male depends on the services of a woman, his dominance is weakened due to the woman's ability to curtail and withhold services (e.g., in the Hadza, Woodburn 1968b). Where women's services are readily obtainable due to an imbalanced sex ratio favoring males, the dominance of the male rises as competition by the females for attention from the males increases. In essence, the male's dominance is rooted in insecurity and the relatively greater need for social reinforcement or assurance. If this concept is valid, males ought to be more sensitive to the social environment during ontogeny than are women, they ought to be more gregarious, more intent on spending time in a peer group where social reinforcement may be had and seeking peer help, and more easily frustrated than females. Clearly, if peer group competition is intense, as is postulated for the agrarian societies, then the males will spend most of their time away from females and will find the demands of females a direct danger to their competitiveness. This, in turn, demands that the woman bother herself as little as possible with the male, and this can be achieved by his keeping her submissive and hemmed in by taboos.

Owing to the high death rates in agricultural, sedentary societies, with the concomitant high birth rates and high population densities, peer groups of children are likely to be large. This is an ideal setting for intense peer competition and trains children in the skills and art of aggression, including overt aggression, from an early age onward (Hutt and Vaizey 1966), but to the detriment of intellectual ability, as suggested by Chance (1969). Other factors that enhance aggression in children are maternal lack of self-esteem, chronic family disorders, punitive discipline, strong belief in supernatural punishment, low performance demand on the children, large families, and a low degree of supervision and care by the parents (Goode 1974, Hinde 1974, pp. 288−9). These are the very factors one would expect in the agricultural surplus societies. Clearly, competition in adulthood, combat, violence, callousness, and bloody ritual ought to come "naturally" to individuals raised in societies that survive by aggression and that have large peer groups. Where these are found, parental influences decrease (Broom and Selznick 1963).

Natural selection for extroversion would lead in the cultural realm to showy, complex, diverse manifestations. Conversely, where introversion is adaptive, we expect cultural displays to be subdued, delicate, and relatively simple. Art would be subdued, the humor gentle and simple, songs and tales would be simple and relatively low in information content. Religious beliefs would be little developed, not only due to introversion, but also because hunters are not nearly as insecure as agriculturists and need not depend for their survival on the whims of weather, or the absence of agents detrimental to the crops (Turnbull 1972). It is agriculturists, not hunters and gatherers, that consider themselves at the mercy of the supernatural. It is not surprising, therefore, that given the chronic insecurity of agriculturists, sorcery is well developed even in primitive agricultural societies (e.g., Meggers 1973). Hence, the function of the witch doctor or medicine man is all-important, not only in reducing suffering brought about by diseases and injuries, but also by fears of sorcery, warfare, and the uncertainties of weather.

In societies with natural selection for extroversion, particularly those that compete strongly for social rank, and where ample time is available for social competition, one can predict great complexity and length of stories, and sharp, caustic—and maybe cynical—humor, because selection favors relatively insensitive persons, and because the very richness in cultural offerings that are competing for attention habituates individuals against all but the extremes. I consider that this prediction has been essentially verified by the study of song and dance style in relation to cultural complexity and productivity. Lomax and Bekowitz (1972) report that the information content of songs and their complexity, as well as the complexity of dancing, varies with the economic productivity of the society; the degree of ornamentation in song increases with social stratification.

Using the rules followed by dominance displays (Chapter 5), we can predict that selection will first be for showiness, complexity, virtuosity of cultural manifestations; later—following the rule of the flip—cultural tastes will swing away to the opposite extreme, to monochromatic, disordered, stark, simple images. Here reduction in complexity will be secondary and will not lead to the art styles of societies of introverted people. It may be added that, in the later societies, individuals will be treated on the basis of individual traits; in extrovert-selected populations, individuals will be treated on the basis of sex and status roles.

Granted an increase in population size, urban centers, social controls, and the rates of change, natural selection will necessarily favor individuals who can cope with the stress of change. Such coping can be brought about by decreasing the individual's sensitivity to the plight of others. Indeed there are studies to show that the type of individuals that survive major dislocations best are those displaying a shallowness of interpersonal relations, a loose attachment to goals, and a lack of concern for means. Such traits are psychopathic traits (Hinkle 1965). It must be evident to anyone acquainted with the profiles of highly creative persons, regardless of vocation (Crosby 1968, Gilchrist 1972), that a social milieu of stressful change will select against the creative individuals. These are usually introverted and sensitive. Intense competition, a fast, hard pace of life, as dictated by market conditions, ought to lead to loss of competence and breakdown in introverted individuals already during ontogeny.

In comparing hunting societies with agricultural ones, another deduction follows. In a hunting society in a marginal habitat, there will be cultural and genetic selection for traits emphasizing energy and effort conservation by the individual. Life is so demanding of individuals that any respite from work helps the individual to restore himself. We can expect persons from such societies to be less overtly energetic than from primitive agrarian societies with a surplus economy. Here intense personal competition ensures that each and every one is on guard, ready to act. Moreover, since the required work output is low, there is incentive for energetic frivolous exercise. We expect that the energetic restless individuals from agrarian cultures can work no more, or even less, than individuals originating from hunting cultures. However, superficially the former will appear more industrious.

In contrast to Paleolithic societies, agricultural societies maximize population size at the expense of individual development in order to create a large labor and soldier force. In a society in which warfare, poor nutrition, crowding, and lack of hygiene increase adult and child mortality, cultural means of population limitation must be abandoned and maximum reproduction encouraged. The milk of domestic animals, furthermore, permits an increase in reproductive rates, since it permits human females to shorten the period of lactation and increase the rate of conceptions per lifetime. There is, however, a limit on how far phenotypic development may be allowed to deteriorate to free resources for maximum population size. If the safety of a tribe depends on competence and skill in combat by individual males rather than on swamping foes with a large number of soldiers, then great physical size, strength, skill, and resourcefulness are at a premium. Moreover, the ability of such males to survive must be maximized, for the weakened tribe may be readily decimated by powerful neighbors. If there is a premium on small-group cooperation, rather than on individual fighting ability alone or massive swamping of the opponents by waves of fighters, then a reversal in cultural selection toward monogamy can be expected. This reversal would set in as individual development is maximized in the social environment of the extended family. Cultural values espousing monogamy will be adopted. Such a system is likely to arise under marginal, rather than good agricultural conditions where there is an emphasis on mixed farming, in order to exploit the landscape's resources.

It is likely that these were the conditions that shaped, for instance, the bodies, behaviors, beliefs, and institutions of the Germanic tribes of Europe prior to their conversion to Christianity. From the sagas of these people, particularly the Icelandic sagas, and the meager historical records the following picture arises.

The Germanic people exploited a forested landscape, usually one of marginal productivity, through plant and animal agriculture, as well as by exploiting the native ecosystem by hunting, fishing, and gathering. They lived on farmsteads, often widely dispersed. This forced self-sufficiency on each family and in so doing raised the woman to a high social role, as we shall see below. Thus, these people were great generalists who derived a living from the land by a great diversity of agricultural and subsistence skills.

The safety of each family and of its possessions was dependent on the reputation of the male members for their competence in combat and also for honorable, fair, and considerate conduct, and on their ability to form alliances with other families. Great skill in arms did not necessarily guarantee safety, for foul personal conduct could bring such disdain that the community would not consider murdering such a misbehaving individual as a punishable evil. Nor would great skill at arms in itself be a desirable trait that would guarantee a man the respect and ready alliance with others. For these reasons the males jealously guarded their reputations and their good names, and could respond fearsomely to any slight. Because there was no judiciary and executive to enforce agreed-upon laws or justice, it was up to the individual to do this. The support of other males at the annual gatherings where cases were heard and judgements were passed was not enough. The individual had to have abilities and resources to punish others who did him and his family wrong. This required great ability in using diverse weapons, and explains the many euphemisms and glorifications of weapons, particularly the sword. Clearly, great physical prowess, and also social competence, were valued. To achieve these, an abundance of resources for ontogenetic development of children was needed, and one way of achieving this was by infanticide practices if living conditions were marginal. The sagas inform us that such was indeed the practice, particularly in poor families. To achieve great physical, intellectual, and social competence of children, a close-knit family life with a supportive milieu for the female is also required. This explains the great value placed on fidelity and loyalty to the family.

Because alliances were terribly important as a protection in strife as well as in law suits at the annual gatherings, a verbal promise was holy and binding. To break an oath, to be devious, or to lie was conduct so unacceptable that the transgressor, and even his relatives, could be persecuted and even killed.

Because reputation and character were so vital in protecting family and possessions, or even acquiring a bride, a man had to seek confrontations with others, whether it were in battle during raids on neighbors or by taking sides in legal disputes at gatherings. Raids and travel to distant lands, the annual gatherings and trips to reinforce bonds of alliance and friendship with allies, necessitated the absence of the males from the household. Women were therefore often left in charge of the estate and had to manage it. This had to include religious functions of blessing the crops and produce of the land. Therefore, women were called upon to perform religious functions; they were often priestesses.

In order to be successful socially, bravery and skill in arms, sport, and combat were mandatory. Such skills could be acquired in large part by hunting. This explains the great love of these people for hunting, and, in particular, their rich lore of hunting ritual and its prominence in their religion. Hunting was good training for warfare, as it required skill in planning, strategy, and tactics. Given the great diversity of physical skills required by crop and animal husbandry in the varied and harsh land Germanic people inhabited, the intellectual skills required to run the affairs of the estate, competence needed in sport and combat, plus the intellectual skills required in the hunt, during raids, and in battle, to learn and recall verbally transmitted laws and ceremonies of religion together with sagas and songs in order to entertain, plus the social skills required to form favorable long-lasting alliances, to bedevil opponents in public debates at law suits during annual meetings, or to cool tempers and appease when strife appears inescapable, is it surprising that phenotypically these people approached those of the late Paleolithic? The physique of these early Middle Age people on average surpassed that of modern Americans, an exceptionally well-developed people physically by comparison with people of other nations (Huber 1968).

Even the structure of the sagas reflects the cultural forces generated by the economic strategy of these people. As discussed by Bertha S. Phillpotts (1932), the sagas are structured in such a way as to tell in great detail and chronologic order observable events and statements made by people. The reader or listener is not informed of inferences that can be made nor are motives ascribed to the actors in the sagas. This is clearly a product of the dangers entailed in publicly making inferences about any person's conduct; observed events and spoken words can be accepted as data but inferences cannot. Granted that individuals jealously guarded their reputations and were ready to repay a slight with the sword, it was safer to tell only what could be seen or heard rather than making inferences about it. The audience was astute enough to make the proper inferences anyway, and the teller of the sagas could never be accused of making false inferences.

At first sight, it is surprising that at the first mention of an important actor in a saga his genealogy is related. On second thought, the surprise vanishes. Given the high esteem in which marital fidelity was held, and the ample opportunity for illicit love affairs in the absence of the males of the household, a recitation of genealogy was proof of a person's legitimacy and the honorable conduct of his family. This protected family reputations, a very important condition to forming favorable alliances and marriages. For the sake of family reputation, it was also necessary to avenge slights incurred by a family member; therefore, the feuds that arose between families, the great importance of the legal wrangles at the annual gatherings, and also the fear of imposing an unjust judgement and being associated with an unjust judgement can be understood.

The foregoing speculative discourse is aimed at elucidating two points. First, strategies of resource exploitation shape both behaviors and biological traits in humans; social systems are a function of a species' adaptive strategies, a point long made by zoologists studying animal societies (Cullen 1957, Brown 1964, Crook 1965, Eisenberg 1966, Lack 1966, Estes 1974, Jarman 1973, Geist 1967, 1971a, 1974a, Wilson 1975). Second, natural selection in post-Pleistocene societies appears to be antagonistic to many human characteristics we greatly value. This leaves but one conclusion: If we value human attributes, if these are enshrined in social values, then we must develop economic strategies compatible with them—but must not allow economic momentum to continue unchecked. We cannot impose arbitrary values on economic strategies, for the latter generate their own rules as to what ought to be done in order to succeed economically. Social values dominated by economic strategy are dominant values—if economic success counts socially. We cannot let economic change proceed blindly, because it will dictate the values we shall ultimately abide by, even if these make us less human (Geist 1978b).

The concept of adaptive strategy in a given ecosystem must not be interpreted simplistically. In particular, adaptive strategy must not be confused with the ecology or with the habitats of a landscape. A given landscape does not give rise deterministically to one standard adaptive strategy and, therefore, to one human social system only. In short, given a desert, one cannot predict the social system of humans exploiting it, although once known, the social system, technology, beliefs, and rituals will probably form a coherent system but one not predictable *a priori*, no matter how closely beliefs, values, ceremonies, rules of conduct, etc., are found to depend on resources. Note the divergent adaptive syndromes of bushmen (Lee 1972, Silberbauer 1972) and Australian aborigines (Tindale 1972), both exploiting deserts.

The point that the evolution of societal structure can be understood as a function of resource exploitation and environmental constraints has been well explored by a number of authors (e.g., Clastres 1972, Turnbull 1972, Watanabe 1972, Hooker 1976). This applies even at the level of state evolution (e.g., Stover 1974, Carneiro 1970, Cox 1973). However, even in the evolution of states, the periglacial environment that was so dominant in the shaping of our species played a decisive, hitherto unrecognized, part. The old civilizations arose largely as an echo of the glaciers, for they developed without exception along powerful rivers whose headwaters rose in the snowfields, icefields, and cirque glaciers of tall mountains. Here the waters gathered the fertile silt and carried it down to the floodplains to deposit it as rich alluvium. Here agriculturists took advantage of the free energy generated by the floodwaters that knocked back the climax vegetation to a productive, early seral stage or cleared the floodplain to be entirely ready for the planting of grains. These mighty rivers confined human populations where they ran through narrow mountain valleys or through deserts, an essential condition to state formation, as elaborated by Carneiro (1970). The Nile, Euphrates, Tigris, Ganges, Mekong, and Yellow Rivers and the Peruvian coastal streams all took their origin in remnant glaciers and snowfields, and owe their fertility to the erosion of rock by ice, snow, and water. It is the silt, the fertile silt, and the meltwaters that ultimately made for great civilizations.

Chapter 15

Health, Professionals, and Creature Comforts

Introduction

I wrote the foregoing chapters not only for academic reasons. My quest is not only for an understanding of that ultimate of concerns, man, but also for a bridge whereby that understanding can cross into practice. I am in quest of a way of life, a lifestyle, that maximizes health, but I want this search to be within the construct of science. I want a verifiable, testable theory of health, one falsifiable by experimental means, standing or falling on the basis of observable, demonstrable, measurable predictions. I want a theory of health within the scientific paradigm, not because the latter is so fashionable, but because I am afraid to delude myself lest verification be a public process. The concern for a way of life that is healthful is so fundamental to any conception of human existence that no social system laying claim to durability can possibly succeed without it. If my conclusions about it are invalid—as they might be—they have to meet their fate by demonstrable means. If I am wrong, the concern still remains and becomes more important with each passing day.

There is no shortage of interest in health in our society; quite the contrary. It is evident in the countless books, advertisements, courses, articles, newspaper columns, and radio and television programs on dieting, exercising, diseases of affluence, pills, and oriental meditation. In the United States, medical costs account for some 8% of the Gross National Product, making it the second largest contributor to that index (Kristein et al 1977); in Canada, the comparable statistic as of 1971 was 7% (Lalonde 1972) and rising. The concern of traditional medicine for health (as opposed to repairs of a damaged body) expresses itself in the disciplines of preventive medicine, epidemiology, and psychiatry. A very active field entitled Environmental Health has arisen (e.g., Eckholm 1977), while new focuses, such as predictive medicine (Cheraskin and Ringsdort 1973) and constitutional medicine (Damon 1970), continue to emerge. Not only are medical doctors and public health officials actors on the national health scene, but so are architects, urban planners, designers, and scientists from diverse backgrounds joining together into associations such as the Environmental Design Research Association or the Association for the Study of Man–Environment Relations. Even older medical disciplines, such as psychiatry, give evidence of internal turmoil and a continued search for a better understanding of and prescription for health, as evidenced by the writings of Duhl (1976).

In such a turmoil, all of which aims at improving health services, it would be surprising if severe critics of the very trend did not appear. They are present. Illich et al (1977) claim that the legitimate interest in health has turned to "healthism," a perversion and sham concocted by professionals to bilk the public, disabling the public in the process, thus, far from detracting from ill health, adding to it. Illich (1975) terms such diseases caused by the medical profession iatrogenic diseases. However, not only the medical professions are to blame, so these critics contend, but professionals as a whole. Moreover, their arguments are not without merit. Dare one speak under such conditions?

The disciplines presently concerned with health, such as preventive medicine in the broad sense, environmental health, predictive and constitutional medicine, as well as psychiatry, have a shortcoming. They lack a theory of health. They are empirical disciplines that have collated experiential information about how to prevent illness from progressing, how to prevent germs reaching human beings, how to avoid pollution, malnutrition, and accidents; in short, how to stay out of trouble. They are collections of lessons on what not to do. However, they do not tell us how to anticipate what not to do; they lack theory. Since they lack theory they cannot readily crystalize their findings and transmit that essence to other professionals, nor can they develop a comprehensive understanding of how to maintain healthy individuals. My particular concern is that a verifiable—that is, scientific—theory of health is needed, to be used by the professional class in daily decision making.

The need for a theory of health has not gone unperceived. Boyden (1972, 1973) saw clearly that a theory of health had to be expressed in a model life style, and that this model life style had to be linked in some manner to our evolutionary history. Human beings had evolved and were adapted to certain environments and could be very healthy only in those encompassed by their biological adaptations. To use another analogy, one does not expect a fish to live healthily on land. It is useful to discuss briefly Boyden's conceptions, for he began on the right road but soon ran into difficulties.

Boyden (1972, 1973) proposed that the model life style for optimum health is to be found in the pre-Neolithic hunter–gatherer societies. I assume he proposed it because the presently accepted paradigm of anthropology conceives the human as evolving gradually in Africa as a hunter–gatherer. Should that paradigm be false, then Boyden's model life style would also be false (unless it predicts correctly improved health, which it does not!). I have shown in the preceding chapters that the presently accepted anthropological view of human evolution is wanting.

Boyden centers his arguments on what he terms the "principle of maladjustment." He argues that species are genetically adapted to a given environment. If they find themselves in an environment different from that, they are likely to suffer maladjustment. Here Boyden makes an error of oversimplification. Species of higher organisms are buffered by epigenetic mechanisms against environmental variation and can suffer maladjustment only where these protective mechanisms, such as culture in human beings, can no longer cope with environmental extremes. Thus, there cannot be for humans *an* environment they are genetically adapted to. Rather, we expect them to cope with a great diversity of environments. Still, the idea of maladjustment is valid, but has to be used very precisely and with reference to epigenetic mechanisms. Suggesting that hunter—gatherers had a life style of optimum health is an act of faith, not a scientific rationale.

Boyden made a peculiar second assumption. He assumed that just because human beings lived in their natural environment such humans must be healthy. A zoologist must find such an assumption about any species of animal untenable. Populations of a species in their natural environment show a range of health, as I discussed in Chapter 6. Therefore, one would assume there would be a spectrum of health in humans living in natural surroundings, depending on the ecological conditions of each population. Moreover, as pointed out in Chapters 1 and 6, Boyden equated natural and evolutionary environments, and this cannot be done; the two concepts are not the same. When evolution acts on populations under maintenance conditions, the environment of individuals is likely to be one responsible for a great amount of disease, deformities, and high mortality. Only the evolutionary environment of species formation during exceptional conditions of superabundant resources is also one that maximizes health. In short, Boyden's pioneering attempt is unfortunately not constructed from valid propositions. Nor does it predict correctly, as hunter—gatherers may in fact be less healthy and less well developed and live significantly shorter lives than individuals from polluted industrialized nations.

In this chapter, I shall turn to outlining a model life style that maximizes health, test it against some empirical data, and discuss its applications. The essence of that life style is shown to be a great amount of diverse, skillful, physical activity, of intense learning of knowledge and skills, of complex interactions with nonpeers, of long-lasting intense social bonds, of developing mastery over a broad range of difficult tasks and a high level of discipline over one's intellect and emotions, and also a life filled with humor, good fellowship, a thorough exercise of bodily pleasures, and a diet both abundant and of high quality. To maximize health is to maximize humanity, a hard and complex task but not without pleasures. I am prepared to accept that the proposed model life style is not likely to prove popular. I am here also pursuing a second objective.

In the preceding chapters, we examined the consequences of the dictates of reproductive fitness to the behavior of individuals. We dealt with the law of least effort, strategies and tactics of resource acquisition, predictability and why individuals strive for it, aggression and its diverse manifestations, the consequences of the dictate to disperse and therefore of phenotype plasticity, and communication between the environment and genes. We did not look at monitoring which must be carried out so that an individual can follow the dictates of reproductive fitness or opt out. I retained this to the last because of its implications for how we see humans— and how we serve them. The implications are particularly relevant to professionals, for they are called upon to help satisfy human needs and wants, and they do require some understanding of the nature of these demands.

I have not addressed myself to urgent problems of this day. Others have done so. For those looking for guidance in environmental management, I point to such a wonderful book as *Man and the Living Envionnment* by the Institute of Ecology (1972), the report of a workshop on global ecological problems, or the handbook on landscape management by Buchwald and Engelhardt (1973). Nor am I concerned with raising anybody's conscience about our environment, for others have done this so eloquently, such as Leopold (1949), Carson (1962), Commoner (1963), White (1967), Hardin (1968), Dubos (1968, 1972), and Ehrlich (1968), and there are others to continue this important task.

It may be noticed that I have avoided eugenics. I did so on purpose to emphasize a conclusion that I reached, at first hesitantly (for the prejudices of a biologist run deep!), namely, that eugenics is a subject so insignificant as to be almost irrelevant. It is a highly relevant subject once humans do develop to the limits of their hereditary potential, for only then can genetic and environmental effects be segregated well. Eugenics can be an excellent distraction from the fact that environment powerfully molds humans during ontogeny and that it is with the environment that we must deal first.

We need a verifiable theory of health. Without a theory of health that generates testable normative criteria for environmental management or design, we must forever be victims of the winds of fashion, following temporary popular trends without guidance as to what is acceptable and what is not. We need such a theory, for human beings are not infinitely malleable nor healthy within much of the range of their adaptability. Without a theory of health there cannot be much resistance to trendiness nor to bureaucratic, political, economic, or special-interest pressures. No corrective feedback prevents unwarranted emphasis on trivia or on damaging decisions. We need a comprehensive theory of health to coordinate the multiplicity of insights generated by specialists, to orchestrate the cacophony of voices that give direction to our political systems, and also to contain malignant economic soccial, or political forces, malignant in their destruction of the very essence that gives us life, let alone humanity. We need a comprehensive theory of health to avoid the pitfalls social science has encountered, as described by Andreski (1972), and to prevent the further isolation of disciplines relevant to an understanding of health (Dreitzel 1972).

What is a theory of health? It is a statement from which one can deduce what to do and what to avoid in maximizing health. We obtained such a theory, briefly, as follows.

(1) For theoretical reasons, we expect that all organisms have the capacity to become a special and rare type of phenotype, one capable of dealing with contingencies, when exposed to unexploited environments. This was termed a "dispersal phenotype."

(2) There is empirical evidence in mammals and birds that such a phenotype does exist under the expected conditions, that is, characterized by exceptional development of tissues and behaviors of low priority. These tissues and behaviors, however, are adaptive in dealing with conspecifics and new environments during dispersal. One correlation noted was that the dispersal phenotype was healthy compared to the maintenance phenotype. Therefore, health is linked to phenotype development.

(3) Using the concepts of C. H. Waddington, it was shown that major evolutionary changes could occur only during dispersal, when resources were abundant. It is then that diagnostic features arise and are in part canalized. From this a definition of the evolutionary environment of innovation or of species formation was derived. This, however, is the very same definition as that which defines the environment of maximum health; it is the environment in which individuals maximize phenotypically the diagnostic features of their species.[1]

[1]Under maintenance conditions, natural selection can also act and change gene frequencies. Here, however, selection reinforces existing adaptations via the evolution of growth priorities (Chapter 6). This type of evolutionary environment *minimizes* health!

(4) Clearly, one needs to know now what are the human diagnostic features and how they arose. If one can decipher this, one consequently arrives at a life style that maximizes health. It is not the only way to the same answer.

(5) One can also generate a model life style that maximizes health by examining the life styles of populations in which individuals show exceptional development of structural tissues of low growth priority; therefore, the concern with people from the upper Paleolithic. I left unexamined a comparison of herders and agriculturists, or upper socioeconomic classes and lower ones; such comparisons ought to support the life style of upper Paleolithic people as one maximizing health. It, in turn, ought to be—in theory—identical to one generated from the theory stated under point 4. The propositions developed have to be testable or falsifiable.

The life style derived from these considerations I have termed the "evolutionary model."

What is gained by living according to this model life style? If followed in principle, it ought to maximize health intrinsically; that is, the evolutionary model aims at developing the potentials inherent in each individual to protect the body against external agents of harm without recourse to drugs or medical treatment. These are the expectations.

I have argued in Chapter 14, as well as in Chapter 6, that the phenotype of individuals is a function of the economic strategy of their population. However, if health is related to phenotype and the phenotype is a function of economic strategy, then the health of individuals can only be maximized by adjusting the economic strategy of the population they dwell in. In short, one must accept a systems perspective. Economic strategies dictate what ought and ought not to be done; they generate values by which people live in order to satisfy material demands. Values, in turn, dictate social structures, such as political institutions, administrative structures, and also the family form. Therefore, health cannot be validly considered outside the framework of a system of economic and social factors of the society. For instance, there is little point urging greater "love and tolerance" when economic dictates reward callous competitiveness, when society lauds those who display evidence of economic success irrespective of the means by which it was achieved. One cannot just dictate that "liberty, love, and empathy" (Esser 1975, p. 3) shall be the accepted values of a society if the dominant economic system of that society mititates against these. If one can identify values that are diametrically opposed to good health, these cannot simply be turned into their opposite by fiat, but only by a judicious adjusting of the socioeconomic system itself.

Of course, we do not always follow social dictates blindly. Occasionally we rebel, and that is termed "exercising our free will." Here we are dealing with ego manifestations, which I discussed in Chapter 5. This is an aspect of very great importance. I have addressed it under Creature Comforts. When there is an appeal to the "free will" of individuals to discipline themselves and take a greater responsibility in matters of personal health, as the Canadian Lalonde document (Lalonde 1972) does; it is, in part, an appeal to oppose economic dictates. One can only go so far with free will toward improving health; beyond that, the system of economic and social forces must give way. That may be the price of improved health.

When reading these lines, the question may be asked whether the concern is not misplaced. Are we in America not healthier today than at any other time? Have not

Americans increased in longevity and body stature, as well as in wealth, due to our past socioeconomic activities? Is it not dangerous to tamper with success? Before euphoria about our wonderful selves engulfs us, we ought to look more closely at some data about ourselves. After all, we are the model others are trying to imitate and excel.

On a given day in 1970, there lived in the United States some 203,235,000 people. Of these, about 2.1 million were society's wards in institutions such as jails, mental hospitals, chronic disease centers, homes for the physically and the mentally handicapped, detention homes and training schools for juvenile delinquents, homes for abandoned children and unwed mothers, and homes for the aged. Of the 2.1 million, 480,000 were children (Gula 1973). Of the uninstitutionalized population, 27.1 million suffered from some sort of handicap which limited their movements and freedom. Of the 3.8 million children of 2.7 million working female heads of families, some 1 million children were left to themselves without supervision, 40,000 of these under 6 years of age (de Huff-Peters 1973). Of the remaining "healthy" population, some 3.6 million were confined to bed due to illness, 86,000 of these with cases serious enough to be admitted to hospital. Some 18,000 of those not ill in bed had been arrested by police; about 4500 people were killed by environmentally linked diseases or social factors such as heart attacks, cardiovascular diseases, cirrhosis of the liver, emphysema, bronchitis, suicide, and murder. So far, these figures are roughly exclusive of one another and we have accounted for close to 34 million out of the 203 million who are handicapped, ill, or detained. There is no way of meaningfully accounting in the casualty figures for the 5.4 million chronic alcoholics, the 20 million individuals with serious hearing, sight, and speech dysfunctions, the 6 million mentally retarded and about twice that number with minimal brain dysfunction, the 9.7 million dependant children and guardians on public assistance or the 2.1 million old people who have to suffer the same fate, nor the 26 million who receive some sort of welfare payment, nor the 3.2 million orphans, nor the 2 million children requiring help to overcome their handicaps, nor the 4 to 5 million (or more) with various congenital deformities (Wallace et al 1973), nor the babies or fetuses that suffered inadequate nutritive environments and died. One could extend this list of figures by citing the statistics from recent texts on preventive medicine (e.g., Sartwell 1973, Wallace et al 1973). The figures, except where indicated otherwise, come from the U.S. Bureau of Census, "Statistical Abstracts of the United States: 1975" (96th ed).

It is significant to note that according to Kristein et al (1977), the medical industry is the second largest industry in the United States, consuming some 8.3% of the Gross National Product in 1975 (compared to 4.6% in 1960). In Canada, according to the Lalonde (1972) report, it is about 7%. There is very good reason to suspect that it is not a cost-effective industry, thus depriving other sectors of society of badly needed tax dollars. Kristein et al (1977) argue for the effectiveness of preventive programs and change of life styles as a means of reducing medical cost. Their figures show that about 25 million people are hypertensive in the U.S., that some 50% of all adults in certain test programs were courting a high risk of death due to obesity, hypertension, smoking, or high blood cholesterol, and that even among children aged 10 to 12 years in a sample of 2500, about 18% were hypertensive. Religious groups dedicated to family unity and good food habits, in contrast to the population norm, had about half the cancer and mortality rates. We cannot get an accurate account of the casualties, but they are probably more than 20% of the population. These figures of misery and destitution speak for themselves. Shall we have the gall to tamper with "success"? Or even question it?

In a discussion about how to increase the health of individuals, we must turn to the large professional class in our society, for it is they who so greatly affect our socioeconomic system. It is they who, by deciding on economic, social, administrative, technological, and jurisdictional matters, shape the milieu we live in.

Professionals and Intervention

By training and by inclination, professionals are interveners in the environment, be they engineers, economists, urban planners, architects, consultants to business or government, medical doctors, lawyers or managers, etc. By their decisions they directly and, more often, indireclty affect the social and physical milieu we live in, whether by generating change or by keeping change from happening. Moreover, they do it daily and usually without understanding that their actions can and do change our environment, except from the limited perspective of their profession and the values by which it abides. They are not normally knowledgeable of what environmental characteristics must not be disturbed lest the disturbance result in loss of the health and competence of the individuals affected. They are not informed of how to create and maintain environments that maximize health and human development, that is, they do not consider a theory of preventive medicine as part of their training, professional work, and responsibility. There are reasons for this, the most obvious being a philosophical *Weltanschauung* that denies the importance of anything but the cultural, moral, and philosophical in the affairs of men.

Professionals, not only in urbansim, architecture, economics, and law, but even those closely allied to social sciences, consider themselves largely beyond biological concerns. To consider such things in their professional practice would be to stoop to vulgarity. Biological and physiological man is of no concern, and the theory of evolution is at best irrelevant, at worst a travesty of humanity. They believe that a human is a unique entity in the universe, fundamentally different from other organisms. It is a cultural, moral, philosophical being, so much so in their view that even a link between these attributes and the biological ones is uninteresting. This is despite the modern embracing of "interdisciplinarity" by professionals—and "systems analysis" by a good many. Within their *Weltanschauung*, political ideals of the eighteenth century have a powerful hold: All humans are created equal and one is as capable as the other, each in his or her own way; humans are born "good," and the rules of society are a constraint on the exercise of this innate goodness; "user satisfaction" is the primary criterion for professional action, since the individual knows best and any interference with him is paternalistic; professional service consists of removing discomforts and difficulties, so that the individuals' energies can be turned to fruitful, productive tasks. The *Weltanschauung* of man carried about by the occidental professional of this age is haunted by the ghosts of Rousseau and the French encyclopedists, by Platonic ideals, and has suffered relatively little distortion by science. In fact, one detects a considerable sentiment against science among professionals.

The sentiment against science is not all that surprising. It has been recognized before. Davis (1972) makes the point that the battle about "evolution" is not won, and so does Leyhausen (1960); the old theological view of man, the supranatural image of man, a view deeply rooted in traditional occidental thought, is today alive and well. Lynn White (1967) espoused the same thoughts in a now-famous essay on the historical roots of our ecological crisis; Hugh Iltis (1970) makes a similar point. Our occidental traditions make us contemptuous of nature, and this is a contempt fed by the glorification of consumer goods, wealth measured as the acquisition of property and the display thereof, and an economic system almost blindly subservient to all, including irrational demands. Not even social scientists, so Davis (1972) points out, have accepted what may be called biological thinking, but are strongly committed to a "cultural" view of man. Lionel Tiger (1975), in an elegant review of somatic factors in human behavior, makes much the same point; his is a review well worth reading, as it shows how much the cultural factor is related to endocrinological, physiological and anatomical factors. I dare say that science has not helped its cause by displaying itself as an enterprise devoted to practical tricks, by pandering to its "scientific method" rather than showing itself as a cultural force developed to understand the world we live in. It is Science's tragedy to be a willing servant to industry, the armament industry, and to totalitarian governments and blind economic enterprise, and to be associated with as tragic a development as atomic power.

I have generalized here, I know. Is this sketch wrong, unfair, myopic? Maybe it is, but surely by erring on the side of generosity. If, to the consternation of humanists, our professionals are guilty of a pedestrian understanding of the teachings articulated in the humanities, they are equally innocent of all but the pedestrian in science.

I propose that, as long as our professional class clings to a crude notion of moral, spiritual, and cultural man, they cannot be instrumental or interested in that great fundamental good we call good health. As long as they fail to perceive the feedback loops between cultural actions and biological expression, they cannot but harm humans through innocent, but still damaging, ignorance. It is very important to recognize that prescribing for health is *not* solely the responsibility of medical doctors to do. They are but one profession affecting our lives. By training they are concerned with repairing damage and possess only inadequate information about how to prevent damage from occurring. Health is not in the purview of medicine! It is in the purview of the professional class as a whole, as well as being a responsibility of each individual. Therefore, a basic understanding of positive health measures, a theory of how environment affects health, should be a common base of all professions, medicine included. Environmental design and preventive medicine in the broad sense should be synonymous. To put it bluntly, the price to society of a world view that is limited to a moral or cultural view of Man the Being, and which ignores post-Darwinian life sciences, is the dismissal by the professional class of any responsibility for health. That is a very big price to pay.

How a conception of man, the moral, responsible, rational agent, manifests itself is illustrated by the Canadian Lalonde document on public health (Lalonde 1972). It hurts to criticize so perceptive a document as Lalonde's and yet it has shortcomings.

Lalonde makes an appeal to individual responsibility. He asks that individuals take steps to improve their health and outlines steps government could take to aid individuals in the task. No explicit or implicit recognition is given to the fact that the wishes of an individual to promote health and the options proposed by Lalonde are irreconcilable with the dictates of our economic system. For instance, in Chapter

12, Lalonde outlines strategies to increase the health of Canadians. He enumerates the essential information on how individuals may improve health. It must be assumed that once the information is available to people, they can readily adjust. Let us assume that the dangers of junk food are explained. How can individuals be expected to heed the information if manufacturers can develop sophisticated advertisements to seduce the consumer into buying junk food? Moreover, these advertisements essentially are paid for by tax dollars. Thus, the well-meaning government will finance both, programs to convince consumers not to buy junk food and, indirectly, programs to seduce the consumer into buying junk food. What government would dare oppose soft drinks? In short, the government, by blindly supporting the economic system within which the individual is the prey of the market, undoes the good it attempts to promote.

The Lalonde document has not attempted to resolve the conflict between "free will" and the machinations of the "economic system." A number of good intentions in the Lalonde document come to nought because of this critical omission. How, one might ask, can a program that tries to keep old or disabled people at home succeed if the presence of old and disabled persons seriously impedes the careers and opportunities to generate a higher income by those who must look after them? Must someone looking after old or disabled people in their families forego promotions because they cannot easily move to another city without causing grief and accepting all the problems of moving and establishing a new network to care for the old and disabled? Must those looking after the old or disabled bear the costs privately which would otherwise be pushed onto society by institutionalizing them? Must someone caring for old or disabled persons forego a social life that is rich in business opportunities or conducive to promotion because such activities are incompatible with the care of those individuals?

How can programs aimed at reducing alcoholism succeed if alcohol or other drugs are a convenient retreat from one's job, family problems, or one's economic and social failures? Who cares about the dangers of alcohol or will listen to an appeal to his responsibility if he is helpless to change his economic role or achieve at least modest success in his social aspirations? How is a greater number of professionals in mental health going to help if they can only placate the symptoms but not the intricate system-generated causes of mental ill health? How can reason and responsibility unaided oppose the onslaught of persuaders gushing from the media?

As long as our socioeconomic system remains our sacred cow, its relationship to health may be glossed over or ignored on purpose. "Free will" on the part of individuals can indeed lead to a few steps toward improving health, but only a few steps, and some steps only at the price of foregoing some personal economic gains (and depressing the Gross National Product). Thereafter the socioeconomic system must be altered so as to encourage values, and through these attitudes and behaviors that enhance health. Without an explicit recognition of this, many of Lalonde's recommendations can be described charitably at best as platitudes.

Of course, a world view of man that ignores the insights of post-Darwinian life science is not the only thing preventing us from readily developing measures to maximize health. The idea of progress is so deeply rooted in occidental thought that change for the sake of change is eminently acceptable. Therefore, practices that are valuable can readily be discarded in favor of questionable, but undoubtedly new alternatives. The sheer quantity of material professionals are expected to master mititates against the inclusion of more for fear that professional competence will be diluted. Moreover, intellectual anesthesia is quite readily available for professionals wishing to ignore the effects of their decisions on human environments. They can always advance the argument that humans are makers of their own environments and, because men are "natural" entities, they cannot produce artificial environments. Arguments as shallow and pointless as this one can be uttered in all seriousness and proffered as fruits of intellectual pursuits! This implies that the strip mines of Appalachia and the acid run-offs would, therefore, be as "natural" as the Rockies and their mountain streams; the throbbing ghettos of Urbania are not artificial, certainly no more so than the camps of the Kung bushmen.

Let me dwell on the fundamental assumptions of professionals a little longer by discussing the "man/environment" field, a rather recent development which aims at producing a good fit between man and physical environments. For an up-to-date introduction to this field I refer to the small, readable book by Irwin Altman (1975). This field—despite claims to the contrary—is nearly divested of medical considerations. This can be quite easily discovered: For instance, find an answer to the question of what benefit it will be to human ontogenetic development and health to concern oneself with "privacy, territoriality, personal space, and crowding." How does one evaluate success or failure of designs that incorporate insights into privacy, territoriality, personal space, and crowding? These questions have not been seriously addressed by the man/environment field precisely because the cultural or spiritual view of man is so very dominant here. Nor has the field examined some very basic assumptions: As in many professions, "user satisfaction" appears to be regarded as a universal good. I shall later argue that this need not be the case. The man/environment field attempts to reduce discomfort and frustration; that this can be detrimental to development and health, as illustrated by Seligman's (1975) book on helplessness, is not recognized, at least not widely enough. The field espouses the view that the "new, dynamic, and changeable environment," as Altman (1975) puts it, is a great good. That excessive change is detrimental to individuals (see Chapter 2), that there is an optimum arousal (Berlyne 1966), and that rapid change and boredom through lack of stimulation are equally detrimental has apparently not been taken seriously. Implied in the writing of this school is the assumption that the individual is always supreme and yet that professional service by the state somehow is preferable to self-help. This is a contradiction. Although it is agreed that the individual must be able to change his physical environment, he must—of course!—be taught by man/environment professionals how to do it (Altman 1975, p. 213).

Since the man/environment field deals very largely with physical spaces, it is not surprising that its apologists have concerned themselves greatly with "territoriality." The tendency of humans in western culture, and also in a great many other cultures, to act as territorial animals, has apparently been raised to the rank of a human requirement. Here the field appears to follow Ardrey (1966). The recognition that territoriality in animals is a consequence of resource defence, and need not be obligatory within a species, and that it vanishes if resources are no longer defendable, has not yet been recognized as applying in the same way to humans. A cultural rather than biological basis for territory does not make territoriality a universal thing—although the man/environment field still treats it as such. There are, of course, political implications to regarding man as innately territorial.

The environment itself is seen in the man/environment field in quite narrow terms. It focuses on the physical environments primarily of adults, and usually on the place of work. The social environment is conceived in spatial terms, as distances and orientation of individuals, as well as an expression of symbols of space claims, be these verbal, behavioral, or cultural. Thus, much of the concern of the man/environment field is how to structure work spaces or public spaces that will permit employees and clients to feel comfortable. There is also some attention given to housing and its spatial layouts. Virtually absent from consideration are the requirements for ontogenetic development and the multitude of consequences of social factors on ontogeny. The rich social behavior of humans is largely bypassed. Without a conception of what criteria signal optimum development and health, the man/environment field has little option but to accept a role as an obedient servant of economic and political dictates. It is not in a position to be an effective critic of society, because it cannot tie down what ought or ought not to be done as sound criteria of health. Make no mistake, there is very genuine humanistic concern in the man/environment field. However, the field suffers from a narrow view of humans and their environment, and from an overdose of intellectual timidity. For an interesting view into the field see also Esser (1975).

The foregoing illustrated why it will be difficult for professionals to accept preventive medicine as part of their responsibility. Yet, because of the rapid changes and increasing artificiality of our environment, it is essential that an easily grasped theory of preventive medicine be generated and accepted as soon as possible. This cannot be left to medicine, due to the perspective of this profession as a science limited to dealing with organic dysfunctions (Kristein et al 1977). It appears that evolutionary biology must take a role in nurturing preventive medicine, just as classical physiology nurtured traditional medicine. We require, not only a collection of empirical facts about what keeps us healthy, but also a broad theory that can be relied upon for guidance when empirical facts are not yet at hand. Professionals require such a theory, for they will at times have to make decisions in the absence of facts. I am proposing such a theory for scrutiny and debate.

Let me emphasize a point of some importance. The only sensible approach to the problem of compatibility between environment and man is to fit the environment to man and not man to the environment. Human beings must not be warped to fit a system by which a society extracts its share of energy and materials from earth. Rather, the system ought to be structured so as to encourage maximum health and development for each individual. Not every exploitation strategy is compatible with this objective and therefore valid. There are some "minds" lamenting that "man" cannot be changed fast enough to fit the realities of the "modern" world. To what end shall one change man to fit an artificial environment he is changing himself? If a maximum of human qualities is desired, such as high intellect, compassion, altruism, and sheer competence, then we had better look at the environments that nurture them, for we may fall woefully short of our genetic potential here. Moreover, changing man basically is changing man genetically. Are those lamenting man's resistance to change advocating eugenics and selective breeding of human beings to fit any fantasmagoria of an environment they opt to design? Are we to be analogous to tough-skinned, resilient, square tomatoes bred to fit mechanical pickers? In short, there is little to be gained from genetic engineering of humans until the potential of the present genetic system is fully utilized; we fall short of this. To "design" humans to fit an environment that humans change only creates a vicious circle without aim or end in sight. Our concern must be to create environments fit for humans, never the reverse.

On Health and a Healthful Environment

Let us now turn to the definition of health and, from it, to an environment that maximizes health. If preventive medicine in the broad sense is to enter into environmental planning and management, if concerns for life styles that maximize health are to become a broad concern of people—not only of professionals—then a clear, unambiguous, definition of health is imperative. A definition proposed by the World Health Organization, "Health is a state of complete physical, mental, and social well-being, and not merely the absence of disease or infirmity," simply will not do. This is not only ambiguous but also avoids facing up to harder criteria on the basis of which health can be evaluated; at the worst, this is a definition of political expediency. It is unacceptable.

If one searches for definitions of health in textbooks of medicine, one does not find definitions that are precise, except in the negative, where health is defined as the absence of disease or infirmity. The converse definition, that health is the maintenance of physiological homeostasis, is, alas, not much better. What is homeostasis, objectively defined? It is an abstract concept, a very valuable one in envisioning the functions of organs, but not any more helpful than "the absence of disease" in defining health. Yet the narrow definition of health (as the absence of disease or infirmity) is a useful unambiguous criterion.

It is recognized that health contains the notion of recovery from disease and injury. The greater the physiological capacity to recover, the healthier the individual. Since vital capacities vary with age, clearly the definition is subject to an age-specific correction. Thus, to maximize health it is quite essential to maximize the age-specific vital capacities of individuals. Alas, vital capacities, the physiological ability to overcome injury, cannot be measured directly. Moreover, we are already skating over psychological factors involved in recovery, for this is not independent of the individual's ability and motivation to help himself, to see hope for the future, etc. (Seligman 1975).

It is also obvious that an individual's physical, intellectual, and social skills, his judgement and experience, and his motivation to stay healthy all enter into the equation as to whether the individual will remain without disease or infirmity. To minimize impairment, therefore, one must maximize not only the physiological vital capacities but also the individual competence to deal with the world at large. Strictly speaking, the state of health, therefore, is the state of maximum adaptation in which a minimum amount of effort is expended on maintenance and a maximum of energy and time are available for activities besides maintenance. However, the conclusion of this must be that health is always an abstract notion and can never be precisely measured; it refers to unrealized potentials unrecognizable as such in individuals. What now?

Rather than searching for a "better" definition of health more will be gained by trying to identify characteristics of individuals who, over a lifespan, suffer the fewest diseases and infirmities. An empirical search for these characteristics can lead to findings such as those collated in textbooks of preventive medicine which, in turn, are not all dissimilar to those found in books on environmental health. By their very admission, authors of texts on preventive medicine focus primarily on how to prevent further deterioration once illness commences, and how to keep recognized dangers to health in check. Only hesitatingly do they enter upon life styles as health promoting. This is most understandable, not only because utterances by experts about how to live find disfavor with a significant segment of the populace, but also because a theory—rather than a collection of empirical facts—is not available as a guide to maximizing the vital capacity of individuals to overcome physiological breakdown of the body.

As long as medicine remains physiology- and laboratory-oriented, it cannot develop a theory about how to maximize health. Such a theory cannot be developed simply by observing and manipulating caged mice, rats, dogs, or monkeys, or even observing patients in hospitals or looking at vital statistics of a population. One must go to natural free-living populations of animals and treat health—or its opposite, the presence of disease, parasitism, congenital deformities, etc.—within the adaptive syndrome of the species and the particular environmental setting of the population. It is then seen that health is closely related to phenotype development, and it is, in turn, a functional response of the genome to the environment encountered. Although my treatment of this matter in this book is far less than exhaustive, it nevertheless leads to the conclusion that health must be at a maximum when the phenotypic expression of a species' diagnostic features are maximized. The environment that does this is the evolutionary environment of a species.

Is this conception superior in principle to that of the World Health Organization? I submit that it is. The desirable state of individuals may now be expressed quite precisely, as the maximum phenotypic development of traits that diagnose us as humans and make us uniquely so. Here is a list of our human diagnostic features that distinguish us from other species within our taxonomic family.

(1) Upright posture with concomitant anatomical peculiarities, such as the vertebral column having an S curve and the joint of the neck in the middle of the base of the skull.
(2) Legs that are longer than arms, also a consequence of bipedality, with concomitant differences in function of legs and arms.
(3) The hands prehensile with a large and strongly opposable thumb.
(4) Toes short, with the first toe frequently longest and not divergent.
(5) Body hair grossly reduced.
(6) Uniquely large brain with enlarged frontal, parietal, and occipital lobes in the cerebral cortex.
(7) Chewing concentrated on premolars with rotary motion, resulting in reduction in canines, recessed chin, and reduction in size of incisors. A further consequence of this is a short jaw with a rounded dental arch and a short vertical face.
(8) Canine teeth no larger than premolars and usually with no gaps in front of or behind the canines.
(9) Unusually high output of sweat and therefore unequalled capability for evaporative cooling.
(10) Unusual manual dexterity, in part a necessity or prerequisite for tool using.
(11) Unusual bodily dexterity, including commonly balancing on one limb; capability of mastering an almost infinite number of activities.
(12) The highest development of intellect.
(13) Vocal mimicry, a prerequisite for language and music.
(14) Language as the major form of communication.
(15) Music, including that produced with the aid of tools.
(16) Visual mimicry, a prerequisite for tool using and manufacture, dancing, sport, and role playing.

(17) Tool using, tool manufacture and the making of tools to make tools.
(18) Dancing.
(19) Sophistication of role playing to a unique extent.
(20) Capacity to act in blind extreme altruism.
(21) Possession of highly developed humor.
(22) The most plastic, diverse system of nonverbal communication.
(23) Extreme control over reflexes and involuntary behavior (self-discipline).
(24) An extended sexual receptivity and very large sexual organs.
(25) Dominance displays largely cultural.
(26) Transmission of very complex traditions.
(27) Exceedingly long ontogeny.
(28) A long life span.
(29) Menarch.

Some of these features are mundane, some are not. Most commonly a list of the morphological features are cited, as can be seen in Simpson (1969), while the physiological, behavioral, and life-historical ones are usually incorrectly and incompletely presented. I have attempted to lump the obviously related features together, but this is not always possible.

Next, we can search for and describe environments that will maximize these diagnostic traits. Even though we may not wish to test ourselves as to how close individually we have come to maximizing traits diagnostic of our species, we cannot withhold information about the environment that achieves it. To do so would be to deprive developing individuals of physical, intellectual, and social competence prior to their reaching adulthood. Do we have the right to impair anybody's development? The policy that avoids this is a policy of maximizing individual development.

I have pointed out that there are clues in the physique and behavior of individuals as to whether they have achieved a development that potentially maximizes health. Morphologically it is highly developed tissues of low growth priority; behaviorally, it is more difficult to be precise, but the manifestations are frequent play with bodily exertion, laughter, preocupation with and demonstrated mastery of many skills, in short, the ability to master problems well. Note that this is what theory predicts; a phenotype maximizing health is one maximizing the ability to deal with contingencies. We have no studies of growth priorities in the human body, nor do we know for sure that the tendency to explore, laugh, play, and do many things is indeed associated with superior health. Nor do we know exactly how to maximize environmentally the behavioral features that distinguish us as uniquely human. We shall have to await detailed studies on life styles as a factor in health. In the meantime, one has to look for short-cuts to verification of the theory, such as examining the adaptive strategies or life styles of people with exceptional phenotypic development as a means of developing a model of how to maximize phenotypic development. That model can be compared to life styles of people characterized by a poor phenotypic development. A detailed comparison of life styles related to human development is still to be made. Nevertheless, one can go a few steps in the process of verifying the hypothesis that life styles maximizing phenotypically the diagnostic features of a species also maximize health. Using constructs from the upper Paleolithic environment, in which we reached our ultimate form as a species, we contrast these against empirical findings of what factors promote health. We construct an evolutionary model of a healthful environment; this is necessary if we are to "design with nature," to quote McHarg (1971).

Evolutionary Model of a Healthful Environment: Its Rationale

In the foregoing chapters, I attempted to reconstruct the origins of human adaptations, as well as the "life style" of upper Paleolithic people, in order to generate a model of human environments that maximize health. If valid, the predictions from this model will coincide with empirical findings. To disprove the model is to demonstrate that it will lead neither to a reduction of disease and infirmity in individuals practicing it in its principle features nor to an enhancement of their physical, intellectual, and social development.

The law of least effort dictates that no more development of a body will take place than is necessary to maximize reproductive fitness. In short, *nothing is developed beyond need*. Furthermore, in the process of phenotypic development, behavior leads, followed by physiology and morphology. This means that *every morphological change reflects a very large amount of goal-directed activity*. Every trait that characterizes a form could *not* have been attained except by having been so important as to maximize reproductive fitness. Therefore, in a human population in which individuals are exceptionally well developed, we are likely to find a life style that—literally—makes one human. One can also conclude *a priori* that such a life style is characterized by excruciating effort on the part of individuals and that for every maximally developed tissue there is a maximum development of related functions. The athletic bodies and exceptionally large brains of people from the upper Paleolithic are therefore highly significant, granted the above perspective from evolutionary biology. Their life style ought not only to maximize health but also reveal how one can enhance the very essence that makes us human.

What are the virtues of an evolutionary model of human life styles that maximize health, or of a theory of health such as proposed here? First, it greatly enlarges the horizons of positive health measures. A scrutiny of the Lalonde report (1972), for instance, shows that two positive health measures are of concern: Good nutrition and exercise. Contrast this with the investigations of how to maximize the many human diagnostic features. Nutrition and exercise are almost lost in the multitude of things that could be done to promote good health. Humor, laughter, dancing, intellectual games, learning how to deal with others, promoting family unity, and many other activities all become agents of good health—so the theory dictates.

Second, the model, if valid, gives a standard against which one can compare existing and imaginary environments and spot likely sources of mismatch and trouble. However, one cannot specify the effects that the deviations will generate; one can only predict an impairment of competence and health. The deviations will probably lead to some underdevelopment of the genetic potential of affected individuals, and this, in turn, may result in shorter life expectancies, greater morbidity, social friction, weak physique—in short, in some cost to the individual and almost certainly to society as well. Even if no disease results, leading a life different from that dictated by the model can be done only at a higher cost of remaining sane and healthy.

Even though the evolutionary model does not spell out the consequences of deviations from its dictates, clearly it permits one to ask and test relevant scientific questions and therefore proceed efficiently in investigating factors that maximize health. In cases when decisions must be made by professionals when relevant facts are missing and an educated guess has to be made, the model can give better guidance than arbitrary economic or political notions that have no basis in human adaptations. Moreover, it permits the professionals or decision makers, such as managers, bureaucrats, or politicians to ask relevant questions of specialists. The evolutionary model can be used to generate quality of life indices that do not depend, as do those used today, on the criteria of user satisfaction. The latter is a dangerous criterion unless used knowledgeably; more of that later.

The model introduces into decision making, quite inadvertently, a time perspective not considered relevant today. Since our very existence, let alone our future, depends on the decisions of our controlling elite, it is they who ought to have not only our short-range, but also our long-range future in mind. A familiarity with the evolutionary model, as well as with the development of human adaptations, must bring the decision makers face to face with the reality that our planet is not stable, and that one cannot plan for a steady state. Massive climatic changes are perfectly natural, natural catastrophies on a small scale, like floods, fires, avalanches, storms, volcanic eruptions, etc., rejuvenate ecosystems, while periodic glaciations on a large scale rejuvenate large areas of the planet and cause enhaned fertility in the old glacial and periglacial zones during the warm interglacials. What effects magnetic reversals have we do not even know. During a glaciation, human life may depend in part on the ability of diverse grazing animals to convert the vegetation adapted to the violent climates of the glacial period into food fit for human beings. Therefore, the conservation of the greatest diversity of plant and animal species is needed to maximize our species' future options. We must also learn, as our distant Paleolithic ancestors evidently did, to keep well below the minimum carrying capacity of our land, rather than try to create a steady state based on an energy-intensive technology that makes our very existence as a species highly precarious. We are probably in for all sorts of surprises from planet Earth, and getting to know human evolutionary history does help to appreciate the earth's instabilities and their consequences.

This model confronts decision makers constantly with the fact of human phenotypic plasticity *vis-à-vis* the decision he is empowered to make. I suspect that it must also dampen euphoric optimism about science and technology, an optimism rooted in the belief that we can readily and cleverly improve upon natural systems. It must also confront the decision maker with the realization that, if one does not act on the model, a variety of economic, political, and technological forces in our society will shape the human environment blindly. Intervention to keep an environment fit for humans is thus not an option but a necessity. Should a decision maker opt for decisions other than those maximizing the physical, intellectual, and social competence of individuals, he is trading off a part of their humanity for some other goal. Let it be quite clear that he or she is then willing to sacrifice a part of the potential development of persons other than himself or herself, their health, longevity, and competitive abilities, for something else, whatever that may be.

The evolution of humans as I described it also introduces a somewhat different perspective to the old nature/nurture argument of concern to academics and professionals alike. In his evolution, *Homo* progesssively dismantled innate controls over his behavior or overode these with an increasingly powerful corticular control system, evidenced morphologically by an increasingly larger cerebral cortex. This new corticular control is so powerful as to override reflexes and, in extreme cases, permit the willful destruction of part of a person's body, and that without the person betraying a sign of agony. However, beneath the cortex rests a store of innate motor patterns, tastes, and aversions, and they can erupt into overt behavior if the cortex permits or if its control vanishes. Humans, thus, are not devoid of innate abilities, but control them to an unusual extent.

If we grant this view, then the cortex can be viewed also as a servant of tastes and emotions that once required ever more refined, diverse, and complex strategies and tactics for satisfaction in the difficult environments of the north. The appetites, aversions, and emotions need change little over time, but the manner of satisfying them and the dictates of the body for homeostasis, sex, or ego satisfaction become more complex as a consequence of adapting to the harsh northern environments. As a class, these complex corticularly controlled means of satisfying biological ends can be termed *culture*. Cultural strategies of resource exploitation become increasingly prevalent. These, however, do mimic biological strategies of resource exploitation. This, in turn, results in strikingly similar adaptive syndromes in animal and human societies, as Wilson (1975) has pointed out. This should not be surprising. The logical relationship of activities in adaptive syndromes based on cultural or on biological mechanisms is, of course, identical; the manner in which they are altered is not. Granted the multitudes of adaptive syndromes of resource exploitation that humans are capable of, whether they are peaceful or warlike does not depend on their genes but on their adaptive strategy of resource exploitation. In short, much of human behavior is not an expression of genetic traits but of culture, and it, in turn, is primarily a product of a population's ecology plus cultural lags and the expression of rules, some of which I discussed in Chapter 5.

Normative Environmental Criteria

What kind of an environment does the evolutionary model predict will maximize the individual's health, as a byproduct of maximizing its physical, intellectual, and, very likely, social competence?

The Social Milieu

The social milieu is formed of an extended family—or its equivalent. The extended family is visualized as being comprised of three generations, the nuclear couple, 2 to 3 siblings spaced 4 to 6 years apart, plus 1 or 2 adults of equal ages or older than the nuclear couple. The extended family should be tied socially to 4 to 6 peer families, to form a social network comprising about 25 to 30 people.

Note that in this structure there are only small groups of peers, few of equal age, and children develop in a group of people composed of predominantly older individuals with few equal-aged companions. This should lead to a low level of aggression among children (see p. 420), maximize supervision by adults, provide maximal opportunity for tutoring, and therefore for superior intellectual development (see p. 370), and bring the child daily into contact with diverse roles based on sex and age, a prerequisite to developing social skills. This mixture of persons aptly demonstrates to the growing child the fate of humans at first hand, from birth to maturity and death.

The extended family, by providing a good social support system, ought not only to protect gestating women and the fetus (Nuckolls et al 1972) and to stimulate good ontogenetic development in children but also to retard senility in the old. The older people, through the activities of the young, ought to be stimulated intellectually and physically with a low frequency of stress-generating inactivity in the daily routine. We are aware that stress and the lack of intellectual and physical exercise accelerate aging (Bourlière 1973b). What we do not know is how much stimulation, and of what kind, promotes optimum development and physical health. The extended family as a model gives us only crude indications. That a low level of social relationships is detrimental to health is discussed by Teele (1970) and in a review by Rabkin and Struening (1976); for the effects of crowding that generate conditions beyond the control of the individual see the discussion by Irwin Altman (1975). Correlates of life in the extended family are a relatively low level of problems related to mental illness and crime (Kyllonen 1967). This finding in turn harmonizes with that of Galle et al (1972) that the formation of delinquent gangs is associated with a high degree of autonomy by the juveniles, in short, with a slippage from enforced rules of conduct. The role of the functional family in preventing sociopathology is discussed by Belfer and Brown (1973).

The extended family can also function in such overlooked areas as the passing on of reasonably sound food habits, acting as an antidote to diets based on haste, diet fads, and television commercials. The young individual is much more likely to acquire sensible food habits here than in a nuclear family, particularly one in which both parents work and children receive minimal supervision. In general, the progression from extended to nuclear to single-parent family increases the workload of child rearing on the adults, increases worries and frustrations, reduces social contact, and increases the chance of suboptimum development of the children. Nevertheless, it cannot be claimed whatsoever that life in an extended family is undemanding and blissful.

Nutrition

Nutrition should be characterized throughout by a high intake of protein, mainly animal protein, a low intake of fats and simple carbohydrates, and a high intake of bulk and fiber. The vitamin and mineral intake ought to resemble that of people subsisting largely on animal products. It also should include a high proportion of living plant or animal tissues, and a low intake of salt and an absence of molecules normally not found in nature. Neither food nor water nor air ought to include molecules other than those found in man's natural environment. This, however, implies that pollution of air with wood smoke and food sprinkled with wood, bone, or fat ash, as found around campfires, is not harmful. The evolutionary model of man does not dictate "clean" air.

The empirical findings about human nutrition are confounded but, on the whole, do support the above dictates. Nutrition is riddled with fads and subject to claims and counterclaims rooted mainly in faith, not in science. The evolutionary model dictates abstention from foods with artificial molecules such as may be included in industrialized foods (Hall 1974), and it clearly leans away from vegetarianism as a means of maximizing ontogenetic development. This is not to deny the value of sophisticated vegetarian diets that do, with minimum addition of animal products, supply abundant nutrients for ontogenetic developments. Nor must the model obscure the fact that our meat industries may offer us health hazards for consumption by selling meats with heavy loads of fat and preservatives.

The importance of good food habits can again hardly be overemphasized for reasons discussed in this book and many others. It promotes both good physical and intellectual development and good health, and it also functions subtly. Thus, tall, well-developed women tend to take more skilled jobs, marry men with more skilled work (Young 1971, p. 242), be more efficient in reproduction because of better lactation performance and a higher percentage of milk fat, as Cuthbertson (1972) points out, and they face fewer difficulties at childbirth (Thomson and Hytten 1973).

Physical Exercise

Physical exercise must include from the earliest age possible the development of diverse motor skills, using many combinations of body muscles and limb−eye coordination. It must include activities enhancing manual dexterity and exercise forcing the body to exertion, causing rapid breathing and forceful heartbeats. Such exercise ought to be performed in part on the initiative of the child, be it in lone exploration or in social activities in part under the guidance of older individuals knowledgeable of skills and able to set a standard of performance by example. Some exercise should be a part of cooperative tasks, and performed in an atmosphere of laughter and music. Humor, merriment, laughter, cooperation, and exercise ought to be performed daily. Ideally, the regime of physical exercise should be part of the normal daily work regime, enhancing physical fitness on one hand and, on the other, maintaining and improving the individual's knowledge about and skill at tasks required for daily life. Small group processes ought to dominate over tasks performed in isolation or in the anonymity of a crowd. To be meaningful, the regime of daily physical exercise must give an opportunity for the display of con-ventional skills in front of others, and standards of performance must be known so that the performer may gain some recognition for his or her efforts. Exercise in the form of jogging is not enough!

The above predictions of the evolutionary model are amply verified by exercise physiology as health promoting, as well as by studies of the effect of music on humans (see Chapter 14). The beneficial effects of exercise on retarding aging are discussed by Bourlière (1973b). Some predictions are untested to my knowledge: Laughter and humor ought to be health promoting, as should the recognition by others (ego satisfaction) of skills performed well. The above sets criteria for tasks to be performed, but does not specify the tasks themselves.

Intellectual Exercises

Intellectual exercises must be frequent daily tasks, including the practice of lan-guage skills—such as rapid scanning for appropriate words, analogies, similes, or metaphors—the practice of musical skills, vocal and instrumental, exercises in strategy and tactics as practiced in discussions (which are social contests), games, sports and exploration, alone or in groups, and in creative activities including art.

This requires daily contact with tutors, performances by individuals with sea-soned intellectual skills, in order to set levels of expectation, social rewards for intellectual achievements and self-discipline, and an occasional change of physical and social setting to stimulate the individual's curiosity, confront him with new problems, force the development of new or rare social skills, enlarge the knowledge base through broadening of experience, and thus develop competence in dealing with the novel and the strange. In the Paleolithic a regular rhythm of movement from seasonal camp to seasonal camp provided the repeated impetus to practice old skills and knowledge and develop new ones. Exercising the intellect for its own sake should, therefore, be health promoting and thus of intrinsic worth.

The relationship between intellectual development and health is not as clear-cut and simple as one might wish, yet intuitively one suspects that it must be present. There is some evidence that it is. Studies by Stott on mentally retarded children showed that they had a higher incidence of illness than normal youngsters (Joffe 1969). Conversely, studies of exceptionally gifted individuals showed them to be a rather healthy lot (Terman and Oden 1959). One can develop health-promoting habits through intellectual insight and the use of reason; one may also gain it on command from others, including the subtle form by an appeal to reason. Maximum intellectual development is a precondition to acquisition of maximum health-promoting habits by individuals.

Education

Education ought to be based on face-to-face instruction, that is, tutoring, in which the peer groups of juveniles should be very small, and always smaller in number than the group of adults. Under tutoring, maximum attention is given to each developing individual with the greatest economy of instruction. Tutor and child ought to know each other well, and that requires a long association. Nor must the tutor frighten the child. Learning should be based on play mimicking adults, which of course requires that the actions of adults be comprehensible to adults and accessible to the observations and questions of children. The children should be able to participate in or observe economic activities, ceremonies, and festivities cherished by adults. Children should have ready access on the one hand to nature and on the other to man-made environments.

The diagnostic feature such education ought to develop is the increase in size of the cerebral cortex in the young, and the maintenance of a well-functioning cerebral cortex in the old through regular exercise. That tutoring is an excellent way to enhance intellectual development and learning there is little doubt. It is also an excellent way of retaining a sense of importance and dignity for the old, and contributes to their physiological and mental health as well. The latter is certainly implied by studies of Timiras and Vernadakis (1972) and Bourlière (1973b).

Keeping Out of Doors in Natural Surroundings

For growing, developing individuals, the out-of-doors serves as a teacher. They become cognizant of the passing of the seasons, the phenomena of weather, the diversity of living thigss, the becoming, growing, declining, fading, dying, decaying, sprouting, blooming, fruiting, coming and going of plants and animals, the texture of natural objects and materials and their properties and potentials, the sensations of their bodies experiencing the tactile, proprioceptive, and sensory stimuli, and experiencing an exercise of emotions from arousal to inspiration and fright. Nature provides *opportunities* to do things, make mud pies, taste berries, throw stones, whittle sticks, and it also provides *privacy* where silly things may be done without peer pressure or interference by elders. Leyhausen (1971) points here to the work of Klimpfinger[2] who showed that in peer groups of children in kindergarten the hostility of the children normally shown (Hutt and Vaizey 1966) was diminished if privacy was granted daily. There is also a positive relation between privacy and intellectual development implied by Chance (1969).

In becoming cognizant of nature, children can develop language skills; in particular, they enrich their store of similes, analogies, and metaphors; they have motivation to talk, harmonize with others on common experiences, and dwell on the entertaining differences. In the process they gain knowledge, as well as intellectual and social skills. Fernandez (1973, 1774) and Geist (1975c) give further account of the importance of nature in language and personality development.

Individuals will gain some of this also in man-made environments by being exposed to the means of production, the customs and rituals of trade and ceremony, seeing the stages of development of diverse enterprises, acquiring the range of conventional skills they may be expected to practice some day, observing the actors in human affairs from the mighty to the pauper, learning the functions, thrills, and dangers of man-made environments, and by this exposure gain an understanding of their own society. Diversity of environment and interacting with it enhances intellectual functioning, and this has been known for some time (Hebb 1966); it builds competence and confidence (Wohlin 1972).

The immediate effect of this would be to phenotypically enlarge the cerebral cortex of the brain, provided the social milieu and nutrition permit it (see p. 369−370). The secondary effect is an increase in physical skill, robustness, and size of body due to the influence of diverse activities.

Yet a daily long exposure to natural environments will probably go further. Variability in the physical milieu stimulates body systems, helps to exercise them, and thereby promotes their effective function and health (see Heiss and Franke 1964). The effects of artificial, as against natural, radiation on diverse physiological processes, as described by John Ott (1973), is at the very least thought provoking, and if confirmed, thoroughly frightening. The ready availability of small air ions in natural landscapes and their depletion in urban environments and buildings argues that the natural environment is best for the individual's health and development (Krueger and Reed 1976). The evolutionary model predicts that a large portion of the day spent out of doors, rather than in the ''caves'' we so steadily occupy, should enhance individual development and health. I see little to disprove this prediction.

The Emotional Regime

There must be security; first, material security and freedom from want of good food and shelter. How to obtain food of high quality and quantity must not be an insoluble problem. There must be the security of solid, dependable, and lasting social bonds among the adults raising the children. This requirement can hardly be overemphasized. Not only does it provide help, comfort, and encouragement to the children and create a thoroughly predictable social milieu, it also generates inadvertently a set of values to emulate and live by; stable, lasting, social bonds within and between families set roles and expectations for children of how to behave as

[2]S. Klimpfinger, *Kinderstudien I and II,* 1952, Film der Bundesstatlichen Hauptstelle für den Unterrichtfilm, Vienna, Austria.

grown-ups. This generates the next requirement, namely, a clear set of social values and the concomitant rules of expected conduct. This must be combined with acceptance and encouragement of intellectual, social, and motor skills—activities that generate identity for the developing individual and a sense of self-worth that give substance to the ability to harmonize with others on common experiences, common knowledge, common skills. We are dealing here with a very fundamental aspect of the human organism. There must also be security for family and self from depredations by other persons; there must be no fear of others attacking and destroying oneself and those one cares for. Only then can the individual regard others not as a threat but as a source of interest, learn to interact with them without the hindrance of excessive prejudices, and develop social competence by dealing with new social situations.

It cannot be stressed enough that individuals require emotional security, which in large part they gain by being competent masters at solving daily problems. Emotional upsets tend to impair the immune system and this, in turn, permits pathogens and apparently malignant tumors to spread, as well as allowing disabilities to develop (Hinkel 1965, Seligman 1975, p. 180−1). It is not at all surprising that individuals with a very high IQ tend to live exceptionally healthy lives (Terman and Oden 1959), as we suspect they are usually masters of the situations they find themselves in.

From the study of how the nuclear family arose and bonding ensued between male and female, it appears that maximizing sensitivity, alertness, humor, and playfulness in adults is a prerequisite to maximizing bonding between partners, and thus family stability. Granted that in the course of human evolution there was a change from the nuclear to the extended family, and that there was a change in the female's role in that she bonded the male by providing an increasing amount of essential services as ever more hostile environments were settled, then—unfortunately—the biological foundation for a stable *nuclear* family in modern man is weak. This conclusion is based on the premise that essential services bonded partners in much of the history of *Homo sapiens* and that there was a concomitant decline in sex and humor as bonding mechanisms. Therefore, the provision of essential services by partners to each other appears to be necessary to strengthen family bonds.

Control over the Environment

The foregoing criteria are such that the individual gains competence and remains continuously master of his or her environment. This point cannot be emphasized enough. By developing diverse physical, intellectual, and social competence, the individual can at once visualize a choice of alternatives to problems, see problems well before they arise, harmonize with a diversity of persons, or with one person in a great diversity of moods and situation; such an individual can feel empathy with others relatively frequently or detect clever selfish strategies by others; he or she can maximize predictability in the social milieu and in the physical environment, but by continuously learning that he or she can exercise control through correct anticipation of, successful preparations for, and execution of tasks the individual remains in high morale and health. Such a person remains physiologically at the peak of possible fitness, is mentally alert, and has a positive attitude toward life with a belief in himself and others, and the immune system remains at maximum efficiency since such persons rarely fall victim to infections, have less chance of developing cancer, have fewer chances of becoming depressed, having accidents, or dying prematurely. We now know that to maintain a person at the peak of health, he or she must have expectations commensurate with his or her abilities so that expectation and actuality can meet and reinforcement—through pattern matching—can take place. There must be the ability to exercise choice, gain information or tools to solve problems, retain access to others to harmonize and to explore, and thus experience the continuous thrills of little discoveries; there must be some breaks from daily routine, and access to peers. Help can only be help if it provides tools, skills, examples, or knowledge for solving problems, but not solutions to problems. One does not help by giving answers to difficult questions, but only questions that lead to answers. Health is generated by individuals under the care of others if they help themselves, remain in social groups, and retain the belief that they are masters of the situation—not an object condemned to helplessness. For an elaboration of this important criterion, I point to the excellent book by Martin Seligman (1975).

The foregoing dealt, in part, with what has been termed "free will," which is, in essence, the ability to choose an alternative that reason would counsel us to reject. We tend to act with enlightened self-interest, except where ego satisfaction makes us act to the contrary, for whatever reason. Yet the option to oppose self-interest is vital for the individual's well-being and health as it is that which generates the sensation of mastery over oneself and one's milieu.

The above model is but a sketch. Much of the discussion justifying it is found in the preceding chapters. In its incompleteness it may hide functions as yet unrecognized as relevant to individual development. Yet, incomplete as it is, the reader is invited to contrast the conditions for a healthy environment, as dictated by the evolutionary model, against environments humans live in today. Whether the evolutionary model is valid or not is a testable proposition. There is a great deal of evidence in favor of it, as marshaled in the preceding chapters, and there is almost certainly a wealth of information in its favor that I am not yet aware of, nor ever will be. The crucial test of its validity is the extent of development of tissues of low growth priority, the incidence and frequency of behaviors not directly related to maintenance as an indicator of exuberance, and the incidence of diverse illnesses. The first two criteria have not yet been developed for humans, and whether they ever will be obviously depends on public interest in a policy of maximizing individual development.

How does this model fare with a comparison to textbooks from preventive medicine? It overlaps and addresses in great part issues other than those addressed by conventional preventive medicine. The latter deals extensively with epidemiology, with the control of viral, bacterial, helminthic diseases and their vectors, of pollution of air and water, sanitation, the control of food for deleterious substances and organisms, occupational health, public health and health support systems, hygiene, and accident prevention and specific diseases greatly affected by environments, e.g., cancer, cardiovascular diseases, dental health, etc. It does not explain to how health should be maximized in individuals through environmental controls during ontogeny. Where it does delve into such factors as nutrition it is an empirical science (Simpson 1970, Sartwell 1973, Wallace et al 1973). The proposed model is thus an enhancement of preventive medicine in that it gives theoretical guidance and is quite explicit in delineating criteria for positive health measures. The evolutionary model addresses itself to the origins of ill health by setting normative criteria for environments that maximize health.

The criteria for maximizing health are also similar to some developed by psychiatrists. For instance, Duhl (1976), despite a different approach, conceptions, and phrasing, articulates conclusions similar to those expressed here. Since psychiatry is an empirical discipline with many of its practitioners keen systematic observers deeply interested in how humans function within social settings, the congruence of conclusions reached may be understandable.

It must be emphasized that the kind of life reflected by the evolutionary model is not one of blissful ease. It is not utopian in the sense of promising some glorious carefree future if we but follow it. The evolutionary model only makes the promise that health is wrung from the environment at the price of toil, sweat, and tears. To develop into a human with full capabilities and health is an arduous task!

Let us now turn to the insights about our environment that can be generated from some of the principles of evolutionary biology rather than the evolutionary model.

Quality of Life

We noted that the evolutionary model identifies as high quality of life that which maximizes phenotypically the diagnostic features of our species. More mundanely, by maximizing the physical, mental, and social competence in individuals during ontogeny, health is also maximized. Unlike the present attempts at measuring the quality of life[3], the quality of life indicators dictated by the evolutionary model should not measure user satisfaction but physical, behavioral, and social indicators such as anthropometric measurements, physical strength, dexterity, verbal fluency, memory, mathematical comptence, musical ability, longevity, diseases indicative of stress, and the extent of social networks. While the evolutionary model permits one to bypass quality of life indicators of an *ad hoc* nature that have no theoretical validity, the concept of "creature comforts" argues *against the notion that emotional satisfactions, discussed below, are a valid guide to our lives in a man-made, highly artificial environment*. It argues that in artificial environments, such as we occupy, severe intellectual control over our actions is essential, since our emotional responses need not be valid guides to the utility of our actions. Furthermore, it can be shown that user satisfaction is based both on sensations once useful in some past environment and on societal dictates that serve the survival of a given economic strategy, on traditions forming the conventional wisdom we grew up with, as well as innovations in the rules of conduct that serve the ego satisfaction of an individual. Therefore, it is dangerous to identify "user satisfaction" as the sole criterion of the "quality of life."

In our daily lives, we are bombarded by the urgings of emotions and sensations from within our body. Most we follow; some we ignore; a few we willfully overcome. We obey some impulses despite good reason not to. We master aversions or appetites because reason tells us we ought to in the interest of one cause or another. Some of these internal messages I should like to label "creature comforts," which will become apparent as we go on. How do these internal messages come about? What is their nature?

---

[3]*The "Quality of Life" Concepts,* a symposium, Environment Protection Agency, US Govt Printing Office, 1972; *Quality of Life Indicators,* a review of the state of the art and guidelines, Environment Protection Agency, US Govt Printing Office, 1973.

Creature Comforts

As was pointed out in the earlier chapter on Cognition, a body must possess monitoring mechanisms to maintain homeostasis.

(1) There are those which maintain control over physiological processes such as regulating the intake of food, water, air, and nutrients, which signal stimuli damaging to the body such as mechanical forces, heat, and cold, and which signal excessive energy expenditures or the need for rest and assimilation of ingested material. Here we are dealing with sensations such as hunger, thirst, choking, pain, craving, tiredness, sleepiness, exertion, or exhaustion. All these are dominant sensations, to ignore which is fraught with danger and which it is very difficult to suppress and control consciously.

(2) A second set of internal sensations can be ascribed to monitoring mechanisms that position the individuals in space, time, or in situations to minimize the cost of maintaining homeostasis. They are due to the monitoring of the external environment. We deal here with the sensations of arousal, fear, and panic, but also with joy, curiosity and the "eureka effect." These sensations, if followed, produce behaviour conforming to the notion that individuals prefer a predictable, responsive environment with an optimum level of arousal.

(3) A third set of internal sensations deal with signaling the degree of matching between visualized and actualized concepts plus signaling the more profitable of two or more alternatives. We have to postulate here not only pattern matching as justified in Chapter 2, but also a comparator that somehow signals which of several alternatives to follow. The internal emotion experienced are those dealing with ego satisfaction. They are desire, yearning, longing, anger, hate, rage, the sensation of a letdown or depression after a slight, insult, or reprimand, or the sensation accompanying laughter in humorous situations. We appear to be dealing here with a monitoring system that is essentially open-ended; it continually generates new situation-dependent goals (patterns) which the individual then attempts to satisfy. This generates the phenomenon of continual self-assertion or self-actualization.

The emotions we term sexual appetites appear to be best placed close to these monitoring systems for they very clearly depend on the operations of the postulated comparator, as well as on the ability to repeatedly satisfy newly posed goals. That is, sexual sensations can hardly operate without internal signals that signal success in ego satisfaction as well as the sensation of identifying the most desirable mate.

We are not entirely uninformed about the operations of the proposed comparator; we identify it at work in the ethological studies on superstimuli. Note the experiments, for instance, in which oyster catchers were presented in different experiments either with an egg larger than their own or with a clutch size larger than their own, and the birds chose "larger" or "more" repeatedly (Manning 1967). Note how unadaptive the response of the birds was to an artificial stimulus not encountered in nature. It suggests that there is no limit on "desirable" alternatives, at least in some cases.

The foregoing sketch served to illustrate the nature of our internal sensations, which call for satisfaction. Clearly, these are not sensations maximizing the individual's ability to live a healthy, long life, but rather sensations that help maximize an individual's reproductive fitness. They are sensations that instruct individuals to follow the rules of life derived from the concept of reproductive fitness described in Chapter 1. Clearly, these are sensations based on the amoral process of the game of the genes, and if this is so then theoretically the well being of the individual is quite secondary to his relative reproductive success. To listen to and obey our internal sensations blindly must, therefore, produce behavior that is highly egocentric and hedonistic, and certainly not behavior that maximizes health. Yet—and this is equally important—to be thwarted in one's wishes is certainly the first step in generating stress, and maybe in reducing health. If following one's urges does result in the individual mastering his milieu, it is health promoting.

Nevertheless, the goals of reproductive fitness are not incompatible with the goals of good health. On the contrary, in a natural environment—so evolutionary theory would dictate—following internal sensations ought to be a reasonable guide to living in safety and comfort. Pleasure, by and large, can be equated here with the useful, aversion with the harmful. Together they permit the adaptive shaping of behavior in a variable environment. We can rightly assume that our internal sensations are the product of a long evolutionary history and that they exist because of their value in maximizing reproductive fitness. Therefore, listening to our internal sensations, or creature comforts, is important to survival and reproduction in some past natural environment. In an *artificial* environment, this is no longer true.

In an artificial environment we are faced by dangers, not only from those stimuli we sense, but also from those we do not sense, as we have not adapted biologically to do so. We have no defence of a biological nature against DDT, high concentrations of heavy metals and carcinogens, harmful antibiotics, and diverse industrial chemicals in our food, water, and air. It is only small comfort to know that various microorganisms and arthropods with short generation times have already adapted biologically to some of the artificial chemicals we have introduced into the biosphere. We cannot detect with our senses radioactivity, radiowaves, and X-rays, but must rely on technological extensions of our senses to detect, monitor, and protect ourselves from them. We continue to produce myriads of these environmental factors that we are not equipped by nature to detect, and consequently we require all available means of detecting them and keeping ourselves free from harm. The factors listed are rightly the concern of environmental scientists (Benarde 1970, Eckholm 1977), and we have little option but to accept their findings and recommendations if we care to protect our bodies from these insidious agents.

Another class of stimuli are readily detected by our senses, found highly pleasing, and are readily indulged in. However, since these substances are rare in nature we have not evolved biological mechanisms to protect ourselves from unhealthy overindulgence in them. Excessive intakes of sugar, salt, and fat, and an excessive indulgence in leisure come to mind at once as harmful, as are also various intoxicants, as discussed by Jones (1974). Sugar, salt, and fat, as well as leisure, are by and large rare and precious to a human in a natural environment; one can hardly get enough of these under natural conditions, and it behoves us to maximize the intake of the substances named, or of rest, whenever it is possible. Since concentrated sources of sugar, salt, and fat are normally hard to come by, we have not evolved a dependable shut-off mechanism akin to to that which prevents excessive intake of a very common substance-water. We tend to drink only as much as we have lost and thereby restore the water balance in our bodies.

In an artificial environment we can also indulge in the use of labor-saving devices, maximize leisure, and indulge in junk foods harming the body thoroughly in the process (Heiss and Franke 1964). We can shut off or remove ourselves from unpleasant sights such as sick and dying persons, or from unpleasant encounters or strenuous confrontation with individuals closely related to us. The outcome of this is self-evident: We escape from developing our bodies, we insult them, and we fail to develop maximum intellectual and social competence. We can, by becoming obedient servants of creature comforts, suffer loss of development, character, and humanity.

The dominance of creature comforts in combination with the opportunism of the market place spawns myriads of "social services" which cater to the creature comforts more of the dominant individuals in society than of the subordinates. The hospitals and old-age homes can be interpreted as being for those who are unwanted, not simply for the service of those not able to help themselves. In addition, the artificiality of our environment spawns disabilities in individuals (Ford 1970, Lave and Seskin 1970), and the market place or the social service agencies respond with support services for the afflicted. The crisis centers and psychiatric care units, as well as nurseries for preschool children and schools themselves to some extent, fall into this category.

Once social service institutions are formed, those employed to supply services have a vested interest in the existence and growth of their facilities. Conversely, the dominant individuals in our society find their creature comforts satisfied by *not* having to care for destitutes, *not* having to be bothered by children, *not* having to reduce the demands of their egos that are satisfied following some dictate of their profession, hobby, or social aspiration. Moreover, we can be quite certain that the *law of least effort* will always ensure that facilities catering to creature comforts of the dominants will attract customers; we see an expression of this in *Jarvis's law* that people in vicinities of medical institutions send more patients there than people residing further away (Shannon and Dever 1974, Chapter 6). In addition, the ego demands of the providers of services ensure that they will lure in customers, so that their service will be full, the degree of disability becoming a secondary matter.

We can expect—alas!—that in institutions "serving" the public, the "services" will be managed in such a fashion as to first satisfy the creature comforts of the "servants." It is here that the accumulation of little transgressions can avalanche into bureaucratic atrocities. Such is the inevitable outcome of conduct based not on the sometimes painful dictates of reason, justice, or principle, but on the emotionally satisfying "easy way out." Here lies a great human tragedy, that of the immature, insecure, or uneducated individual who is subject to the remorseless workings of the innate ego that seduces his reason into rationalization, rather than being subject to and disciplined by reason. It follows that the "servant" in his domain will try to act as the dominant and superior *vis-à-vis* the institutionalized individual. A condescending insulting attitude can be the consequence unless confronted by a person with a strong character. It also follows that institutions may in fact conflict with their primary goal, such as hospitals *increasing* morbidity by generating helplessness in patients, robbing them of control over their surroundings, reducing predictability so vital to any organism's health, robbing them of morale, and hastening, or at times causing, death. Here I refer to a most enlightening discussion of this subject by Seligman (1975). This author also reviews the experiences from orphanages in which management was structured to conform to the wishes of staff rather than to the needs of children, with nightmarish results. If this concept were invalid would Sommer (1972), for instance, have stressed the need for hospital staff to simulate the experiences of patients as generated by the treatment they received and facilities of the hospital?

It is inevitable that institutions spawn employees who attempt to dominate those they serve, and if they find no resistence to their actions they develop a condescending, callous, domineering behavior, which in extreme cases graduates into the behavior of excesses well known through that of guardians and administrators of concentration camps of Germany in World War II, or those in Russia described in Solzehnitzin's books. Secondly, institutions serve the creature comforts of their employees first, and this, too, grows into excesses if not resisted. Thirdly, to satisfy the creature comfort of ego satisfaction by becoming prestigious, administrators of social service institutions aim at increasing their budgets, a measure of importance in our society. Fourthly, if institutions do meet resistance from those they serve or supervise, they turn into servants of those supervised, of the special-interest group, and give symbolic assurance only to the public which foots the bill. This is a most important point, repeatedly observed and fully recognized by social scientists, as has been pointed out by Crowe (1969). On all four points, the employees of a bureaucracy, a social service institution, follow the dictates of creature comforts. These are expressions of our animal ancestry, not of our humanity and culture.

Display as a means of ego satisfaction is readily fostered by the free market, and energy and materials are diverted in exceedingly great amounts to that end. There are industries that cater to the visual and olfactory enhancement of our biological display organs, permitting those who have the wherewithal a unique sophistication of decor, dress, and make-up far beyond what individual effort could produce. Consumer goods are primarily a product of creature comforts, and their first function often is as ego displays, their second for specific work to be done, and their third to create a tickle of excitement and maybe a saving of labor. The car dealer, the merchant in furniture, weapons, sporting equipment, and delicatessen, the operator of massage parlors, and the friendly pornographer—all provide services catering to creature comforts of diverse kinds. Creature comforts, the guardian angels of environments past, turn quite easily into the opposite unless they are rationally controlled.

However, we are not only subject to basic biological drives but also subject to the rules dictated by society. We are readily subject to such rules because we have such a powerful cortical control over our behavior; self-control is the legacy of the dispersal phenotype, as I have repeatedly stated in this book. We have also been structured by natural selection to have great faith in culturally transmitted rules (see p. 241). Societal rules, I argued, are a product of the economic strategy of a society. They are also subject to a lag in the retention of old cultural values, traditions that may conflict with the existing economic strategy. In addition, the drive for ego satisfaction ensures opposition to the rules of conduct, be they rules relating directly to our economic strategy or to the conventional wisdom we grew up with. Clearly, humans are creatures condemned to perpetual internal conflict.

Please note that the rules generated by an economic strategy, or by tradition, or by ego satisfaction in no way need be compatible with good health. In a society, prevailing myths may very well serve a chosen few who control society, as Stover's (1974) work on the cultural ecology of China convincingly illustrates. People can be, and are, conditioned to desires and behaviors that do not serve their interests at all. The rules of society serve to maintain a society, not the individual. Thus, listening to creature comforts or to society's dictates is not the road to an environment that maximizes health, nor necessarily to one enhancing the quality of life.

The dominance of creature comforts, plus the damage they can inflict on individuals and the turmoil they can cause in a highly organized society, make it imperative that some of society's codes of conduct deal with the control of creature comforts. Religions, customs, moral codes, and laws contain the means of controlling them. How tough these creature comforts are is betrayed by the complexity of the social institutions prescribing behavior, as well as by the diversity of means to enforce them—the jails, the instruments of torture, the sophisticated means of execution, the public rituals of administrating shame or coercion, and the symbolic degradation of transgressors even after their death by corpse mutilation and banishment from sacred soil. The "original sin" one can identify as creature comforts. The more artificial the environment, the greater the civilization, the more they must be controlled lest their free exercise leads to a weakening of the community, making it less ready and capable of defending itself or preying on others, and thus becomes prey to neighbors better able to control the creature comforts of their citizens. Clearly, the closer a cultural adaptive strategy approaches tha used by humans in a natural environment, that is, living by hunting from undefendable resources, the less control is needed over creature comforts.

Yet creature comforts need to be satisfied, even in civilizations. They are expressed in festivities in which conduct is condoned that would elsewhere be unbecoming, such as gluttony, drunkenness, clowning, insulting of superiors, and open flirting, and even promiscuity. Civilizations may institutionalize the satisfaction of creature comforts, so that the logical companion of moral codes and tortures must be festivals and bordellos. It also follows from the foregoing that the class most dominant, or the sex most dominant, will by and large be able to satisfy their creature comforts more frequently, use this as a symbol of their rank, and create a double standard, One can thus interpret civilizations, in part, as a means of controlling creature comforts of the citizens while making them gain dominance over other people and safeguarding means of economic production toward greater dominance. Since creature comforts can be expected to be much the same in different people, it need not surprise us that great civilizations shared much in common, while their differences can be traced, in part, to differences in the adaptive strategies that generated their wealth.

The foregoing serves to illustrate that an uncritical plunge into the concept of quality of life and nonchalant acceptance of "user satisfaction" as a guiding principle in shaping our environment is not likely to produce an environment supporting health.

(1) Our internal sensations at best serve to maximize reproductive fitness, not health, provided the individual lives in some bygone natural environment rather than in our modern artificial one.

(2) The individual's wishes to comply with social and economic myths of current and past origins serves at best to preserve a given economic strategy and conventional wisdoms, not to maximize the individual's health.

(3) The dictates of ego satisfaction ensure a continuous stream of wishes—some counter to the prevailing social myths—which ensure a change in social values. This may or may not be compatible with an environment that supports health. To do this, we must have some understanding of what environmental parameters do support health. We must have a model of an environ-

ment that maximizes health that is derived quite independently of the idea of user satisfaction. The evolutionary model is one such attempt.

Note that in our discussion of health, the idea of physical health is closely linked to physical, intellectual, and social competence. A closer examination reveals that these three related competences are based on maximizing environmentally our species' diagnostic features. Physical, intellectual, and social competence appear to be universal values, that is, they are desirable and regarded as a universal good. The reverse of physical, intellectual, and social competence is clearly unacceptable, and individuals causing others to be underdeveloped robs them of part of their humanity. It may be noted that this articulation is very close to the notion of ethics as developed by C. H. Waddington (1960). Without belaboring the point, conduct by individuals or society that deprives others of the development of our species' diagnostic features is unethical by Waddington's criteria since it ultimately threatens the very existence of our species. My criterion for judging whether certain conduct leads toward greater health or not and Waddington's criterion for judging whether a given conduct is ethical or not happen to be identical.

Testing the Evolutionary Model

One test of the evolutionary model of environmental criteria that maximize health is to generate predictions from the model and compare them with empirical results. To disprove the model is the test. I have shown earlier that, in fact, the predictions of the model conform to known facts about what maximizes health, but it clashes with the assumption that life styles of aboriginal people, existing today or in the recent past, are a valid guide to a healthful environment. Here we shall examine some rather mundane tests.

An example of the validity of the evolutionary model, as tested against empirical research, is illustrated by the unfortunate fad of bottle feeding babies. An evolutionary biologist would expect that deprivation of mother's milk was not likely to be adequately substituted for by cow's milk or some synthetic fluid. Granted the differences in milk composition between species of mammals, one would expect that the milk composition would be selected to impart a maximum of growth and development to the infant. Children on cow's milk and calves on human milk ought not to do as well as each infant on its own mother's milk. This proposition is, of course, testable, and somewhat humorous it would be—save for the results.

Unless absolutely necessary, bottle feeding should be dispensed with in favor of breast feeding. Bottle feeding is not the way to maximize babies' ontogenetic development and health—breast feeding is. Bottle feeding reduces the intake of critical amino acids such as vitamin C; it increases the waste of nutrients and thus the metabolic cost of development; it reduces the availability of lipids needed for brain development, deprives the child of specific immunoglobulins, and thereby increases morbidity and the risk of mortality by reducing the child's resistence to infections; it increases the risk of illness unless care is taken to ensure that the bottle feeding is hygenic, and even then bottle feeding is associated with higher infant morbidity and mortality; it can lead to high rates of inner ear infections and consequent deafness frequently to those affected; it increases the monetary cost of raising the baby as well as the inconvenience—the latter gladly accepted for the sake of being "modern"—and it is a very real drain on the economies of developing nations which must import cow's milk to satisfy a harmful fad. The interested may pursue this rather sad subject, including the blatant irresponsibility of industry in promoting bottle feeding for the sake of profits, or the effects on Canadian Eskimos recently "civilized" to bottle feeding and other of our civilization's gifts, in the symposium proceedings edited by Jeliff and Jeliff (1971) and in Schaefer (1971a) and Hildes and Schaefer (1973); an excellent discourse on the value of milk is found in Brody (1945, pp. 794—95). Our hypothesis, derived from the evolutionary model, that mother's milk is superior to cow's milk or synthetic milk substitute fed in baby bottles can be regarded as verified.

A more critical test would be to verify the prediction that extending nursing over the first three years of ontogeny leads to more superior phenotypic development than curtailing it after one year or less and substituting cow's milk thereafter. I am not aware of pertinent data to test this prediction.

There is more to the relationship of an infant at its mother's breast than suckling. There is a feedback between infant and mother, and we have reason to suspect that this feedback maximizes babies' health. To deprive a baby of mother is certain to result in devastating consequences to the baby, even if it survives the trauma (see Seligman 1975). If this feedback at the mother's breast is reduced, certainly the baby's experience of being able to control the world around him must be reduced somewhat, his mimicry must be reduced and, unless we know better, we have no business assuming that nothing is impaired in the baby.

The evolutionary model predicts that individuals living closest to its dictates benefit more by enhanced physical, intellectual, and social development, by better health and increased longevity. This in practice is not readily testable, but there is evidence compatible with this view. In modern society those individuals having a relatively great contact with nature, who practice a diversity of sports and cultural pursuits while laying emphasis on intellectual competence, and having the where-withal to obtain tutors, traveling, and supplying an excellent diet, are the upper classes. If this is so, individuals from these classes ought to be characterized by relatively large bodies, large brains, dolichocephaly, and superior intellectual competence. A comparison of social classes from Great Britain suggests that this is so; individuals from upper classes are on average considerably taller and have higher average IQs than those of other classes (Tanner 1962, 1966, Young 1971, Deutsch 1973). Huber (1968) also reports on a correlation between body size and class inferred from the distribution of grave goods in men from a medieval population. If the correlation between body size and IQ indicated is valid, then the progressive return toward normal body size, as experienced over the last century in America (Tanner 1962, Damon 1968, Huber 1968) ought to be paralleled by an increase in intellectual competence. Surprisingly, there is some evidence for this hypothesis (Sears 1975, p. 54). The notion that greater competence is correlated with (phenotypic!) body size is supported by the finding of Krapinos (in Huber 1968) that men inducted into the United States army were taller by about one half inch than the population average. In short, there is reason to believe, and scattered empirical evidence supports it, that the evolutionary model for good physical, intellectual, and probably social development during ontogeny does do what it is expected to do. Moreover, the optimum environment for ontogenetic development is more often ecountered in families with wealth and traditions, a finding amply verified (see also reviews in Caldwell and Ricciuti 1973, Horowitz 1975). In practice, a national policy of maximizing individual development should increase the frequency of individuals of development comparable to that of the upper classes.

One could develop many more examples here showing that the predictions of the evolutionary model, when tested against available data, vindicate the model, although disproof and not proof was the aim. In every instance, the test is reduced injury, morbidity and mortality, or an increase in health. Note that an environmental increase in IQ is correctly predicted by the model, as is the relationship of nutrition to health, as is the amount of vitamin C readily metabolized by our bodies—and there are other examples in the preceding chapter. The evolutionary model's predictions are vindicated also by examining the diverse causes of morbidity, illness, and death as described, for instance, for America's poor by Roger Hurley (1971). This is a disturbing account of how absence of material security, social bonding, cognitive stimulation, poor nutrition, and the dehumanizing treatment of institutions combine to inflict misery on those not competent to respond adequately. To continue further testing of the evolutionary model by myself could lead to advocacy. But advocacy is not my intention, not advocacy but tests to determine its validity, tests of whether it does generate more competence and health for individuals during ontogeny. We must now look at a different aspect of the evolutionary model and its use in decision making by professionals.

The Evolutionary Model in Decision Making: Examples

How does one use the evolutionary model in decision making? Let us turn for an illustration to the matter of noise pollution. The evolutionary model predicts that continuous exposure to noise above the level encountered, for instance, in a windy forest or steppe ought to lead to some sort of grief. The model does not predict what kind of grief; only empirical research can show it. Here are the findings on the effects of noise pollution, as discussed by Farr (1967), Boggs and Simon (1968), Rosen (1970), Arvay (1970), Geber (1970), Jensen and Rasmussen (1970), Junghans and Nitschkoff (1975), and Magrab (1975): Hearing loss sets in, affecting people mainly over 40 years of age. Hearing loss is gradual and quite painless. Excessive noise leads to an inability to concentrate or learn, it reduces work output, increases arousal and the frequency of mistakes, deprives individuals of sleep, leads to chronic stress and hypertension, increases the work load on the heart, increases metabolism, is associated with coronary heart disease and arteriosclerosis, increases susceptibility to infections and audiogenic seizures, worsens the conditions of those who suffer from duodenal ulcers, affects the unborn fetus, and increases the frequency of congenital deformities; during early ontogeny it leads to a reduced rate of language acquisition and stuttering and to histologically determinable damage to the inner ear and central nervous system.

However, we turn to a respected textbook in preventive medicine, such as that of Sartwell (1973 ed), and find in Chapter 24 that Baetjer quotes approvingly (p. 864) from K. D. Kryter that ''Quantitative evidence that regular or for that matter irregular environmental noise causes any physiological or mental ill health appears to be completely lacking''. Baetjer does point out, though, that the matter of noise and its relation to ill health is controversial.

Who does one listen to when experts disagree? One needs guidance from an independent source. The predictions of the evolutionary model are clearly in conflict with Baetjer's account in Sartwell's textbook. Therefore, we ought to design our environments to mimic noise levels of common outdoor environments and reject the "all is well unless disproven" approach of Baetjer in this case. Let us not forget the claims of some scientists that no evidence exists that smoking could possibly cause ill health, that pesticides could not possibly be anything but a boon to mankind, etc.

In professional decision making, one must ensure that the very elementary needs of humans as dictated by the evolutionary model are satisfied, e.g., that cities have clean air. Housing and traffic must be located and built so as to permit natural ventilation and filtration of the air, so that inversions are precluded and toxic pollutants are contained. An urban policy maximizing profit from land and built primarily to accommodate vehicles will not satisfy the basic need of humans for fresh air. We are not designed to run on pavement. Why do we insist on building hard surfaces to walk upon? We need bulky fibrous food with a high content of diverse proteins, vitamins, essential oils, and minerals, not the convenience foods produced for the convenience of the manufacturers (Hall 1974). What about the social realm? Let us turn to a recent book by Oscar Newman (1972) on "defensible space." His main argument is that to reduce crime in buildings housing a large number of people, it is essential to design spaces in such a fashion as to reduce anonymity. Rather than double-loaded corridors of the conventional high-rise, one ought to cluster 5 to 8 living units about a common semiprivate space, which in turn is connected to a public space acting as a roadway. The public, private, and semiprivate spaces have to be physically well defined. Windows should be positioned to overview the semiprivate spaces, permitting surveillance. The buildings themselves should house about 50 families for natural surveillance to operate; common petty criminals prefer the anonymity of larger buildings for their endeavours. The location of buildings must be such as to minimize the vulnerability of residents, for instance, by locating them in areas with ready access to police or neighborly help. Newman emphasizes that the building type matters less with middle-income families which maintain better control over children and may be able to afford a doorman.

We recognize in Newman's recommendations a number of attributes of human sociality that reflect old adaptations: Living clusters of 5 to 8 peer families gives rise to 20 to 40 individuals divided into peer groups of 5 to 8. The latter is the ideal group for discourse and socialization; the former is close to the typical peer family network of Paleolithic and primitive cultures, as discussed earlier. Newman's design recommendations reduce *anonymity* of residents by subjecting a limited number of them to frequent viewing once they are on semiprivate spaces. A reduction in *anonymity* is equivalent to increasing *predictability* in the social milieu (see Chapter 2). Strangers are readily recognized and can be challenged. "Watchmen," be they volunteers or hired help, and closed-circuit television achieve similar results. However, one can go beyond Newman's recommendations, which essentially state that we ought to cluster living spaces much as one finds in primitive cultures.

Anonymity can be reduced by diverse means; let us total all means, beginning with those proposed by Newman.

(1) Cluster residence by 5 to 8 peer families around a surveyable semipublic space.
(2) Keep building capacity to 50 families or less.

(3) Increase surveyability of spaces public and semipublic by various means, including closed-circui television monitoring.

(4) Define spaces unambiguously by means of conspicuous borders. This requirement does not serve some kind of a human territorial "instinct," which we do not possess; rather, it satisfies principles of perception that permit vertebrates to orient themselves by conspicuous lines.

(5) Place, where possible, a watchman at the entrance to the residence. These can be hired or volunteers, or accidental watchmen such as residents enjoying each other's company in front of the door or in a foyer.

(6) Provide a great mix of age classes to reduce the size of residential peer groups and thereby enhance bonding between peers.

(7) Increase the number of older residents bonded to nuclear families. These older residents provide surveillance and exercise some control over the activities of children and juveniles.

(8) Provide within walking distance of the residence a diversity of public meeting centers, such as coffee houses, pubs, enclosed gardens for winter visits, parks, small stores, playgrounds, mail pick-up locations, etc. These meeting centers facilitate establishment and maintenance of contact between residents, thereby reducing anonymity.

(9) Within a residential area some odd, and thereby striking structure should be provided, or some structural uniqueness given to buildings so that they begin to function as symbols. Unique structures stimulate discussion, increase harmonizing, facilitate bonding, and, in the long run, tie individuals to what they perceive is a unique place of residence. The unique characterizing feature may also be a natural feature left in place during residential development of the area—a cliff, a beautiful creek valley, a pond, a glacial erratic, a prominent hill, etc. In time, residents may feel like erecting a monument themselves to one of their worthy fellow residents, as it once was common to do.

(10) Provide for residential festivities for some to participate in and for all to talk about. Such festivals also become symbols. Some such activities as open house in the local school can be enlarged. Ethnic groups would have the easiest time generating such festivities.

(11) Permit residents to cluster by background to increase their common heritage and thus facilitate cooperation, harmony, and mutual help. This is one means of enhancing the development of a strong social fabric.

(12) Devise noncoercive means of reducing residential turnover rates.

(13) Limit access to the total housing area to reduce the number of spontaneous visits by strangers, and increase the frequency of residents looking at a stranger. Should such a stranger enter with sinister motives, the many curious glances meeting him will be experienced as rather discouraging.

(14) Generate self-government by residents and maximize the opportunity for participation in various community functions.

(15) Create crises. Maybe these need not be created, as diverse interests are likely to create them anyway. They can and do function to unite the community in cooperation and foster cohesion and a sense of belonging.

(16) Reduce the opportunity for the display of wealth in the community to reduce the apparent discrepancy of wealth in the residential area. Display of wealth as a status symbol can lead to envy, to ridicule by those unable to keep up,

or, conversely, to an *a priori* drop in social value of residents with a lower income. Conversely, there ought to be the opportunity for displaying conventional skills, even to the poorest, such as patches of ground for flower gardening or lawns to be kept trim and beautiful.

All the above factors singly or collectively reduce anonymity and enhance the formation of strong viable social networks that develop in individuals a liking and loyalty for their community, reduce the turnover rates of residents, and enforce acceptable behavior on community members.

What are the Costs of Maximizing Individual Development?

If our socioeconomic system were static, the answer would be easy. It is not. We have to contend with the inertia of population growth, with the dictates of nuclear technology already inflicted on eons of generations to come, with the drastic reduction in availability of liquid fuels for much of our energy needs and therefore the dictates of future energy technologies. For instance, should solar heating of homes become very important in our energy budgets, then cities as we know them are likely to disappear. Instead, we shall see low houses well spaced out so that none shadows another robbing it of sun; there will be no tall trees unless their shadows do not interfere with the solar-energy-collecting panels of buildings; the north slopes of hills and mountains will be left unbuilt upon, as will be floodplains in deep valleys. Cities will be spread out, probably clustered, with densities of people probably low. Should this come about, I for one am optimistic that we shall have environments much more closely fitting our biological potential than those we are now creating. In designing our environments, our determining factor ought not to be technological accidents but a policy of maximizing individual development as a means of maintaining a healthy, capable population.

What are the costs of *not* doing this? The first consequence of denying a policy that maximizes individual development is an affirmation that doing violence to those incapable of protecting themselves is just and proper; to dehumanize fellow citizens is acceptable; to deprive them of health, competence, and life is justified. Granted adequate economic growth, there will be enough money to build hospitals, wards, and detention centers to support failing bodies and minds, as well as keeping failing individuals out of circulation. Sick people are a great stimulus to the free market because they force ingenious support systems to be developed. We need not worry, therefore, how to maximize individual competence, since our technology and social services will support failing human parts and individuals. Think of the possibilities in sophisticated drug therapies, in technological aids to restore breathing, heartbeat, and muscle function, the artificial organs which can be installed, including plastic penises. *Homo sapiens* will become *Homo cyborgensis,* but what a boon to the medical sciences, the economy, the electronics industry, the service industries, to our instiuttions of higher learning, and even to basic sciences, etc., etc.

Something is wrong in the above argument for a perpetuation of values from the *status quo,* is it not? It is an argument for complex service systems full of frailty to support our culture and a burgeoning number of individuals incapable of looking after themselves, while affirming that limited competence in individuals is quite all right.

Of course, one can develop such a society and with a *laissez-faire* attitude we shall almost certainly reach it.

The costs of maximizing individual development are diverse and far reaching. A policy of maximizing individual development would lead to a reduction in the world population so as to maximize resources per individual. It would probably lead to a reduction in the size of extractive and manufacturing industries, because of the need to increase the size of complex ecosystems and the emphasis on self-help instead of bureaucratic service. This would lead to a reduction in the need for diverse social services, including those of education, welfare, policing, and entertainment. Inevitably, it would also reduce the amount of service each individual would perform within strictly economic institutions. One would move closer to economies as envisioned by Schumacher (1973). Urbanization would almost certainly be decreased. By shifting functions and responsibilities, now held by institutions and the state, closer to the individual one would necessarily reduce the rate of cultural evolution. The "costs" would thus probably be a severe reduction in the size, complexity, and rate of development of our socio-economic system. Increasing the decision-making powers of individuals would probably lead to increasing the power of local governing bodies. There would also probably be a reduction in the size of nation-states. However, to pursue a policy of maximizing individual development requires security, and that ultimately implies world peace—as yet an unattainable goal.

References

Addicott, W. O.: Tertiary climate change in the marginal northeast Pacific ocean. *Science* **165,** 583−585 (1969)

Agranoff, B. W.: Memory and protein synthesis. *Sci Am* **216(6),** 115−122 (1967)

Akiskal, H. S., McKinney, W. T.: Depressive disorders: Towards a unified hypothesis. Science **182,** 20−29 (1973)

Alcock, J.: The evolution of the use of tools by feeding animals. *Evolution* **26,** 467−473 (1972)

Aleksiuk, M., Baldwin, J.: Temperature dependence of tissue metabolism in monotremes. *Can J Zool* **51,** 17−19 (1973)

Alexander, R. D.: The evolution of social behaviour. *Annu Rev Ecol Syst* **5,** 325−383 (1974)

Allee, W. C.: *Cooperation Among Animals.* New York: Schuman, 1951

Allee, W. C., Guhl, A. M.: Concerning the group survival value of the social peck order. *Anat Rec* **84(4),** 497−498 (1942)

Allport, G.: *The Nature of Prejudice.* Reading, Mass.: Addison-Wesley, 1954

Allred, G. L., Baker, L. R., Bradley, W. G.: Additional studies of anomalies of the skull in desert bighorn sheep. *Trans Cal Nev Sec Wildl Soc* 40−47 (1966)

Altman, I.: *The Environment and Social Behaviour.* Monterey: Brooks-Cole, 1975

Altmann, G.: *Die Orientierung der Tiere im Raum.* Neue Brehm-Bücherei 369. Wittenberg-Lutherstadt: Ziemsen, 1966

Altmann, M.: Social behaviour of elk, *Cervus canadensis nelsoni,* in the Jackson Hole area of Wyoming. *Behaviour* **4,** 116−143 (1953)

Altmann, S. A.: A field study of the sociobiology of rhesus monkeys, *Macaca mulatta. N Y Acad Sci* **102(2),** 338−435 (1962)

Altmann, S. A.: The structure of primate social communication. In: *Social Communication Among Primates.* Altmann, S. A. (ed). Chicago: Univ. Chicago Press, 1967, 325−362

Altmann, S. A.: Baboons, space, time and energy. *Amer Zool* **14(1),** 221−248 (1974)

Altmann, S. A., Altmann, J.: *Baboon Ecology.* Chicago: *Univ. Chicago Press,* 1970

Amadon, D.: Birds of the Congo and Amazon forest: A comparison. In: *Tropical Forest Ecosystems in Africa and South America.* Meggers, B. J., Ayensu, E. S., Duckworth, W. D. (eds). Washington: Smithsonian Inst, 1973, 267−277

Anderson, D. E., Brady, J. V.: Preavoidance blood pressure elevations accompanied by heart rate decrease in the dog. *Science* **172,** 595−597 (1971)

Anderson, E.: *Plants, Man and Life*. Berkeley: Univ. California Press, 1952

Anderson, P. K.: Ecological structure and gene flow in small mammals. In: *Variation in Mammalian Populations*. Berry, R. J., Southern, H. N. (eds). *Symp Zool Soc London* **26**, 299—325. London: Academic Press, 1970

Andreski, S.: *Social Science as Sorcery*. Harmondsworth, Middx.: Penguin, 1972

Andrew, R. J.: The origins and evolution of calls and facial expressions of primates. *Behaviour* **20**, 1—109 (1963)

Angel, J. L.: Ecological aspects of paleodemography. In: *The Skeletal Biology of Earlier Human Populations*. Brothwell, D. R. (ed). Oxford: Pergamon, 1968, 263—270

Annis, R. C., Frost, B.: Human visual ecology and orientation anisotropies in acuity. *Science* **182**, 729—731 (1973)

Antonius, O.: Über Symbolhandlungen und Verwandtes bei Säugetieren. *Z Tierpsychol* **3**, 263—278 (1939)

Archer, J.: Effects of population density on behaviour in rodents. In: *Social Behaviour in Birds and Mammals*. Crook, J. H. (ed). London: Academic Press, 1970, 169—210

Ardrey, R.: *African Genesis*. New York: Dell, 1963

Ardrey, R.: *The Territorial Imperative*. New York: Atheneum, 1966

Armelagos, G. J.: Disease in ancient Nubia. *Science* **163**, 225—259 (1969)

Arnheim, R.: *Entropy and Art*. Berkeley: Univ. California Press, 1971

Arvay, A.: Effects of noise during pregnancy upon foetal viability and development. In: *Physiological Effects of Noise*. Welch, B. L., Welch, A. S. (eds). New York: Plenum, 1970, 91—115

Asmussen, E.: Growth in muscular strength and power. In: *Physical Activity*. Rarick, G. L. (ed). New York: Academic Press, 1973, 60—80

Astrand, P.: *Experimental Studies of Physical Working Capacity in Relation to Sex and Age*. Copenhagen: Munksgaard, 1952

Atz, J. W.: The application of the idea of homology to behaviour. In: *Development and Evolution of Behaviour*. Aronson, L. R., Tobach, E., Lehrman, D. S., Rosenblatt, J. S. (eds). San Francisco: Freeman, 1970, 53—74

Axelrod, D. I., Bailey, H. P.: Cretaceous dinosaur extinction. *Evolution* **22**, 595—611 (1968)

Azrin, N. H.: Aggressive responses of paired animals. In: *Symposium on the Medical Aspects of Stess in the Military Climate*. Washington: Reed Army Inst Res, 1964, 329—352

Azrin, N. H., Hutchinson, R. R., McLaughlin, R.: The opportunity for aggression as an operant reinforcer during aversive stimulation. *J Exp Anal Behav* **8**, 171—80 (1965)

Backhaus, D.: Experimentelle Untersuchungen über die Sehschärfe und das Farbensehen einiger Huftiere. *Z Tierpsychol* **16**, 445—467 (1959)

Backhaus, D.: *Beobachtungen an Giraffen in Zoologischen Gärten und freier Wildbahn*. Brussels: Inst Parcs Nat Congo et Ruanda-Urundi, 1961

Baker, J. R.: Cro-Magnon man, 1868—1968. *Endeavour* **27**, 87—90 (1968)

Baker, P. T.: Human adaptation to high altitude. *Science* **163**, 1149—1156 (1969)

Bakker, R. T.: Dinosaur physiology and the origin of mammals. *Evolution* **25**, 636—658 (1971)

Bakker, R. T.: Anatomical and ecological evidence of endothermy in dinosaurs. *Nature (Lond)* **238**, 81—85 (1972)

Balakci, A.: The Natsiluk Eskimos, adaptive processes. In: *Man the Hunter*. Lee, R. B., de Vore, I. (eds). Chicago: Aldine, 1968, 78—82

Bandy, O. J., Butler, E. A., Wright, R. C.: Alaska upper Miocene glacial deposits and the *Tuberotalia pachyderma* datum plane. *Science* **166**, 607—609 (1969)

Banfield, A. W. F.: Preliminary investigation of the barren ground caribou: Part ii. life history, ecology and utilization. *Wildl Mangt Bull* Series 1, 10B. Ottawa: Can Wildl Service, 1954

Banfield, A. W. F.: A revision of the reindeer and caribou genus *Rangifer*. Ottawa: *Nat Mus Can Bull* **177**, 1961

Banks, E. M., Willson, M. F. (eds): Ecology and evolution of social organization. *Amer Zool* **14(1)**, 7−264 (1974)

Bannikov, A. G., Zhirnov, L. V., Lebedeya, L. S., Fandeev, A. A.: *Biology of the Saiga*. Trans. from Russian. Springfield: US Dept of Commerce, 1961

Barash, D. P.: The evolution of marmot societies. *Science* **185**, 415−420 (1974)

Barghusen, H. R.: A review of fighting adaptations in dinocephalians (Reptilia, Therapsida). *Paleobiology* **1**, 295−311 (1975)

Barghusen, R. H., Hopson, J. A.: Dentary-squamosal joint and the origin of mammals. *Science* 168, 573−575 (1970)

Barlow, H. B., Narasimhan, R., Rosenfeld, A.: Visual pattern analysis in machines and animals. *Science* **177**, 567−575 (1972)

Barnes, R. M.: *A History of the Regiments and Uniforms of the British Army*. London: Sphere (6th ed), 1967

Barnett, S. A.: Competition among wild rats. *Nature (Lond)* **175**, 126 (1955)

Barnett, S. A.: Social behaviour in wild rats. *Proc Zool Soc Lond* **130**, 107−152 (1958)

Barnett, S. A.: *Instinct and Intelligence*. London: MacGibbon & Kea, 1967

Barrette, C.: The social behaviour of captive muntjacs *Muntiacus reevesi* (Ogilby 1839). *Z Tierpsychol* **43**, 188−213 (1977a)

Barrette, C.: Some aspects of the behaviour of muntjacs in Wilpattu National Park. *Mammalia* **41(1)**, 1−34 (1977b)

Barrette, C.: Fighting behaviour of muntjac and the evolution of antlers. *Evolution* **31(1)**, 169−176 (1977c)

Bartholomew, G. A., Casey, T. M.: Endothermy during terrestrial activity in large beetles. *Science* **195**, 882−883 (1977)

Bartholomew, G. A., Tucker, V. A.: Control of changes in body temperature, metabolism and circulation by the agamid lizard *Amphibolurus barbatus*. *Physiol Zool* **36**, 199−218 (1963)

Bartholomew, G. A., Tucker, V. A.: Size, body temperature, thermal conductance, oxygen consumption and heart rate in Australian varanid lizards. *Physiol Zool* **37(4)**, 341−354 (1964)

Baskin, L. M.: Reindeer, their Ecology and Behaviour. Unedited trans. from Russian. Ottawa: *Can Wildl Service*, 1970

Baskin, L. M.: Management of ungulate herds in relation to domestication. In: *The Behaviour of Ungulates and its Relation to Management*. Geist, V., Walther, F. R. (eds). Morges: IUCN Publications New Series **24**, 1974, 530−541

Batcheler, C. L.: Compensatory responses of artificially controlled mammal populations. *Proc N Z Ecol Soc* **15**, 25−30 (1968)

Bateson, G.: Redundancy and coding. In: *Animal Communication*. Sebeok, T. A. (ed). Bloomington: Indiana Univ. Press, 1968, 614−626

Battistini, R., Vérin, P.: Ecological changes in prehistoric Madagascar. In: *Pleistocene Extinctions*. Martin, P. S., Wright, H. E. (eds). New Haven: Yale Univ. Press, 1967, 407−427

Baxi, P. G.: A natural history of childbearing in the hospital class of women in Bombay. *J Obstet Gynecol India* **811**, 26−51 (1957)

Bayless, T. M., Christopher, N. L.: Disaccharides deficiency. *Amer J Clin Nutr* **22**, 181−190 (1969)

Beck, H.: Minimal requirements for a biobehavioural paradigm. *Behav Sci* **16**, 442−455 (1971)

Beck, S. L., Gavin, D. L.: Susceptibility of mice to audiogenic seizures is increased by handling their dams during gestation. *Science* **193**, 427−428 (1976)

Begleiter, H., Porjesz, B., Yerre, C., Kissin, B.: Evoked potential correlates of expected stimulus intensity. *Scince* **179**, 814−816 (1973)

Bekoff, M.: Social play and play-soliciting by infant canids. *Amer Zool* **14(1)**, 323–340 (1974)

Belfer, M. L., Brown, B. S.: Juvenile delinquency. In: *Maternal and Child Health Practices*. Wallace, H. M., Gold, E. M., Lis, E. F. (eds). Springfield, Ill.: Thomas, 1973, 868–883

Bell, R. H. V.: A grazing ecosystem in the Serengeti. *Sci Amer* **225**, 86–93 (1971)

Bellairs, A. d'A.: *Reptiles*. London: Hutchinson, 1968

Belmont, L., Marolla, F. A.: Birth order, family size and intelligence. *Science* **182**, 1096–1101 (1973)

Benarde, M. A.: *Our Precarious Habitat*. New York: Norton, 1970

Beninde, J.: Zur Naturgeschichte des Rothirsches. *Monog. Wildsäugetiere* **4**. Leipzig: Schöps, 1937.

Bennett, A. F., Dalzell, B.: Dinosaur physiology. *Evolution* **27(1)**, 1–26 (1973)

Bennett, E. L., Diamond, M. C., Krech, D., Rosenzweig, M. R.: Chemical and anatomical plasticity of the brain. *Science* **146**, 610–619 (1964)

Benzie, D., Gill, J. C.: Radiography of the skeletal and dental condition of the Soay sheep. In: *Island Survivors*. Jewell, P. A., Milner, C., Boyd, J. M. (eds). London: Athlone, 1974, 326–337

Berg, A.: *The Nutrition Factor*. Washington: Brookings Inst, 1973.

Bergerud, A. T.: The role of the environment in the aggregation, movement and disturbance behaviour of caribou. In: *The Behaviour of Ungulates and its Relation to Management*. Geist, V., Walther, F. (eds). Morges: IUCN Publications New Series **24**, 1974, 552–584

Berlyne, D. E.: Curiosity and exploration. *Science* **153**, 25–33 (1966)

Bertram, B. C.: The social system of lions. *Sci Amer* **232(5)**, 54–61 (1975)

Beschel, R. E.: The diversity of tundra vegetation. In: *Productivity and Conservation in Cimcumpolar Lands*. Fuller, W. A., Kevan, P. G. (eds). Morges: IUCN Publications New Series **16**, 1970, 85–92

Bibikov, D. I.: Die Murmeltiere. *Neue Brehm Bücherei* **388**. Wittenberg-Lutherstadt: Ziemsen, 1968

Binford, L. R.: Post-Pleistocene adaptations. In: *New Perspectives in Archeology*. Binford, S. R., Binford, L. R. (eds). Chicago: Aldine, 1968a, 313–341

Binford, S. R.: Ethnographic data and understanding the Pleistocene. In: *Man the Hunter*. Lee, R. B., De Vore, I. (eds). Chicago: Aldine, 1968b, 274–275

Binford, S. R.: Late middle paleolithic adaptations and their possible consequences. *Bioscience* **20**, 280–283 (1970)

Birch, H. G.: The relation of previous experience to insightful problem solving. *J Comp Psychol* **38**, 367–383 (1945)

Birdsell, J. B.: Some environmental and cultural factors influencing the structuring of Australian aboriginal populations. *Amer Nat* **87**, 171–207 (1953)

Birdsell, J. B.: Some predictions for the Pleistocene based on equilibrium systems among recent hunter-gatherers. In: *Man the Hunter*. Lee, R. B., De Vore, I. (eds). Chicago: Aldine, 1968, 229–240

Birdsell, J. B.: *Human Evolution*. Chicago: Rand McNally, 1972

Birdwhistell, R. L.: Kinesis and communication. In: *Exploration in Communication*. Carpenter, E., McLuhan, M. (eds). Boston: Beacon, 1960

Bissonette, T. H.: Influence of light on the hypophysis. *Endocrinology* **22**, 92 (1938)

Bitterman, M. E.: The comparative analysis of learning. *Science* **188**, 699–709 (1975)

Blain, E. H.: Elevated arterial blood pressure in an asymptotic population of meadow voles (*Microtus pennsylvanicus*). *Nature (Lond)* **242**, 135 (1973)

Blair, W. F.: Amphibians and reptiles. In: *Animal Communication*. Sebeok, T. A. (ed). Bloomington: Indiana Univ Press, 1968, 289–310

Blanc, A. C.: Some evidence for the ideologies of early man. In: *Social Life of Early Man*. Washburn, S. L. (ed). Chicago: Aldine, 1961, 119–136

Blaxter, K. L.: Energy utilization in the ruminant. In: *Digestive Physiology and Nutrition of the Ruminant*. Lewis, D. (ed). London: Butterworth, 1960, 183−197

Blaxter, K. L.: The fasting metabolism of adult wether sheep. *Brit J Nutr* **16**, 615−626 (1962)

Blaxter, K. L., Wainman, F. M., Wilson, R. S.: The regulation of food intake by sheep. *Anim Prod* **3**, 51−61 (1961)

Boggs, D. H., Simon, J. R.: Differential effect of noise on tasks of varying complexity. *J Appl Psychol* **52**, 148−153 (1968)

Bonatti, E.: North Mediterranean climate during the last Würm glaciation. *Nature (Lond)* **209**, 984−985 (1966)

Bonnefille, R.: Palynological evidence for an important change in the vegetation of the Omo Basin between 2.5 and 2 million years BP. In: *Earliest Man and Environments in the Lake Rudolf Basin*. Coppens, Y., Howell, F. C., Isaac, G. L., Leakey, R. E. F. (eds). Chicago: Univ. Chicago Press, 1976, 421−431

Bordes, F.: *A Tale of Two Caves*. New York: Harper & Row, 1972

Boren, J. H.: *When in Doubt, Mumble*. New York: Van Nostrand Reinhold, 1972

Borns, H. W., Goldthwait, R. F.: Late-Pleistocene fluctuations of Kaskawulsh Glacier, southwestern Yukon Territory, Canada. *Amer J Sci* **264**, 600−619 (1966)

Bostock, H. S.: Physiography of the Canadian cordillera, with special reference to the area north of the fifty-fifth parallel. *Geol Surv Can Memoir* **247**. Ottawa: Mines and Resources, 1948

Bostock, H. S.: Geology of northwest Shakwak valley, Yukon Territory. *Geol Surv Can Memoir* **267**. Ottawa: Mines & Technical Surveys, 1952

Bostock, H. S. (ed): Yukon Territory: Selected field reports of the Geological Survey of Canada 1898−1933. *Geol Surv Can Memoir* **284**. Ottawa: Mines & Technical Surveys, 1957

Boulding, K. E.: Am I a man or a mouse—or both? In: *Man and Aggression*. Ashley Montagu, M. F. (ed). Oxford: Oxford Univ. Press, 1968, 88−90

Bourlière, F.: Density and biomass of some ungulate populations in eastern Congo and Rwanda, with notes on population structure and lion/ungulate ratios. *Zool Afr* **1**, 199−207 (1965)

Bourlière, F.: The comparative ecology of rainforest mammals in Africa and tropical America. In: *Tropical Ecosystems in Africa and South America*. Meggers, B. J., Ayensu, E. S., Duckworth, W. D. (eds). Washington: Smithsonian Inst, 1973a 279−292

Bourlière, F.: Ecology of human senescence. In: *Textbook of Geriatric Medicine and Gerontology*. Brockenhurst, J. C. (ed). London: Churchill, 1973b, 60−74

Bourlière, F., Hadley, M.: The ecology of tropical savannas. *Annu Rev Ecol Syst* **1**, 125−152 (1970)

Bowerman, W. G.: *Studies in Genius*. New York: Philosophical Library, 1947

Boyd, J. M., Jewell, P. A.: The Soay sheep and their environment. In: *Island Survivors*. Jewell, P. A., Milner, C., Boyd, J. M. (eds). London: Athlone, 1974, 360−373

Boyden, S.: Biological determinants of optimum health. In: *Human Biology of Environmental Change*. Vorster, D. J. M. (ed). London: Int Biol Prog 1972, 3−11

Boyden, S.: Evolution and health. *Ecologist* **3**, 304−309 (1973)

Brace, C. L.: Cultural factors in the evolution of the human dentition. In: *Culture and the Evolution of Man*. Montagu, M. F. A. (ed). New York: Oxford Univ. Press, 1962, 343−354

Brace, C. L.: Structural reduction in evolution. *Amer Nat* **97**, 39−49 (1963)

Brace, C. L.: The fate of the classic Neanderthals: A consideration of hominid catastrophism. *Curr Anthropol.* **5(1)**, 3−43 (1964)

Brace, C. L.: Environment, tooth form and size in the Pleistocene. *J Dent Res* **46**, 809−816 (1967)

Brace, C. L.: Sexual dimorphism in human evolution. In: *Man in Evolutionary Perspective*. Brace, C. L., Metress, J. (eds). New York: Wiley, 1973, 238−254

Brace, C. L., Livingstone, R. B.: On creeping Jensenism. In: *Man in Evolutionary Perspective*. Brace, C. L., Metress, J. (eds). New York: Wiley, 1971, 426−437

Brace, C. L., Mahler, P. E.: Post-Pleistocene changes in human dentition. *Amer J Phys Anthropol.* **34**, 191−203 (1971)

Brady, J. V.: Ulcers in executive monkeys. *Sci Amer* **199(4)**, (1958)

Brain, C. K.: An attempt to reconstruct the behaviour of australopithecines: The evidence for interpersonal violence. *Zool Afr* **7(1)**, 379−401 (1972)

Bramwell, C. D., Fellgett, P. B.: Thermal regulation in sail lizards. *Nature (Lond)* **242**, 203−205 (1973)

Brattstrom, B. H.: The evolution of reptilian social behaviour. *Amer Zool* **14**, 35−49 (1974)

Bray, J. R.: Solar-climate relationships in the post-Pleistocene. *Science* **171**, 1242−1243 (1971)

Brereton, J. L.: A self-regulating density-independent continuum in Australian parrots and its implication to ecological management. In: *The Scientific Management of Animal and Plant Communities for Conservation*. Duffey, E., Watt, A. S. (eds). London: Blackwell, 1971, 207−221

Brink, A. S.: Speculations on some advanced mammalian characteristics in the higher mammal-like reptiles. *Paleontol Afr* **4**, 77−96 (1956)

Brody, S.: *Bioenergetics and Growth*. New York: Reinhold, 1945

Bromley, P. T.: Pregnancy, birth, behavioural development of the fawn and territoriality in the pronghorn (*Antilocapra americana* Ord) on the National Bison Range, Moise, Montana. Unpubl. MA Thesis, Univ. Montana, 1967

Bromley, P. T.: Aspects of the behavioural ecology and sociobiology of the pronghorn *Antilocapra americana*. Unpubl. PhD Thesis, Univ. Calgary, 1977

Bronowski, J., Bellugi, U.: Language name and concept. *Science* **168**, 669−677 (1970)

Bronson, F. H., Eleftheriou, B. E.: Chronic physiological effects of fighting in mice. *Gen Comp Endocrinol.* **4**, 9−14 (1964)

Bronson, F. H., Eleftheriou, B. E.: Relative effects of fighting in mice: Separation of physical or psychological causes. *Science* **147**, 627−628 (1965)

Broom, L., Selznick, P.: *Sociology*. New York: Harper (3rd ed), 1963

Brothwell, D.: Diet, economy and biosocial change in late Pleistocene Europe. In: *Economy and Settlement in Neolithic and Early Bronze Age Britain and Europe*. Simpson, D. D. A. (ed). New York: Humanities Press, 1971

Brown, J. K.: A note on the division of labour by sex. *Amer Anthropol* **72**, 1073−1078 (1970)

Brown, J. L.: The evolution of diversity in avian territorial systems. *Wilson Bull* **76(2)**, 160−169 (1964)

Brown, J. L.: Alternate routes to sociality in jays—with a theory for the evolution of altruism and communal breeding. In: *Ecology and evolution of social organization*. Banks, E. M., Willson, M. F. (eds). *Amer Zool* **14(1)**, 1974, 63−80

Brown R. E.: Decreased brain weight in malnutrition and its implications. *E Afr Med J* **42**, 584−595 (1965)

Brozek, J. M.: Food as an essential: Experimental studies on behavioural fitness. In: *Food and Civilization*. Voice of America Forum Lectures. Springfield, Ill.: Thomas, 1966, 29−60

Brunner-Orne, M., Orne, F. E.: Englische Handglocken und ihre Verwendung in der psychiatrischen Klinik. In: *Musik in der Medizin*. Teirich, H. R. (ed). Stuttgart: Fischer, 1958, 98−103

Bryan, A. L.: Paleoenvironments and cultural diversity in late Pleistocene South America. *Quat Res* **3**, 237−256 (1973)

Bryson, R. A.: A perspective on climatic change. *Science* **184**, 753−760 (1974)

442   References

Buchwald, K., Engelhardt, W. (eds): *Landschaftspflege und Naturschutz in der Praxis*. Munich: Bayerischer Landwirtschaftsverlag, 1973

Buechner, H. K.: Territorial behaviour in the Uganda kob. *Science* **133**, 698–699 (1961)

Buechner, H. K.: Territoriality as a behavioural adaptation to environment in the kob. *Proc 14th Internat Cong Zool* **3**, 59–63 (1963)

Buechner, H. K., Schloeth, R.: Ceremonial mating behaviour in Uganda kob (*Adenota kob thomasi* Neumann). *Z Tierpsychol* **22**, 209–225 (1965)

Bützler, W.: Kampf- und Paarungsverhalten, soziale Rangordnung und Activitätsperiodik beim Rothirsch. *Adv Ethol* **16**, Berlin: Parey, 1974

Bygott, J. D.: Cannibalism among wild chimpanzees. *Nature (Lond)* **238**, 410–411 (1972)

Caldwell, B. M., Ricciuti, H. N. (eds): Child development and social policy. *Rev Child Devel Res* **3**. Chicago: Univ. Chicago Press, 1973

Caldwell, B. M., Elardo, R., Elardo, P.: Enrichment of early childhood experience. In: *Maternal and Child Health Practices*. Wallace, H. M., Gold, E. M., Lis, E. F. (eds). Springfield, Ill.: Thomas, 1973, 776–794

Calhoun, J. B.: Mortality and movement of brown rats *Rattus norwegicus* in artificially supersaturated populations. *J Wildl Manage* **12**, 167–172 (1948)

Calhoun, J. B.: A behavioural sink. In: *Roots of Behaviour*. Bliss, E. (ed). New York: Hoeber, 1962, 295–315

Calhoun, J. B.: The social use of space. In: *Physiological Mammalogy*. Mayer, W. V., van Gelder, R. G. (eds). New York: Academic Press, 1963a, 2–188

Calhoun, J. B.: The ecology and sociology of the Norway rat. Public Health Publication 1008. Washington: US Dept. Health, Education, Welfare, 1963b

Campbell, B., Sarawar, M., Petersen, W. E.: Mechanism of immunization of infant by mother's milk. *Science* **125**, 932 (1957)

Campbell, B. G.: *Human Evolution*. Chicago: Aldine (2nd ed), 1974

Campbell, D. T.: *Pattern Matching as an Essential in Distal Knowing*. New York: Holt, Rinehart, Winston, 1966

Carbyn, L. N.: Wolf predation and behavioural interactions with elk and other ungulates in an area of high prey diversity. Unpubl. PhD thesis, Univ. Toronto, 1975

Carey, F. G.: Fishes with warm bodies. *Sci Amer* **228(2)**, 36–44 (1973)

Carey, F. G., Teal, J. M.: Heat conservation in tuna fish muscle. *Proc Nat Acad Sci* **56**, 1464–1469 (1966)

Carl, E. A.: Population control in arctic ground squirrels. *Ecology* **52**, 395–413 (1971)

Carneiro, R. L.: A theory of the origin of the state. *Sci* **169**, 733–738 (1970)

Carothers, S. W.: Population structure and social organization of southwestern riparian birds. *Amer Zool* **14**, 97–108 (1974)

Carpenter, C. R.: Sexual behaviour of free ranging Rhesus monkeys, *Macaco mulatta*: Specimens, procedures and behavioural characteristics of estrus. *J Comp Psychol* **33**, 113–142 (1942)

Carpenter, C. R.: Naturalistic behaviour of nonhuman primates. University Park, Pa.: Pennsylvania State Univ Press, 1964

Carpenter, C. R.: The howlers of Barro Colorado Island. In: *Primate Behaviour*. De Vore, I. (ed). New York: Holt, 1965, 250–291

Carr, C. J.: Plant ecological variation and pattern in the lower Omo Basin. In: *Earliest Man and Environments in the Lake Rudolf Basin*. Coppens, Y., Howell, F. C., Isaac, G. L., Leakey, R. E. F. (eds). Chicago: Univ. Chicago Press, 1976, 432–467

Carson, D. H.: Population concentrations and human stress. In: *Explorations in the Psychology of stress and Anxiety*. Rourke, B. P. (ed). Don Mills, Ontario: Longmans Canada, 1969, 27–42

Carson, R.: *Silent Spring*. Boston: Houghton Mifflin, 1962

Cartmill, M.: Rethinking primate origins. *Scince* **184**, 436–443 (1974)

Cassel, J.: Health consequences of population density and crowding. In: *Rapid Population Growth: Consequences and Policy Implications*. Baltimore: Johns Hopkins Press, 1971

Cavalli-Sforza, L. L., Bodmer, W. F.: *The Genetics of Human Populations*. San Francisco: Freeman, 1971

Chance, M. R. A.: Towards the biological definition of ethics. In: *Biology and Ethics*. Ebling, F. J. (ed). London: Academic Press, 1969

Chandra, R. K.: Antibody formation in first and second generation offspring of nutritionally deprived rats. *Science* **190**, 289−290 (1975)

Chang, K.: New evidence on fossil man in China. *Science* **136**, 749−760 (1962)

Chapell, J.: Astronomical theory of climatic changes. *Quat. Res.* **3**, 221−236 (1973)

Chapman, C. B., Mitchell, J. H.: The physiology of exercise. *Sci Amer* **212(5)**, 88−96 (1965)

Chard, C. S.: Archeology in the Soviet Union. *Science* **163**, 774−779 (1969)

Chatterjee I. B.: Evolution and the biosynthesis of ascorbic acid. *Science* **182**, 1271−1272 (1973)

Chavaillon, J.: Evidence for the technical practices of early Pleistocene hominids. In: *Earliest Man and Environments in the Lake Rudolf Basin*. Coppens, Y., Howell, F. C., Isaac, G. L., Leakey, R. E. F. (eds). Chicago: Univ. Chicago Press, 1976, 565−573

Cheraskin, E., Ringsdorf, W. M.: *Predictive Medicine*. New Canaan: Keats, 1973

Cherry, C.: *On Human Communication*. New York: Science Editions, 1957

Cheyne, J. A., Foster, W. M., Spence, J. B.: The incidence of disease and parasites in the Soay sheep population of Hirta. In: *Island Survivors*. Jewell, P. A., Milner, C., Boyd, J. M. (eds). London: Athlone, 1974, 388−359

Chitty, D., Southern, N. H.: *Control of Rats and Mice*. Oxford: Clarendon, 1954

Chitty, H.: Variations in the weight of adrenal glands of field voles *Microtus agrestis*. *J Endocrinol*. **22**, 387−393 (1961)

Christian, J. J.: Endocrine adaptive mechanisms and the physiologic regulation of population growth. In: *Physiological Mammalogy, Vol 1, Mammalian Populations*. Mayer, W., Van Gelder, R. (eds). New York: Academic Press, 1963, 189−381

Christian, J. J.: Social subordination, population density and mammalian evolution. *Science* **168**, 84−90 (1970)

Clapperton, J. L.: The energy metabolism of sheep walking on the level and on gradients. *Brit J Nutr* **18**, 47−54 (1964)

Clark, C.: Cultural differences in reactions to discrepant communication. *Human Organ* **27** 125−131 (1968)

Clark, G.: *The Earlier Stone Age Settlement of Scandinavia*. Cambridge: Cambridge Univ. Press, 1975

Clark, J. D.: *The Prehistory of Africa*. Southampton: Thames & Hudson, 1970

Clark, J. G. D.: *Archeology and Society*. Cambridge, Mass.: Harvard Univ. Press, 1957

Clark, J. G. D.: Primitive man in Egypt, Western Asia and Europe in Mesolithic times. In: *Primitive Man in Egypt, Western Asia and Europe*. Garrod, D. A. E., Clark, J. G. D. Cambridge: Cambridge Univ. Press, 1965, 23−61

Clark, J. G. D.: *The Stone Age Hunters*. New York: McGraw-Hill, 1967

Clark, J. G. D.: *World Prehistory*. Cambridge: Cambridge Univ. Press (2nd ed), 1969

Clarke, S. H., Brander, R. Radiometric determination of porcupine surface temperature under two conditions of overhead cover. *Physiol Zool* **46**, 230−237 (1973)

Clastres, P.: The Guayaki. In: *Hunters and Gatherers Today*. Bicchieri, M. G. (ed). New York: Holt, Rinehart Winston, 1972, 138−174

Clemens, W. A.: Origin and early evolution of marsupials. *Evolution* **22**, 1−18 (1968)

CLIMAP Project members: The surface of the ice-age Earth. *Science* **191**, 1131−1137 (1976)

Cohen, J. E.: *Casual Groups of Monkeys and Men*. Cambridge, Mass.: Harvard Univ. Press, 1971

Colbert, E. H.: *Evolution of the Vertebrates*. New York: Wiley, 1955

Colbert, E. H.: *Dinosaurs*. New York: Dutton, 1961

Cole, La Mont. C.: Man's ecosystem. *Bioscience* **16**, 243−248 (1966)

Collias, N. E.: Aggressive behaviour among vertebrate animals. *hysiol Zool* **17(1)**, 83−123 (1944)

Commoner, B.: *Science and Survival*. New York: Viking, 1963

Cook, R.: The general nutritional problems of Africa. *African Aff* **65**, 329−340 (1966)

Cook, S. F.: Aging of and in populations. In: *Developmental Physiology and Aging*. Timiras, P. S. (ed). New York: Macmillan, 1972, 581−606

Cooke, H. B. S.: The fossil mammal fauna of Africa. In: *Evolution, Mammals and Southern Continents*. Keast, A., Erk, F. C., Glass, B. (eds). Albany: State Univ. New York Press, 1972, 89−139

Cooke, H. B. S.: Pleistocene chronology: Long or short? *Quat Res* **3**, 206−220 (1973)

Coon, C. S.: *The Origin of Races*. New York: Knopf, 1962

Cornwall, I.: *Ice Ages, Their Nature and Effects*. London: Baker, 1970

Corruccini, R. S.: Multivariant analysis of *Gigantopithecus* mandibles. *Amer J Phys Anthropol.* **42**, 167−170 (1975)

Coulson, J. C.: Differences in the quality of birds nesting in the centre and on the edges of a colony. *Nature (Lond)* **217**, 478−479 (1968)

Count, E. W.: Brain and body weight in man: Their antecedents in growth and evolution. *Ann N Y Acad Sci* **46(10)**, 993−1122 (1947)

Cowles, R. B.: Missiles, clay pots and mortality rates in primitive man. *Amer Nat* **97**, 29−37 (1963)

Cox, B. (ed): *Cultural Ecology*. Toronto: McClelland & Stewart Carleton, 1973

Craig, J. V., Baruth, R. A.: Inbreeding and social dominance ability in chickens. *Anim Behav* **13**, 109−113 (1965)

Craig, J. V., Bhagwat, A. L.: Agonistic and mating behaviour of adult chickens modified by social and physical environments. *Appl Anim Ethol* **1**, 57−65 (1974)

Cravioto, J., De Licardie, E. R.: Nutrition and behaviour and learning. In: *World Review of Nutrition and Dietetics, Vol. 16*. Bourne, G. H. (ed). New York: Karger (1973)

Cronin, T. J.: Influence of lactation upon ovulation. *Lancet* **2**, 422−424 (1968)

Croog, S. H.: The family as a source of stress. In: *Social Stress*. Levine, S., Scotch, N. A. (eds). Chicago: Aldine, 1969, 19−53

Crook, J. H.: The adaptive significance of avian social organization. *Symp Zool Soc Lond* **14**, 181−218 (1965)

Crook, J. H.: The socioecology of primates. In: *Social Behaviour in Birds and Mammals*. Crook, J. H. (ed). London: Academic Press, 1968a, 103−166

Crook, J. H.: The nature and function of territorial aggression. In: *Man and Aggression*. Ashley Montagu, M. F. (ed). Oxford: Oxford Univ. Press, 1968b, 141−178

Crook, J. H.: Social organization and the environment: Aspects of contemporary social ethology. *Anim Behav* **18(2)**, 197−209 (1970a)

Crook, J. H. (ed): *Social Behaviour in Birds and Mammals*. New York: Academic Press, 1970b

Crosby, A.: *Creativity and Performance in Industrial Organizations*. London: Tavistock, 1968

Crowe, B. L.: The tragedy of the commons revisited. *Science* **166**, 1103−1107 (1969)

Crowcroft, P.: *Mice All Over*. London: Foulis, 1966

Cullen, E.: Adaptations of the kittiwake to cliff-nesting. *Ibis* **99**, 275−302 (1957)

Cumming, D. H. M.: A field study of the ecology and behaviour of warthog. *Mus Mem* **7**. Salisbury, Rhodesia: Nat Mus Monum Rhodes (1975a)

Cummins, R. A., Walsh, R. N., Budtz-Olsen, O. E., Konstantinos, T., Horsfall, C. R.: Environmentally-induced changes in the brain of elderly rats. *Nature (Lond)* **243**, 516−518 (1973)

Cunningham, M.: *Intelligence: Its Organization and Development*. New York: Academic Press, 1972

Cureton, T. K.: *The Physiological Effects of Exercise Programs on Adults*. Springfield, Ill.: Thomas, 1969

Curry-Lindahl, K.: *Conservation for Survival: An Ecological Strategy*. New York: Morrow, 1972a

Curry-Lindahl, K.: *Let Them Live: A Worldwide Survey of Animals Threatened with Extinction*. New York: Morrow, 1972b

Curry-Lindahl, K.: Nature conservation and water resources aspects of the highlands of East Africa. Proc. Tech Conf. on Agroclimatology of the Highlands of East Africa, Nairobi, 1973. Geneva: World Meteorological Organization, 1974

Cuthbertson, D. P.: The world setting. In: *The Biology of Affluence*. Smith, G., Smythe, J. C. (eds). Edinburgh: Oliver & Boyd, 1972, 109–126

Dahlberg, A. A.: Dental evolution and culture. *Human Biol* **35**, 237–249 (1963)

Damon, A.: Secular trends in height and weight within old American families at Harvard, 1870–1965, within twelve four-generation families. *Amer J Phys Anthropol*. **29**, 45–50 (1968)

Damon, A.: Constitutional medicine. In: *Anthropology and the Behavioural and Health Sciences*. Mering, O., Kasdan, L. (eds). Pittsburg: Univ. Pittsburgh Press, 1970, 179–195

Daniels, F., van der Leun, J. C., Johnson, B. E.: Sunburn. *Sci Amer* **219**, 38–46 (1968)

Dansgaard, W., Tauber, A.: Glacier oxygen-18 content and Pleistocene ocean temperatures. *Science* **166**, 499–502 (1969)

Darlington, P. J.: *Zoogeography: The geographic distribution of animals*. New York: Wiley, 1957

Dart, R. A.: The predatory implimental technique of *Australopithecus*. *Amer J Phys Anthropol*. **7**, 1–38 (1949)

Dart, R. A.: The predatory transition from ape to man. *Inter Anthropol. Ling Rev* **1(4)**, 201–218 (1953)

Dart, R. A.: The osteodontokeratic culture of *Australopithecus prometheus*. *Transvaal Mus Mem* **10**, 1957

Dart, R. A.: *Adventures with the Missing Link*. London: Hamish Hamilton, 1959

Darwin, C.: *The Expression of the Emotions in Man and Animals*. Chicago: Univ. Chicago Press (1872)

Darwin, C.: *Origin of Species*. New York: Carlton House (6th ed), 1972

Dasmann, R. F., Taber, R. D.: Behaviour of Columbia black-tailed deer with reference to population ecology. *J Mammal* **37(2)**, 143–164 (1956)

Davenport, J. W., Gonzales, L. M., Carey, J. C., Bishop, S. B., Hagquist, W. W.: Environmental stimulation reduces learning deficits in experimental cretinism. *Science* **191**, 578–579 (1976)

Davids, A., de Vault, S.: Maternal anxiety during pregnancy and childbirth anomalies. *Psychosom Med* **24**, 464–470 (1962)

Davis, D. D.: The Giant Panda: A morphological study of evolutionary mechanisms. Fieldiana: *Zool Mem* **3**. Chicago: Chic Nat Hist Mus, 1964

Davis, D. E.: The physiological analysis of aggressive behaviour. In: *Social Behaviour and Organization Among Vertebrates*. Etkin, W. (ed). Chicago: Univ. Chicago Press, 1964, 53–74

Davis, K.: Sociological aspects of genetic control. In: *Genetics and the Future of Man*. Roslansky, I. D. (ed). New York: Appleton-Century-Crofts, 1965, 173–204

Davis, P. M. C.: Man the mythic earth-tree. *Month* **5**, 210–216 (1972)

Davison, A. N., Dobbing, J.: Myelination as a vulnerable period in brain development. *Brit Med Bull* **22**, 40–44 (1968)

Dawson, J.: Urbanization and mental health in a West African community. In: *Magic, Faith and Healing: Studies in Primitive Psychiatry Today*. Kiev, A. (ed). London: Free Press of Glencoe, Collier-Macmillan, 1964, 305–342

Dawson, T. J., Hulbert, A. J.: Standard energy metabolism of marsupials. *Nature (Lond)* **221**, 383 (1969)

Dawson, W. R., Bartholomew, G. A.: Temperature regulation and water economy of desert birds. In: *Desert Biology*. Browning, G. W. (ed). London: Academic Press, 1968

Day, M. H.: Hominid postcranial remains from the East Rudolf succession. In: *Earliest Man and Environments in the Lake Rudolf Basin*. Coppens, Y., Howell, F. C., Isaac, C. L., Leakey, R. E. F. (eds). Chicago: Univ. chicago Press, 1976, 507–521

Day, R. H.: Visual spatial illusions: A general explanation. *Science* **175**, 1335–1340 (1972)

De Bock, E. A.: On the behaviour of the mountain goat (*Oreamnos americanus*) in Kootenay National Park. Unpubl. MSc thesis, Univ. Alberta, Edmonton, 1970

Deevey, E. S.: Discussion of Part V. In: *Man the Hunter*. Lee, R. B., de Vore, I. (eds). Chicago: Aldine, 1968, 248–249

De Huff-Peters, A.: Day care. In: *Maternal and Child Health Practices*. Wallace, H. M., Gold, E. M., Lis, E. F. (eds). Springfield, Ill.: Thomas, 1973, 744–760

De Long, A. J.: The communication process: A generic model for man–environment relations. *Man-Environ Syst* **2(5)**, 263–313 (1972)

De Montellano, B. O.: Empirical Aztec medicine. *Science* **188**, 215–220 (1975)

Denenberg, V. H., Ottinger, D. R., Stephens, M. W.: Effects of maternal factors upon growth and behaviour of the rat. *Child Devel* **33**, 65–71 (1962)

Denenberg, V. H., Rosenberg, K. M.: Nongenetic transmission of information. *Nature (Lond)* **216**, 549–550 (1967)

Denton, G. H., Karlén, W.: Holocene climatic variations—Their pattern and possible cause. *Quat Res* **3**, 155–205 (1973)

Denton, G. H., Stuiver, M.: Late Pleistocene glacial stratigraphy and chronology, Northeastern St Elias Mountains, Yukon Territory, Canada. *Geol Soc Amer Bull* **78**, 485–510 (1967)

Denton. G. H., Armstrong, R. L., Stuiver, M.: The late Cenozoic glacial history of Antarctica. In: *Late Cenozoic Glacial Ages*. Turekian, K. (ed). New York: Yale Univ. Press, 1971, 267–306

De Ricqlès, A.: Vers une histoire de la physiologie thermique: Les données histologiques et leur interprétation fonctionelle. *Acad Sci Paris CR* Series D. **275**, 1745–1748 (1972a)

De Ricqlès, A.: Vers une histoire de la physiologie thermique: L'apparition de l'endothermie et le concept de Reptile. *Acad Sci Paris CR* Series D. **275**, 1875–1878 (1972b)

De Ricqlès, A.: Nature et signification des 'surface épiphysaires' chez les tétrapodes fossiles. *Aycad Sci Paris CR* Series D. **274**, 3084–3087 (1972c)

De Ricqlès, A., de Bonis, L., Lebeau, M.-O.: Etude de la répartition des types de tissus osseux chez les vertébrés tétrapodes au moyen de l'analyse factorielle des correspondances. *Acad Sci Paris CR* Series D. **274**, 3527–3530 (1972)

Des Meules, P.: The influence of snow on the behaviour of moose. *Quebec Service Faune Rapport* **3**, 51–73 (1964)

Destunis, G., Seebandt, R.: Beitrag zur Frage der Musikeinwirkung auf die Zwischenhirngesteuerten Funktionen des Kindes. In: *Musik in der Medizin*. Teirich, H. R. (ed). Stuttgart: Fischer, 1958, 34–42

Deutsch, C. P.: Social class and child development. In: *Child Development and Social Policy, Vol. 3*. Caldwell, B. M., Ricciuti, H. N. (eds). Chicago: Univ. Chicago Press, 1973, 233–282

Deutsch, J. A.: The cholinergic synapse and the site of memory. *Science* **174**, 788–794 (1971)

De Vries, H. A.: Physiology of Exercise. Dubuque, Iowa: Brown, 1966

Diamond, I. T., Hall, W. C.: Evolution of neocortex. *Science* **164**, 251–261 (1969)

Dickeman, M.: Demographic considerations of infanticide in man. *Ann Rev Ecol Syst* **6**, 107–137 (1975)

Diebold, A. R.: Anthropological perspectives: Anthropology and the comparative psychology of communicative behaviour. In: *Animal Communication*. Sebeok, T. A. (ed). Bloomington: Indiana Univ. Press, 1968, 525–571

Dodds, D. O.: Observations of prerutting behaviour of Newfoundland moose. *J Mammal* **39**, 412–416 (1958)

Dornstreich, M. D.: Food habits of early man: Balance between hunting and gathering. *Science* **179**, 306 (1973)

Dreitzel, H. P.: Introduction. In: *The Social Organization of Health*. Dreitzel, H. P. (ed). Recent Sociology **3**. New York: Macmillan, 1972

Dreizen, S., Spirakis, C. N., Stone, R. E.: Undernutrition slows growth of adolescents. *J Pediat* **70**, 256–263, (1967)

Drozdz, A., Osiecki, A.: Intake and digestibility of natural feeds by roe deer. *Acta Theriol* **18**, 81–91 (1973)

Dru, D., Walker, J. P., Walker, J. B.: Self-produced locomotion restores visual capacity after striate lesions. *Science* **187**, 265–266 (1975)

Drury, W. H.: Birds of the Saint Elias quadrangle in the southwestern Yukon Territory. *Can Field Nat* **67**, 103–128 (1953)

Dubos, R.: *Medical Utopias, The Dreams of Reason*. New York: Columbia Univ. Press, 1961

Dubos, R.: *So Human an Animal*. New York: Scribner, 1968

Dubos, R.: *A God Within*. New York: Scribner, 1972

Dubost, G.: Les niches écologiques des forêts tropicales Sudamericaines et Africaines, sources de convergences remarquables entre Rongeurs et Artiodactyles. *La Terre et al Vie* **22**, 3–28 (1968)

Duhl, L. J.: The Process of Re-Creation: The health of the "I" and the "us". Ethics in Science and Medicine, **3**, 33–63 (1976)

Dumond, D. E.: The limitation of human population: A natural history. *Science* **187**, 713–721 (1975)

Dunbar, M. J.: The scientific importance of the circumpolar region, and its flora and fauna. In: *Productivity and Conservation in Circumpolar Lands*. Fuller, W. A., Kevan, P. G. (eds). Morges: IUCN Publications New Series **16**, 1970, 71–77

Dunn, F. L.: Epidemiological factors: Health and disease in hunter–gatherers. In: *Man the Hunter*. Lee, R. B., de Vore, I. (eds). Chicago: Aldine, 1968, 221–228

Durham, W. H.: The adaptive significance of cultural behaviour. *Human Ecol* **4(2)**, 89–121 (1976)

Eaton, R. L.: *The Cheetah*. New York: Van Nostrand Reinhold, 1974

Eckhardt, R. B.: Population genetics and human origins. *Sci Amer* **226(1)**, 94–103 (1972)

Eckholm, E. P.: *The Picture of Health*. Norton and Company, New York, 256 pp. (1977)

Edds, M. V.: Hypertrophy of nerve fibres to functionally overloaded muscles. *J Comp Neurol* **93**, 259–275 (1950)

Edwards, H. P., Barry, W. R., Wyspianski, J. O.: Effects of differential rearing on photic evoked potentials and brightness discrimination in the albino rat. *Devel Psychobiol* **2**, 133–138 (1969)

Edwards, R. J., Ritcey, R. W.: The migration of a moose herd. *J Mammal* **37**, 486–494 (1956)

Edwards, W. E.: The late-Pleistocene extinction and diminution in size of many mammalian species. In: *Pleistocene Extinctions*. Martin, P. S., Wright, H. E. (eds). New Haven: Yale Univ. Press, 1967, 141–154

Egorov, O. V.: *Wild Ungulates of Yakutia*. Trans. from Russian. Springfield: US Dept. Commerce, 1967

Ehrlich, P.: *The Population Bomb*. New York: Ballantine, 1968

Eibl-Eibesfeldt, I.: Angeborenes und Erworbenes im Nestbauverhalten der Wanderratte. *Naturwiss* **42**, 633–634 (1955)

Eibl-Eibesfeldt, I.: Angeborenes und Erworbenes im Verhalten einiger Säuger. *Z Tierpsychol* **20**, 705–754 (1963)

Eibl-Eibesfeldt, I.: *Grundrisse der Vergleichenden Verhaltensforschung, Ethologie*. Munich: Pieper, 1967

Eibl-Eibesfeldt, I.: *Liebe und Hass*. Munich: Pieper, 1970

Eichwald, H. F., Fry, P. G.: Nutrition and learning. *Science* **163**, 644–648 (1969)

Eisenberg, J. F.: Studies on the behaviour of *Permyscus maniculatus gambelii* and *Peromyscus californicus parasiticus, Behaviour* **19**, 177−207 (1962)

Eisenberg, J. F.: The social organization of mammals. *Handbuch der Zoologie, Vol. 8,* **10(7)**, 1−92 (1966)

Eisenberg, J. F.: Communication mechanisms and social integration in the black spider monkey, *Ateles fusciceps robustus,* and related species. *Smithsonian Contrib Zool* **213**, 1976

Eisenberg, J. F., Lockhart, M.: An ecological reconnaissance of Wilpattu National Park, Ceylon. *Smithsonian Contrib Zool* **101**, 1972

Eisenberg, J. F., Muckenhirn, N. A., Rudran, R.: The relation between ecology and social structure in primates. *Science* **176**, 863−873 (1969)

Elkin, D.: Perceptual development in children. *Amer Sci* **63**, 533−541 (1975)

El-Najjar, M. Y., Robertson, A. L.: Spongy bones in prehistoric America. *Science* **193**, 141−143 (1976)

Elton, C.: *The Ecology of Invasion by Animals and Plants.* London: Methuen, 1958

Emiliani, C.: Pleistocene paleotemperatures. *Science* **168**, 822−824 (1970)

Emlen, S. T.: Territoriality in the bullfrog *Rana catesbeiana. Copeia* **1968(2)**, 240−243 8)

Ertl, J. P., Schafer, E. W. P.: Brain response correlates of psychometric intelligence. *Nature (Lond)* **223**, 421−422 (1969)

Esser, A. H. (ed): *Health and the Built Environment: Stanley House 1974 Workshop.* Ottawa: Health & Welfare Canada, 1975

Estes, R. D.: Territorial behaviour of the wildebeest (*Connochaetes taurinus* Burchell 1823). *Z Tierpsychol* **26**, 284−370 (1969)

Estes, R. D.: Social organization of the African Bovidae. In: *The Behaviour of Ungulates and its Relation to Management.* Geist, V., Walther, F. R. (eds). Morges: IUCN Publications New Series **24**, 1974, 166−205

Etkin, W.: Social behaviour and the evolution of man's mental faculties. *Amer Nat* **88**, 129−142 (1954)

Ewer, R. F.: *Ethology of Mammals.* London: Logos, 1968

Fagen, R. M., George, K.: Play behaviour and exercise in young ponies (*Equus caballus* L). *Behav Ecol Sociobiol* **2**, 267−269 (1977)

Fairbridge, R. W.: Shellfish-eating preceramic Indians in coastal Brazil. *Science* **191**, 353−359 (1976)

Farmer, F. A., Neilson, H. R.: The caloric value of meat and fish of northern Canada. *J Can Diet Assoc* **28(4)**, 174−178 (1967)

Farmer, F. A., Ho, M. L., Neilson, H. R.: Analyses of meats eaten by humans or fed to dogs in the Arctic. *J Can Diet Assoc* **32(3)**, 137−141 (1971)

Farnham, A. B.: *Home Tanning and Leather Guide.* Tuscon: Sincere, 1950

Farr, L. E.: Medical consequences of environmental noise. *JAMA* **202**, 171−174 (1967)

Feduccia, A.: Dinosaurs as reptiles. *Evolution* **27(1)**, 166−169 (1973)

Fengler, F. A.: Die systematische spezielle Musiktherapie im Dienste der Sprachheilarbeit. *Deutsch Ges-Wes* **5**, 1488 (1950)

Fenna, D., Mix, L., Schaefer, O., Gilbert, J. A. L.: Ethanol metabolism in various racial groups. *Can Med Assoc J* **105**, 472−475 (1971) ·

Fernandez, J. W.: Analysis of ritual: Metaphoric correspondence as elementary forms. *Science* **182**, 1366−1367 (1973)

Fernandez, J. W.: The mission of metaphor in expressive culture. *Curr Anthropol.* **15(2)**, 119−145 (1974)

Fesbach, S., Fesbach, N.: Influence of the stimulus object upon the complementary and supplementary projections of fear. *J Abnorm Soc Psychol* **66**, 498−502 (1963)

Festinger, L.: *A Theory of Cognitive Dissonance.* Stanford: Stanford Univ. Press, 1957

Field, R. A., Smith, F. C., Hepworth, W. G.: The pronghorn antelope carcass. *Agric Exp Stat Bull* **575**. Laramie: Univ. Wyoming, 1972

Field, R. A., Smith, F. C., Hepworth, W. G.: The mule deer carcass. *Agric Exp Stat Bull* **589**. Laramie: Univ. Wyoming, 1973a.

Field, R. A., Smith, F. C., Hepworth, W. G.: The elk carcass. *Agric Exp Stat Bull* **594**. Laramie: Univ. Wyoming, 1973b

Fisher, R. A.: *The Genetic Theory of Natural Selection.* London: Oxford Univ. Press, 1930

Fitch, F. J., Miller, J. A.: Radioisotropic age determinations of Lake Rudolf artifact site. *Nature (Lond)* **226**, 226−228 (1970)

Flannery, K. V., Marcus, J.: Formative Oaxaca and the Zapotec cosmos. *Amer Sci* **64**, 374−383 (1976)

Flavell, J. H.: *The Developmental Psychology of Jean Piaget.* New York: Van Nostrand, 1963

Ford, A. B.: Casualties of our time. *Science* **167**, 256−263 (1970)

Formozov, A. N.: *Snow Cover as an Integral Factor of the Environment and its Importance to the Ecology of Mammals and Birds.* English trans. Edmonton: Univ. Alberta, 1946

Foster, J. B., Coe, M. J.: The biomass of game animals in Nairobi National Park, 1960−1966. *J Zool Lond* **155**, 413−425 (1968)

Fox, M. W., Andrews, R. V.: Physiological and biochemical correlates of individual differences in behaviour of wolf cubs. *Behaviour* **46**, 129−140 (1973)

Frädrich, H.: Das Verhalten der Schweine (Suidae, Tayassuidae) und Flusspferde (Hippopotamidae). *Handbuch der Zoologie, Vol. 8.* Berlin: De Gruyter, 1967

Frädrich, H.: A comparison of behaviour in the Suidae. In: *The Behaviour of Ungulates and its Relation to Management.* Geist, V., Walther, F. R. (eds). Morges: IUCN Publications New Series **24**, 1974, 133−143

Frankel, G. S., Gunn, D. L.: *The Orientation of Animals.* London: Clarendon, 1940

Freeman, R. D., Thibos, L. N.: Electrophysiological evidence that abnormal early visual experience can modify the human brain. *Science* **180**, 876−878 (1973)

French, C. E., McEwan, L. C., Magruder, N. D., Ingram, R. H., Swift, R. W.: Nutritional requirements of white-tailed deer for growth and antler development. *Penn State Univ Agric Exp Stat Bull* **600**. University Park: Pennsylvania State Univ., 1955

Frenzel, B.: The Pleistocene vegetation of northern Eurasia. *Science* **161**, 637−648 (1968)

Freuchen, P.: *Book of the Eskimos.* Greenwich, Conn: Fawcett-Premier, 1961

Frisch, K. von: *Aus dem Leben der Bienen.* Berlin. Springer-Verlag, 1959

Frisch, K. von: *Animal Architecture.* New York: Harcourt Brace Jovanovich, 1974

Fuller, W. A., Kevan, P. G. (eds): Productivity and Conservation in Northern Circumpolar Lands. Morges: IUCN Publications New Series **16**, 1970

Gable, G.: *Stone Age Hunters of the Kafu.* Boston: Boston Univ. Press, 1965

Gage, N. L., Berliner, D. C.: *Educational Psychology.* Chicago: Rand McNally, 1975

Galbraith, J. K.: *The Affluent Society.* New York: Houghton Mifflin, 1958

Galbraith, J. K.: *The New Industrial State.* New York: Houghton Mifflin, 1967

Galef, B. G.: Target novelty elicits and directs shock-associated aggression in wild ruts. *J Comp Physiol Psychol* **71**, 87−91 (1970)

Galle, O. R., Grove, W. R., McPherson, J. M.: Population density and pathology: What are the relations for man. *Science* **176**, 23−30 (1972)

Gallup, G. G.: Chimpanzees: Self-recognition. *Science* **167**, 87−87 (1970)

Galton, P. M.: A primitive dome-headed dinosaur (*Ornithischia pachycephalosauridae*) from the lower Cretaceous of England and the function of the dome of pachycephalosaurids. *J Paleontol* **45(1)**, 40−47 (1971)

Gardarsson, A., Moss, R.: Selection of food by Icelandic ptarmigan in relation to its availability and nutritive value. In: *Animal Populations in Relation to their Food Resources.* Watson, A. (ed). Oxford: Blackwell, 1970, 47−72

Gardner, R. A., Gardner, B. T.: Teaching sign language to a chimpanzee. *Science* **165**, 664−672 (1969)

Garn, S. M.: Culture and the direction of human evolution. *Human Biol* **35(3)**, 221−236 (1963)

Garstang, W.: The theory of recapitulation: A critical restatement of the biogenetic law. *Linn Soc J Zool* **35,** 81–101 (1922)

Garutt, W. E.: *Das Mammut. Neue Brehm-Bücherei.* Wittenberg-Lutherstadt: Ziemsen, 1964

Gaston, E. T. (ed): *Music in Therapy.* New York: Macmillan, 1968

Gaulin, S. J. C., Kurland, J. A.: Primate predation and bioenergetics. *Science* **191,** 314–314 (1976)

Gauthier-Pilters, H.: The behaviour and ecology of camels in the Sahara, with special reference to nomadism and water management. In: *The Behaviour of Ungulates and its Relation to Management.* Geist, V., Walther, F. R. (eds). Morges: IUCN Publications New Series **24,** 1974, 542–551

Geber, W. F.: Cardiovascular and teratogenic effects of chronic intermittant noise stress. In: *Physiological Effects of Noise.* Welch, B. L., Welch, A. S. (eds). New York: Plenum, 1970, 85–90

Geist, V.: Feral goats in British Columbia. *Murrelet* **41,** 1–7 (1960)

Geist, V.: On the behaviour of the North American moose (*Alces alces andersoni* Peterson 1950) in British Columbia. *Behaviour* **20(3–4),** 377–416 (1963)

Geist, V.: On the rutting behaviour of the mountain goat. *J Mammal* **45(4),** 551–568 (1965)

Geist, V.: The evolution of hornlike organs. *Behaviour* **27,** 175–215 (1966a)

Geist, V.: On the behaviour and evolution of North American mountain sheep. Unpubl. PhD thesis, Univ. British Columbia, 1966b

Geist, V.: Ethological observations on some North American cervids. *Zool Beitrage* **12,** 219–250 (1966c)

Geist, V.: The evolutionary significance of mountain sheep horns. *Evolution* **20(4),** 558–566 (1966d)

Geist, V.: On fighting injuries and dermal shields of mountain goats. *J Wildl Manage* **31(1),** 192–194 (1967a)

Geist, V.: A consequence of togetherness. *Nat Hist* **76(8),** 24–31 (1967b)

Geist, V: On the interrelation of external appearance, social behaviour and social structure of mountain sheep. *Z Tierpsychol* **25,** 194–215 (1968a)

Geist, V.: On delayed social and physical maturation in mountain sheep. *Can J Zool* **46,** 899–904 (1968b)

Geist, V.: Mountain Sheep: A Study in Behaviour and Evolution. Chicago: Univ. Chicago Press, 1971a

Geist, V.: On the relation of social evolution and dispersal in ungulates during the Pleistocene, with emphasis on the old-world deer and the genus *Bison. Quat Res* **1,** 283–315 (1971b)

Geist, V.: A behavioural approach to the management of wild ungulates. In: *The Scientific Management of Animal and Plant Communities for Conservation:* Duffey, E., Watt, A. S. (eds). Oxford: Blackwell, 1971c, 413–424

Geist, V.: An ecological and behavioural explanation of mammalian characteristics and their implication to Therapsid evolution. *Z Säugetierk* **37,** 1–5 (1972)

Geist, V.: On the relationship of social evolution and ecology in ungulates. *Amer Zool* **14,** 205–20 (1974a)

Geist, V.: On fighting strategies in animal combat. *Nature (Lond)* **250,** 354 (1974b)

Geist, V.: On the evolution of reproductive potential in moose. *Le Naturaliste Canadien* **101,** 527–537 (1974c)

Geist, V.: On the relationship of ecology and behaviour in the evolution of ungulates: Theoretical considerations. In: *The Behaviour of Ungulates and its Relation to Management.* Geist, V., Walther, F. R. (eds). Morges: IUCN Publications New Series **24,** 1974d, 235–246

Geist, V.: *Mountain Sheep and Man in the Northern Wilds.* Ithaca: Cornell Univ. Press, 1975a

Geist, V.: On life in the sight of glaciers. *Nature Canada* **4(3),** 10–16 (1975b)

Geist, V.: Wildlife and people in an urban environment: The biology of cohabitation. In:

Proceedings of the Symposium of Wildlife in Urban Canada. Euler, D., Gilbert, F., McKeating, G. (eds). Guelph, Ontario: Univ. Guelph, 1975c, 36−47

Geist, V.: A comparison of social adaptations in relation to ecology in gallinaceous bird and ungulate societies. In: *Annual Review of Ecology and Systematics* **8.** Johnston, R. F., Frank, P. W., Michener, C. D. (eds). Palo Alto: Annual Reviews, 1977, 193−207

Geist, V.: On weapons, combat and ecology. In: *Aggression, Dominance and Individual Spacing. Vol. 4, Advances in the Study of Communication and Effect,* Krames, L. (ed). New York: Plenum, 1978a

Geist, V.: Adaptive strategies, late Paleolithic life styles: Towards a theory of preventive medicine. In: *Proceedings of the Symposium on Population Control by Social Behaviour, London 1977.* London: Inst. Biology, 1978b, 245−260

Geist, V.: Adaptive strategies in the behaviour of elk. In: *The Ecology and Management of the North American Elk.* Thomas, J. W. (ed). Washington: Wildlife Manage Inst, 1978c

Geist, V.: Behaviour patterns of mule deer. In: *Mule Deer and Blacktailed Deer,* Wallmo, O. C. (ed). Washington: Wildlife Society, in press.

Geist, V., Karsten, P.: The wood bison (*Bison bison athabascae* Rhodes) in relation to hypotheses on the origin of the American bison (*Bison bison* Linnaeus). *Z Säugetierkunde* **42(2),** 119−127 (1977)

Geist, V., Walther, F. R. (eds): *The Behaviour of Ungulates and its Relation to Management.* Morges: IUCN Publications New Series **24,** 1974

Geschwind, N.: The organization of language and the brain. *Science* **170,** 940−944 (1970)

Gibson, E. J.: The development of perception as an adaptive process. *Amer Sci* **58,** 98−107 (1970)

Gidion, S.: *Mechanization Takes Command.* New York: Oxford Univ. Press Norton, 1948

Gilbert, B. K.: The influence of foster rearing on adult social behaviour in fallow deer (*Dama dama*). In: *The Behaviour of Ungulates and its Relation to Management.* Geist, V., Walther, F. R. (eds). Morges: IUCN Publications New Series **24,** 1974, 247−273

Gilchrist, M.: *The Psychology of Creativity.* Carleton, Australia: Melbourne Univ. Press, 1972

Gilliard, E. T.: Evolution of bowerbirds. *Sci Amer* **209(2),** 38−46 (1963)

Gilula, M. F., Daniels, D. N.: Violence and man's struggle to adapt. *Science* **164,** 396−405 (1969)

Ginsburg, B. E.: Genotypic factors in the ontogeny of behaviour. In: *Animal and Human, Vol. XII.* Masserman, J. H. (ed). New York: Grune & Stratton, 1968, 12−17

Giterma, R. E., Golubeva, L. V.: Vegetation of eastern Siberia during the Anthropogene period. In: *The Bering Land Bridge.* Hopkins, D. M. (ed). Stanford: Stanford Univ. Press, 1967, 232−244

Glaser, H. H., Heagarty, M. C., Bullard, D. M., Pivchik, E. C.: Physical and psychological development of children with early failure to thrive. *J Pediat* **73,** 690−698 (1968)

Glass, B., Ericson, D. B., Heezen, B. C., Opdyke, N. D., Glass, J. A.: Geomagnetic reversals and Pleistocene chronology. *Nature (Lond)* **216,** 437−442 (1967)

Glickman, S. E., Sroges, R. W.: Curiosity in zoo animals, *Behaviour* **27,** 151−188 (1966)

Goldberg, S.: *The Inevitability of Patriarchy.* New York: Morrow, 1973

Goldthwait, R. P.: Evidence from Alaska glaciers of major climatic changes. In: *World Climate from 8000 to 0 BC.* Sawyer, J. S. (ed). London: Royal Meteorol. Soc, 1966, 40−51

Golub, A. M., Masiarz, F. R., Villars, T., McConnell, J. V.: Incubation effects in behaviour induction in rats. *Science* **168,** 392−395 (1970)

Goodall, J. van Lawick: The behaviour of free-living chimpanzees in the Gombe Stream Reserve. *Anim Behav Monog* **1(3),** 161−311 (1968)

Goodall, J. van Lawick: *In the Shadow of Man.* London: Collins, 1971

Goode, W. J.: Force and violence in the family. In: *Intimacy, Family and Society.* Skolnick, A., Skolnick, J. H. (eds). Boston: Little Brown, 1974, 72−92

Gordon, M. S., Bartholomew, G. A., Grinnell, A. D., Jorgensen, C. B., White, F. N.:

Animal Function: Principles and Adaptations. New York: Macmillan, 1968

Gordon, W. J. J.: *Synectics*. New York: Collier Macmillan, 1961

Gorecki, A.: Metabolic rate and energy budget in bank vole. *Acta Theriol* **13(20)**, 341–365 (1968)

Gottschlich, H. J.: Biotop und Wuchsform—eine craniometrischallometrische Studie an europäischen Populationen von *Cervus elaphus*. *Beiträge zur Jagd- und Wildforschung* **78**, 83–101 (1965)

Gough, K.: The origin of the family. In: *Intimacy, Family and Society*. Skolnick, A., Skolnick, J. H. (eds). Boston: Little Brown, 1974, 41–60 First published 1971.

Graham, N. McC: Measurement of the heat production of sheep: The influence of training and of a tranquilizing drug. *Proc Austral Soc Anim Prod* **4**, 138–144 (1962)

Graham, N. McC.: Energy costs of feeding activities and energy expenditure of grazing sheep. *Austral J Agric Res* **15(6)**, 969–973 (1964)

Graham, N. McC.: The net energy value of artificially dried subterranean clover harvested before flowering. *Austral J Agric Res* **20**, 365–373 (1969)

Gray, J.: *The Psychology of Fear and Stress*. New York: McGraw-Hill, 1971

Grayson, D. K.: Pleistocene avifaunas and the overkill hypothesis. *Science* **195**, 691–693 (1977)

Green, A. E., Lazell, J. D., Wright, R. W.: Anatomical evidence for a countercurrent heat exchange in leatherback turtle *(Dermochelys coriaced)*. *Nature (Lond)* **244**, 181 (1973)

Green, D. L.: Environmental influence on Pleistocene hominid dental evolution. *Bioscience* **20**, 276–279 (1970)

Greenbie, B. B.: An ethological approach to community design. In: *Environmental Design Research, Vol. 1*. Preiser, F. E. (ed). Stroudsburg, Pa.: Dowden Hutchinson & Ross, 1973, 14–23

Greenough, W. T.: Experimental modification of the developing brain. *Amer Sci* **63**, 37–46 (1975)

Grinnell, A. D.: Sensory physiology. In: *Animal Function: Principles and Adaptations*. Gordon, M. S., Bartholomew, G. A., Grinnell, A. D., Jorgensen, C. B., White, F. N. (eds). New York: Macmillan, 1968, 396–460

Grobstein, P., Chow, K. L., Spear, P. D., Mathers, L. H.: Development of rabbit visual cortex: Late appearance of a class of receptive fields. *Science* **180**, 1185–1187 (1973)

Grodums, E. I., Dempster, G.: Passive neutralizing antibodies in experimental Coxsackie B-1 virus infection. *Can J Microbiol* **10**, 53–61 (1964)

Grubb, P.: Population dynamics of Soay sheep. In: *Island Survivors*. Jewell, P. A., Milner, C., Boyd, J. M. (eds). London: Athlone, 1974, 242–272

Grüsser, O. Y., Grüsser-Cornehls, U., Bullock, T. H.: Functional organization of movement detecting neurons in the frog retina. *Pflügers Arch Physiol* **279**, 88–93 (1964)

Grzimek, B. V.: Versuche über das Farbensehen von Pflanzenessern 1: das farbige sehen (und die Sehschärfe) von Pferden. *Z Tierpsychol* **9**, 289 (1952)

Guhl, A. M.: The frequency of mating in relation to social position in small flocks of white leghorns. *Anat Rec Suppl* **81**, 113(1941)

Guilday, J. E.: Differential extinction during late Pleistocene and Recent times. In: *Pleistocene Extinctions*. Martin, P. S., Wright, H. E. (eds). New Haven: Yale Univ. Press, 1967, 121–140

Gula, M.: Child caring institutions. In: *Maternal and Child Health Practices*. Wallace, H. M., Gold, E. M., Lis, E. F. (eds). Springfield, Ill.: Thomas, 1973, 1267–1279

Gundlach, H.: Brutfürsorge, Brutpflege, Verhaltensontogenese und Tagesperiodik beim europäischen Wildschwein *(Sus scrofa L.)*. *Z Tierpsychol* **25**, 955–995 (1968)

Guthrie, R. D.: The extinct wapiti of Alaska and the Yukon Territory. *Can J Zool* **44**, 47–57 (1966)

Guthrie, R. D.: Pleoecology of the large-mammal community in interior Alaska during the late Pleistocene. *Amer Midl Nat* **79(2)**, 346–363 (1968)

Guthrie, R. D.: Evolution of human threat display organs. *Evol Biol* **1**, 257–302 (1970a)

Guthrie, R. D.: Bison evolution and zoogeography in North America during the Pleistocene. *Quart Rev Biol* **45**, 1−15 (1970b)

Guyton, A. C.: *Basic Human Physiology: Normal Function and Mechanisms of Disease.* Philadelphia: Saunders, 1971

Hafez, E. S. E., Signoret, J. P.: The behaviour of swine. In: *The Behaviour of Domestic Animals.* Hafez, E. S. E. (ed). London: Ballière, Tindall & Cassell (2nd ed), 1969, 349−390

Hagen, V. W. von: *World of the Maya.* New York: Mentor, 1960

Hagen, V. W. von: *Das Reich der Inka.* Frankfurt: Fischer Bücherei, 1967

Hahn, H.: Baumschliefer, Buschschliefer, Kippschliefer. Die Neue Brehm-Bücherei **246.** Wittenberg-Lutherstadt: Ziemsen, 1959

Halas, E. S., Hanlon, M. J., Sandstead, H. H.: Intrauterine nutrition and aggression. *Nature (Lond)* **257**, 221−222 (1975)

Hall, E. T.: *The Silent Language.* New York: Fawcett (1959)

Hall, E. T.: A system for the notation of proxemic behaviour. *Amer Anthropol.* **65**, 1003−1026 (1963)

Hall, K. R. L.: Aggression in monkey and ape societies. In: *The Natural History of Aggression.* Carthy, J. D., Ebling, F. J. (eds). London: Academic Press, 1964, 51−64

Hall, R. H.: Food for Naught. Hagerstown, Md.: Harper & Row, 1974

Hall, W. K.: Natality and mortality of white-tailed deer (*Odocoileus virginianus dacotensis* Goldman and Kellogg) in Camp Wainwright, Alberta. Unpubl. MSc Thesis, Univ. Calgary, 1973

Halloran, A. F.: American bison weights and measurements from the Wichita Mountain Wildlife Refuge. *Proc Oklahoma Acad Sci* **41**, 212−218 (1960)

Hamilton, W. D.: The genetical theory of social behaviour. *J Theoret Biol* **7(1)**, 1−52 (1964)

Hamilton, W. R.: The lower Miocene ruminants of Gebel Zelten, Libya. *Bull Brit Mus Nat Hist* **21(3)**, 1973

Hammond, J.: *Farm Animals.* London: Arnold, 1960

Hansen, G.: Bighorn sheep populations of the Desert Game Range. *J Wildl Manage* **31**, 693−706 (1967)

Hanson, H., Jones, R. L.: The biogeochemistry of blue, snow and Ross geese, appendix 2. *Ill. Nat Hist Soc Spec Publ* **1.** Carbondale: S. Illinois Univ. Press, 1976

Hardin, G.: The tragedy of the commons. *Science* **162**, 1243−1248 (1968)

Harkness, R. D.: Mechanical properties of skin in relation to its biological function and its chemical components. In: *Biophysical Properties of Skin.* Elden, H. R. (ed). New York: Wiley, 1971, 393−436

Harlow, H. F., Harlow, M. K., Svomi, S. J.: From thought to therapy: Lessons from a primate laboratory. *Amer Sci* **59(5)**, 539−549 (1971)

Harlow, H. F., Zimmerman, R. R.: Affectionate responses in the infant monkey. *Science* **130**, 421−432 (1959)

Harrer, G.: Das 'Musikerlebnis' im Griff des naturwissenschaftlichen Experiments. In: *Grundlagen der Musiktherapie und Musikpsychologie.* Harrer, G. (ed). Stuttgart: Fischer, 1975, 3−47

Hart, Liddell: *Strategy.* New York: Praeger (2nd rev ed), 1967

Hayes, W. N., Saiff, E. I.: Visual alarm reactions in turtles. *Anim Behav* **15**, 102−106 (1967)

Haynes, G. V.: Elephant hunting in North America. *Sci Amer* **214(6)**, 104−112 (1966)

Hays, J. D., Imbrie, J., Shackleton, N. J.: Variations in the Earth orbit: Pacemaker of the ice ages. *Science* **194**, 1121−1132 (1976)

Healey, M. C.: Aggression and self-regulation of population size in deer mice. *Ecology* **48**, 377−392 (1967)

Heath, J. E.: The origin of thermoregulation. In: *Evolution and Environment.* Drake, E. T. (ed). New Haven: Yale Univ. Press, 1968, 259−278

Hebb, D. O.: The nature of fear. *Psychol Rev* **53**, 259−276 (1946)

Hebb, D. O.: *A Textbook of Psychology*. Philadelphia: Saunders, 1966

Hecht, K.: Zur Rolle der zentralnervalen Regulation in der Organismus-Umwelt-Bezienung. In: *Im Mittelpunkt der Mensch*. Lohs, K., Döring, S., (eds). Berlin: Akademie Verlag, 1975, 151−173

Heck, L.: *Der Rothirsch*. Berlin: Parey, 1956

Hediger, H.: *The Psychology and Behaviour of Animals in Zoos and Circuses*. New York: Dover, 1955

Heimer, W. E., Smith, A. C.: Ram horn growth and population quality: their significance to Dall sheep management in Alaska. *Wildl Tech Bull* **5** (1975)

Heinzelin, J. de: Ishango. *Sci Amer* **206(6)**, 105−116 (1962)

Heiss, F., Franke, K. (eds): *Der Vorzeitig verbrauchte Mensch*. Stuttgart: Enke, 1964

Heizer, R. F., Napton, L. K.: Biological and cultural evidence from prehistoric human coprolites. *Science* **165**, 563−568 (1969)

Henley, E. D., Moisset, B., Welch, B. L.: Catecholamine uptake in cerebral cortex: Adaptive changes induced by fighting. Science **180**, 1050−1052 (1973)

Henry, F. M.: Specificity vs. generality in learning motor skills. *Proc Coll Phys Educ Assoc,* 126−128 (1958)

Henry, H. M.: Reply to Kay 1973. *Science* **182**, 396 (1973)

Henry, J.: Normal and abnormal behaviour. In: *Anthropology and the Behavioural and Health Sciences*. Mering, O., Kasdan, L. (eds). Pittsburgh: Univ. Pittsburgh Press, 1970, 128−140

Henry, F. M., Rogers, D. E.: Increased response latency for complicated movements and a 'memory drum' theory of neuromotor reaction. *Res Quart* **31**, 448−458 (1960)

Heptner, W. G., Nasimovitsch, A. A.: *Der Elch. Neue Brehm-Bücherei* **386.** Wittenberg-Lutherstadt: Ziemsen, 1968

Heptner, W. G., Nasimovitsch, A. A., Bannikov, A. G.: *Säugetiere der Sowjet Union*. Jena: Gustav Fischer, 1961

Herrero, S.: Aspects of evolution and adaptation in American black bear (*Ursus americanus* Pallas) and brown and grizzly bears (*U. arctos* Linné) of North America. In: *Bears: Their Biology and Management*. Herrero, S. (ed). Morges: IUCN Publications New Series **23**, 1972

Hess, E. H.: Attitude and pupil size. *Sci Amer* **212(4)**, 46−54 (1965)

Hester, J. J.: The agency of man in animal extinctions. In: *Pleistocene Extinctions*. Martin, P. S., Wright, H. E. (eds). New Haven: Yale Univ. Press, 1967, 169−192

Hester, J. J.: Ecology of the North American Paleo-Indians. *Bioscience* **20**, 213−217 (1970)

Hibbard, C. W.: Pleistocene vertebrates from the Upper Becerra. Univ. *Michigan Mus Paleontol Contrib* **12**, 47−96 (1955)

Hildes, J. A., Schaefer, O.: Health of Igloolik eskimos and changes with urbanization. *J Human Evol* **2**, 241−246 (1973)

Hinde, R. A.: *Animal Behaviour*. New York: McGraw-Hill (2nd ed), 1970

Hinde, R. A.: *Biological Bases of Human Social Behaviour*. New York: McGraw-Hill, 1974

Hinde, R. A., Spencer-Booth, Y.: Effects of brief separation from mother on Rhesus monkeys. *Science* **173**, 111−118 (1971)

Hinde, R. A., Stevenson, J. G.: Goals and response control. In: *Development and Evolution of Behaviour*. Aronson, L. R., Tolbach, E., Lehrman, D. S., Rosenblatt, J. S. (eds). San Francisco: Freeman, 1970, 216−237

Hinkle, L. E.: Studies of human ecology in relation to health and behaviour. *Bioscience* **Aug,** 517−520 (1965)

Hoar, W. S.: *General and Comparative Physiology*. Engelwood Cliffs: Prentice-Hall, 1966

Hobbs, P. V., Harrison, H., Robinson, E.: Atmospheric effects of pollutants. *Science* **183,** 909−915 (1974)

Hochstrasser, D. L., Trap, J. W.: Social medicine and public health. In: *Anthropology and the Behavioural and Health Sciences*. Mering, O. von, Kasdan, L. (eds). Pittsburgh: Univ. Pittsburgh Press, 1970, 242−261

Hockett, C. F., Altmann, S. A.: A note on design features. In: *Animal Communication*.

Sebeok, T. A. (ed). Bloomington: Indiana Univ. Press, 1968, 61−72

Hodes, H. L., Berger, R., Ainbender, E., Hevizy, M. M., Zepp, H. D., Kochwa, S.: Proof that colostrum polio antibody is different from serum antibody. *J. Pediat.* **65,** 1017−1018 (1964)

Hoebel, E. A.: Song duels among the Eskimo. In: *Law and Warfare.* Bohannan, P. (ed). New York: Natural History Press, 1966, 256−262

Hoefs, M.: Ecological investigation in Kluane National Park, Yukon Territory. Mimeo. report to Can Wildl Serv, Ottawa, 1973

Hoefs, M.: Food selection by Dall sheep (*Ovis dalli dalli* Nelson). In: *The Behaviour of Ungulates and its Relation to Management.* Geist, V., Walther, F. R. (eds). Morges: IUCN Publications New Series **24,** 1974, 759−786

Hoefs, M.: Ecological investigation of Dall sheep (*Ovis dalli dalli* Nelson) and their habitat on Sheep Mountain, Kluane National Park, Yukon Territory, Canada. Unpubl. PhD thesis, Univ. British Columbia, 1975

Hoefs, M., Thomson, J. W.: Lichens from the Kluane Game Sanctuary, southwest Yukon Territory. *Can Field Nat* **86,** 249−252 (1972)

Hoefs, M., Cowan, I. McT., Krajina, V. J.: Phytosociological analysis and synthesis of Sheep Mountain, southwest Yukon Territory, Canada. *Syesis* **8, Supp. 1,** 125−228 (1975)

Hofer, H.: Prolegomena primatologiae. In: *Die Sonderstellung des Menschen.* Hofer, H., Altner, G. (eds). Stuttgart: Fischer, 1972, 113−148

Hoffman, R. S., Taber, R. D.: Origin and history of Holarctic Tundra ecosystems with special reference to their vertebrate faunas in Arctic and Alpine environments. In: Arctic and Alpine Environments. Wright, H. E., Osburn, W. H. (eds). Bloomington: Indiana Univ. Press, 1968, 143−170

Hokanson, J. E., Burgess, M.: The effects of three types of aggression on vascular processes. *J Abnorm Soc Psychol* **64,** 446−449 (1962)

Hokanson, J. E., Shelter, S.: The effect of overt aggression on vascular processes. *J Abnorm Soc Psychol* **63,** 446−448 (1961)

Holmes, T. A.: Multidisciplinary studies in tuberculosis. In: *Personality, Stress and Tuberculosis.* Sparer, P. J. (ed). New York: Internat. Univ. Press, 1956, 376−412

Holst, E. von: Relations between the central nervous system and the peripheral organs. *Brit J Anim Behav* **2,** 89−94 (1954)

Holst, E. von, Mittelstaedt, H.: Das Reafferenzprinzip. *Naturwiss* **37,** 464−476 (1950)

Hooker, C.: Cultural form, social institution, physical system: Remarks towards a systematic theory. In: *Man and his Environment, Vol. 2.* Mohtadi, M. F. (ed). Oxford: Pergamon, 1976, 169−182

Hopkins, C. D.: Electric communication in fish. *Amer Sci* **62,** 426−437 (1974)

Hopson, J. A.: The origin of the mammalian middle ear. *Amer Zool* **6,** 437−450 (1966)

Hopson, J. A.: The origin and adaptive radiation of mammal-like reptiles and nontherian mammals. *Ann N Y Acad Sci* **167,** 144−216 (1969)

Hopson, J. A.: Postcanine replacement in the gomphodont cynodont *Diademodon.* In: *Early Mammals.* Kermack, D. H., Kermack, K. A. (eds). London: Academic Press, 1971, 1−21

Hopson, J. A.: The evolution of cranial display structures in hadrosaurian dinosaurs. *Paleobiology* **1,** 21−43 (1975)

Horejsi, B. L.: Behavioural differences in bighorn lambs (*Ovis canadensis canadensis* Shaw) during years of high and low survival. Proc. Northern Wild Sheep Council Symp, Hinton, Alberta, 1972, 51−73

Horejsi, B. L.: Suckling and feeding behaviour in relation to lamb survival in bighorn sheep (*Ovis canadensis canadensis* Shaw). Unpubl. PhD thesis, Univ. Calgary, 1976

Horn, G., Rose, S. P. R., Bateson, P. P. G.: Experience and plasticity in the central nervous system. *Science* **181,** 506−514 (1973)

Hornocker, M. G.: An ecological study of the mountain lion. Unpubl. PhD thesis, Univ. British Columbia, 1967

Hornocker, M. G.: An analysis of mountain lion predation upon mule deer and elk in Idaho. *Wildl Monog* **21**. Washington: Wildlife Soc., 1970

Horowitz, F. D. (ed): *Review of Child Development Research 4*. Chicago: Univ. Chicago Press, 1975

Horowitz, F. D., Paden, L. Y.: Effectiveness of environmental intervention programs. In: *Child Development and Social Policy 3*. Caldwell, B. M., Ricciuti, H. N. (eds). Chicago: Univ. Chicago Press, 1973, 331–402

Howarth, E., Skinner, N. F.: Salivation as a physiological indicator of introversion. *J Psychol* **73**, 123–128 (1969)

Howell, R. C., Coppens, J.: An overview of Hominidae from the Omo succession, Ethiopia. In: *Earliest Man and Environments in the Lake Rudolf Basin*. Coppens, Y., Howell, F. C., Isaac, G. L., Leakey, R. E. F. (eds). Chicago: Univ. Chicago Press, 1976, 522–531

Howells, W.: *Evolution of the Genus Homo*. Reading, Mass.: Addison-Wesley, 1973a

Howells, W.: *The Pacific Islanders*. London: Weidenfeld & Nicolson, 1973b

Hrdy, B. L.: Male–male competition and infanticide amont the langurs (*Presbytis entellus*) of Abu, Rajastan. *Fol Primatol* **22**, 19–58 (1974)

Hubel, D. H.: The visual cortex of the brain. *Sci Amer* **109(5)**, 54–62 (1963)

Huber, N. M.: The problem of stature increase: Looking from the past to the present. In: *The Skeletal Biology of Earlier Human Populations*. Brothwell, D. R. (ed). Oxford: Pergamon, 1968, 67–102

Huey, R. B.: Behavioural thermoregulation in lizards: Importance of associated costs. *Science* **184**, 1001–1003 (1974)

Hughes, C. C., Hunter, J. M.: Disease and 'development' in Africa. In: *The Social Organization of Health*. Dreitzel, H. P. (ed). New York: Macmillan, 1971, 150–214. First publ 1970

Hughes, O. L., Campbell, R. B., Muller, J. E., Wheeler, I. O.: Glacial limits and flow patterns, Yukon Territory, south of 65 degrees north latitude. Geological Survey of Canada Paper 68–34. Ottawa: Energy, Mines and Resources, 1969

Hull, T. G.: *Diseases Transmitted from Animals to Man*. Springfield, Ill.: Thomas (4th ed), 1955

Hulse, F. S.: *The Human Species*. New York: Random House, 1963

Hultén, E.: *Flora of Alaska and Neighboring Territories*. Stanford: Stanford Univ. Press, 1968

Hunt, J. McV.: *Intelligence and Experience*. New York: Ronald, 1961

Hunter, R. F.: Home range behaviour in hill sheep. In: *Grazing in Terrestrial and Marine Environments*. *Symp. Brit. Ecol. Soc. 4*. Crisp, D. J. (ed), 1964, 155–171

Hurley, R.: The health crisis of the poor. In: *The Social Organization of Health*. Dreitzel, H. P. (ed). New York: Macmillan, 1971, 83–122. First publ 1969

Hutchinon, V. H., Dowling, H. G., Vinegar, A.: Thermoregulation in a brooding female Indian python, *Python moluris biviatatus*. *Science* **151**, 694–696 (1966)

Hutt, C., Vaizey, M. J.: Differential effects of group density on social behaviour. *Nature. (Lond)* **209**, 1371–1372 (1966)

Igić, R., Stern, P., Basagic, E.: Changes in emotional behaviour after application of chlorinesterase in the septal and amygdala region. *Neuropharmacology* **9**, 73–75 (1970)

Illich, I.: *Medical Nemesis: The Expropriation of Health*. London: Calder & Boyar, 1975

Illich, I., Zola, J. K., McKnight, J., Caplan, J., Shaiken, J.: *Disabling Professions*. London: Boyars, 1977

Iltis, H.: Flowers and human ecology. In: *New Movements in the Study and Teaching of Biology*. Selmes, C. (ed). London: Maurice Temple Smith, 1974, 289–317

Iltis, H. H.: The population explosion, the conservation crisis and the Catholic church. *Scarracenia* **12**, May, 39–50 (1970)

Institute of Ecology: *Man and the Living Environment*. Madison: Univ. Wisconsin Press, 1972.

Isaac, G. L.: Traces of Pleistocene hunters: An East African example. In: *Man the Hunter*. Lee, R. B., de Vore, I. (eds). Chicago: Aldine, 1968, 253–261.

Isaac, G. L.: Plio-Pleistocene artefact assemblages from East Rudolf, Kenya. In: *Earliest Man and Environments in Lake Rudolf Basin*. Coppens, Y., Howell, F. C., Isaac, G. L., Leakey, R. E. F. (eds). Chicago: Univ. Chicago Press, 1976, 552–564

Isaac, G. L., Leakey, R. E. F., Behrensmeyer, A. K.: Archeological traces of early hominid activities east of Lake Rudolf, Kenya. *Science* **173**, 1129–1134 (1971)

Isaac, G. L., Harris, J. W. K., Crader, D.: Archeological evidence from the Koobi Fora formation. In: *Earliest Man and Environments in Lake Rudolf Basin*. Coppens, Y., Howell, F. C., Isaac, G. L., Leakey, R. E. F. (eds). Chicago: Univ. Chicago Press, 1976, 533–551

Itô, Y.: Groups and family bonds in animals in relation to their habitat. In: *Development and Evolution of Behaviour*. Aronson, L. R., Tobach, E., Lehrman, D. S., Rosenblatt, J. S. (eds). San Francisco: Freeman, 1970, 389–415

Jackson, R. L., Westerfeld, R., Flynn, M. A., Kimball, E. R., Lewis, R. B.: Growth of 'well-born' American infants fed human and cow's milk. *Pediatrics* **33**, 642–652 (1964)

Jaco, E. G.: Mental illness in response to stress. In: *Social Stress*. Levine, S., Scotch, N. A. (eds). Chicago: Aldine, 1970, 210–227

Jacobson, H. N.: Nutrition and pregnancy. In: *Maternal and Child Health Practices*. Wallace, H. M., Gold, E. M., Lis, E. F. (eds). Springfield, Ill.: Thomas, 1973, 311–331

Jaedicke, H. G.: Über Musiktherapie im psychotherapeutischen Heilplan. *Z Psychother Med Psychol* **4**, 93–98 (1954)

Jaeger, J. J., Wesselman, H. B.: Fossil remains of micromammals from the Omo group deposits. In: *Earliest Man and Environments in Lake Rudolf Basin*. Coppens, Y., Howell, F. C., Isaac, G. L., Leakey, R. E. F. (eds). Chicago: Univ. Chicago Press, 1976, 351–369

Jansky, L.: Nonshivering thermogenesis and its thermoregulatory significance. *Biol Rev* **48**, 85–132 (1973)

Jarman, P. J.: The development of a dermal shield in impala. *J Zool Lond* **166**, 349–356 (1972)

Jarman, P. J.: The social organization of antelope in relation to their ecology. *Behaviour* **48(3–4)**, 215–266 (1973)

Jelliffe, D. B., Jelliffe, E. F. (eds): The uniqueness of human milk. *Amer J Clin Nutr* **24**, 968–1024 (1971)

Jenkins, F. A.: Chimpanzee bipedalism: Cineradiographic analysis and implications for the evolution of gait. *Science* **178**, 877–879 (1972)

Jensen, M. M., Rasmussen, A. F.: Audiogenic stress and susceptibility to infection. In: *Physiological Effects of Noise*. Welch, B. L., Welch, A. S. (eds). New York: Plenum, 1970, 7–20

Jerison, H. J.: Quantitative analysis of evolution of the brain in mammals. *Science* **121**, 1012–1014 (1961)

Jerison, H. J.: Brain evolution and dinosaur brains. *Amer Natur* **103**, 575–588 (1969)

Joffe, J. M.: *Parental Determinants of Behaviour*. Oxford: Pergamon, 1969

Joffe, J. M., Rawson, R. A., Mulick, J. A.: Control of their environment reduces emotionality in rats. *Science* **180**, 1383–1384 (1973)

Johansen, K., Hanson, D.: Functional anatomy of the hearts of lung-fishes and amphibians. *Amer Zool* **8**, 191–210 (1968)

John, E. R.: Switchboard versus statistical theories of learning and memory. *Science* **177**, 850–864 (1972)

Johnson, A., Bezeau, L. M., Smoliaks, S.: Chemical composition and in vitro digestibility of alpine tundra plants. *J Wildl Manage* **32**, 773–777 (1968)

Johnston, D. H., Beauregard, M.: Rabies epidemiology in Ontario. *Bull Wildl Dis Assoc* **5(3)**, 357–370 (1969)

Jolly, A.: *The Evolution of Primate Behaviour*. New York: Macmillan, 1972

Jones, H. B.: The measurement of health and human life values. In: *Science and Absolute Values*. Tarrytown: Inter. Cult. Found., 1974, 241–255

Jonkel, C. F., Cowan, I. McT.: The black bear in the spruce-fir forest. *Wildl Monog* **27**, 1971

Jordan, P. A., Shelton, P. C., Allen, D.: Number turnover and social structure of the Isle Royal wolf population. *Amer Zool* **7**, 233–252 (1967)

Joyce, J. P., Blaxter, K. L., Park, C.: The effect of natural outdoor environments on the energy requirements of sheep. *Res. Vet. Sci.* **7**, 342–359 (1966)

Junghans, R., Nitschkoff, S.: Einfluss des Lärms auf Organismus—notwendige Schritte zur Lärmbekämpfung. In: *Im Mittelpunkt der Mensch*. Lohs, K., Döring, S. (eds). Berlin: Akademie Verlag, 1975, 174–186

Jungius, H.: The biology and behaviour of the reed buck (*Redunca arundinum* Boddaert 1785) in the Kruger National Park. *Mammalia Depicta*. Berlin: Parey, 1971

Kagan, J.: Attention and psychological change in the young child. *Science* **170**, 826–832 (1970)

Kagan, J.: Emergent themes in human development. *Amer Sci* **64**, 186–196 (1976)

Kagan, J., Klein, R. E.: Cross-cultural perspectives on early development. *Amer Psychol* **28**, 947–961 (1973)

Karmel, B. Z.: Complexity, amounts of contour and visually dependent behaviour in hooded rats, domestic chicks and human infants. *J Comp Physiol Psychol* **69(4)**, 649–657 (1969)

Katz, A. H.: The social causes of disease. In: *The Social Organization of Health*. Dreitzel, H. P. (ed). New York: Macmillan, 1971, 5–14. First publ 1959.

Katz, D.: *Gestaltpsychologie*. Basel: Schwabe (2nd ed), 1948

Kaufman, J. H.: Habitat use and social organization of nine sympatric species of macropodid marsupials. *J Mammal* **55**, 66–80 (1974a)

Kaufman, J. H.: The ecology and evolution of social organization in the kangaroo family. *Amer Zool* **14**, 51–62 (1974b)

Kay, R. F.: Humerus of robust Australopithecus. *Science* **182**, 396 (1973)

Keast, A.: Continental drift and the evolution of the biota on southern continents. In: *Evolution, Mammals and Southern Continents*. Keast, A., Erk, F. C., Glass, B. (eds). Albany: State Univ. New York Press, 1972, 23–87

Keast, A., Erk, F. C., Glass, B. (eds): *Evolution, Mammals and Southern Continents*. Albany: State Univ. New York Press, 1972

Keeley, K.: Prenatal influences on behaviour of offspring of crowded mice. *Science* **135**, 44 (1962)

Kellogg, W. N.: Communication and language in home-raised chimpanzee. *Science* **162**, 423–427 (1968)

Kelsall, J. P.: *The Caribou*. Ottawa: Queen's Printer, 1968

Kerr, F. A.: Ice fields of Western Canada. *Can Geog J* **7**, 234–246 (1933)

Kesner, R. P., Conner, H. S.: Independence of short- and long-term memory: A neural system analysis. *Science* **176**, 432–434 (1972)

Kimble, G. A., Perlmuter, L. C.: The problem of volition. *Psychol Rev* **77(5)**, 361–383 (1970)

King, M. C., Wilson, A. C.: Evolution at two levels in humans and chimpanzees. *Science* **188**, 107–116 (1975)

Kinsey, A. C., Pomeroy, W. B., Martin, C. E.: *Sexual Behaviour in the Human Male*. Philadelphia: Saunders, 1948

Kitchen, D. W.: Social behaviour and ecology of the pronghorn. *Wildl Monog* **38**, 1974

Kitchen, D. W., Bromley, P. T.: Agonistic behaviour of territorial pronghorn bucks. In: *The Behaviour of Ungulates and its Relation to Management*. Geist, V., Walther, F. R. (eds). Morges: IUCN Publications New Series **24**, 1974, 365–381

Kleiber, M.: *The Fire of Life*. New York: Wiley, 1961

Klein, D. R.: Range-related differences in growth of deer reflected in skeletal ratios. *J Mammal* **45**, 226–235 (1964)

Klein, D. R.: The introduction, increase and crash of reindeer on St. Matthew Island. *J Wildl Manage* **32**, 350−367 (1968)

Klein, D. R.: Food selection by North American deer and their response to overutilization of preferred plant species. In: *Animal Populations in Relation to their Food Resources.* Watson, A. (ed). Oxford: Blackwell, 1970a, 25−46

Klein, D. R.: Tundra ranges north of the boreal forest. *J Range Manage* **23(1)**, 8−14 (1970b)

Klein, D. R., Strandgaard, H.: Factors affecting growth and body size of roe deer. *J Wildl Manage* **36**, 64−79 (1972)

Klein, R. G.: Mousterian cultures in European Russia. *Science* **165**, 257−264 (1969)

Klein, R. G.: The late Quaternary mammalian fauna of Nelson Bay Cave (Cape Province, South Africa): Its implications for megafaunal extinctions and environmental and cultural change. *Quat Res* **2**, 135−142 (1972)

Klein, R. G.: *Ice-Age Hunters of the Ukraine.* Chicago: Univ. Chicago Press, 1973.

Kleine, H. O.: Die Gefährdung der Frau im Zeitalter der Technik. In: *Der Vorzeitig verbrauchte Mensch.* Heiss, F., Franke, K. (eds). Stuttgart: Enke, 1964, 386−405

Klingel, H.: Sociale Organisation und Verhalten freilebender Steppenzebras. *Z Tierpsychol* **24**, 580−624 (1967)

Klingel, H.: A comparison of the social behaviour of the Equidae. In: *The Behaviour of Ungulates and its Relation to Management.* Geist, V., Walther, F. R. (eds). Morges: IUCN Publications New Series **24**, 1974, 124−132

Klose, H.: Musik als Heilmittel in der Chirurgie. *Die Medizinische* **37**, 1252−1256 (1954)

Kluger, M. J., Ringler, D. H., Anver, M. R.: Fever and survival. *Science* **188**, 166−168 (1975)

Knaus, W., Schröder, W.: Das Gamswild. Berlin: P. Parey (2nd ed), 1975

Koehler, O.: Vom Unbenannten Denken. *Zool Anz Suppl.* **16**, 202−211 (1952)

Koenig, O.: Kultur und Verhaltensforschung. Munich: Deutsch. Taschenbuch, 1970

Koestler, A.: *The Act of Creation.* London: Hutchinson, 1964

Koestler, A.: Humour and wit. In: *Encyclopedia Britannica, Vol. 9.* Chicago: Encycl. Brit. (15th ed), 1974, 5−11

Köhler, W.: *The Mentality of Apes.* London: Routledge Kegan Paul, 1927

Komarek, E. V.: The natural history of lightning. Proc. 3rd Tall Timbers Fire Ecol. Conf., Tallahassee, Florida, 139−187 (1964)

Komarek, E. V.: Fire ecology—grasslands and man. Proc. 4th Tall Timbers Fire Ecol. Conf., Tallahassee, Florida, 169−2200 (1965)

Komarek, E. V.: Fire—and ecology of man. Proc. 6th Tall Timbers Fire Ecol. Conf., Tallahassee, Florida, 143−170 (1967)

Kortland, A.: Experimentation with chimpanzees in the wild. In: *Neue Ergebnisse der Primatologie.* Starck, D., Schneider, R., Kuhn, H. J. (eds). Stuttgart: Fisher, 1967, 208−224

Kortland, A., van Zon, J. C. J.: The present state of research on the dehumanization hypothesis of African ape evolution. Proc. 2nd Inter. Cong. Primatology 3, 1969, 10−13

Kramer, A.: Soziale Organisation und Sozialverhalten einer Gemspopulation (*Rupicapra rupicapra* L) der Alpen. *Z Tierpsychol* **26**, 889−964 (1969)

Krebs, C. J., Gains, M. S., Keller, B. L., Myres, J. H., Tamarin, R. H.: Population cycles in small rodents. *Science* **179**, 35−41 (1973)

Krech, D. M., Rosenzweig, M. R., Bennett, E. L.: Effects of environmental complexity and training on brain chemistry. *J Comp Physiol Psychol* **53**, 509−519 (1960)

Krekorian, C. O., Vance, V. J., Richardson, A. H.: Temperature dependent maze learning in desert iguana *Dipsosaurus sorsalis. Anim Behav* **16**, 429−436 (1968)

Krinsley, D. B.: Pleistocene geology of the southwest Yukon Territory, Canada. *J Glaciol* **5**, 385−397 (1965)

Kristein, M. M., Arnold, C. B., Wynder, E. L.: Health economics and preventive care. *Science* **195**, 457−462 (1977)

Krueger, A. P., Reed, E. J.: Biological impact of small air ions. *Science* **193**, 1209−1213 (1976)

Krumbiegel, J.: *Biologie der Säugetiere, Vol. 1.* Krefeld: Agis, 1954

Kruska, D.: Vergleichend cytoarchitektonische Untersuchungen an Gehirnen von Wild und Hausschweinen. *Z Anat Entwick Gesch* **131**, 291−324 (1970a)

Kruska, D.: Uber die Evolution des Gehirns in der Ordnung Artiodactyla Owen 1848, ins besondere der Teilordnung Suina Gray 1868. *Z Säugetierk* **35**, 214−238 (1970b)

Kruuk, H.: *The Spotted Hyena.* Chicago: Univ. Chicago Press, 1972

Kühme, W.: Freilandstudien zur Soziologie des Hyänenhundes (*Lycaon pictus lupinus* Thomas 1902). *Z Tierpsychol* **22(5)**, 495−541 (1965)

Kuhn, O.: Die Säugetierähnlichen Reptilien. *Neue Brehm-Bücherei* **423**. Wittenberg-Lutherstadt: Ziemsen, 1970

Kuhn, O.: Die Vorzeitlichen Vögel. *Neue Brehm-Bücherei* **435**. Wittenberg-Lutherstadt: Ziemsen, 1971

Kühne, W. G.: The evolution of a synorgan, nineteen stages concerning teeth and dentition from the pelycosaur to the mammalian condition. *Bull Group Int Rech Sc Stomat* **16**, 293−325 (1973)

Kukla, G. J., Kukla, H. J.: Increased surface albedo in the Northern Hemisphere. *Science* **183**, 709−714 (1974)

Kummer, H.: Tripartite relations in hamadryas baboons. In: *Social Communication among Primates.* Altmann, S. A. (ed). Chicago: Univ. Chicago Press, 1967

Kummer, H.: *Social Organization of the hamadryas baboon.* Chicago: Univ. Chicago Press, 1968

Kurtén, B.: *Pleistocene Mammals of Europe.* London: Weidenfeld & Nicolson, 1968a

Kurtén, B.: *The Age of Dinosaurs.* New York: McGraw-Hill, 1968b

Kurtén, B.: *Not from the Apes.* New York: Random House, 1972

Kurtén, B.: *The Cave Bear Story.* New York: Columbia Univ. Press, 1976

Kurten, B.: *Time and hominid brain size.* Commentationes Biologicae **36:** 1−8, 1971

Kyllonen, R. L.: Crime rates versus population density in the United States. *Yearbk Soc Gen Syst Res* **12**, 137−145 (1967)

La Barre, W.: *The Ghost Dance: The Origins of Religion.* New York: Doubleday Delta, 1970

Lack, D.: *Population Studies of Birds.* London: Oxford Univ. Press, 1966

Lal, P.: The tribal man in India: A study in the ecology of primitive communities. In: *Ecology and Biogeography in India.* Mani, M. S. (ed). The Hague: W. Junk, 1974, 281−329

Lalonde, M.: *A New Perspective on the Health of Canadians.* Ottawa: Queen's Printer, 1972

Lambourn, L. J.: Relative effects of environment and live weight upon the feed requirements of sheep. *Proc N Z Soc Anim Prod* **21**, 92−108 (1961)

Lamprecht, F., Eichelman, B., Thoa, N. B., Williams, R. B., Kopin, I. J.: Rat fighting behaviour: Serum dopamine-.4b-hydroxylase and hypothalamic tryosine hydroxylase. *Science* **177**, 1214−1215 (1972)

Larsen, T.: Discussion. In: *Bears—Their Biology and Management.* Herrero, S. (ed). Morges: IUCN Publications New Series **23**, 1972, 253

Larson, R. L.: Physical activity and the growth and development of bone and joint structures. In: *Physical Activity.* Rarick, G. L. (ed). New York: Academic Press, 1973, 33−59

Lasker, G. W.: Human biological adaptability. *Science* **166**, 1480−1486 (1969)

Laughlin, W. S.: Acquisition of anatomical knowledge by ancient man. In: *Social Life of Early Man.* Washburn, S. L. (ed). Chicago: Aldine, 1961, 150−175

Laughlin, W. S.: Eskimos and Aleuts: Their origins and evolution. *Science* **142**, 633−645 (1963)

Laughlin, W. S.: Human migration and permanent occupation in the Bering Sea area. In: *The Bering Land Bridge.* Hopkins, D. M. (ed). Stanford: Stanford Univ. Press, 1967, 409−450

Laughlin, W. S.: Hunting: An integrating biobehaviour system and its evolutionary importance. In: *Man the Hunter*. Lee, R. B., de Vore, I. (eds). Chicago: Aldine, 1968, 304–320

Laughlin, W. S.: Aleuts: Ecosystem, holocene history and Siberian origin. *Science* **189**, 507–515 (1975)

Lave, L. B., Seskin, E. P.: Air pollution and human health. *Science* **169**, 723–732 (1970)

Lawick, H. von, Lawick-Goodall, J. von: *Innocent Killers*. Boston: Houghton Mifflin, 1971

Laws, R. M.: Behaviour, dynamics and management of elephant populations. In: *The Behaviour of Ungulates and its Relation to Management*. Geist, V., Walther, F. R. (eds). Morges: IUNC Publications New Series **24**, 1974, 513–529

Layzer, D.: Heritability analyses of IQ scores: Science or numerology? *Science* **183**, 1259–1266 (1974)

Leach, E.: Don't say "boo" to a goose. In: *Man and Aggression*. Montagu, M. F. A. (ed). London: Oxford Univ. Press, 1966, 65–73

Leakey, M. D.: Fauna and artefacts from a new Plio-Pleistocene locality near Lake Rudolf in Kenya. *Nature (Lond)* **226**, 223–224 (1970)

Leakey, R. E. F.: Evidence for an advanced Plio-Pleistocene hominid from East Rudolf, Kenya. *Nature (Lond)* **242**, 448–450 (1973a)

Leakey, R. E. F.: Further evidence of lower Pleistocene hominids from East Rudolf, North Kenya, 1972. *Nature (Lond)* **242**, 170–173 (1973b)

Leakey, R. E.: Hominids in Africa. *Amer Sci* **64**, 174–178 (1976a)

Leakey, R. E.: An overview of the Hominidae from East Rudolf, Kenya. In: *Earliest Man and Environments in the Lake Rudolf Basin*. Coppens, Y., Howell, F. C., Isaac, G. L., Leakey, R. E. F. (eds). Chicago: Univ. Chicago Press, 1976b, 476–483 (1976b)

Lee, B. C., de Vore, I.: Problems in the study of hunter-gatherers. In: *Man the Hunter*. Lee, R. B., de Vore, I. (eds). Chicago: Aldine, 1968, 3–12. See also Discussion, 245–248.

Lee, D. H. K.: Studies of heat regulation in sheep with special reference to the Merino. *Aust J Agric Res* **1**, 200–216 (1950)

Lee, R. B.: What hunters do for a living, or, how to make out on scarce resources. In: *Man the Hunter*. Lee, R. B., de Vore, I. (eds). Chicago: Aldine, 1968, 30–48

Lee, R. B.: The Kung Bushman of Botswana. In: *Hunters and Gatherers Today*. Biccheri, M. G. (ed). New York: Holt Rinehart Winston, 1972, 326–368

Leeds, A.: Reindeer herding and Chukchi social institutions. In: *Man, Culture and Animals*. Leeds, A., Vayda, A. P. (eds). Washington: *AAAS Publ* **78**, 1965, 87–128

Le Gros Clark, W. E.: *The Fossil Evidence for Human Evolution*. Chicago: Univ. Chicago Press (rev ed), 1964

Le May, M.: The language capability of Neanderthal man. *Amer J Phys Anthropol.* **42**, 9–14 (1975)

Lemkov, P. V.: Mental health services. In: *Preventive Medicine and Public Health*. Sartwell, P. E. New York: Appleton-Century-Crofts (10th ed), 1973, 547–590

Lenneberg, E. H.: On explaining language. *Science* **164**, 635–643 (1969)

Lent, P. C.: Rutting behaviour in a barren-ground caribou population. *Anim Behav* **13**, 259–264 (1965)

Lent, P.: Mother–infant relationships in ungulates. In: *The Behaviour of Ungulates and its Relation to Management*. Geist, V., Walther, F. R. (eds). Morges: IUCN Publications New Series **24**, 1974, 14–55

Leopold, A.: *A Sand County Almanac*. London: Oxford Univ. Press, 1949

Leopold, A. C., Ardrey, R.: Toxic substances in plants and the food habits of early man. *Science* **176**, 511–513 (1972)

Leopold, A. C., Ardrey, R.: Reply to food habits of early man: Balance between hunting and gathering. *Science* **179**, 306–307 (1973)

Leroi-Gourhan, A.: The evolution of Paleolithic art. *Sci Amer* **218(2)**, 58–70 (1968)

Lethmate, J., Dücker, G.: Untersuchungen zum Selbsterkennen im Spiegel bei Orang-Utans und einigen anderen Affinarten. *Z Tierpsychol* **33**, 248–269 (1973)

Leuthold, W.: Variations in territorial behaviour of the Uganda kob (*Adenota kob thomasi* Neumann 1896). *Behaviour* **27**, 214−251 (1966)

Leuthold, W.: Observations on home range and social organisation of lesser Kudux *Tragelaphus imberbis* in V. Geist and F. Walther (eds.) *The Behaviour of Ungulates and its Relation to Management*. IUCN. Publication No. 24. New series pp. 206–234. Morges, Switzerland 1974.

Levine, S., Scotch, N. A.: Social Stress. In: *Social Stress*. Levine, S., Scotch, N. A. (eds). Chicago: Aldine, 1970, 1−16

Levitsky, D. A., Barnes, R. H.: Nutritional and environmental interactions in the behavioural development of rat: Long-term effects. *Science* **176**, 68−71 (1972)

Levitt, J.: The traceological method: An introduction and case study analysis of some late prehistoric hide scraping tools from the Northwest plains. Unpubl. MA thesis, Univ. Calgary, 1976

Leyhausen, P.: Verhaltensstudien bei Katzen. *Z Tierpsychol Beiheft* **2**, 1956

Leyhausen, P.: Charles Robert Darwin—cine Irrlehre. *Orion* **1960(8)**, 655−660 (1960)

Leyhausen, P.: Dominance and territoriality as complements in mammalian social structure. In: *Behaviour and Environment*. Esser, A. H. (ed). New York: Plenum, 1971, 22−32

Liddell, H. S.: Sheep and goats: The psychological effects of laboratory experiences of deprivation and stress upon certain experimental animals.In: *Beyond the Germ Theory*. Gladstone, I. (ed). New York: *N Y Acad Med* 1954, 106−119

Liddell, H. S.: A biological basis for psychopathology. In: *Problems of Addiction and Habituation*. Hook, P. H., Zubin, J. (eds). New York: Grune & Stratton, 1958, 120−133

Liddell, H. S.: Contribution of conditioning in the sheep and goat to an understanding of stress, anxiety and illness, In: *Western Psychiatric Institute and Clinic Lectures on Experimental Psychiatry*. Pittsburg: Univ. Pittsburg Press, 1961, 227−255

Lieberman, J. (ed): Animal disease and human health. *Ann N Y Acad Sci* **70**, 277−766 (1958)

Lieberman, M. W.: Early developmental stress and later behaviour. *Science* **141**, 824−825 (1963)

Lieberman, P., Crelin, E. S.: On the speech of Neanderthal man. *Ling Inq* **2**, 203−222 (1971)

Lincoln, G. A.: The role of antlers in the behaviour of red deer. *J Exp Zool* **182(2)**, 233−250 (1972)

Lincoln, G. A., Youngson, R. W., Short, R. V.: The way in which testosterone controls the social and sexual behaviour of the red deer stag. *Horm Behav* **3**, 375−396 (1972)

Lindsey, C. C.: Proglacial lakes and fish dispersal in southwestern Yukon Territory. *Verhandl Int Ver Limnol* **19**, 2364−2370 (1975)

Livingston, D. A.: Late Quaternary climatic change in Africa. *Ann Rev Ecol Syst* **6**, 249−280 (1975)

Lofland, L.: *A World of Strangers: Order and Action in Urban Public Spaces*. New York: Basic Books, 1973

Lomax, A., Berkowitz, N.: The evolutionary taxonomy of culture. *Science* **177**, 228−239 (1972)

Long, C. A.: The origin and evolution of mammary glands. *Bioscience* **19**, 519−523 (1969)

Loomis, W. F.: Skin-pigment regulation of vitamin-D biosynthesis in man. *Science* **176**, 501−506 (1967)

Lorenz, K.: *Das Sogenannte Böse*. Vienna: Borotha-Schoeler, 1963

Lorenz, K.: *On Aggression*. London: Methuen, 1966

Lorenz, K.: *Die Rückseite des Spiegels*. Munich: Pieper, 1973

Lorenz, K. Z.: Analogy as a source of knowledge. *Science* **185**, 229−234 (1974)

Louw, G. N., Malan, M. E. (eds): Animal behavior symposium. *Zool Afr* **7(1)**, 1−412 (1972)

Lowenstein, F. W.: Some considerations of biological adaptation by aboriginal man to the tropical rain forest. In: *Tropical Forest Ecosystems in Africa and South America*. Meggers, B. J., Ayensu, E. S., Duckworth, W. D. (eds). Washington: Smithsonian Institution, 1973, 293−310

Lozek, V.: Zum Problem der Zahl der quartären Klimaschwankungen. *Quartär* **22**, 1−16 (1971)

Lubell, D., Hassan, F. A., Gautier, A., Ballais, J. L.: The Capsian Escargotières. *Science* **191**, 910−920 (1976)

Lyman, S. M., Scott, M. B.: Territoriality: A neglected sociological dimension. *Soc Prob* **15**, 236−249 (1967)

Lytle, L. D., Messing, R. B., Fisher, L., Phebus, L: Effects of long-term corn consumption on brain seratonin and the response to electric shock. *Science* **190**, 692−694 (1975)

MacArthur, R.: Sex differences in field dependence for the Eskimo. *Inter J Psychol* **2**, 139−140 (1967)

MacNeish, R. S.: Investigations in southwest Yukon: Archeological excavation, comparisons and speculations. Papers of the R.S. Peabody Found for Archeology **6(2)**, 201−488 (1964)

MacNeish, R. S.: Early man in the Andes. *Sci Amer* **224(4)**, 36−46 (1971)

MacNeish, R. S.: Early man in the New World. *Amer Sci* **64**, 316−327 (1976)

Maegher, M. M.: The bison of Yellowstone National Park. Washington: *US Nat Park Serv Sci Monog* **1**, 1973

Maglio, V. J.: Vertebrate faunas and chronology of hominid-bearing sediments east of Lake Rudolf, Kenya. *Nature (Lond)* **239**, 379−385 (1972)

Magoun, H. W.: Advances in brain research with implications for learning. In: *On the Biology of Learning*. Pribram, K. H. (ed). New York: Barcourt Brace & World, 1969, 169−190

Magrab, E. B.: *Environmental Noise Control*. New York: Wiley, 1975

Mandl, A. M., Zuckerman, S.: Factors influencing the onset of puberty in albino rats. *J Endocrinol* **8**, 357−364 (1952)

Mania, D., Toepfer, V.: *Königsave, Gliederung, Ökologie und mittelpaläolithische Funde der letzten Eiszeit*. Berlin: VEB Deut. Verlag. Wissens, 1973

Manning, A.: *An Introduction to Animal Behaviour*. London: Arnold, 1967

Margalef, R.: On certain unifying principles in ecology. *Amer Natur* **97**, 357−374 (1963)

Margolis, S. V., Kennett, J. P.: Antarctic glaciation during the Tertiary recorded in subantarctic deep-sea cores. *Science* **170**, 1085−1187 (1970)

Maringer, J., Bandi, H. G.: *Art in the Ice Age*. New York: Praeger, 1953

Markert, C. L., Shaklee, J. B., Whitt, G. S.: Evolution of a gene. *Science* **189**, 102−114 (1975)

Markov, K. K.: The pleistocene history of Antarctica. In: *The Periglacial Environment*. Péwé, T. L. (ed). Montreal: McGill-Queen's, 1969, 263−270

Marler, P.: The logical analysis of animal communication. *J Theor Biol* **1(3)**, 295−317 (1961)

Marler, P.: Visual systems. In: *Animal Communication*. Sebeok, T. A. (ed). Bloomington: Indiana Univ. Press, 1968, 103−126

Marler, P., Hamilton, W. J.: *Mechanisms of Animal Behaviour*. New York: Wiley, 1966

Marshack, A.: Upper Paleolithic notation and symbol. *Science* **178**, 817−827 (1972a)

Marshack, A.: *The Roots of Civilization*. New York: McGraw-Hill, 1972b

Marshack, A.: Implications of the Paleolithic symbolic evidence for the origin of language. *Amer Sci* **64**, 136−145 (1976)

Martin, P. S.: Prehistoric overkill. In: *Pleistocene Extinctions*. Martin, P. S., Wright, H. E. (eds). New Haven: Yale Univ. Press, 1967, 75−120

Martin, P. S.: The discovery of America. *Science* **179**, 969−974 (1973)

Martin, P. S., Wright, H. E.: *Pleistocene Extinctions*. New Haven: Yale Univ. Press, 1967

Maslow, A. H.: *Motivation and Personality*. New York: Harper, 1954

Mason, J. W.: Central nervous factors in the regulation of endrocrine secretion. *Rec Prog Horm Res* **15**, 345−389 (1959)

Massey, A.: Agonistic aids and kinship in a group of pigtail macaques. *Behav Ecol Sociobiol* **2**, 31−40 (1977)

Masters, W. H., Johnson, V. E.: *Human Sexual Response*. Boston: Little Brown, 1966

Masuda, M., Holmes, T. H.: Magnitude estimations of social readjustment. *J Psychosom Res* **11**, 213−218 (1967)

Mata, L. J., Wyatt, R. G.: Host resistance to infection. *Amer J Clin Nutr* **24**, 976−986 (1971)

Matsumoto, Y. S.: Social stress and coronary heart disease in Japan. In: *The Social Organization of Health*. Dreitzel, H. P. (ed). New York: Macmillan, 1971, 123−149

Matthew, W. D.: Climate and evolution. *Ann N Y Acad Sci* **24**, 171−318 (1915)

Matthews, L. H.: Overt fighting in mammals. In: *The Natural History of Aggression*. Carthy, J. D., Ebling, F. J. (eds). New York: Academic Press, 1964, 23−32

Mattingly, I. G.: Speech cues and sign stimuli. *Amer Sci* **60**, 327−337 (1972)

Matthissen, P.: *Under the Mountain Wall*. New York: Viking/Ballantine, 1962

Maynard Smith, J., Price, G. R.: The logic of animal conflict. *Nature (Lond)* **246**, 15−18 (1973)

Mayr, E.: *Systematics and the Origin of Species*. New York: Columbia Univ. Press, 1942

Mayr, E.: *Animal Species and Evolution*. Cambridge, Mass.: Belknap Harvard, 1966

McBride, G., Arnold, G. W., Alexander, G., Lynch, J. J.: Ecological aspects of the behaviour of domestic animals. *Proc Ecol Soc Austral* **2**, 133−165 (1967)

McCay, C. M., Maynard, A., Sperling, G., Barnes, L. L.: Retarded growth, life span, ultimate body size and age changes in the albino rat after feeding diets restricted in calories. *J Nutr* **18**, 1−13 (1939)

McCrea, W. H.: Ice ages and the Galaxy. *Nature (Lond)* **255**, 607−609 (1975)

McCulloch, J. S. G., Talbot, L. M.: Comparison of weight estimation methods for wild animals and domestic livestock. *J App Ecol* **2**, 59−69 (1965)

McGhee, R.: A current interpretation of central Canadian Arctic prehistory. *Inter Nord* **13−14** Dec, 171−180 (1974)

McHarg, I. L.: *Design with Nature*. New York: Doubleday, 1971

McHugh, T.: Social behaviour of the American buffalo *(Bison bison bison)*. *Zoologica* **43(1)**, 1−40 (1958)

McHugh, T.: *The Time of the Buffalo*. New York: Knopf, 1972

McKusick, V. A.: *Human Genetics*. Englewood Cliffs: Prentice-Hall (2nd ed), 1969

McLean, F. C., Urist, M. R.: Bone. Chicago: Univ. Chicago Press, 1968

McNab, B.: Bioenergetics and the determination of home range size. *Amer Natur* **97**, 133−140 (1963)

Mech, D. L.: *The Wolf*. New York: Natural History Press, 1970

Meddis, R.: On the function of sleep. *Anim Behav* **23**, 676−691 (1975)

Meggers, B. J.: Some problems of cultural adaptation in Amazonia, with emphasis on the preEuropean period. In: *Tropical Forest Ecosystems in Africa and South America*. Meggers, B. J., Ayensu, E. S., Duckworth, W. D. (eds). Washington: Smithsonian, 1973, 311−320

Meggitt, M. J.: The association between Australian aborigines and dingoes. In: *Man, Culture and Animals*. Leeds, A., Vayda, A. P. (eds). Washington: AAAS Publ, 1965, 7−26

Menzel, E. W.: Chimpanzee spatial memory organization. *Science* **182**, 943−945 (1973)

Merrick, H. V., Merrick, J. P. S.: Archeological occurrences of earlier Pleistocene Age from Skungura Formation. In: *Earliest Man and Environments in the Lake Rudolf Basin*. Coppens, Y., Howell, F. C., Isaac, G. L., Leakey, R. E. F. (eds). Chicago: Univ. Chicago Press, 1976, 574−583

Merton, R. K.: *Social Structure and Social Theory*. London: Collier-Macmillan Free Press of Glencoe, 1957

Mettler, F. A.: Culture and the structural evolution of the neural system. In: *Culture and the Evolution of Man*. Montagu, M. F. A. (ed). New York: Oxford, 1962, 155−201. First Publ 1955.

Metzgar, L. H.: An experimental comparison of screech owl predation on resident and transient white-footed mice *(Peromyscus leucopus)*. *J. Mammal* **48(3)**, 387−391 (1967)

Michael, E. D., Cureton, T. K.: Effect of physical training on cardiac output and at 15,000 simulated altitude. *Res Quart* **24**, 446−452 (1953)

Michael, R. P., Bonsall, R. W., Warner, P.: Human vaginal secretions: Volatile fatty acid content. *Science* **186**, 1217−1219 (1974)

Michael, R. P., Keverne, E. B.: Pheromones in the communication of sexual status in primates. *Nature (Lond)* **218**, 746−749 (1968)

Michael, R. P., Keverne, E. B., Bonsall, R. W.: Pheromones: Isolation of male sex attractants from a female primate. *Science* **172**, 964−966 (1971)

Mileski, M., Black, D. J.: The social organization of homosexuality. *Urb Life Cult* **1(2)**, 187−199 (1972)

Milgram, S.: The experience of living in cities. *Science* **167**, 1461−1468 (1970)

Miller, D. A.: Evolution of pumatt chromosomes. *Science* **198**, 1116−1124 (1977)

Miller, L.: Discussion. In: *Bears—Their Biology and Management*. Herrero, S. (ed). Morges: IUCN Publications New Series **23**, 1972, 254

Miller, R. S.: Patterns and processes in competition. In: *Advances in Ecological Research, Vol. 4*. Cragg, J. B. (ed). London: Academic Press, 1967, 1−74

Milner, B.: Psychological defects produced by temporal lobe excision. *Res Pubs Assoc Res Nerv Ment Dis* **36**, 244−257 (1956)

Mitchell, G. D.: Persistent behaviour pathology in Rhesus monkeys following early social isolation. *Folia Primat* **8**, 132−147 (1968)

Mize, R. R., Murphy, E. H.: Selective visual experience fails to modify receptive field properties of rabbit striated cortex neurons. *Science* **180**, 320−322 (1973)

Modha, M. L.: The ecology of the Nile crocodile on Central Island, Lake Rudolf. *E Afr Wildl J* **5**, 74−95 (1967)

Moen, A. J.: Energy exchange of white-tailed deer, western Minnesota. *Ecology* **49**, 676−682 (1968)

Moen, A. N.: *Wildlife Ecology*. San Francisco: Freeman, 1973

Mohr, E.: Das Urwildpferd. *Neue Brehm-Bücherei* **249**. Wittenberg-Lutherstadt: Ziemsen, 1970

Morkridin, V. P.: Control of harmful predators. In: *Reindeer Husbandry*. Zhigunov, P. S. (ed). Trans. from Russian. Springfield: US Dept. Commerce, 1961, 311−325

Molnar, S.: Some functional interpretations of tooth wear in prehistoric and modern man. Unpubl. PhD Thesis, Univ. California, Santa Barbara, 1968

Mongait, A. L.: *Archeology in the USSR*. Gloucester, Mass.: Smith, 1955

Montague, M. F. A.: Time, morphology and neoteny in the evolution of man. In: *Culture and the Evolution of Man*. Montagu, M. F. A. (ed). New York: Oxford Univ. Press, 1960, 324−342

Moore, A. U.: Effects of modified maternal care in the sheep and goats. In: *Early Experience and Behaviour*. Newton, G., Levine, S. (eds). Springfield, Ill.: Thomas, 1968, 481−529

Morris, D.: *The Naked Ape*. New York: McGraw-Hill, 1967

Morrison, B. J., Hill, W. F.: Socially facilitated reduction of the fear response in rats raised in groups or in isolation. *J Comp Physiol Psychol* **63**, 71−76 (1967)

Morrison, R. R., Ludvigson, H. W.: Discrimination by rats of conspecific odours of reward and nonreward. *Science* **167**, 904−905 (1970)

Morton, E. S.: Ecological sources of selection on avian sounds. *Amer Natur* **109**, 17−34 (1975)

Moulton, D. G.: Olfaction in mammals. *Amer Zool* **7**, 421−429 (1967)

Movius, H. L.: A wooden spear of third interglacial age from lower Saxony. *Southwest J*

Anthropol **6,** 139–142 (1950)

Moynihan, M.: Control, suppression, decay, disappearance and replacement of displays. *J Theoret Biol* **29,** 85–112 (1970)

Mueller, H. C.: Oddity and specific search image more important than conspicuousness in prey selection. *Nature (Lond)* **233,** 345–346 (1971)

Muller, E. A.: Physiology of muscle training. *Rev Can Biol* **21,** 303–313 (1962)

Muller, J. E.: Kluane Lake map area. Geological Survey of Canada Memoir 340. Ottawa: Energy Mines and Resources, 1967

Müller-Schwarze, D., Müller-Schwarze, C.: Olfactory imprinting in a precocial mammal. *Nature. (Lond)* **229,** 55 (1971)

Müller-Schwarze, D. Pheromones in black-tailed deer (Odocaileus/hemionus columbianus). *Animal Behaviour* **19 (1)** 141–152 (1971).

Mulvaney, D. J.: The prehistory of the Australian aborigine. *Sci Amer* **214(3),** 84–93 (1966)

Munro, T.: *Form and Style in the Arts.* Cleveland, Ohio: Case Western Reserve Univ., 1970

Murray, B. M., Murray, D. F.: Notes on mammals in alpine areas of the northern St Elias Mountains, Yukon Territory and Alaska. *Can Field Natur* **83,** 331–338 (1969)

Murton, R. K.: Some predator–prey relationships in bird damage and population control. In: *The Problem of Birds as Pests.* Murton, R. K., Wright, E. N. (eds). New York: Academic Press, 1968, 157–169

Musgrave, J.: How dextrous was Neanderthal man? *Nature (Lond)* **233,** 538–541 (1971)

Mykytowycz, R.: Further observations on the territorial function and histology of the submandibular cutaneous (chin) glands in the rabbit, *Oryctolagus cuniculus* (L). *Anim Behav* **13,** 400–412 (1965)

Mykytowycz, R.: Observations on odoriferous and other glands in the Australian wild rabbit *Oryctolagus cuniculus* (L) and the hare *Lepus europaeus* P: The anal gland. *CSIRO Wildl Res* **11,** 11–29 (1966a)

Mykytowycz, R.: Observations on odoriferous and other glands in the Australian wild rabbit *Oryctolagus cuniculus* (L) and the hare *Lepus europaeus* P: The inguinal glands. *CSIRO Wildl Res* **11,** 49–64 (1966b)

Mykytowycz, R.: Observations on odoriferous and other glands in the Australian wild rabbit *Oryctolagus cuniculus* (L) and the hare *Lepus europaeus* P: Harder's, lachrymal and submandibular glands. *CSIRO Wildl Res* **11,** 65–90 (1966c)

Mykytowycz, R., Dudzinski, M. L.: A study of the weight of odoriferous and other glands in relation to social status and degree of sexual activity in the wild rabbit *Oryctolagus cuniculus* (L). *CSIRO Wildl Res.* **11,** 31–47 (1966)

Napier, J.: The evolution of the hand. *Sci Amer* **207(6),** 56–62 (1962)

Napier, J.: The antiquity of human walking. *Sci Amer* **216(4),** 56–66 (1967)

Nasimovich, A. A.: *The Role of Snowcover Conditions in the Life of Ungulates in the USSR.* English trans. Ottawa: Canadian Wildlife Service, 1955

Neel, J. V.: Lessons from a primitive people. *Science* **170,** 815–822 (1970)

Newbigging, P. L.: The perceptual reintegration of words which differ in connotative meaning. *C J Psychol* **15,** 133–141 (1961)

Newman, O.: *Defensible Space.* New York: Macmillan, 1972

Newman, R. W.: Why man is such a sweaty and thirsty naked animal. *Human Biol* **42,** 12–27 (1970)

Newton, I.: Eruptions of crossbills in Europe. In: *Animal Populations in Relation to their Food Resources.* Watson, A. (ed). Oxford: Blackwell, 1970, 337–353

Newton, M.: Human lactation. In: *Milk: The Mammary Gland and its Secretion, Vol. 1.* Kon, S. K., Cowie, A. T. (eds). New York: Academic Press, 1961, 281–320

Newton, M.: Psychological differences between breast and bottle feeding. *Amer J Clin Nutr* **24,** 993–1004 (1971a)

Newton, M.: Uniqueness of human milk: Mammary effects. *Amer J Clin Nutr* **24,** 987–990 (1971b)

Newton, N. R., Newton, M.: Relation of the let-down reflex to the ability to breast feed. *Pediatrics* **5**, 726−733 (1950)

Ney, F. I., Carlson, J., Garrett, G.: Family size, interaction, effect and stress. *J Marr Fam* **32**, 216−226 (1970)

Nickerson, N. H., Rowe, W. H., Richter, E. A.: Native plants in the diet of North Alaskan Eskimos. In: *Man and his Foods*. Smith, C. E. (ed). Tuscaloosa, Ala.: Univ. Alabama Press, 1973, 3−27

Niemeyer, W.: *Legasthenic und Milieu*. Hanover: Schroeder, 1974

Nievergelt, B.: *Der Alpensteinbock*. *Mammalia Depicta*. Berlin: Parey, 1966

Nikolaevskii, L. D.: Reindeer hygiene. In: *Reindeer Husbandry*. Zhigunov, P. S. (ed). Trans. from Russian. Springfield: US Dept. Commerce, 1961, 57−77

Nishikawa, J., Hafez, E. S. E.: Reproduction of horses. In: *Reproduction in Farm Animals*. Hafez, E. S. E. (ed). Philadelphia: Lea & Febiger, 1968, 289−300

Noble, G. K.: Sexual selection among fishes. *Biol Rev* **13**, 133−158 (1938)

North, F. A.: The influence of poverty on maternal and child health. In: *Maternal and Child Health Practices*. Wallace, H. M., Gold, E. M., Lis, E. F. (eds). Springfield, Ill.: Thomas, 1973, 155−167

Noton, D., Stark, L.: Eye movements and visual perception. *Sci Amer* **224(6)**, 35−43 (1971)

Nuckolls, C., Cassel, J., Kaplan, B.: Psychological assets, life crisis and the prognosis of pregnancy. *Amer J Epidemiol* **95**, 431−440 (1972)

Nye, F. I., Carlson, J., Garrett, G.: Family size, interaction, effect and stress. *J Marr Fam* **32**, 216−226 (1970)

Oakley, K. P.: *Man the Tool-Maker*. Chicago: Univ. Chicago Press, 1949

Oakley, K. P.: On man's use of fire, with comments on toolmaking and hunting. In: *Social Life of Early Man*. Washburn, S. L. (ed). Chicago: Aldine, 1961, 176−193

Oboussier, H.: Quantitative und morphologische Studien am Hirn der Bovidae, ein Beitrag zur Kenntnis der Phylogenie. *Gegenbauers Morph Jahrb* **117(2)**, 162−168 (1971)

Oboussier, H., Schliemann, H.: Hirn-Körpergewichtbeziehungen bei Boviden. *Z Säugetierk* **31**, 464−471 (1966)

Olds, J.: Emotional centres in the brain. *Science J* **3(5)**, 87−92 (1967)

Oloff, H. B.: *Zur Biologie und Ökologie des Wildschweines: Beiträge zur Tierkunde und Tierzucht, Vol. 2*. Frankfurt: Schöps, 1951

Olson, E. C.: The evolution of mammalian characters. *Evolution* **13**, 344−353 (1959)

Olson, E. C.: Community evolution and the origin of mammals. *Ecology* **47**, 291−302 (1966)

O'Gara, B. W.: Unique aspects of reproduction in the female pronghorn. (*Antilocapra americana* Ord). *Am. J. Anatomy* **125(2)**, 217−231 (1969)

Ooshima, A., Fuller, G., Cardinale, G., Spector, S., Udenfriend, S.: Collagen biosynthesis in blood vessels of brain and other tissues of the hypertensive rat. *Science* **190**, 898−900 (1975)

Orians, G. H.: On the evolution of mating systems in birds and mammals. *Amer Natur* **103**, 589−603 (1969)

Osborn, J. W.: The evolution of dentitions. *Amer Sci* **61(5)**, 548−559 (1973)

Osofsky, H. J., Rajan, R.: The adolescent pregnancy. In: *Maternal and Child Health Practices*. Wallace, H. M., Gold, E. M., Lis, E. F. (eds). Springfield, Ill.: Thomas, 1973, 884−889

Ostrom, J. H.: Cranial morphology of the hadrosaurian dinosaurs of North America. *Amer Mus Nat Hist* **122**, 37−186 (1961)

Ostrom, J. H.: A reconsideration of the paleoecology of hadrosaurian dinosaurs. *Amer J Sci* **262**, 975−997 (1964)

Ostrom, J. H.: Functional morphology and evolution of the ceratopsian dinosaurs. *Evolution* **20**, 290−308 (1966)

Ott, J. N.: *Health and Light*. Old Greenwich, Conn: Devin-Adair, 1973

Otte, D.: Effect and functions in the evolution of signal systems. *Ann Rev Ecol Syst* **5**, 385–417 (1974)

Owen-Smith, N.: The social system of the white rhinoceros. In: *The Behaviour of Ungulates and its Relation to Management.* Geist, V., Walther, F. R. (eds). Morges: IUCN Publications New Series **24**, 1974, 341–351

Oxnard, C.: *Uniqueness and Diversity in Human Evolution.* Chicago: Univ. Chicago Press, 1975

Ozertskovskaya, N. N., Romanova, V. I., Alekseeva, M. I., Pereverseva, E. V., Uspenskii, S. M.: Human trichinosis in the Soviet Arctic and the characteristics of the strain of Arctic Trichinella. In: *Productivity and Conservation in Northern Circumpolar Lands.* Fuller, W. A., Kevan, P. G. (eds). Morges: IUCN Publications **16**, 1970, 133–142

Parke, R. D., Collmer, C. W.: Child abuse: An interdisciplinary analysis. In: *Review of Child Development Research 5.* Hetherington, E. M. (ed) Chicago: Univ. Chicago Press, 1975, 509–590

Parizkova, J.: Body composition and exercise during growth and development. In: *Physical Activity.* Rarick, G. L. (ed). New York: Academic Press, 1973, 98–124

Parker, G. R.: Biology of the Kaminuriak Population of Barren Ground Caribou. Ottawa: CWS Report Series 20, 1972

Parkinson, C. N.: *Parkinson's Law.* Boston: Houghton Mifflin, 1957

Passingham, R. E.: The brain and intelligence. *Brain Behav Evol* **11**, 1–15 (1975)

Patterson, B., Pascual, R.: The fossil mammal fauna of South America. In: *Evolution, Mammals and Southern Continents.* Keast, A., Erk, F. C., Glass, B. (eds). Albany: State Univ. New York Press, 1972, 247–310

Patton, R. G., Gardner, L. I.: *Growth Failure in Maternal Deprivation.* Springfield, Ill.: Thomas, 1963

Pauling, L. C.: *Vitamin C and the Common Cold.* San Francisco: Freeman, 1970

Pauling, L. C.: That man Pauling! *Nutr Today* **6(2)**, 21–24 (1971)

Pauling, L. C.: The new medicine. *Nutr Today* **7(5)**, 18–23 (1972)

Payne, N.: Discussion. In: *Bears—Their Biology and Management.* Herrero, S. (ed). Morges: IUCN Publications New Series **23**, 1972, 254

Pearson, A. M.: Population characteristics of the northern interior grizzly in the Yukon Territory, Canada. In: *Bears—Their Biology and Management.* Herrero, S. (ed). Morges: IUCN Publications New Series **23**, 1972, 32–35.

Penfield, W.: The interpretive cortex. *Science* **129**, 1719–1725 (1959)

Perin, C.: *With Man in Mind.* Cambridge, Mass.: MIT Press, 1970

Peter, L. J.: *The Peter Prescription.* New York: Bantam, 1972

Peters, H. F.: A feedlot study of bison, cattolo and Hereford calves. *Can J Anim Sci* **38**, 87–90 (1958)

Petocz, R. G.: The effect of snow cover on the social behaviour of bighorn rams and mountain goats. *Can J Zool* **51**, 987–993 (1973)

Petrides, G. A., Swank, W. G.: Population densities and range carrying capacity for large mammals in Queen Elizabeth National Park, Uganda. *Zool Afr* **1(1)**, 209–225 (1965)

Petropoulos, E. A., Vernadakis, A., Timiras, P. S.: Developmental changes in the offspring of rats electroshocked during gestation. *Exp Neurol* **20**, 481–495 (1968)

Petropoulos, E. A., Lau, C., Liao, C. L.: Neurochemical changes in the offspring of rats subject to stressful conditions during gestation. *Exp Neurol* **37**, 86–99 (1972)

Pettigrew, J. D., Freeman, R. D.: Visual experience without lines: Effect on developing cortical neurons. *Science* **182**, 599–601 (1973)

Péwé, T. L.: Permafrost and its effect on life in the North. Corvallis: Oregon State Univ. Press, 1966

Pfeiffer, J. E.: *The Emergence of Man.* New York: Harper & Row (2nd ed), 1972

Phillpotts, B. Z.: *Edda and Saga.* London: Thornton Butterworth (Folcraft Library ed), 1973. First publ 1932

Pilbeam, D., Gould, S. J.: Size and scaling in human evolution. *Science* **186**, 892–901 (1974)

Poglayen-Neuwall, I.: On the marking behaviour of the kinkajou (*Potos flavus* Schreber). *Zoologica* **51(4)**, 137–141 (1966)

Popov, S. P.: Reindeer in haulage and commercial hunting. In: *Reindeer Husbandry*. Zhigunov, P. S. (ed). Trans. from Russian. Springfield: US Dept. Commerce, 1961, 189–214

Porsild, A. E.: Contribution to the flora of the southwestern Yukon Territory. *Nat Mus Can Bull* **216**, 1–86 (1966)

Portmann, A.: *Einführung in die vergleichende Morphologie der Wirbeltiere*. Basel: Schwabe, 1959

Post, R. H.: Population differences in red and green colour vision deficiencies. In: *Man in Evolutionary Perspective*. Brace, C. L., Metress, J. (eds). New York: Wiley, 1971, 387–401

Potter, R. G., New, M. L., Wyon, J. B.: Applications of field studies to research on the physiology of human reproduction; lactation and its effect upon birth intervals in elevan Punjab villages, India. *J Chron Dis* **18**, 1125–1140 (1965)

Poulter, T. C.: Marine mammals. In: *Animal Communication*. Sebeok, T. A. (ed). Bloomington: Indiana Univ. Press, 1968, 405–465

Powell, G. F., Basel, A. J., Blizzard, R. M.: Emotional deprivation and growth retardation simulating idiopathic hypopituitarism. I: Clinical evaluation of the syndrome. *N Engl J Med* **276**, 1271–1278 (1967a)

Powell, G. F., Basel, J. A., Raiti, S., Blizzard, R. M.: Emotional deprivation and growth retardation simulating idiopathic hypopituitarism. II: Endocrinologic evaluation of the syndrome. *N Engl J Med* **276**, 1279–1283 (1967b)

Premack, D.: Language and intelligence in ape and man. *Amer Sci* **64(6)**, 674–683 (1976)

Preobrazhenskii, B. V.: Management and breeding of reindeer. In: *Reindeer Husbandry*. Zhigunov, P. S. (ed). Trans. from Russian. Springfield: US Dept. Commerce, 1961, 78–128

Pressey, A. W.: An assimilation theory of geometric illusions. In: *Readings in General Psychology: Canadian Contributions*. Pressey, A. W., Zubek, J. P. (eds). Toronto: McClelland & Stewart, 1970, 178–185

Preston, E. M.: Computer simulated dynamics of a rabies-controlled fox population. *J Wildl Manage* **37**, 501–512 (1973)

Pribram, K. H.: A neuropsychological model. In: *Expression of the Emotions in Man*. Knapp, P. H. (ed). Nes York: International Univ. Press, 1963, 209–229

Pribram, K. H.: The new neurology and the biology of emotion. *Amer Psychol* **22**, 830–838 (1967)

Prideaux, T.: *Cro-Magnon Man*. New York: Time-Life Books, 1973

Priester, E. S.: Musik im Dienste von Medizin und Zahnheilkunde. *Das Deut Zahnärz* **16**, 564–568 (1962)

Protsch, R. R.: The dating of upper Pleistocene sub-Sahara fossil hominids and their place in human evolution: With morphological and archeological implications. Unpubl. PhD thesis, Univ. California, Los Angeles, 1973

Quay, W. B.: Microscopic structure and variation in the cutaneous glands of the deer *Odocoileus virginianus*. *J Mammal* **40**, 114–128 (1959)

Quay, W. B., Müller-Schwarze, D.: Functional histology of integumentary glandular regions in black-tailed deer (*Odocoileus hemionus columbianus*). *J Mammal* **51**, 675–694 (1970)

Quay, W. B., Müller-Schwarze, D.: Relations of age and sex to inteumentary glandular regions in Rocky Mountain mule deer (*Odocoileus hemionus hemionus*). *J Mammal* **52(4)**, 670–685 (1971)

Rabkin, J. G., Struening, E. L.: Life events, stress and illness. *Science* **194**, 1013–1020 (1976)

Raesfeld, F. von: *Das Deutsche Weidwerk*. Frevert, W. (ed). Berlin: Parey (7th ed), 1952

Rahe, R. R., McKean, J. D., Arthur, R. J.: A longitudinal study of life-change and illness patterns. *J Psychosom Res* **10**, 355–366 (1967)

Ralls, K.: Mammalian scent marking. *Science* **171**, 443–449 (1971)

Rampton, V. N.: Neoglacial fluctuations of the Natazhat and Klutlan glaciers, Yukon Territory, Canada. *Can J Earth Sci* **7(5)**, 1236–1263 (1970)

Rarick, G. L.: Stability and change in motor ability. In: *Physical Activity*. Rarick, G. L. (ed). New York, Academic Press, 1973a, 201–224

Rarick, G. L.: Motor performance of mentally retarded children. In: *Physical Activity*. Rarick, G. L. (ed). New York: Academic Press, 1973b, 225–256

Rauseh, R. A.: Some aspects of population ecology of wolves in Alaska. *Amer Zool* **7**, 253–266 (1967)

Read, D. W.: Hominid teeth and their relationship to hominid phylogeny. *Amer J Phys Anthropol.* **42**, 105–126 (1975)

Reed, C. A.: Extinction of mammalian megafauna in the old world late Quaternary. *Bioscience* **20**, 284–288 (1970)

Reid, R. L., Miles, S. C.: Studies on the carbohydrate metabolism of sheep: The adrenal response to psychological stress. *Austral J Agric Res* **13**, 282–295 (1962)

Rensberger, J. M.: Sanctimus (Mammalia, Rodentia) and the phyletic relationships of the large arikareean geomyoids. *J Paleontol* **47**, 835–853 (1973)

Reynolds, V., Reynolds, R.: Chimpanzee of the Bundongo Forest. In: *Primate Behaviour*. de Vore, I. (ed), New York: Holt Rinehart Winston, 1965, 368–424

Riney, T.: Evaluating condition of free-ranging red deer (*Cervus elaphus*) with special reference to New Zealand. *N Z J Sci Tech* **36**, 429–463 (1954)

Riney, T.: The impact of introductions of large herbivores on the tropical environment. In: *The Ecology of Man in the Tropical Environment*. Elliott, H. F. L. (ed). Morges: IUCN Publications New Series **4**, 1964, 261–273

Robel, R. J.: Possible role of behaviour in regulating greater prairie chicken populations. *J Wildl Manage* **34**, 306–312 (1970)

Robins, E.: *Africa's Wildlife: Survival or Extinction*. London: Odhams, 1963

Robinson, F. N.: Vocal mimicry and the evolution of bir song. *Emu* **75(1)**, 23–27 (1975)

Robinson, J. T.: Early hominid posture and locomotion. Chicago: Univ. Chicago Press, 1972a

Robinson, J. T.: The bearing of East Rudolf fossils on early hominid systematics. *Nature (Lond)* **240**, 239–240 (1972b)

Robinson, P. L.: A problem of fauna replacement on permotriassic continent. *Paleontology* **14(1)**, 131–153 (1971)

Rodbard, S.: Warm-bloodedness. *Sci Month* **77**, 137–142 (1953)

Roe, F. G.: The North American Buffalo. Univ. Toronto Press, Toronto (2nd ed), 1970

Romer, A. S.: *The Vertebrate Body*. Philadelphia: Saunders (3rd ed), 1962

Romer, A. S.: *Vertebrate Paleontology*. Chicago: Univ. Chicago Press (3rd ed), 1966

Rosen, S.: Noise, hearing and cardiovascular function. In: *Physiological Effects of Noise*. Welch, B. L., Welch, A. S. (eds). New York: Plenum, 1970, 57–66

Rosenmayer, L.: The quality of ageing. In: *Science and Absolute Values*. Tarrytown: Inter. CULT. Found., 1975, 287–318

Ross, M. H., Bras, G.: Food preference and length of life. *Science* **190**, 165–167 (1975)

Rotter, I.: Die Warane (Varanidae). *Neue Brehm-Bücherei* **325.** Wittenberg-Lutherstadt: Ziemsen, 1963

Rowan, W.: On photoperiodism, reproductive periodicity and the annual migration of birds and certain fishes. *Proc Boston Soc Nat Hist* **38**, 147–189 (1926)

Rowell, T. E., Hinde, R. A.: Responses of Rhesus monkeys to mildly stressful situations. *Anim Behav* **11**, 235–243 (1963)

Rowlands, M. J.: Defence: A factor in the organization of settlements. In: *Man Settlement and Urbanism*. Ucko, P. J., Tringham, T., Dimbley, G. W. (Eds). London: Duckworth, 1972, 447–462

Ruddelle, R.: Chiefs and commoners: Nature's balance and the good life among the Nootka. In: *Cultural Ecology*. Cox, B. (ed). Toronto: Carleton Library 65, McClelland & Stewart, 1973, 254–265

Russell, R.: Discussion. In: *Bears—Their Biology and Management*. Herrero, S. (ed). Morges: IUCN Publications New Series **23**, 1972, 253

Ruttern, N. W., Geist, V., Shackleton, D. M.: A bighorn sheep skull 9280 years old from British Columbia. *J Mammal* **53**, 641−644 (1972)

Ruwet, J. C.: *Introduction to Ethology*. New York: Inter. Univ. Press, 1972

Ruyle, E. E., Cloak, F. T., Slobodkin, L. B., Durham, W. H.: The adaptive significance of cultural behaviour: Comments and reply. *Human Ecol* **5(1)**, 49−68 (1977)

Sadleir, R. M. F.: *The Ecology of Reproduction in Wild and Domestic Mammals*. London: Methuen, 1969

Sahlins, M. D.: Notes on the original affluent society. In: *Man the Hunter*. Lee, R. B., de Vore, I. (eds). Chicago: Aldine, 1968, 85−89

Salzen, E. A.: Imprinting and environmental learning. In: *Development and Evolution of Behaviour*. Aronson, L. R., Tobach, E., Lehrman, D. S., Rosenblatt, J. S. (eds). San Francisco: Freeman, 1970, 158−178

Sancetta, C., Imbrie, J., Kipp, N. G.: Climatic record of past 130,000 years in North Atlantic deep-sea core V23−82: Correlation with the terrestrial record. *Quat Res* **3**, 110−116 (1973)

Santayana, G.: *A Sense of Beauty*. New York: Random House (Modern Library ed), 1955. First publ 1896

Sartwell, P. E. (ed): *Preventive Medicine and Public Health*. New York: Appleton-Century-Crofts (10th ed), 1973

Savalev, D. G.: Control of warble flies and blood-sucking diptera. In: *Reindeer Husbandry*. Zhigunov, P. S. (ed). Trans. from Russian. Springfield: US Dept. Commerce, 1961, 294−310

Sayler, A., Salmon, M.: Communal nursing in mice: Influence of multiple mothers on the growth of young. *Science* **164**, 1309−1310 (1969)

Scarr-Salapatek, S.: Race, social class and IQ. *Science* **174**, 1285−1295 (1971a)

Scarr-Salapatek, S.: Unknowns in the IQ equation. *Science* **174**, 1223−1228 (1971b)

Schaefer, O.: Nutrition problems in the Eskimo. *Can Nutr Notes* **20(8)**, 85−89 (1964)

Schaefer, O.: Carbohydrate metabolism in Eskimos. *Arch Envir Health* **18**, 142−147 (1969)

Schaefer, O.: Pre- and postnatal growth acceleration and increased sugar consumption in Canadian Eskimos. *Can Med Assoc J* **103**, 1059−1068 (1970)

Schaefer, O.: Otitis media and bottle-feeding. *Can J Public Health* **62**, 478−489 (1971a)

Schaefer, O.: When the Eskimo comes to town. *Nutr Today* **6(6)**, 8−16 (1971b)

Schaefer, O.: Vigorous exercise and coronary heart disease. *Lancet* **1** (7807), 840 (1973)

Schaefer, O.: Hildes, J. A., Greidanus, P., Leung, D.: Regional sweating in Eskimos compared to Caucasians. *Can J Physiol Pharmacol* **52**, 960−965 (1974)

Schafer, E. W. P., Marcus, M. M.: Self-stimulation alters human sensory brain responses. *Science* **181**, 175−177 (1973)

Schaller, G.: *The Mountain Gorilla*. Chicago: Univ. Chicago Press, 1963

Schaller, G.: *The Deer and the Tiger*. Chicago: Univ. Chicago Press, 1967

Schaller, G.: *The Serengeti Lion*. Chicago: Univ. Chicago Press, 1972

Schaller, G., Lowther, G.: The relevance of carnivore behaviour to the study of early hominids. *S W J Anthropol*. **25(4)**, 307−341 (1969)

Schaller, G., Mirza, Z. B.: On the behaviour of Punjab urial *Ovis orientalis punjabiensis*. In: *The Behaviour of Ungulates and its Relation to Management*. Geist, V., Walther, F. R. (eds). Morges: IUCN Publications New Series **24**, 1974, 306−323

Schenkel, R.: On sociology and behaviour in impala (*Aepycerus melampus suara* Matschie). *Z Säugetierk* **31**, 177−205 (1966a)

Schenkel, R.: Play, exploration and territoriality in the wild lion. In: *Play, Exploration and Territoriality in Mammals*. Jewell, P. A., Loizos, C. (eds). Symp. Zool. Soc. London 18. London: Academic Press, 1966b, 11−22

Schenkel, R.: Submission: Its features and function in the wolf and dog. *Amer Zool* **7**, 319−329 (1967)

Schenkel, R., Schenkel-Hulliger, L.: Ecology and behaviour of the black rhinoceros (*Di-*

ceros bicornis L). *Mammalia Depicta*. Berlin: Parey, 1969

Schiff, W.: Perception of impending collision: A study of visually directed avoidant behaviour. *Psychol Monog* **79**, no. 11, 1965

Schiff, W., Caviness, J. A., Gibson, J. J.: Persistent fear responses in Rhesus monkeys to the optical stimulus looming. *Science* **136**, 982−983 (1962)

Schiller, P. H.: Manipulative patterns in the chimpanzee. In: *Instinctive Behaviour*. Schiller, C. H. (ed). New York: International Universities Press, 1957, 264−287

Schleidt, W. M.: Reaktion von Truthühnern auf fliegende Raubvögel und Versuche zur Analyse ihrer AAHs. *Z Tierpsychol* **18**, 534−560 (1961)

Schleidt, W. M.: Die historische entwicklung des Begriffe "Angeborenes auslösendes Schema" und "Angeborener Auslösemachanismus" in der Ethologie. *Z Teirpsychol* **19**, 697−722 (1962)

Schloeth, R.: Das Sozialleben des Camargue Rindes. *Z Tierpsychol* **18**, 574−627 (1961)

Schmidt-Nielsen, K.: *How Animals Work*. Cambridge: Cambridge Univ. Press, 1972a

Schmidt-Nielsen, K.: Locomotion: Energy cost of swimming, flying and running. *Science* **177**, 222−223 (1972b)

Schnierla, T. C.: Aspects of stimulation and organization in approach/withdrawal processes underlying vertebrate behavioural development. In: *Advances in the Study of Behaviour, Vol. 1*. Lehrman, D. S., Hinde, R. A., Shaw, E. (eds). New York: Academic Press, 1965, 1−74

Schoenauer, N.: *Introduction to Contemporary Indigenous Housing*. Montreal: Reporter Books, 1973

Scholander, P. F., Hock, R., Walters, V., Irving, L.: Body insulation of some arctic and tropical mammals and birds. *Biol Bull* **99**, 225−236 (1950)

Schull, W. J., Neel, J. V.: *The Effects of Inbreeding on Japanese Children*. New York: Harper & Row, 1965

Schultz, A. H.: Some factors influencing the social life of primates in general and or early man in particular. In: *Social Life of Early Man*. Washburn, S. L. (ed). Chicago: Aldine, 1961, 60−90

Schultze-Westrum, T.: Innerartliche verständigung durch Düfte beim Gleitbeutler *Petaurus breviceps paperanus* Thomas (Marsupialia, Phalangeridae). *Z Vergleich Physiol* **50**, 151−220 (1965)

Schumacher, E. F.: *Small is Beautiful*. New York: Harper & Row, 1973

Schwabe, C.: *Musik-Therapie*. Jena: Fischer, 1969

Schwartz, G. E., Higgins, J. D.: Cardiac activity preparatory to overt and covert behaviour. *Science* **173**, 1144−1146 (1971)

Scott, J. H.: *Dentofacial Development and Growth*. Oxford: Pergamon, 1967a

Scott, J. P.: *Aggression*. Chicago: Univ. Chicago Press, 1958

Scott, J. P.: Agonistic behaviour of mice and rats. *Amer Zool* **6**, 683−701 (1966)

Scott, J. P.: The evolution of social behaviour in dogs and wolves. *Amer Zool* **7**, 373−381 (1967b)

Scott, J. P., Fredricson, E.: The causes of fighting in mice and rats. *Physiol Zool* **24**, 273−309 (1951)

Scott, W. B.: *A History of Land Mammals in the Western Hemisphere*. New York: Macmillan, 1937

Scotter, G. N. and Simmons, N. M.: Mortality of Dall Sheep within a cave. *J Mammal,* **57**, 387−89 (1976)

Scrimshaw, N. S.: Infant malnutrition and adult learning. *Sat Rev* **51(11)**, 64−66 (1968)

Scrimshaw, N. S., Gordon, J. E. (eds): *Malnutrition, Learning and Behaviour*. Cambridge, Mass.: MIT Press, 1968

Scuttles, G.: *The Social Construction of Communities*. Chicago: Univ. Chicago Press, 1972

Sears, R. R.: Your ancients revisited: A history of child development. In: *Review of Child Development Research 5*. Hetherington, E. M. (ed). Chicago: Univ. Chicago Press, 1975, 1−73

Sebeok, T. A.: Coding in the evolution of signalling behaviour. *Behav Sci* **7(4)**, 430–442 (1962)

Sebeok, T. A. (ed): *Animal Communication*. Bloomington: Indiana Univ. Press, 1968

Seifter, E., Cohen, M. H., Riley, V.: Of stress, vitamin A and tumours. *Science* **193**, 74–75 (1976)

Seitz, E.: Die Bedeutung geruchlicher Orientierung beim Plumplori *Nicticebus coucang* Boddaert 1785 (Prosimii, Lorisidae). *Z Tierpsychol* **26(1)**, 73–103 (1969)

Seligman, M. E. P.: *Helplessness: On Depression, Development and Death*. San Francisco: Freeman, 1975

Selye, H.: *Stress*. Montreal: Acta, 1950

Selye, H.: *Stress Without Distress*. New York: Signet, 1974

Shackleton, D. M.: Population quality and bighorn sheep (*Ovis canadensis canadensis*). Unpubl. PhD thesis, Univ. Calgary, 1973

Shank, C. C.: Some aspects of social behaviour in a population of feral goats (*Capra hircus* L). *Z Tierpsychol* **30**, 488–528 (1972)

Shannon, G. W., Dever, G. E. A.: *Health Care Delivery: Spatial Perspectives*. New York: McGraw-Hill, 1974

Shapiro, R. J.: Creative Research Scientists. *Psychol Afr Monog Suppl* **4**, 1968

Sharp, R. P.: The Wolf Creek glaciers, St. Elias Range, Yukon Territory. *Geog Rev* **37**, 26–52 (1947)

Sharp, R. P.: Studies of superglacial debris on valley glaciers. *Amer J Sci* **247**, 289–315 (1949)

Sharp, R. P.: Accumulation and ablation on the Seward-Malaspina glacier system, Canada-Alaska. *Bull Geol Soc Amer* **62**, 725–744 (1951a)

Sharp, R. P.: Glacial history of Wolf Creek, St. Elias Range, Canada. *J Geol* **59**, 97–117 (1951b)

Sharp, R. P.: *Glaciers*. Eugene: Univ. Oregon Press, 1960

Shaw, C. E.: The male combat dance of some crotalid snakes. *Herpetology* **4**, 137–145 (1948)

Sherp, H. W.: Dental caries: Prospects for prevention. *Science* **173**, 1199–1205 (1971)

Shneour, E.: *The Malnourished Mind*. Garden City, N.Y.: Anchor, 1974

Sikes, S. K.: *The Natural History of the African Elephant*. London: Weidenfeld & Nicolson, 1971

Silberbauer, G. B.: The G/Wl Bushmen. In: *Hunters and Gatherers Today*. Bicchieri, M. G. (ed). New York: Holt Rinehart Winston, 1972, 271–326

Simmons, K. E. L.: Ecological determinants of breeding adaptations and social behaviour in two fish-eating birds. In: *Social Behaviour in Birds and Mammals*. Crook, J. H. (ed). London: Academic Press, 1968

Simon, H.: How big is a chunk. *Science* **183**, 482–488 (1974)

Simpson, G. G.: Horses. New York: Oxford Univ. Press, 1951

Simpson, G. G.: *Biology and Man*. New York: Harcourt Brace Jovanovich, 1969

Simpson, J.: *An Introduction to Preventive Medicine*. London: Heinemann Medical, 1970

Singer, J. L.: Daydreaming and the stream of thought. *Amer Sci* **62**, 417–425 (1974)

Smith, D. E., King, M. B., Hoebel, B. G.: Lateral hypothalamic control of killing: Evidence for a cholinoceptive mechanism. *Science* **167**, 900–901 (1970)

Smith, G. W.: Arctic pharmacognosia. *Arctic* **26(4)**, 324–333 (1973)

Smith, H.: *Man and his Gods*. New York: Grosset & Dunlap, 1952

Smith, H. C.: Music in relation to employer attitudes, piecework production and industrial accidents. *App Psychol Monog* **14**, 1–59 (1947)

Smith, W. J.: Message-meaning analysis. In: *Animal Communication*. Sebeok, T. A. (ed). Bloomington: Indiana Univ. Press, 1965, 44–60

Smith, W. J.: Messages of vertebrate communication. *Science* **165**, 145–150 (1969)

Sommer, R.: *Personal Space*. Englewood Cliffs: Prentice-Hall, 1969

Sommer, R.: *Design Awareness*. San Francisco: Holt Rinehart Winston, 1972

Southwick, C. H.: Conflict and violence in animal societies. In: *Animal Aggression*. Southwick, C. H. (ed). New York: Van Nostrand Reinhold, 1970a, 1−13

Southwick, C. H.: Genetic and environmental variables influencing animal aggression. In: *Animal Aggression*. Southwick, C. H. (ed). New York: Van Nostrand Reinhold, 1970b, 213−229

Sowls, L. K.: Social behaviour of the collared peccary *Dicotyles tajacu* in V. Geist & F. Walther (eds.) *The Behaviour of Ungulates and its Relation to Management*. IUCN Publication No. 24. New series pp. 144−165.

Spain, D. M., Nathan, D. J., Gellis, M.: Weight, body type and the prevalence of coronary heart diseases in males. *Amer J Med Sci* **245**, 1963, 63−68

Sparks, B. W., West, R. G.: *The Ice Age in Britain*. London: Methuen, 1972

Sparrow, R. D., Springer, P. F.: Seasonal activity patterns of white-tailed deer in eastern South Dakota. *J Wildl Manage* **34**, 420−431 (1970)

Speer, A.: *Inside the Third Reich*. New York: Macmillan, 1970

Speth, J. D., Davis, D. D.: Seasonal variability in early hominid predation. *Science* **192**, 441−445 (1976)

Spinage, C. A.: Horns and other bony structures of the skull of the giraffe, and their functional significance. *E Afr Wildl J* **6**, 53−61 (1968)

Spitz, R.: Ontogenesis: The proleptic function of emotion. In: *Expression of the Emotions in Man*. Knapp, P. H. (ed). New York: International Universities Press, 1963

Stanley, S. M.: An explanation of Cope's rule. *Evolution* **27**, 1−26 (1973)

Stanley, S. M.: Relative growth of the titanothere horn. *Evolution* **28**, 447−457 (1974)

Stare, F. J.: Not quite cricket. *Nutr Today* **6(1)**, 18−20 (1971)

Steel, R.: Die Dinosaurier, *Neue Brehm-Bücherei* **432.** Wittenberg-Lutherstadt: Ziemsen, 1970

Stein, L., Wiese, C. D.: Possible etiology of schizophrenia: Progressive damage to the noradrenergic reward system by 6-hydroxydopamine. *Science* **171**, 1032−36 (1971)

Stein, M., Schiavi, R. C., Camerino M.: Influence of brain and behaviour on the immune system. *Science* **191**, 435−440 (1976)

Stent, G. S.: Limits to the scientific understanding of man. *Science* **187**, 1052−1057 (1975)

Stevenson, S. S.: Comparison of breast and artificial feeding. *J Amer Diet Assoc* **25**, 752−756 (1949)

Steward, J. H.: Causal factors and processes in the evolution of prefarming societies. In: *Man the Hunter*. Lee, R. B., de Vore, I. (eds). Chicago: Aldine, 1968, 321−334

Stewart, T. D.: Form of the pubic bone in Neanderthal man. *Science* **131**, 1437−1438 (1960)

Stewart, T. D.: Neanderthal scapula with special attention to the Shanidar Neanderthals from Iraq. *Amer J Phys Anthropol.* **21**, 409−410 (1963)

Stock, C.: *Rancho La Brea*. Los Angeles: Los Angeles County Mus. Nat. Hist. (6th ed), 1956

Stock, M. B., Smythe, P. M.: The effect of undernutrition during infancy on subsequent brain growth and intellectual development. *S Afr Med J* **41**, 1027−1030 (1967)

Stokes, W. M. L., Condie, K. C.: Pleistocene bighorn sheep from the Great Basin. *J Paleontol* **35**, 598−609 (1961)

Stokvis, B.: Psychosomatische Gedanken über Musik. In: *Musik in der Medizin*. Teirich, H. R. (ed). Stuttgate: Fischer, 1958, 43−53

Stover, L. E.: *The Cultural Ecology of the Chinese Civilization*. New York: Mentor, 1974

Strauss, G., Sayles, L. R.: *Personnel*. Engelwood Cliffs: Prentice-Hall, 1972

Struhsaker, T. T.: Behaviour of vervet monkeys and other cercopithecines. *Science* **156**, 1197−1203 (1967)

Strum, S. C.: Primate predation. *Science* **187**, 755−757 (1975)

Sugiyama, Y.: On the social change of hanuman langurs (*Presbytis entellus*) in their natural condition. *Primates* **6**, 381−418 (1965)

Sutcliff, A. J.: Spotted hyaena: Crusher, gnawer, digester and ollector of bones. *Nature (Lond)* **227**, 1110−1113 (1970)

Sutermeister, H. M.: Psychosomatik des Musikerlebens, Prolegomena zur Musiktherapie. *Acta Psychotheor* **12**, 91 (1964)

Sweet, L. E.: Camel pastoralism in North Arabia and the minimal camping unit. In: *Man, Culture and Animals*. Leeds, A., Vayda, A. P. (eds) Washington: AAAS Publ **78**, 1965, 129–152

Talbot, L. M., Talbot, M. H.: The high biomass of wild ungulates on the East African savannah. *Trans. 28 N. Amer. Wildl. Nat. Resources Conf, Detroit*. Trefethen, J. B. (ed). Washington: Wildl. Mangt. Inst, 1963, 465–476

Tanner, J. M.: *Growth at Adolescence*. Oxford: Blackwell (2nd ed), 1962

Tanner, J. M.: Galtonian eugenics and the study of growth. *Eugen Rev* **58**, 122–135 (1966)

Tattersall, I., Eldredge, N.: Fact, theory and fantasy in human paleontology. *Amer Sci* **65(2)**, 204–211 (1977)

Tavolga, W. N.: Fishes. In: *Animal Communication*. Sebeok, T. A. (ed). Bloomington: Indiana Univ. Press, 1968, 271–288

Taylor, C. R., Rowntree, V. J.: Running on two or on four legs: Which consumes more energy. *Science* **179**, 186–187 (1973)

Taylor, R. E.: The Marquis de Sade and the first *psychopathia sexualis*. In: *An Analysis of the Kinsey Report on Sexual Behaviour of the Human Male and Female*. Geddes, D. P. (ed). New York: Mentor, 1954, 193–210

Teele, J. E.: Social pathology and stress. In: *Social Stress*. Scotch, N. A. (ed). Chicago: Aldine, 1970, 228–256

Teirich, H. R.: Musik im Ramen einer nervenärzlichen Praxis. In: *Musik in der Medizin*. Teirich, H. R. (ed). Stuttgart: Fischer, 1958, 119–137

Teleki, G.: The predatory behaviour of wild chimpanzees. Lewisburg, Pa.: Bucknell Univ. Press, 1973

Tembrock, G.: Das Verhalten des Rotfuchses. *Handb Zool* **8(9)**, 1–20 (1957)

Tembrock, G.: Beobachtungen zur Fuchsranz unter besonderer Berücksichtigung der Lautgebung. *Z Tierpsychol* **16(3)**, 351–368 (1959)

Templeton, J. R.: Reptiles. In: *Comparative Physiology of Thermoregulation. Vol. 1. Invertebrates and Nonmammalian Vertebrates*. Whittow, G. C. (ed). New York: Academic Press, 1970, 167–221

Tener, J. S.: *Muskoxen*. Ottawa: Queen's Printer, 1965

Thenius, E., Hofer, H.: *Stammesgeschichte der Säugetiere*. Berlin: Springer, 1960

Terasmae, J.: Notes on quaternary paleoecological problems in the Yukon Territory and adjacent regions. Geol. Survey Can. Paper 67–46. Ottawa: Energy Mines and Resources Canada, 1967

Terman, L. M., Oden, M. H.: *The Gifted Group at Midlife*. Stanford: Stanford Univ. Press, 1959

Thoma, A.: On Vérteszöllös man. *Nature (Lond)* **236**, 464–465 (1972)

Thompson, D'A. W.: *On Growth and Form*. London: Cambridge Univ. Press (abr ed), 1961. First publ 1917.

Thompson, W. R.: Influence of prenatal maternal anxiety on emotionality in young rats. *Science* **125**, 698–699 (1957)

Thomson, A. M., Hytten, F. E.: Nutrition during pregnancy. *World Rev Nutr Diet* **16**, 22–45 (1973)

Thorpe, W. H.: *Learning and Instinct in Animals*. London: Methuen, 1963

Tiger, L.: *Men in Groups*. London: Nelson, 1969

Tiger, L.: Somatic factors and social behaviour. In: *Biosocial Anthropology*. Fox, R. (ed). New York: Wiley, 1975, 115–132

Tiger, L., Fox, R.: *The Imperial Animal*. Toronto: McClelland Stewart, 1971

Timiras, P. S.: Development and plasticity of the nervous system. In: *Developmental Physiology and Aging*. Timiras, P. S. (ed). New York: Macmillan, 1972, 129–165

Timiras, P. S., Vernadakis, A.: Structural, biochemical and functional aging of the nervous system. In: *Developmental Physiology and Aging*. Timiras, P. S. (ed). New York:

Macmillan, 1972, 502–526

Tinbergen, N.: *The Study of Instinct*. London: Oxford Univ. Press, 1951

Tinbergen, N.: *Social Behaviour in Animals*. London: Methuen, 1953

Tinbergen, N.: *Curious Naturalist*. Country Life Ltd. London, 280 pp. (1958)

Tinbergen, N.: On war and peace in animals and man. *Science* **160,** 1411–1418 (1968)

Tindale, N. B.: The Pitjandjara. In: *Hunters and Gatherers Today*. Bicchieri, M. G. (ed). New York: Holt Rinehart Winston, 1972, 217–268

Tinkle, D. W.: The concept of reproductive effort and its relation to the evolution of life histories of lizards. *Amer Nat* **103,** 501–516 (1969)

Tobach, E.: Some guidelines to the study of the evolution and development of emotion. In: *Development and Evolution of Behaviour*. Aronson, L. R., Tobach, E., Lehrman, D. S., Rosenblatt, J. S. (eds). San Francisco: Freeman, 1970, 238–253

Tobias, P. V.: Implications of the new age estimates of the early South African hominids. *Nature (Lond)* **246,** 79–83 (1973)

Toffler, A.: *Future Shock*. New York: Random House, 1970

Townsend, R.: *Up the Organization*. New York: Knopf Fawcett, 1970

Tränkel, W.: Über die anregende und entspannende Wirkung von Musik. In: *Musik in der Medizin*. Teirich, H. R. (ed). Stuttgart: Fischer, 1958, 54–67

Traxel, W., Wrede, G.: Hautwiederstandtmessungen bei Musikdarbeitung. *Z Exper Angew Psychol* **6,** 293–309 (1959)

Tringham, R.: Territorial demarcation of prehistoric settlements. In: *Man, Settlement and Urbanism*. Ucko, P. J., Tringham, R., Dimbley, G. W. (eds). London: Duckworth, 1972, 463–475

Trivers, R. L.: The evolution of reciprocal altruism. *Quart Rev Biol* **46(4),** 35–57 (1971)

Trivers, R. L.: Parental investment and sexual selection. In: *Sexual Selection and the Descent of Man*. Campbell, B. (ed). Chicago: Aldine, 1972, 136–179

Trivers, R. L.: Parent-offspring conflict. *Amer Zool* **14,** 249–264 (1974)

Trivers, R. L., Willard, D. E.: Natural selection of parental ability to vary the sex ratio of offspring. *Science* **179,** 90–92 (1973)

Troyer, W. A., Hensel, R. J.: Structure and distribution of a Kodiak bear population. *J Wildl Manage* **28,** 769–772 (1964)

Tsukamoto, J. Y.: Experimental Farm Mile 1019, Alaska Highway, Yukon Territory Progress Report 1953–1959. Ottawa: Canada Dept. Agriculture, 1963.

Tucker, V. A.: Oxygen transport by the circulatory system of the green iguana (*Iguana iguana*) at different body temperatures. *J Exp Biol* **44,** 77–92 (1966)

Turnbull, C. M.: *The Mountain People*. New York: Simon & Schuster Touchstone, 1972

Tuttle, R. H.: Knuckle-walking and the problem of human origins. *Science* **166,** 953–961 (1969)

Tyler, L. E.: *The Psychology of Human Differences*. New York: Appleton-Century-crofts (2nd ed), 1956

Tyszka, H. von: Das Grosshirnfurchenbild als Merkmal der Evolution. Untersuchungen an Boviden I. *Mitt Hamburg Zool Mus Inst* **63,** 121–158 (1966)

Uhazy, L. S., Holmes, J. C., Stelfox, J. G.: Lungworms in the Rocky Mountain bighorn sheep of western Canada. *Can J Zool* **51,** 817–824 (1973)

Ulrich, R. E., Azrin, N. H.: Reflexive fighting in response to aversive stimulation. *J Exp. Anal Behav* **5,** 511–520 (1962)

Ulrich, R. E., Hutchinson, R. R., Azrin, N. H.: Pain-elicited aggression. *Psychol Rec* **15,** 111–126 (1965)

Uspenskii, S. M.: Problems and forms of fauna conservation in the Soviet arctic and subarctic. In: *Productivity and Conservation in Circumpolar Lands*. Fuller, W. A., Kevan, P. G. (eds). Morges: IUCN Publications **16,** 1970, 199–207

Valentine, J. W., Campbell, C. A.: Genetic regulation and the fossil record. *Amer Sci* **63,** 673–680 (1975)

Vallois, H. V.: The social life of early man: The evidence of skeletons. In: *Social Life of Early Man*. Washburn, S. L. (ed). Chicago: Aldine, 1961, 214–235

Vandenberg, J. G.: Esinophil response to aggressive behaviour in CFW albino mice. *Anim Behav* **8**, 13−18 (1960)

Vanderwolf, C. H.: Limbic-diencephalic relations in the higher level control of movement. Dept. Psychol. Physiol. Res. Bull. 126. London: Univ. Western Ontario, 1969

Van Valen, L.: Therapsids as mammals. *Evolution* **14**, 304−313 (1960)

Van Valen, L.: Body size and numbers of plants and animals. *Evolution* **27**, 23−35 (1973)

Van Valen, L., Sloan, R. S.: Ecology and the extinction of the dinosaurs. *Evol Theory* **2**, 37−64 (1977)

Vereshchagin, N. K.: Primitive hunters and Pleistocene extinction in the Soviet Union. In: *Pleistocene Extinctions*. Martin, P. S., Wright, H. E. (eds). New Haven: Yale Univ. Press, 1967, 365−398

Vereshchagin, N. K.: The mammoth horizon. *Animals,* Feb, 483−441 (1970)

Vereshchagin, N. K.: The mammoth 'cemetries' of northeast Siberia. *Polar Rec* **17**, 3−12 (1974)

Vernadakis, A., Valcana, T., Curry, J. J., Maletta, G. J., Hudson, D., Timiras, P. S.: Alterations in growth of brain and other organs after electroshock in rats. *Exp Neurol* **17**, 505−516 (1967)

Verner, J., Willson, M. F.: The influence of habitats on mating systems in North American passerine birds. *Ecology* **47**, 143−147 (1966)

Vibe, C.: The arctic ecosystem influenced by fluctuations in sunspots and drift-ice movements. In: *Productivity and Conservation in Northern Circumpolar Lands*. Fuller, W. A., Kevan, P. G. (eds). Morges: IUCN Publications New Series **16**, 1970, 115−120

Viereck, L. A.: Plant succession and soil development on gravel outwash of the Muldrow Glacier, Alaska. *Ecol Monog* **36**, 161−199 (1966)

Villiers, H. de: The tablier and steatopygia in Kalahari bushwomen. *S Afr J Sci* **57**, 223−227 (1964)

Vogel, V. J.: *American Indian Medicine*. New York: Ballantine, 1970

Waddington, C. H.: *The Strategy of the Genes*. London: George Allen & Unwin, 1957

Waddington, C. H.: *The Ethical Animal*. Chicago: Univ. Chicago Press, 1960

Waddington, C. H.: *The Evolution of an Evolutionist*. Ithaca, N.Y.: Cornell Univ. Press, 1975

Walker, A.: Remains attributable to *Australopithecus* in the East Rudolf succession. In: *Earliest Man and Environments in the Lake Rudolf Basin*. Coppens, Y., Howell, F. C., Isaac, G. L., Leakey, R. E. F. (eds). Chicago: Univ. Chicago Press, 1976, 484−489

Walker, E. P.: *Mammals of the World*. Baltimore: Johns Hopkins Press, 1964

Wallace, H. M.: Factors associated with perinatal mortality and morbidity. In: *Maternal and Child Health Practices*. Wallace, H. M., Gold, E. M., Lis, E. F. (ed), 1973, 500−531

Wallace, H. M., Gold, E. M., Lis, E. F. (eds). *Maternal and Child Health Practices*. Springfield, Ill.: Thomas, 1973

Walther, F. R.: Zum Kampf- und Paarungsverhalten einiger Antilopen. *Z Tierpsychol* **15**, 340−380 (1958)

Walther, F. R.: Entwicklungszüge im Kampf- und Paarungsverhalten der Horntiere. *Jahrb G Opel Freigehege* **3**, 90−115 (1960)

Walther, F. R.: Verhaltensstudien an der Gattung *Tragelaphus* de Blainville (1816) in Gefangenschaft unter besonderen Berücksichtigung des Sozialverhaltens. *Z Tierpsychol* **21**, 393−467 (1964a)

Walther, F. R.: Einige Verhaltensbeobachtungen an Thomsongazellen (*Gazella thomsoni* Günther 1884) im Ngorongoro-Krater. *Z Tierpsychol* **21**, 871−890 (1964b)

Walther, F. R.: Verhaltensstudien an der Grantgazelle (*Gazella granti* Brooke 1872) im Ngorongoro-Krater. *Z Tierpsychol* **22**, 167−208 (1965)

Walther−F. R.: *Mit Horn und Huf*. Berlin: Parey, 1966

Walther, F. *Verhalten der Gazellen*. Neue Brehm—Bücherei No. **373**. Ziemsen Verlag, Wittenberg Lutherstadt. 144 pp. (1968)

Walther, F. R.: Some reflections on expressive behaviour in combats and courtship of certain horned ungulates. In: *The Behaviour of Ungulates and its Relation to Management*.

Geist, V., Walther, F. R. (eds). Morges: IUCN Publications New Series **24**, 1974, 56–106

Ward, I. L.: Prenatal stress feminizes and demasculinizes the behaviour of males. *Science* **175**, 82–84 (1972)

Warren, R. J., Lepow, M. L., Bartsche, G. E., Robbins, F. C.: The influence of breast milk on intestinal infection with Sabin Type 1 polio virus vaccine. *Amer J Dis Child* **102**, 685–686 (1961)

Washburn, A. L.: *Periglacial Processes and Environments*. New York: St Martin's Press, 1973

Washburn, S. L., Lancaster, C. S.: The evolution of hunting. In: *Man the Hunter*. Lee, R. B., de Vore, I. (eds). Chicago: Aldine, 1968, 293–303

Watanabe, H.: Subsistence and ecology of northern food gatherers with special reference to the Ainu. In: *Man the Hunter*. Lee, R. B., de Vore, I. (eds). Chicago: Aldine, 1968, 69–77

Watanabe, A.: The Ainu. In: *Hunters and Gatherers Today*. Bicchieri, M. G. (ed). New York: Holt Rinehart Winston, 1972, 448–484

Waterbolk, H. T.: Food production in prehistoric Europe. *Science* **162**, 1093–1102 (1968)

Watson, A., Moss, R.: A current model of population dynamics in red grouse. Proc. 15th Inter. Ornithol. Cong. Leiden: Brill, 1972, 134–149

Watt, K. E. F.: *Principles of Environmental Science*. New York: McGraw-Hill, 1973

Watts, C. R., Stokes, A. W.: *The social order of turkeys*. Sci Amer **224(6)**, 112–118 (1971)

Webb, S. D.: Extinction-origination equilibrium in Late Cenozoic land mammals of North America. *Evolution* **23**, 688–702 (1969)

Webster, A. J. F.: Prediction of heat loss from cattle exposed to cold environments. *J Appl Physiol* **30**, 684–690 (1971)

Webster, A. J. F., Blaxter, K. L.: The thermal regulation of two breeds of sheep exposed to air temperatures below freezing point. *Res Vet Sci* **7**, 466–479 (1966)

Weidenreich, F.: On the earliest representatives of modern mankind recovered on the soil of east Asia. *Peking Nat Hist Bull* **13**, 161–175 (1939a)

Weidenreich, F.: The duration of life of fossil man in China and the pathological lesions found in his skeleton. *Chin Med J* **55**, 34–44 (1939b)

Welch, B. L., Welch, A. S.: Some aspects of brain biochemistry correlated with general nervous reactivity and aggressivenes. In: *Animal Aggression*. Southwick, C. H. (ed). New York: Van Nostrand Reinhold, 1970, 187–200

Welles, R. E., Welles, F. B.: The Bighorn of Death Valley. Fauna and Flora of National Parks of the US **6**. Washington: US Dept. Interior, 1961

Wellington, W. G.: Individual differences as a factor in population dynamics. *Can J Zool* **35**, 163–292 (1957)

Wellington, W. G.: Qualitative changes in natural populations during changes in abundance. *Can J Zool* **38**, 289–314 (1960)

Weltman, A. S., Sackler, A. M., Sparber, S. B.: Endocrine, metabolic and behavioural aspects of isolation stress in female albino mice. *Aerospace Med* **37**, 804–810 (1966)

Wendorf, F.: Early man in the New World: Problems of migration. *Amer Nat* **100**, 253–270 (1966)

Wendorf, F., Romvald, S., Rushiti, S., Vance, H. C., Achiel, G., Michal, K.: The prehistory of the Egyptian Sahara. *Science* **193**, 103–114 (1976)

Wendt, H.: Musik und Schlaftherapie. In: *Musik in der Medizin*. Stuttgart: Fischer, 1958, 104–112

Went, F. W.: The size of man. *Amer Sci* **56**, 400–413 (1968)

Wernick, R.: Danubian minicivilization bloomed before ancient Egypt and China. *Smithsonian* **5(12)**, 34–40 (1975)

White, F. N.: Circulation. In: *Animal Functions: Principles and Adaptations*. Gordon, M. S., Bartholomew, G. A., Grinnell, A. D., Jorgensen, C. B. (eds). New York: Macmillan, 1968, 152–229

White, L.: The historical roots of our ecological crisis. *Science* **155**, 1203−1207 (1967)

White, R. W.: Ego and reality in psychoanalytic theory. *Psychol Iss* **3**, no. 3, 1963

Whyte, W. H.: *The Organization Man*. New York: Simon and Schuster, 1956

Wickler, W.: Die biologische Bedeutung auffallenden farbiger, nackter Hautstellen und innerartliche Mimikry der Primaten. *Die Naturwissens.* **50**, 481−482 (1963)

Wickler, W.: Ursprung und biologische Deutung des Genitalpräsentierens männlicher Primaten. *Z Tierpsychol* **23**, 422−437 (1966)

Wickler, W.: Sociosexual signals and their intraspecific imitation among primates. In: *Primate Ethology*. Morris, D. (ed). London: Weidenfeld & Nicolson, 1967

Wickler, W.: *Mimicry in Plants and Animals*. World University Library, New York: McGraw-Hill, 1968

Wickler, W.: *The Sexual Code*. Garden City, N.Y.: Doubleday Anchor, 1969

Wickler, W.: Umfunktionierung als Evolutionsprinzip. *Naturwiss Med* **7(34)**, 3−14 (1970)

Wickler, W.: Uber stammes- und kulturgeschichtliche Semantisierung des männlichen Genitalpräsentierens. *Akt Fragen Psychiat Neurol* **11**, 122−137 (1971)

Wickler, W.: Biologische Grundlagen menschlichen Verhaltens. *Heidelberger Tasch* **121**, 169−181 (1973)

Wiley, R. H.: Evolution of social organization and life history patterns among grouse. *Quart Rev Biol* **49**, 201−227 (1974)

Wilkens, H.: Genetic interpretation of regressive evolutionary processes: Studies on hybrid eyes of two astyanax cave populations (Characidae, Pisces). *Evolution* **25**, 530−544 (1971)

Wilkie, F., Eisdorfer, C.: Intelligence and blood pressure in the aged. *Science* **172**, 959−962 (1971)

Willerman, L., Naylor, A. F., Myrianthopoulos, N. C.: Intellectual development of children from interracial matings. *Science* **170**, 1329−1331 (1970)

Williams, G. C.: *Adaptation and natural selection*. New York: Princeton Univ. Press, 1966

Williams, R. B., Eichelman, B.: Social setting: Influence of the physiological response to electric shock in the rat. *Science* **174**, 613−614 (1971)

Williams, R. J.: *Nutrition Against Disease*. New York: Pitman, 1971

Wilson, E. O.: *Sociobiology*. Cambridge, Mass: Harvard Univ. Belknap Press, 1975

Winick, M., Meyer, K. K., Harris, R. C.: Malnutrition and environmental enrichment by early adoption. *Science* **190**, 1173−1175 (1975)

Wohlin, H.: *Freiflächen für Kinder*. Munich: Callewey, 1972

Wolpoff, M. H.: "Telanthropus" and the single-species hypothesis. *Amer Anthropol* **70**, 477−493 (1968)

Wood, A. J., Cowan, I. McT., Nordan, H. C.: Periodicity of growth in ungulates as shown by deer of the genus *Odocoileus*. *Can J Zool* **40**, 594−603 (1962)

Wood, B. A.: Remains attributable to *Homo* in the East Rudolf succession. In: *Earliest Man and Environments in the Lake Rudolf Basin*. Coppens, Y., Howell, F. C., Isaac, G. L., Leakey, R. E. F. (eds). Chicago: Univ. Chicago Press, 1976, 490−506

Woodburn, J.: An introduction to Hadza ecology. In: *Man the Hunter*. Lee, R. B., de Vore, I. (eds). Chicago: Aldine, 1968a, 49−55

Woodburn J.: Stability and flexibility in Hadza residential groupings. In: *Man the Hunter*. Lee, R. B., de Vore, I. (eds). Chicago: Aldine, 1968b, 103−110

Worsley, T. R.: Terminal Cretaceous events. *Nature (Lond)* **230**, 318−320 (1971)

Wyler, A. R., Masuda, M., Holmes, T. H.: Seriousness of illness rating scale. *J Psychosom Res* **11**, 363−374 (1968)

Wynne-Edwards, V. C.: *Animal Dispersion in Relation to Social Behaviour*. Edinburgh: Oliver & Boyd, 1962

Yerkes, R. M.: Social behaviour of chimpanzees: Dominance between mates in relation to sexual status. *J Comp Psychol* **30**, 147−186 (1940)

Young, B. A.: Application of the carbon dioxide entry rate technique to measurement of energy expenditure by grazing cattle. In: *Fifth Symposium on the Energy Metabolism of Farm Animals*. Zurich: Joris Druck, 1971, 237−241

Young, B. A., Corbett, J. L.: Energy requirement for maintenance of grazing sheep measured by calorimetric techniques. *Proc Aust Soc Anim Prod* **1**, 327−334 (1968)

Young, E., Bronkhorst, P. I. L.: Overstraining disease in game. *Afr Wildl* **25**, 51−52 (1971)

Young, J. Z.: *An Introduction to the Study of Man*. Oxford: Clarendon, 1971

Young, M., Benjamin, B., Wallis, C.: The mortality of widowers. *Lancet* **2**, 454−456 (1963)

Zajonc, R. B.: Family configuration and intelligence. *Science* **192**, 227−236 (1976)

Zangwill, O. L.: Lashley's concept of cerebral mass action. In: *Current Problems in Animal Behaviour*. Thorpe, W. H., Zangwill, O. L. (eds). Cambridge: Cambridge Univ. Press, 1961, 59−86

Zeuner, F. E.: Geschichte der Haustiere. Trans. rev. by Boessneck, J., Haltenorth, T. Munich: Bayrischer Landwirtschaft Verlag, 1967

Zhigunov, P. S. (ed): *Reindeer Husbandry*. Trans. from Russian. Springfield: US Dept. Commerce, 1961

Zimmerberg, B., Glick, S. D., Jerussi, T.: Neurochemical correlate of a spatial preference in rats. *Science* **185**, 623−625 (1974)

Zimmerman, C. C., Broderick, C. B.: Nature and role of informal family groups. *Marr Fam Living* **16**, 107−111 (1954)

Zimmerman, C. C., Cervantes, L. F.: *Successful American Families*. New York: Pageant, 1960

Zimmermann, R. R., Strobel, D. A., Maguire, D., Steere, R. R., Hom, H. L.: The effects of protein deficiency on activity, learning, manipulative tasks, curiosity and social behaviour of monkeys. In: *Play*. Bruner, J. S., Jolly, A., Sylva, K. (eds). New York: Penguin, 1976, 496−511. First publ 1973

Zipf, G. K.: *Human Behaviour and the Principle of Least Effort*. Cambridge, Mass.: Addison-Wesley, 1949

Index

A Springer-Verlag Journal

Behavioral Ecology and Sociobiology

Managing Editor: Hubert Markl, University of Konstanz

Editors: John H. Crook, Bristol; Bert Holldobler, Cambridge, Mass.; Hans Kummer, Zurich; Edward O. Wilson, Cambridge, Mass.

Behavioral Ecology and Sociobiology presents original contributions and short communications on the analysis of animal behavior on both the individual and population levels. Special emphasis in this international journal is given to the functions, mechanisms, and evolution of ecological adaptions of behavior.

The journal, based on a philosophy developed and nurtured for more than a half a century in the original *Zeitschrift fur Vergleichende Physiologie* (now *Journal of Comparative Physiology*), covers all important aspects of animal behavior. Areas most frequently examined include orientation in space and time, communication and all other forms of social and interspecific behavioral interaction, including predatory and antipredatory behavior, origins and mechanisms of behavioral preferences and aversions (with respect to food, locality, and social partners), behavioral mechanism of competition and resource partitioning, and population physiology and evolutionary theory of social behavior.
Contact Springer-Verlag for sample copies and subscription information.

Topics in Environmental Physiology and Medicine
Edited by **K.E. Schaefer**

Zoophysiology and Ecology
Coordinating Editor: D.S. Farner
Editors: W.S. Hoar, B. Hoelldobler, H. Langer, M. Lindauer